Advances in Global Change Research

Volume 65

More information about this series at http://www.springer.com/series/5588

Mirjana Šijačić-Nikolić • Jelena Milovanović
Marina Nonić
Editors

Forests of Southeast Europe Under a Changing Climate

Conservation of Genetic Resources

Springer

Editors
Mirjana Šijačić-Nikolić
Faculty of Forestry
University of Belgrade
Belgrade, Serbia

Jelena Milovanović
Faculty of Applied Ecology FUTURA
Singidunum University
Belgrade, Serbia

Marina Nonić
Faculty of Forestry
University of Belgrade
Belgrade, Serbia

ISSN 1574-0919 ISSN 2215-1621 (electronic)
Advances in Global Change Research
ISBN 978-3-319-95266-6 ISBN 978-3-319-95267-3 (eBook)
https://doi.org/10.1007/978-3-319-95267-3

Library of Congress Control Number: 2018963838

Cover illustration: View from National Park "Djerdap", Serbia
Author: Marina Nonić

This Springer imprint is published by the registered company Springer Nature Switzerland AG
The registered company address is: Gewerbestrasse 11, 6330 Cham, Switzerland

Preface

Forest genetic resources represent a source of variability which is a prerequisite for the adaptation and survival of different types of forest trees and hence of forest ecosystems in altered climatic conditions. They can also satisfy the different needs of mankind if we consider them as the starting material in the process of breeding or the already existing raw material. Bearing in mind that we are witnesses of climate change that adversely affect the overall biodiversity, and therefore the forest genetic resources, it is necessary to intensify the work on their conservation and further directed use. The aim of this book is to point out the significance and potential of forest genetic resources in the countries of Southeastern Europe, bearing in mind that this is an area with pronounced biodiversity and significant forest cover. The book elaborates current state and legal and institutional framework for forest genetic resources conservation, in situ and ex situ conservation case studies from Southeast Europe, as well as climate change aspects in forest genetic resources conservation. Looking at the current state, with the desire to improve it, concluding remarks and recommendations have been defined as a separate chapter which consists of objectives, priorities, and measures for conservation of forest genetic resources of Southeast Europe under a changing climate in the future. We believe that the book will be of benefit to both professionals and researchers as well as to students, teachers, and all others interested in forest genetic resources and nature conservation.

The book is partly supported through the following projects: "Establishment of forest plantations to increase the afforested areas in Serbia" (TR31041); project "Studying climate change and its influence on the environment: impacts, adaptation and mitigation" (43007); and project "Ecoremediation of degraded areas through agro-energy crops cultivation" (TR31078) financed by the Ministry of Education, Science and Technological Development of the Republic of Serbia.

Our special thanks go to the 66 authors of different chapters of the book who represent 28 various scientific institutions from 9 countries of Southeast Europe (Bosnia and Herzegovina, Bulgaria, Croatia, Greece, Macedonia, Montenegro, Romania, Serbia, and Slovenia) and contributed significantly to the quality of the book. We sincerely hope that the book will find its readers and contribute to the awareness raising about the significance of forest genetic resources conservation in the context of climate change.

Belgrade, Serbia

Mirjana Šijačić-Nikolić
Jelena Milovanović
Marina Nonić

Introduction

The climate change we witness is a serious threat to people, the living world, and our environment. No one can foresee long-term consequences, while we are daily witnessing the absence of common weather, and extreme events such as drought, floods, forest fires, hurricanes, and the like. All current data and information, as well as developed models, indicate that the ecosystems, products, and services they provide, as well as the biodiversity that is their integral part, are becoming degraded and less able to function. Within global change, in the European context, it is quite clear that Southeast Europe will be especially hard hit in terms of its biodiversity whose significant part is in forests. Climate change represents particular challenges for both forests and individual trees and forest genetic resources which are the basis for adaptation and an adequate response to climate change.

In the countries of Southeast Europe, including Greece, Macedonia, Montenegro, Serbia, Bulgaria, Romania, Bosnia and Herzegovina, Croatia, and Slovenia, climate change can lead to an increase in average temperatures, an increase in the frequency of extreme weather conditions, an increase in coastal erosion, impact on marine biodiversity, river level increase due to extreme rainy periods, more intensive and frequent flooding, high pressure on water sources, changes in structure and layout of arable land including poor harvesting, changes in habitat conditions and species schedules, problems with foreign invasive species, and the reduction of the richness of biodiversity. These events will have an impact on local and regional communities that depend on services provided by the ecosystem in the form of food, drinking water, fuel, raw materials sources, etc., which currently provides acceptable living conditions and well-being for most of the population in this area. The fact is that many Southeast European regions belong to most vulnerable areas to climate change and where the most pronounced impacts of climate change on ecosystems can be expected in the future, including forests.

Within the overall biodiversity of Southeast Europe, forest ecosystems occupy a significant place. Forests represent a significant natural resource that fulfills economic, cultural, and spiritual needs of man. These needs are related to forest products and services, such as wood and wood products, water, human and animal food, medical services, fuel, recreation, wildlife habitat protection, and landscape diversity. Forests can also be considered as a significant source of biological and genetic diversity. Lately, forests have an indispensable role in mitigating climate change caused by human activity. Forests, characterized by a wide range of production and environmental functions, meet the different needs of mankind. Forest trees and other woody plants provide the living space of many other organisms, building complex mechanisms of genetic diversity. Genetic variability, both between and within species, has multiple fundamental values. Thanks to it, the trees and shrubs are adapting to the new environment conditions, even if they were under the negative influence of pests, diseases, or climate change. The diversity of forest ecosystems is an important basis in the processes of evolution, selection, and breeding in order to meet the needs of mankind for biomass.

Within forest resources, a significant place, perhaps the most significant, is occupied by hundreds of trees that carry a different genetic structure (DNA). This genetic variability of tree species, which has a potential or real value for man, is designated as *forest genetic resources.* In the context of climate change and increasing exploitation of forest genetic resources, the scientific and professional public is aware of the necessity of conserving forest genetic resources. The basic international legal and political framework for forest genetic resources in Europe is located within EUFORGEN (European Forest Genetic Resources Program), which was established in 1994 and represents a common mechanism for European countries in promoting conservation and targeted use of forest genetic resources.

Most of the countries of Southeast Europe are members of EUFORGEN, so cooperation in the field of conservation of forest genetic resources has been established and activities are coordinated. Authors of this book intended to present the state and the current activities for conservation of forest genetic resources in the countries of Southeast Europe. According to the current state, goals, priorities, and concrete measures were defined as guidelines for improving this area.

Contents

Contributors

Jelena M. Aleksić Institute of Molecular Genetics and Genetic Engineering, University of Belgrade, Belgrade, Serbia

Alexander H. Alexandrov Forest Research Institute, Bulgarian Academy of Sciences, Sofia, Bulgaria

Paraskevi G. Alizoti School of Forestry and Natural Environment, Aristotle University of Thessaloniki, Thessaloniki, Greece

Vlatko Andonovski Faculty of Forestry in Skopje, University Ss. Cyril and Methodius, Skopje, Macedonia

Ivan Andrić Faculty of Forestry, University of Zagreb, Zagreb, Croatia

Ina Y. Aneva Institute of Biodiversity and Ecosystem Research, Bulgarian Academy of Sciences, Sofia, Bulgaria

Ecaterina Nicoleta Apostol National Institute for Research and Development in Forestry "Marin Drăcea" Ilfov County, Voluntari, Romania

Filippos A. Aravanopoulos School of Forestry and Natural Environment, Aristotle University of Thessaloniki, Thessaloniki, Greece

Evangelia V. Avramidou Laboratory of Silviculture, Forest Genetics and Biotechnology, Institute of Mediterranean Forest Ecosystems and Forest Products Technology, Athens, Greece

Marko Bajc Slovenian Forestry Institute, Ljubljana, Slovenia

Jasmina Balijagić Biotechnical Faculty, University of Montenegro, Podgorica, Montenegro

Dalibor Ballian Faculty of Forestry, University of Sarajevo, Sarajevo, Bosnia and Herzegovina

Evangelos Barbas Laboratory of Forest Genetics and Tree Breeding, Faculty of Agriculture, Forestry and Natural Environment, Aristotle University of Thessaloniki, Thessaloniki, Greece

Peter Bebi WSL Institute for Snow and Avalanche Research SLF, Davos, Switzerland

Mariya Belovarska Executive Forest Agency, Ministry of Agriculture, Food and Forestry, Sofia, Bulgaria

Svilena Bozhinova Forest Seed Legislation Station, Sofia, Bulgaria

Gregor Božič Slovenian Forestry Institute, Ljubljana, Slovenia

Robert Brus Department for Forestry and Renewable Forest Resources, Biotechnical Faculty, University of Ljubljana, Ljubljana, Slovenia

Marius Budeanu National Institute for Research and Development in Forestry "Marin Drăcea", Voluntari, Romania

Dijana Čortan Faculty of Education, University of Novi Sad, Sombor, Serbia

Milić Čurović Biotechnical Faculty, University of Montenegro, Podgorica, Montenegro

Alexandru Lucian Curtu Department of Forest Sciences, Transilvania University of Braşov, Braşov, Romania

Branislav Cvjetković Faculty of Forestry, University of Banja Luka, Banja Luka, Bosnia and Herzegovina

Vanja Daničić Faculty of Forestry, University of Banja Luka, Banja Luka, Bosnia and Herzegovina

Jovana Devetaković Faculty of Forestry, University of Belgrade, Belgrade, Serbia

Rumen D. Dobrev Forest Research Institute, Bulgarian Academy of Sciences, Sofia, Bulgaria

Ivan Evtimov University of Forestry, Sofia, Bulgaria

Anna-Maria Farsakoglou Laboratory of Forest Genetics and Tree Breeding, Faculty of Agriculture, Forestry and Natural Environment, Aristotle University of Thessaloniki, Thessaloniki, Greece

Velichko Gagov University of Forestry, Sofia, Bulgaria

Vladislava Galović Institute of Lowland Forestry and Environment, University of Novi Sad, Novi Sad, Serbia

Georgi Gogushev Regional Forestry Directorate – Blagoevgrad, Blagoevgrad, Bulgaria

Kostas Ioannidis Institute of Mediterranean Forest Ecosystems and Forest Products Technology, Hellenic Agricultural Organization "DEMETER", Athens, Greece

Vladan Ivetić Faculty of Forestry, University of Belgrade, Belgrade, Serbia

Kristjan Jarni Department for Forestry and Renewable Forest Resources, Biotechnical Faculty, University of Ljubljana, Ljubljana, Slovenia

Dušan Jokanović Faculty of Forestry, University of Belgrade, Belgrade, Serbia

Miodrag Jovančević Biotechnical Faculty, University of Montenegro, Podgorica, Montenegro

Davorin Kajba Faculty of Forestry, University of Zagreb, Zagreb, Croatia

Ino-Vasileia Korompoki Laboratory of Forest Genetics and Tree Breeding, Faculty of Agriculture, Forestry and Natural Environment, Aristotle University of Thessaloniki, Thessaloniki, Greece

Vasiliki-Maria Kotina Laboratory of Forest Genetics and Tree Breeding, Faculty of Agriculture, Forestry and Natural Environment, Aristotle University of Thessaloniki, Thessaloniki, Greece

Branislav Kovačević Institute of Lowland Forestry and Environment, University of Novi Sad, Novi Sad, Serbia

Hojka Kraigher Slovenian Forestry Institute, Ljubljana, Slovenia

Aleksandar Lučić Institute of Forestry, Belgrade, Serbia

Zoran Maksimović SE "Srbijašume", Belgrade, Serbia

Ermioni Malliarou Laboratory of Forest Genetics and Tree Breeding, Faculty of Agriculture, Forestry and Natural Environment, Aristotle University of Thessaloniki, Thessaloniki, Greece

Milan Mataruga Faculty of Forestry, University of Banja Luka, Banja Luka, Bosnia and Herzegovina

Jelena Milovanović Faculty of Applied Ecology FUTURA, Singidunum University, Belgrade, Serbia

Jelena Nedeljković Faculty of Forestry, University of Belgrade, Belgrade, Serbia

Vesna Nikolić Faculty of Forestry, University of Belgrade, Belgrade, Serbia

Dragan Nonić Faculty of Forestry, University of Belgrade, Belgrade, Serbia

Marina Nonić Faculty of Forestry, University of Belgrade, Belgrade, Serbia

Saša Orlović Institute of Lowland Forestry and Environment, University of Novi Sad, Novi Sad, Serbia

Momchil Panayotov University of Forestry, Sofia, Bulgaria

Denitsa Pandeva Executive Forest Agency, Ministry of Agriculture, Food and Forestry, Sofia, Bulgaria

Andrej Pilipović Institute of Lowland Forestry and Environment, University of Novi Sad, Novi Sad, Serbia

Flaviu Popescu National Institute for Research and Development in Forestry "Marin Drăcea", Simeria, Romania

Emil Popov Forest Research Institute, Bulgarian Academy of Sciences, Sofia, Bulgaria

Vladan Popović Institute of Forestry, Belgrade, Serbia

Dragoş Postolache National Institute for Research and Development in Forestry "Marin Drăcea", Cluj-Napoca, Romania

University of Agricultural Sciences and Veterinary Medicine, Cluj-Napoca, Romania

Mirjana Šijačić-Nikolić Faculty of Forestry, University of Belgrade, Belgrade, Serbia

Neculae Şofletea Faculty of Silviculture and Forest Engineering, Transilvania University of Braşov, Braşov, Romania

Srđan Stojnić Institute of Lowland Forestry and Environment, University of Novi Sad, Novi Sad, Serbia

Rifet Terzić Faculty of Natural Sciences, University of Tuzla, Tuzla, Bosnia and Herzegovina

Nikolaos Tourvas Laboratory of Forest Genetics and Tree Breeding, Faculty of Agriculture, Forestry and Natural Environment, Aristotle University of Thessaloniki, Thessaloniki, Greece

Evgeni Tsavkov University of Forestry, Sofia, Bulgaria

Nickolay Tsvetanov University of Forestry, Sofia, Bulgaria

Nikolco Velkovski Faculty of Forestry in Skopje, University Ss. Cyril and Methodius, Skopje, Macedonia

Marjana Westergren Slovenian Forestry Institute, Ljubljana, Slovenia

Stefan Yurukov University of Forestry, Sofia, Bulgaria

Petar Zhelev University of Forestry, Sofia, Bulgaria

Part I
State of Forest Genetic Resources Conservation in Southeast Europe

Chapter 1
State of Forests in Bosnia and Herzegovina: Ecological and Vegetation Distribution, Management and Genetic Variability

Milan Mataruga, Dalibor Ballian, Rifet Terzić, Vanja Daničić, and Branislav Cvjetković

Abstract According to the data obtained after the *Second National Forest Inventory in Bosnia and Herzegovina* (2006–2009) was carried out, the total area of forests and forest land in Bosnia and Herzegovina (BiH) is 3,231,500 ha or 63.08%, whilst the area covered by forest takes up 2,904,600 ha, or 56.7% of the total Bosnia and Herzegovina land area The land area of BiH is divided into four ecological vegetation areas (Pripanon, Inner Dinarides, Transitional Illyrian-Moesiac area and Mediterranean area). The spatial distribution of the main vegetation types is influenced by horizontal zonation and altitudinal belts. These areas are then followed by the mountainous belt with beech forests, beech and fir, beech and fir with spruce and the subalpine beech forest belt. There is a great number of endemic species and habitats and relic locations – a terrestrial flora and fauna shelter, which are now preserved under specific paleo-climate conditions. BiH is one of the countries in Europe with the highest diversity of both, plants and animals species, which are endangered by climate change. The abundance of diversity is reflected in a large number of plant and animal species within the region (about 4500 higher plants, 600 taxa mosses and about 80 ferns, around 250 species of forest trees and shrubs). This is why the position of BiH is high in the list of European

M. Mataruga (✉) · V. Daničić · B. Cvjetković
Faculty of Forestry, University of Banja Luka, Banja Luka, Bosnia and Herzegovina
e-mail: milan.mataruga@sf.unibl.org; vanja.danicic@sf.unibl.org; branislav.cvjetkovic@sf.unibl.org

D. Ballian
Faculty of Forestry, University of Sarajevo, Sarajevo, Bosnia and Herzegovina
e-mail: balliandalibor9@gmail.com

R. Terzić
Faculty of Natural Sciences, University of Tuzla, Tuzla, Bosnia and Herzegovina
e-mail: rifet.terzic@untz.ba

M. Šijačić-Nikolić et al. (eds.), *Forests of Southeast Europe Under a Changing Climate*, Advances in Global Change Research 65,
https://doi.org/10.1007/978-3-319-95267-3_1

countries of a high biodiversity level. The variability has been investigated by molecular markers for the most important species. The variability is high and it is important for meeting climate changes. Afforestation trends in BiH indicate that the afforestation volume decreased. Establishihng of new forests are not at the satisfactory level due to the lack of financial resources and insufficient attention to the aforested areas.

Keywords Ecology · Vegeation · Forests · Bosnia and Herzegovina

1.1 Introduction

With an area of 51,129 km^2 Bosnia and Herzegovina (abbreviation BiH) comprises two entities and the Brčko District, a self – governing administrative unit. Federation of Bosnia and Herzegovina (FBiH) takes up 50% of the BiH land area and Republika Srpska (RS) takes up 49%. Brčko District takes up the remaining 1% (Agency for Statistics of Bosnia and Herzegovina 2014).

According to preliminary results of the census, Bosnia and Herzegovina has a population of 3,791,622 people, with 2,371,603 people living in FBiH, 1,326,991 people living in Republika Srpska and 93,028 people living in Brčko District (Agency for Statistics of Bosnia and Herzegovina 2014).

Bosnia and Herzegovina lies between latitudes 42°26′ and 45°15′ N, and longitudes 15°45′ and 19°41′ E. Out of the total land area of Bosnia and Herzegovina, 5% is covered by lowlands, 24% by hills, 42% by mountains, and 29% by karst (NEAP 2003). The extremely high diversity of landscapes throughout Bosnia and Herzegovina resulted out of the synergy from the long-term coexistence and the high level of interaction between biological and geological diversity, as well as the human impact on nature. In addition, the effects of the specific orography, geological structure, hydrology and climate deliver their influence, too. Therefore, the BiH landscape varies from the mountainous in the central part, through the cultivable land in the north, all the way to the Mediterranean in the south (Fig. 1.1). Most city areas are mainly located in the valleys.

The war that struck Bosnia and Herzegovina during the 1990s and the economic recovery in the post-war period, which was mainly based on the use of natural resources, produced certain negative impacts on forest resources in BiH. The extensive population migrations due to the war activities, as well as to the urbanization processes, resulted in a spontaneous change of the land use, converting it from the agricultural type of land to the forest type of land, directly affecting the areas coverd by forests in BiH.

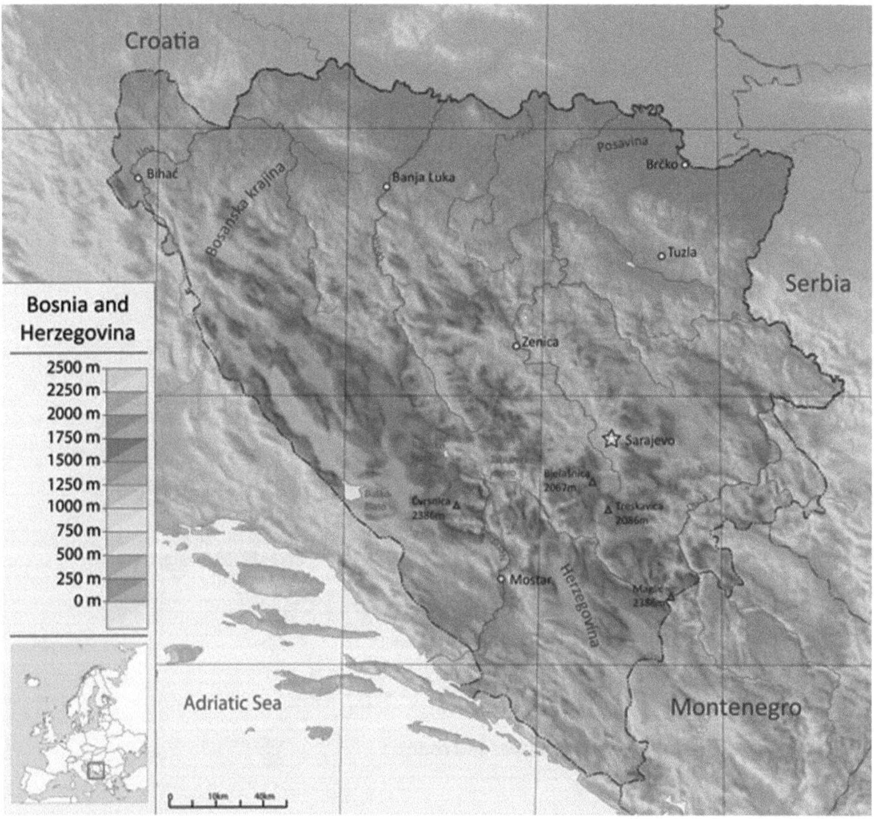

Fig. 1.1 Bosnia and Herzegovina, physical map. (Source: http://www.vidiani.com/physical-map-of-bosnia-and-herzegovina/ (on February 13th 2017))

1.2 Ecological Vegetation Zonation in Bosnia and Herzegovina

The land area of BiH is divided into four ecological vegetation areas (Pripanon, Inner Dinarides, Transitional Illyrian-Moesiac area and Mediterranean area). The spatial distribution of the main vegetation types is influenced by horizontal zonation and altitudinal belts. The lowest base area is covered by the oak forests belt throughout the lowlands and hilly areas of the Illyrian and Transitional Illyrian-Moesiac provenances. These areas are then followed by the mountainous belt with beech forests, beech and fir, beech and fir with spruce and the subalpine beech forest belt (Fig. 1.2). There is a particular spruce forests belt, which is then followed by mugo pine forests of a distinct floristic diversity characterized by numerous endemic and relict species.

Although small in size, Bosnia and Herzegovina is a country of a high bio-ecological diversity potential. It is located in the southeastern part of Europe on the Balkan peninsula and is one of the "hot spots" of the world's biodiversity (Myers et al. 1999). All the way from the south to the north, the diversity of the wildlife is

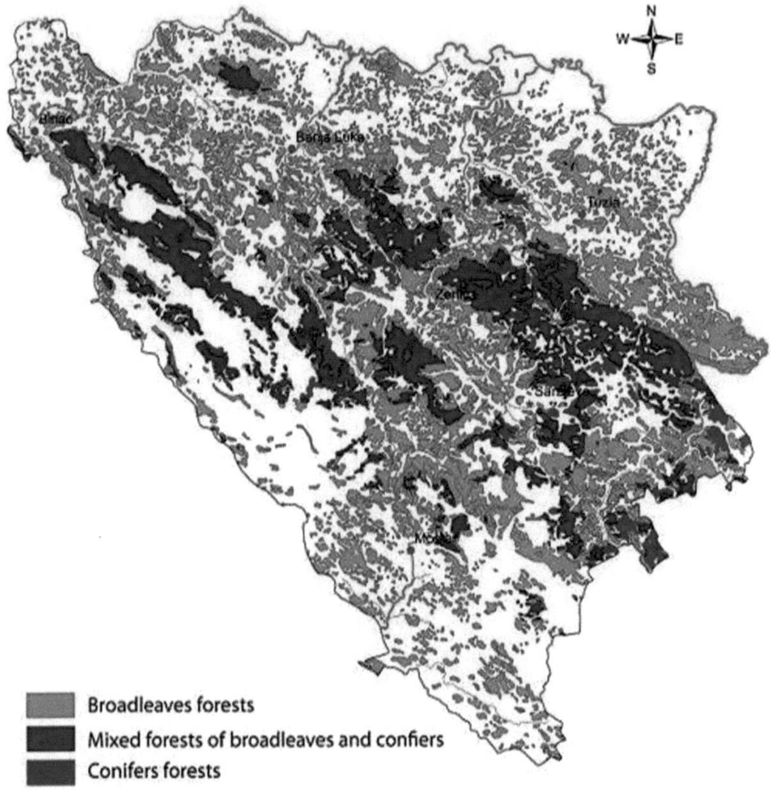

Fig. 1.2 Broadleaf and coniferous forest distribution in BiH

conditioned by the proximity of the Adriatic Sea on the south, then, by the Dinaric mountain chain in the central part, as well as by the Pannonian Plain in the north of the country (Fig. 1.2).

There is a great number of endemic species and habitats and relic locations – a terrestrial flora and fauna shelter, which are now preserved under specific paleo-climate conditions. BiH is one of the countries in Europe with the highest diversity of both, plants and animals species, which are endangered by climate change. It is necessary to emphasize the need to allocate larger areas to be governed by certain protection systems, in order to preserve the forest ecosystems affected by excessive felling activities which led to soil degradation, caused landslides, etc. These steps are to include drought management and desertification, increase capacities for carbon dioxide absorption, thus, provide greater climate stability.

The findings of the European CORINE Land Cover (or CLC2000) project gave its contribution to the research activities of the forest complex at the national level, since the project itself plays a significant role in the activities related to the ecosystem protection, the prevention of biodiversity loss, as well as keeping track of climate change impacts, assessing the development of the agriculture and the implementation of the Water Framework Directive. The CLC2000 is a set of data

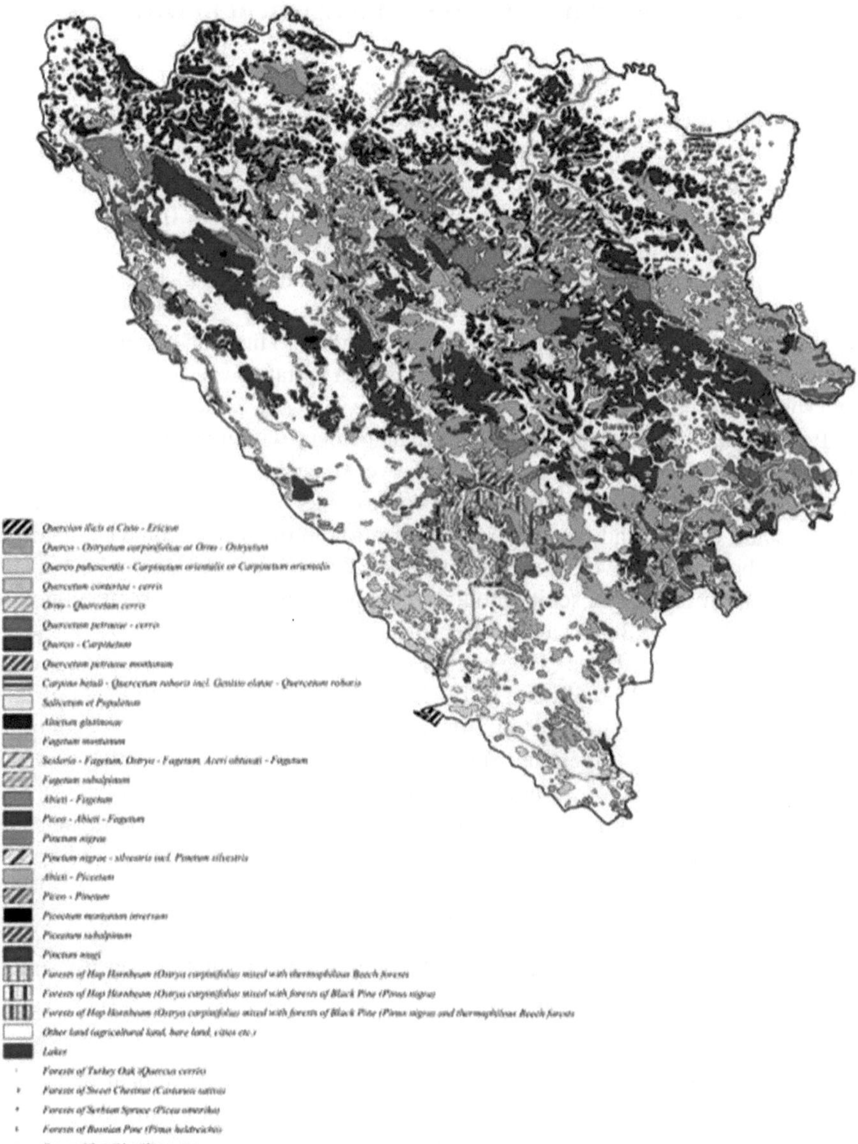

Fig. 1.3 Land area covered by forests. (Source: CLC 2006)

significant for the implementation of key priorities set in *the Sixth Environment Action Programme of the European Community*. The aforementioned Project on the land cover, had its goal to define the structure of the earth's biophysical cover utilization, and put the focus on the forest cover. The CLC data for BiH were further updated during 2006, and deployed to create a GIS map of the land use with the accompanying database (Fig. 1.3).

1.3 Forest Condition and Forest Management in BiH

According to the data obtained after the *Second National Forest Inventory in BiH* (2006–2009) was carried out, the total area of forests and forest land in BiH is 3,231,500 ha or 63.08%, whilst the area covered by forest takes up 2,904,600 ha or 56.7% of the total Bosnia and Herzegovina land area (Fig. 1.2). Thus, it is obvious that forests are also a significant source of the employment force (in forestry, wood industry) and of energy (firewood, mostly forest waste). Nevertheless, forests make a significant source of non-wood forest products (mushrooms, forest fruits and herbs and other herbs) and a spacious wildlife habitat.

With respect to the ownership structure, according to the data obtained from the *Second National Forest Inventory*, 70% of the area is state owned and run by Public Enterprises, while 30% is privately owned.

Professional development and forest management within the forestry sector is governed by traditional systems. In general, ever since the war started in Bosnia and Herzegovina (1992–1995), the forestry sector has faced significant structural changes and the demand for strong modernization in order to become competitive on the global market. Even though the forest and wood industry do not have a significant role in creating a social product, they are noticed as important economic resources, moreover, Bosnia and Herzegovina's natural resources.

Forest sector contributes to the overall employment situation more than it contributes to GDP (in RS 6.5% and in FBiH 4.6% in 2008). Although the contribution of forestry sector to GDP in BiH was only 0.83% in 2010, this economic activity is of strategic importance due to its export activities and creating new job positions. State owned forests are managed by public enterprises at entity level and are under the control of relevant ministry departments and entity parliaments. The legal and institutional framework relevant to forestry sector is structured across two entities.

Statistical data of the forest resources have two key elements:

- Statistical information based on the data collected from public forestry enterprises and statistical agencies,
- National Forest Inventory (NFI).

Along with the data obtained from the 2006 to 2009 BiH National Inventory, there are also significant data obtained from the National Inventory carried out from 1964 to 1968. Questions and dilemmas whether the information of the forest condition are reliable and whether they can be deployed as the basis for planning and creating the forestry policy in BiH, arouse since those data have been used in national and international reports on forest resources.

The *Second NFI* was carried out during 2006–2009, but official results haven't been reported yet. However, some of the first findings of this inventory were partially made public (UNDP 2014; FAO 2015).

Table 1.1 Areas covered by forests in BiH

Forest type	Regular forests management system (commercial)	Low commercial value forests (coppice and degraded)	Protected areas	Special purpose	Inaccessible areas	Total
	Hectars					
1. High forests	1,329,500	46,300	5200	8800	262,600	1,652,400
2. Sprout forests	843,200	158,700	1600	2400	246,300	1,252,200
Total (1 + 2)	2,172,700	205,000	6800	11,200	508,900	2,904,600
3. Degraded forests	52,700	41,100	0	100	36,700	130,600
4. Bare ground	55,700	88,400	800	3400	38,900	187,200
Total (3 + 4)	108,400	129,500	800	3500	75,600	317,800
5. Other forest area	3300	3100	–	100	2600	9100
FAO forests (1 + 2 + 3 + 5)	2,228,700	241,600	6800	11,400	548,200	3,035,700
Forests and forest land	2,284,400	337,600	7600	14,800	587,100	3,231,500

Source: Possibilities of using biomass from forestry and wood industry in Bosnia and Herzegovina, UNDP in Bosnia and Herzegovina, Sarajevo, October 2014

The latest findings of the *Second NFI* are presented in Table 1.1. In economic sense, forests are defined as forests managed for economic purposes. Protected forests are defined as forests that primarily protect land, water, settlements, for economic purposes or against natural disasters and forests which are established as belts in order to protect forests which grow at higher vegetation altitudes. Forests for special purposes are those which are rarely found in nature or of a particular cultural, religious or historical significance, including national parks, nature parks, or forests intended for: leisure activities, sports, recreation, education, science and hunting. In addition to the above mentioned, there are also forest areas of special interest for the defense of the country, seed sources. A special forest segment represents inaccessible forests under minefields.

Out of the total area under forests, 1,652,400 ha are high forests and 1,252,200 ha are sprout forests. The rest is characterized as forest land and includes shrubs, infertile forest land and other types of forest areas.

A significant increase of the forest areas in all categories (more than 15%) is evident if compared to the *First NFI* (total area of forests and forest land was 2734 million hectares).

The current traditional management system based on natural renewal activities, which have been in practice for centuries in BiH and which contributed to the cre-

Table 1.2 The structure of forests and forest land by forest tree species

Forest type	FB&H (ha)	RS-B&H (ha)	Brčko district (ha)	B&H (ha)	%
1. Beech forest	453,000	544,000	2200	999,200	30.92
2. Coniferous and mixed forests of coniferous and broadleaf forests within the beech and fir forest (with spruce) area of ditribution	443,100	319,700	–	762,800	23.61
3. Pine forests	146,000	70,400	400	216,800	6.71
4. Common oak forests	25,800	41,100	4000	70,900	2.19
5. Sessile oak forests	185,900	269,300	3500	458,700	14.19
6. Thermophilous oak forests	334,400	205,000	400	539,800	16.70
7. Willow, poplar and alder forests	14,200	24,600	2000	40,800	1.26
8. Pioneer forest communities	23,300	30,500	400	54,200	1.68
9. Introduced tree species plantations	11,700	21,100	100	32,900	1.02
10. Secondary beech forests	55,300	100	–	55,400	1.71
Total (ha)	1,692,700	1,525,800	13,000	3,231,500	100.00

ation of a significant diversity of forest cover and its intensive renewal, as well as the application of today's increasingly recognizable practice "forests managed in a harmony with nature" is essential for the continuous natural development of the forestry fund. For that reason today in BiH there are 93% of natural and only 7% of the planted forests. Plantations with selectable fast growing clones and intensive agrochemicals are almost gone.

Pure beech forests dominate in Bosnia and Herzegovina, since they cover more than 30% of the area, oak forests cover about 30%, and mixed coniferous-broadleaf forests slightly over 23% (Table 1.2). With regard to the forest composition, BiH is dominated by: beech forests with about 31%, sessile oak forests 14% and common oak 2%, thermophilous oak forests 17%, willow, poplar and alder forests 1%, then coniferous forests, and mixed coniferous and broadlef forests 23%, while pine forests account for about 7%. The rest are mainly planted with autochthonous (4%) or allochtonous species (1%) (FAO 2015). In addition to the fact that most of the forests are of commercial benefits, the role played by forests in order to preserve the biodiversity in BiH is enormous.

It should be noted that there are several reasons for this increase of the forest area if it is to be compared to the data from the 1960s *First National Inventory*. Two main reasons may be:

- The change in the calculation methodology and different statistical approach, which both make the two data largely uncomparable,
- Increase of the forest area due to reforestation (less) and natural grow of trees on the abandoned land (more).

Table 1.3 The increase of the wood stock of all accessible and inaccessible forests by vegetation type

Vegetation	BiH		
	ha	(m^3/ha)	(m^3)
1. High forests	1,652,400	6,85	11,316,477
2. Coppice forests	1,252,200	2,27	2,848,362
Total of forest (1 + 2)	2,904,600	4,89	14,216,803
Accesable, econimic forests	2,225,200	5,13	11,421,336
3. Brushwood	130,600	–	–
4. Bare land	187,200	–	–
Other wooded land (3 + 4)	317,800	–	–
5. other non productive area	9100	–	–
Total forest and other forest land	3,044,300	4,67	14,202,196

The forest certification that has been carried out in Bosnia during the last decade is of a great significance for the quality management of the forest fund. More specifically, three public forest enterprises have undergone an international inspection in order to obtain Forest Stewardship Council (FSC) certificates, while several other enterprises are preparing for the same procedure. The implementation of this system significantly contributes to a more efficient management approach from the aspect of a greater contribution to the protection and improvement of all forest functions, all the way from economic sustainability, social responsibility to the environmental sustainability.

According to the data presented in Table 1.3, the total growth of the annual wood stock (not including branchwood, as well as stumps and undreground parts) in all forests in BiH is slightly over 14 million cubic meters, which could be characterized as a significantly higher increase in the felling volume.

At the same time, the felling volume in BiH has increased during the last 5–6 years (Figs. 1.4 and 1.5). Thus, the calculation shows that felling activities annually grow by 9% in relation to the previous year. Consequently, for the last 5 years in BiH the felling volume has increased by 15.22% if it is compared to 2010.

The capacities of mechanical wood processing exceed habitat potentials of our forests (Musić et al. 2013), and arise conflicts between forestry and wood processing sectors (Marić et al. 2012). Forestry faces various problems: lack of a long-term strategy and the development of protected areas (Stupar 2011) and forests policies, the over-estimated economic and social functions endanger the ecological function, inefficient organization, as well as the practices that are not aligned with the EU practices. All these factors have put forestry sector in a difficult position, for it reflects inadequate breeding and protection activities, biotic and abiotic vulnerability, forest land decrease, biodiversity loss and poor access to forests (FAO 2015). This synergistic impact makes Bosnia and Herzegovina forests and forest systems vulnerable.

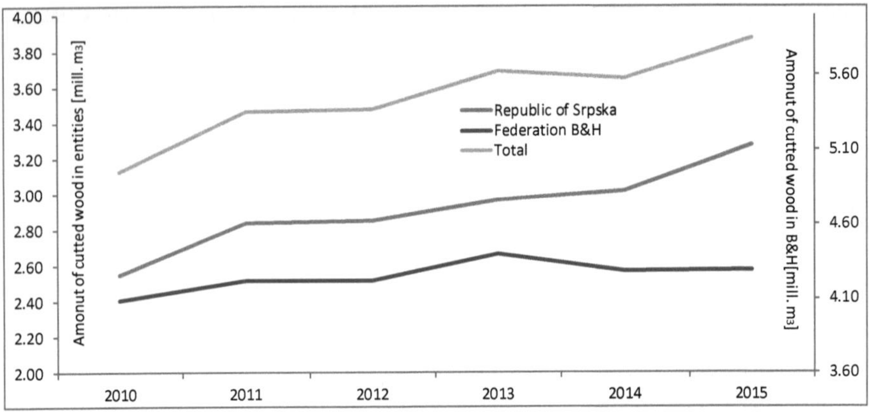

Fig. 1.4 Felling volume (millions of cubic meters) in entities and total in BiH during 2010–2013. (Source: Statistical yearbooks published by Agency for Statistics of Republika Srpska and Federation of B&H)

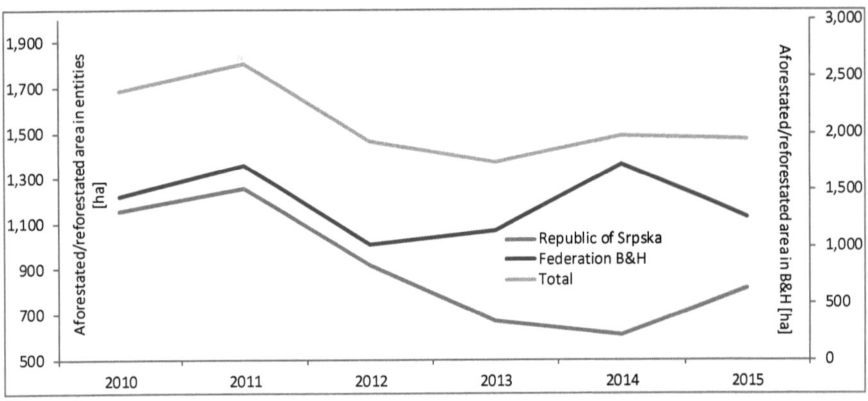

Fig. 1.5 Felling volume in entities and total in BiH during 2010–2013. (Sorce: Statistical yearbooks published by Agency for Statistics of Republika Srpska and Federation of Bosnia and Herzegovina)

1.4 Forest Restoration

Afforestation trends in BiH indicate that the afforestation volume decreased (Fig. 1.5) during the last 5 years. Apart from the fact that the afforestation volume in 2011 increased by 9,15%, due to the significant decrease of the afforestation volume, in regard to 2010 when the percent of the afforested land was 26,18%, making the afforestation volume less for almost a quarter. The afforestation problem in Bosnia and Herzegovina is because of the poor ways of making records of the afforested land and then in the way that land is then displayed to public. Namely, the records of the afforested land are entered a year after the seedlings were planted. At that moment most of the afforested areas are recorded as "successfully afforested". At the same time, due to the lack of financial resources and insufficient attention to the forested areas (left behind to competing vegetation), these areas are completely covered by weed, in most cases 5–10 years later, and the success of afforestation is severely depleted. Therefore, this data on the increase of the afforested area in BiH through afforestation activities should not be taken for granted.

If the measures, defined in the 1970s and 1980s, which were then incorporated into legal provisions, were implemented, the Federation of BiH should, on the basis of available felling volume (NEAP 2003), annually afforest more than 5000 ha, that is to produce over 15,000,000 seedlings in nurseries. On the basis of the same measures, in Republika Srpska, over 4000 ha should be annually afforested, or 10,000,000 seedlings produced. The total amount of seedlings to be produced for the afforestation activities in order to cover the logged surface is about 25,000,000 seedlings.

In addition, according to statistical data, about 3000 ha of forests are annually destroyed in forest fires. Fire hazards have considerably increased over the past few decades due to the rise in average and extreme air temperatures, especially in the mountainous areas of BiH.

When it comes to the forest restoration with the planting stock coming from the local resources, in BiH there are 210 seed objects and 25 nurseries producing forest tree stock material. Out of these, 18 ones are in the Federation of BiH and 7 are in Republika Srpska. Nursery production in Republika Srpska offers its nursery products to Public Forest Enterpise and other customers through its five nurseries, which are part of the modernly equipped Ṣeedlings and Nursery Center, seated in Doboj, providing services for cone, fruit and seeds processing, as well as a line for conveyor production of the latest generation purchased from BCC. The total area of nursery is 132 ha meanwhile only one third of the area is used to produce forest tree planting stock. The capacities of plant processing activities and nurseries in BiH exceed the demands for planting stock of these public forest enterprises (Mekić 2003). At the same time, the range of planting stock is unfavorable because over 90% of them are conifers (Ballian 2000; Mekić 2003; Mataruga et al. 2012).

1.5 Previous Research Work of the Forest Tree Species and Shrubs Diversity in BiH

The abundance of diversity is reflected in a large number of plant and animal species within the region (about 4500 higher plants, 600 taxa mosses and about 80 ferns, around 250 species of forest trees and shrubs – Brujić 2011). This is why the position of BiH is high in the list of European countries of a high biodiversity level. This diversity provides forest ecosystems with a better starting position to adapt to climate change, but at the same time put them in risk of losing rare and unique species. The main types of trees are: fir, spruce, white and black pine, beech, various oak species and a large number of broadleaf tree species (such as maple, elm, ash) along with fruit trees (cherry, apple, pear).

The species diversity is defined at the ecosystem, species and gene level. The species diversity in BiH is characterized by the highest degree of endemicity in Europe. It is believed that there are many more endemic species in the Balkan mountains than in other mountains that are geographically close to the Mediterranean area, but there are fewer phylogeographic research activities compared to other mountains (Hewit 2011). According to the IUCN criteria, there are over 600 taxa of higher plants of different conservation status, 250 vertebrate (fish, amphibians, reptiles, birds and mammals), dozens of invertebrate species, and 50 fungi species and lichens. In addition to the specific and structural diversity of forests in BiH, there is also an extremely high degree of landscape diversity which integrated all forms of geological and biological diversities in a wider sense.

According to preliminary findings, out of 5000 taxa of vascular plants, 10% of them are endangered (Red List of Protected Flora and Fauna of Republika of Srpska 2012; Red List of Flora of Federation BiH 2013). Due to the intensive habitat conversion, it is necessary to constantly monitor to what extent the natural gene pool has declined. The abundance of the gene pool is determined by differences which occur within the ecological habitat conditions in BiH and at the same time are of a high variability and conditioned by specific bio-ecological conditions, therefore they are organized into appropriate ecological vegetation units.

The basis for the adaptation to climate change lies in the variability of the species itself, ie the survival of resistant populations. This requires a research work of genetic variability in order to predict the level of adaptability of the species to negative impacts of the environment. Previous research activities which were based on a molecular level may be in the function of defining the conservation measures of the forest genetic resources in BiH, although it should be kept in mind that variations at the molecular level are not necessarily related to variations of biological functions (Ballian 2012). Along with the GIS data, they contribute to a better understanding of variation patterns for the purposes of improving management practices and tracking forest genetic resources flow. Analyses of molecular markers for several significant species were performed in Bosnia and Herzegovina. The preview is in Table 1.4.

As confirmed by all research findings in BiH, differences between populations or within populations are visible and clear when genetic parameters for differences were deployed. Apart from the natural selection, other possible causes of diversity may be anthropogenic activities, developmental factors or adaptation processes to specific ecological conditions.

The Dinarid area is very specific in terms of climatic, edaphic and orographic conditions which occur within a relatively small area, and at the same time directly

Table 1.4 Species and markers deployed for variation study

Species	Investigations by markers				References
	Biochemical	nSSR	cp DNK	mt DNK	
Silver fir	Yes	No	Yes	Yes	Ballian (2003a, b, 2009) and Gömöry et al. (2004)
Omorika	Yes	Yes	Yes	Yes	Ballian (2006), Ballian et al. (2006a, b, c), and Gömöry et al. (2007)
Norway spruce	Yes	Yes	Yes	Yes	Ballian (2007), Ballian et al. (2007a, b, 2009), Mataruga et al. (2014), and Cvjetković et al. (2017)
Austrian pine	Yes	No	No	No	Mataruga et al. (2007)
Ash	No	Yes	Yes	No	Ballian et al. (2008a, b)
Pendiculate oak	No	Yes	Yes	No	Slade et al. (2008) and Ballian et al. (2010a, b)
Sessile oak	No	No	Yes	No	Ballian et al. (2006a, b, c) and Slade et al. (2008)
Downy oak	No	No	Yes	No	Slade et al. (2008) and Ballian et al. (2010a, b)
Italian oak	No	No	Yes	No	Slade et al. (2008)
Schots pine	Yes	No	No	No	Ballian and Božič (2004) and Ballian et al. (2005)
Yew	Yes	Yes	Yes	Yes	Ballian et al. (2008a, b)
Croatia sibiraea	Ne	Yes	Yes	Yes	Ballian et al. (2006a) and Potter et al. (2007)
Black poplar	Ne	Yes	Yes	Yes	Kajba et al. (2004) and Hase et al. (2007)
Chesnut	Ne	Yes	Ne	Ne	Fussi et al. (2016)
Wild cheery	No	Yes	No	No	Ballian (2004)

Table is an extract from Ballian (2012) paper and updated with other research findings

affect the differentiation of various ecotypes. Numerous domestic and foreign experts believe that forest trees of Dinarid species show great variability if they are compared to the same species on the north. The influence of orographic factors, i.e. genetic adaptation to altitude and the formation of ecotypes, can be seen from physiological-genetic researches that have been carried out on many species in the world and therefore can be expected to occur in our country.

Therefore, the following standouts can be made on the basis of the obtained results

- The obtained variability is not typical of western and central Europe, only for the Balkans.
- Many of the studied gene loci of forest trees, with the exception of *Picea omorika*, show clonal variability, which therefore indicate adaptivity to certain ecological conditions, but there is a possibility that this is due to the colonization

effects. Haplotypes show high polymorphism in all populations, which may be the result of a disjuncted distribution area of the studied species in Bosnia and Herzegovina

- Some of the research activities in Bosnia and Herzegovina referred to the apenine origin of some populations. This is especially true for oak and silver fir which reveal specific qualities of western Dinarides.

Small and isolated populations, and in this case primarily those in the southern and western parts of Bosnia and Herzegovina, and some in the Posavina area, do not show any differences to the populations of the central Dinarids at the molecular genetic level. All gene loci that characterize this species have been isolated in the population, although the inbreeding load is noticed, for there is low heterozygosity and constant genetic drift phenomena. Also, in small populations the number of haplotypes is smaller than in large ones, resulting from the isolation and poor exchange of genetic material with other populations. Thus, in small and isolated populations, the influence of anthropomorphic changes in the last 2000 years has been noticed.

Based on the above mentioned information, it can be concluded that there are differences between populations growing in different ecological niches, these differences in habitat ecology are caused by genetic differentiation among populations and can be recorded by molecular and biochemical markers. This has been confirmed by a number of proven experiments conducted in Europe, including in BiH provenances. It is evident that specific selection processes are found in each population.

References

Agencija za statistiku Bosne i Hercegovina. (2014). Preliminarni rezultati popisa stanovništva u BiH.

Ballian, D. (2000). *Kvaliteta sadnog materijala u rasadnicima u Federaciji Bosne i Hercegovine. Seminar "Sjemensko-rasadnička proizvodnja u BiH – Aktuelno stanje i perspektive"* (pp. 76–78). Brčko. *Zbornik radova.*

Ballian, D. (2003a). Procjena genetičke varijabilnosti obične jele (*Abies alba* Mill.) analizom izoenzima u dijelu prirodnih populacija Bosne i Hercegovine i Hrvatske. *Šumarski List, 3–4*, 135–151.

Ballian, D. (2003b). Procjena genetičke varijabilnosti obične jele (*Abies alba* Mill.) analizom cpDNA u dijelu prirodnih populacija Bosne i Hercegovine i Hrvatske. *Šumarski List, 7–8*, 347–357.

Ballian, D. (2004). Varijabilnost mikrosatelitne DNK u populacijama divlje trešnje iz središnje Bosne. *Šumarski List, 11–12*, 649–654.

Ballian, D. (2006). Kontrola podrijetla Pančićeve omorike (*Picea omorika*/Panč./purk.) iz plantaže kod Kaknja pomoću izoenzimskih biljega. *Šumarski List, 7–8*, 295–304.

Ballian, D. (2007). Genetička struktura smreke (*Picea abies* Karst.) na planini Vlašić. *Radovi HDZU, 9*, 211–220.

Ballian, D. (2009). Genetičke strukture obične jele (*Abies alba* Mill.) iz zapadne i istočne Bosne. *Šumarski List, 9/10*, 501–512.

Ballian, D. (2012). *Genetic variability of forest trees in Bosnia and Herzegovina and important for their method preservation* (pp. 97–110). Academy of Sciences and Arts of Bosnia and Herzegovina Second International Colloqium "Biodiversity – Theoretical and practical aspects", 3 Dec 2010, Sarajevo, Bosnia and Herzegovina, Special Editions CXLVIII.

Ballian, D., & Božič, G. (2004). Kontrola morfološke identifikacije klonova iz sjemenske plantaže bijelog bora "Koziji grm" pomoću izoenzimskih markera. *Radovi Šumarski Fakultet U Sarajevu, 1*, 47–56.

Ballian, D., Bogunić, F., & Konnert, M. (2005). Usporedba molekularno genetičkih svojstava sjemenskih plantaža običnog bora (*Pinus sylvestris* L.) u Bosni i Hercegovini. *Radovi Šumarski Institute Jastrebarsko, 41*(2), 7–16.

Ballian, D., Grebenc, T., Božič, G., Melnik, V., Wraber, T., & Kraigher, H. (2006a). History, genetic differentiation and conservation strategies for disjunct populations of Sibiraea species from South-Eastern Europe and Asia. *Conservation Genetics, 7*(6), 895–907.

Ballian D., Ivanković M., & Slade D. (2006b). The distribution of the populations of sessile oak (*Quercus petraea* (Matt.) liebl) in Bosnia and Herzegovina by cloroplasts dna (cpDNA) and its application in forestry. In: Bohnens, J., & Rau, H.-M. (Hrsg.). Forstliche Genressourcen als Produktionsfaktor. 26. Tagung der Arbeitsgemeinschaft Forstgenetik und Forstpflanzenzüchtungvom 20. bis 22. Oktober 2005 in Fuldatal (pp. 61–66, 121–131). Hann. Münden: Hessen-Forst.

Ballian, D., Longauer, R., Mikić, T., Paule, L., Kajba, D., & Gömöry, D. (2006c). Genetic structure of a rare European conifer, Serbian spruce (*Picea omorika* (Panč.) Purk.). Plant Sistematics and Evolution, 260, 53–63 Springer.

Ballian, D., Bogunić, F., Konnert, M., Kraigher, H., Pučko, M., & Božič, G. (2007a). Genetička diferenciranost subpopulacija obične smreke (*Picea abies* (L.) Karst.) na planini Igman. *Šumarski List, 1/2*, 13–23.

Ballian, D., Bogunić, F., & Božič, G. (2007b). Genetička varijabilnost obične smreke (*Picea abies* /L./ H. Karst.) u bosanskom dijelu Dinarida. *Šumarski List, 6/7*, 237–246.

Ballian, D., Giersberg, B., & Tröber, U. (2008a). Genetička varijabilnost obične tise (*Taxus baccata* L.) u Bosni i Hercegovini. *Šumarski List, 9*(10), 431–443.

Ballian, D., Monteleone, I., Ferrazzini, D., Kajba, D., & Bellett, P. (2008b). Genetic characterization of common ash (*Fraxinus excelsior* L.) population in Bosnia and Herzegovina. *Periodicum Biologorum, 110*(4), 323–328.

Ballian, D., Bogunić, F., & Božič, G. (2009). Genetic Research of Norway spruce (*Picea abies* (L.) Karst.) from extreme areas of *Spaghno-Piceetum* in Slovenia & Bosnia and Herzegovina. *Dendrobiology, 61*(Supplement), 137–144.

Ballian, D., Ivanković, M., Gračan, J., Perić, S., Marjanović, H., Bobinac, M., & Slade, D. (2010a). Analysis of pubescent oak (Quercus pubescens Willd.) Ba means of chloroplast DNA (cpDNA) in the western part of Balkan peninsula. *Acta Societatis Botanicorum Poloniae, 79*(3), 189–195.

Ballian, D., Belletti, P., Ferrazzini, D., Bogunić, F., & Kajba, D. (2010b). Genetic variability of pedunculate oak (*Quercus robur* L.) in Bosnia and Herzegovina. *Periodicum Biologorum, 112*(3), 353–362.

Brujić, J. (2011). Zaštićena područja, rijetke i ugrožene vrste u RS i prioriteti u zaštiti šumskih genetičkih resursa. Predavanja na temu "Program očuvanja šumskih genetičkih resursa Republike Srpske". *Ljubić-Han Kram*, 01–02 Sept 2011.

CLC. (2006). Technical guidelines EEA Technical report No 17/2007.

Cvjetković, B., Konnert, M., Fussi, B., Mataruga, M., Šijačić-Nikolić, M., Daničić, V., & Lučić, A. (2017). Norway spruce (*Picea abies* karst.) variability in progeny tests in Bosnia and Herzegovina. *Genetika, 49*(1), 259–272.

FAO. (2015). The forest sector in Bosnia and Herzegovina preparation of IPARD forest and fisheries sector reviews in Bosnia and Herzegovina (pp. 1–146).

Fussi, B., Petropoulos, S., Daničić, V., Aravanopoulos, F., & Konnert, M. (2016). Genetic diversity of Sweet chestnut (*Castanea sativa* Mill.) populations of Central and South- eastern European origin. IUFRO. Forest genomics and Forest Tree Genetics, *Book of abstracts* (p. 49).

Gömöry, D., Longauer, D., Liepelt, S., Ballian, D., Brus, R., Kraigher, H., Parpan, V. I., Stupar, P. I., Paule, L., & Ziegenhagen, B. (2004). Variation patterns of mitochondrial DNA of *Abies alba* Mill. in suture zones of postglacial migration in Europe. *Acta Societatis Botanicorum Poloniae, 73*(3), 203–206.

Gömöry, D., Paule, L., & Ballian, D. (2007). Genetic aspect of the conservation of rare endem tree species: A case study of *Picea omorika* (Panč.) Purk. *Geobiocenologicke Spisy, Brno, 12*, 73–82.

Hase, I., Heinze, B., Arthofer, W., Kajba, D., & Ballian, D. (2007, January 13–17). Candidate genes for hairiness. In *Poplar. Plant & animal genomes XV conference*. San Diego: Town & Country Convention Center. Books of abstracts.

Hewit, G. (2011). Mediterranean Peninsulas: The evolution of hotspots. In F. E. Zachos & J. Christian (Eds.), *Habel biodiversity hotspots distribution and protection of conservation priority areas* (pp. 123–147). Heidelberg/Dordrecht/London/New York: Springer.

http://www.lib.utexas.edu/maps/europe/bosnia_herzegovina_land_1993.jpg (On February 23rd 2017).

http://www.vidiani.com/physical-map-of-bosnia-and-herzegovina/ (On February 13th 2017).

Kajba, D., Ballian, D., Heinze, B., Idzojtic, M., & Bogdan, S. (2004). *Populus nigra* ssp. *caudina* and its importance for forest tree improvement and conservation of poplar genetic resources. In *IPC 2004 – Chile, 22nd session, International Poplar Commission*, Santiago – Chile, 28 November – 9 December 2004. Book of abstracts.

Marić, B., Avdibegović, M., Blagojević, D., Bećirović, D. Ž., Brajić, A., Mutabdžija, S., Delić, S., & Pezdevšek Malovrh, Š. (2012). Conflicts between forestry and wood-processing industry in Bosnia and Herzegovina: Reasons, actors and possible solutions. *South-East European Forestry, 1*, 41–48.

Mataruga, M., Isajev, V., Konstantinov, K., Mladenović-Drinić, S., & Daničić, V. (2007). Protein in seed and seedlings of selected austrian pine (*Pinus nigra* Arnold) trees as genetic markers tolerant to drought. *Genetika, 39*(2), 259–272.

Mataruga, M., Isajev, V., Balotić, P., Rose, R., & Wu, S. (2012). Forest seed and seedling production in Republic of Srpska (Bosnia & Herzegovina) – Current status and future development. International scientific conference "Forestry science and practice for the purpose of sustainable development of forestry – 20 years of Faculty of Forestry in Banja Luka". In *Proceedings* (pp. 621–634).

Mataruga, M., Galović, V., Isajev, V., Orlović, S., Cvjetković, B., Daničić, V., & Balotić, P. (2014, September 28–October 2). Genetic characterization of Norway spruce (*Picea abies*/L./Karst.). in Bosnia and Herzegovina. In *Fifth congress of the Serbian Genetic Society*, Belgrade. Book of abstracts (p. 248).

Mekić, F. (2003). Dinamička trajnost načelo temelja šumarske politike. Stanje i perspektive proizvodnje sadnog materijala u rasadnicima Federacije Bosne i Hercegovine, Sarajevo 18 Dec 2003. Šumarski fakultet Univerziteta u Sarajevu, Katedra za uzgajanje šuma i urbanog zelenila (pp. 7–168).

Musić, J., Obućina, M., Gurda, S., & Halilović, V. (2013). Odnosi šumarstva i drvne industrije u Federaciji BiH. *Nova Mehanizacija Šumarstva, 34*, 51–60.

Myers, N., Mittermeier, R. A., Mittermeier, C. G., da Fonseca, G., & Kent, J. (1999). Biodiversity hotspots for conservation priorities. *Nature, 403*, 853–858.

NEAP – Akcioni plan za zaštitu okoliša BiH. (2003). Federalno ministarstvo prostornog uređenja i okoliša i Ministarstvo za urbanizam, stambeno-komunalne djelatnosti, građevinarstvo i ekologiju Republike Srpske (pp. 6–123).

Potter, D. S., Still, M., Grebenc, T., Ballian, D., Božič, G., Franjić, J., & Kraigher, H. (2007). Phylogenetic relationships in tribe *Spiraeeae* (*Rosaceae*) inferred from nucleotide sequence data. *Plant Sistematics and Evolution, 266*, 105–118 Springer.

Red List of Protected Flora and Fauna of Republika of Srpska. (2012). Official Gazette of Republic of Srpska no. 124/12.

Red List of Flora of Federation BiH. (2013). Official Gazette of Federation BiH No. 7/14 (2014).

Slade, D., Skvorc, Z., Ballian, D., Gracan, J., & Papes, D. (2008). The chloroplast DNA polymorphisms of White Oaks of section *Quercus* in the Central Balkans. *Silvae Genetica, 57*(4–5), 227–234.

Stupar, V. (2011). *Nature conservation in Bosnia and Hercegovina – Problems and perspective* (pp. 7–14). Bulletin of Faculty of Forestry University of Banja Luka.

UNDP. (2014). *Possibilities of using biomass from forestry and wood industry in Bosnia and Hercegovina* (pp. 1–21).

Chapter 2
Wild Fruit Tree Species of Montenegrin Forests

Milić Ćurović, Miodrag Jovančević, and Jasmina Balijagić

Abstract Wild fruit woody species without any doubt present an essential element of biodiversity of Montenegrin forests. The aim of this paper was to examine the geographical distribution of these species as a basic precondition for their monitoring and protection. According to the National Forest Inventory in Montenegro, there are seven wild fruit tree species present: chestnut (*Castanea sativa* Mill.), Turkish hazelnut (*Corylus colurna* L.), walnut (*Juglans regia* L.), wild apple (*Malus sylvestris* Mill.), wild cherry (*Prunus avium* L.) and wild pears (*Pyrus communys* L. and *Pyrus amigdaliformis* Vill.). Seven analyzed species are spread over an area of 9498 ha which takes up 0,69% of the territory of Montenegro. Most common wild fruit tree species in Montenegro are wild pear, wild cherry and wild apple. The highest values of average timber dimensions show wild cherry, chestnut and Turkish hazelnut. Further analysis of the distribution according to the altitude shows which species are located mainly on terrains of lower altitude and which species spread mainly at altitudes above 800 m and how altitude influenced the diversity of these species. Distribution and main characteristics of Montenegrin wild fruit tree species shows potential for establishment of collecting site areas as first step of in situ conservation.

Keywords Wild fruit · Forest · Tree species · Montenegro

2.1 Introduction

Intensive forest use followed by management and silviculture measures oriented to reach the maximum economic forest valorization, resulted in establishment of more homogeneous forest stands and made wild fruit trees scarce. Importance of preservation of wild fruit tree species is not only in increasing the forest ecosystem biodiversity but also in obtaining of valuable wood, to have unique generative rootstock

M. Ćurović (✉) · M. Jovančević · J. Balijagić
Biotechnical Faculty, University of Montenegro, Podgorica, Montenegro
e-mail: curovic@t-com.me; mijojovancevic@gmail.com; jas.be@t-com.me

M. Šijačić-Nikolić et al. (eds.), *Forests of Southeast Europe Under a Changing Climate*, Advances in Global Change Research 65,
https://doi.org/10.1007/978-3-319-95267-3_2

as well as possibilities to use fruits. Sustainable growing and use of wild fruit tree species could reduce wood import from tropical regions. Tending to meet the demand for walnut (*Juglans regia* L.) and wild cherry (*Prunus avium* L.) hardwood timber from national production, Italian forest authorities have started with developing tree-farming programmes (Ducci et al. 2006).

The fruit trees often have dual-purpose and produce a timber as end-product, especially wild cherry (*Prunus avium* L.), chestnut (*Castanea sativa* Mill.), walnut (*Juglans regia* L.), pear (*Pyrus communis* L.) and apple (*Malus sylvestris* Mill.). Silvoarable systems based on fruit production (including nuts) covered extensive tracts in central Europe as recently as the last century (Eichhorn et al. 2006).

One of the very important elements of preserving the genetic diversity of Montenegrin forests is certainly the preservation and protection of wild fruit tree species. A comprehensive knowledge about the spatial pattern of genetic diversity is necessary for conservation of forest genetic resources (Jolivet et al. 2011). Spatial and species-specific variability of tree fruit in European forests have been poorly studied (Fernández-Martínez et al. 2017).

Research on wild cherry occurrence in forest stands in Czech Republic resulted with plus trees certification and establishment of seed orchards and clone archives (Kobliha 2002). Wild cherry occurrence is connected with fertile, warm sites and stand edges in Germany (Kleinschmit et al. 1996) but it occurs in the whole territory of Spain but only as individual trees (Miranda 1996).

It is also belived that walnut (*Juglans regia* L.) appeared in Europe around 5000 BC (Huntley and Birks 1983). It was utilized for fruit, timber, coloring and as a source of many medical compounds. In the Balkan region natural populations of walnut are characterized by a large polymorphism, but they are dominated by trees with small fruits (Korać et al. 1990).

European chestnut (*Castanea sativa* Mill.) is highly valued, and therefore broadly cultivated, for its fruit and timber. It proved to be important not only for rural diversification, local economies and cultures, but also for its direct contribution to the landscape and environment. The genetic research encompassed 82 populations that were selected in France, Greece, Italy, Spain and the United Kingdom (Villani et al. 2006).

The wild apple (*Malus sylvestris* Mill.) is more important as generative rootstock than for fruit use. In terms of generative rootstock production, wild apple with generally smaller fruit mass is more commercial in comparison to the autochthonous apple varieties. Topical issue is how to preserve wild apples. Nevertheless, germplasma of wild apple in less urban areas in Montenegro is rather preserved (Šebek 2013).

Forest genetic resources are unique resource whose conservation is urgent. The genetic resources of Montenegrin forests are not introduced in an appropriate manner internationally, and it should be an imperative to include Montenegro in global initiatives in the field of genetic resource preservation (Lazarevic 2012). The forest genetic resources preservation refers to a set of actions and strategies that ensure the continuous existence, expansion and availability of these resources for the present and future generations (Šijačić-Nikolić and Milovanović 2007). First steps in the preservation of genetic resources of fruit tree species are to explore, collect, store, and protect genetic materials of wild species (Gixhari et al. 2014).

In this paper we analyzed data of spatial distribution of seven main wild fruit tree species of Montenegrin forests: chestnut (*Castanea sativa* Mill.), Turkish hazelnut (*Corylus colurna* L.), walnut (*Juglans regia* L.), wild apple (*Malus sylvestris* Mill.), wild cherry (*Prunus avium* L.) and wild pears (*Pyrus communys* L. and *Pyrus amigdaliformis* Vill.). Beside that we analyzed values of basic measurement elements: Area of occurrence, mean diameter (Ds), mean high (Hs) and timber volume per ha (V/ha) on different altitude according data of the First National forest inventory (NFI) in Montenegro (MARD 2013).

2.2 Material and Methods

Council Directive 1999/105/EC demands each EU country to establish a provenance regions system (PRs) for regulation of the movement and use of propagation materials. PRs for forest species should be the basic management units for forest genetic resources. According to Ducci and Vannuccini (2006), the accurate forest inventory data are of utmost importance as species' geographical distribution is essential information when defining species-specific PRs. Base of this research were data obtained from the First National forest inventory (NFI) in Montenegro, data of regular Inventory and field research on several locations.

In the NFI of Montenegro, information is assessed for sample plots which are distributed over the land area of Montenegro as follows: The clusters network with the sample plots is spread through entire territory of Montenegro and is based on a basic raster – a regular 2 × 2 km square raster, set up by *Gauss-Krüger's coordinate system and projection* – zone 6. On each square raster, there is the cluster comprising four sample plots, positioned in the tops of the square whose leg length is 200 m. The sample plot is made up of four concentric circles The radius measurements are conducted and the following values are obtained: the radius of the first circle is $r_1 = 5.64$ m ($p_1 = 1$ *acre* on plain terrain), of the second circle is $r_2 = 7.98$ m ($p_2 = 2$ *acres*), of the third circle $r_3 = 12.62$ m ($p_3 = 5$ acres) and the radius of the fourth circle is $r_4 = 17.84$ m ($p_4 = 10$ *acres*). Diameter of the first concentric circle is measured on all trees above 10 cm *at* high forests, while taxation limit for coppice woodland is 5 cm. Diameters of the second concentric circle are measured on all trees above 15 cm, the third circle diameters were measured on all trees above 30 cm and the measurement of the fourth circle diameters were conducted on the trees above 50 cm.

In a regular inventory data were obtained by collecting in a network of permanent and non permanent points which are placed on grid 100 × 100 m at a level of Management units. Field research was done by total measurement in simple plots of 10 acres.

2.3 Distribution and Characteristics

According to NFI data, seven wild tree fruit species are determined: Chestnut (*Castanea sativa* Mill.), Turkish hazelnut (*Corylus colurna* L.), Walnut (*Juglans regia* L.), Wild apple (*Malus sylvestris* Mill.), Wild cherry (*Prunus avium* L.) and Wild pears (*Pyrus communys* L. and *Pyrus amigdaliformis* Vill.) and their distribution is shown on the map (Fig. 2.1).

Montenegrin wild fruit tree populations are distinguished by its excellent properties and are adapted to ecological conditions. The hills are abundant with the biotypes of wild apples, pears and cherries. The wild species of these fruit trees grow here in the parish, hilly and mountainous regions of altitude up to 1500 m, and have them at all exposures and inclinations where the deciduous trees grow. Apples and pears are an integral part of the forest, and grow in associations with: beech, white oak, birch, and then along the borders, next to the fields, in the meadows and slopes of the forest, and are located mostly along the streams.

Surface on which occur analized seven species (Table 2.1) is 9498 ha which takes up 0,69% of the territory of Montenegro. Most common wild fruit tree species in Montenegro are wild pear, wild cherry and wild apple. These three species make 85,7% of the area under wild fruit tree species in Montenegro.

In some earlier reports of the presence of Noble hardwood tree species in Montenegro, it was noted that only single species of wild cherry (*Prunus avium*), wild apple (*Malus sylvestris*) and wild pear (*Pyrus pyraster*) were present in the woods of sessile oak–hornbeam, mountainous beech, maple and ash (Orlovic et al. 2006).

The smallest representation of wild fruit tree species has almond-leaved pear and chestnut. Chestnut is recorded in just a couple locations in the southern part of Montenegro. The largest populations of chestnut are located in Boka Kotorska and on the southeastern slopes of Mt. Rumija to the vicinity of Skadar Lake on area of 300 ha of very heterogeneous genotypes of chestnut. Population genetic variability is very high and gene expression is highly affected (Odalovic et al. 2013).

The highest values of medium tree height were recorded of wild cherry (14,2 m), Turkish hazelnut (12,6 m) and chestnut (11,8 m), while highest values of average stem diameter show chestnut (57,1 cm), Turkish hazelnut (41,0 cm) and then wild cherry (29,0 cm). The average dimensions of the trees are a clear indicator of the timber volume. These three species shows the highest values of timber volume.

Further analysis of the distribution according to the altitude (Table 2.2) shows that chestnut is located exclusively on terrains of less than 800 m above sea level. Similar is to walnut, where only 12% of the occurence are above the altitude of 800 m. The spread of Turkish hazelnuts, wild apples, wild pears and wild cherries is mainly at altitudes above 800 m.

In an area with a relief, such as in Montenegro, altitude is the main factor affecting the tempo of plant development. Altitude conditions the vertical change of climatic elements. Stronger winter frosts and shorter vegetation at higher altitudes could be limited factor for fruit trees.

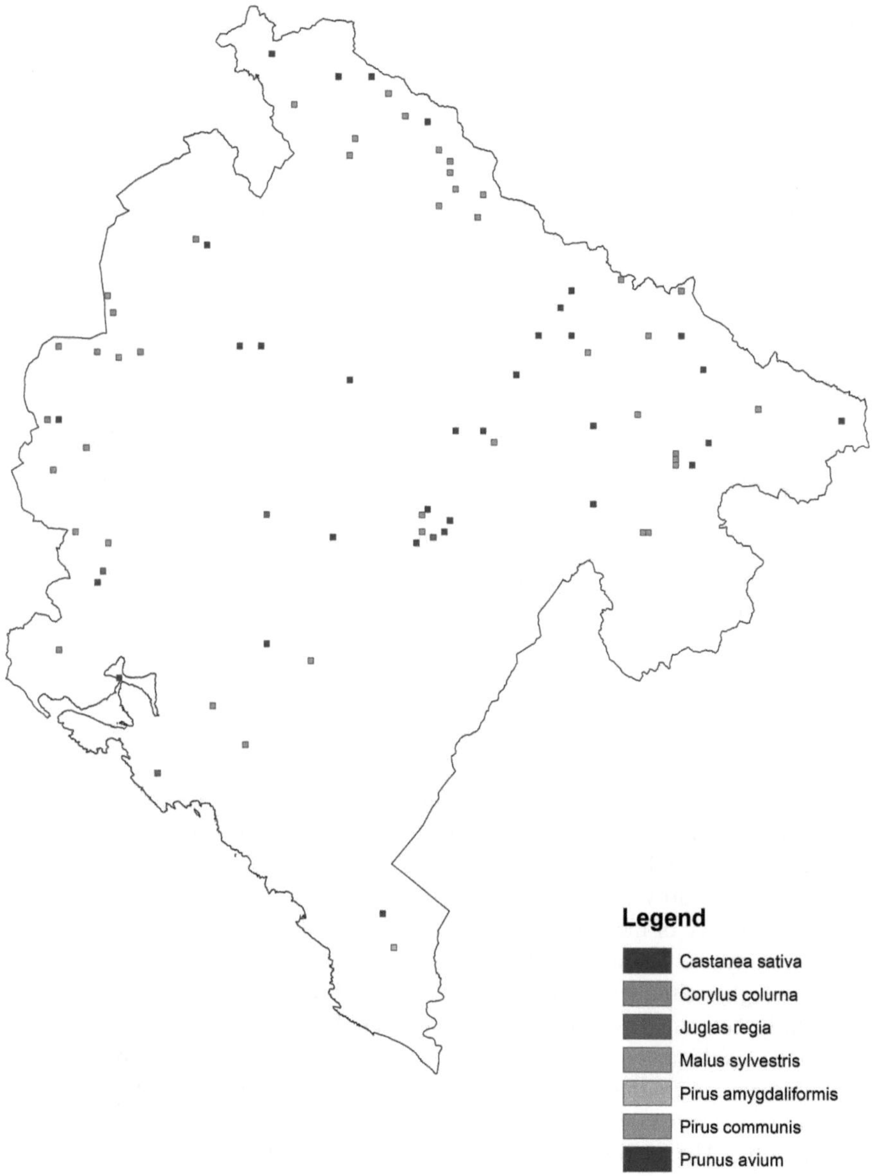

Fig. 2.1 Geographical distribution of wild fruit tree species according to the Montenegrin Forest Inventory

Climate change will affect forest ecosystems primarily in moving of the vegetation vertical layout belts. In Montenegrin forests climate changes could have the greatest impact on the species with narrow ecological valence and the smallest spread areas such as wild fruit tree species. Resistance of the forests to climate

Table 2.1 Area of occurrence, mean stem diameter (Ds), mean high (Hs) and timber volume per ha (V/ha) of Montenegrin wild fruit tree species

Species	Area (ha)	Ds (cm)	Hs (m)	V (m^3/ha)
Castanea sativa Mill.	193,39	57,1	11,8	120,0
Corylus colurna L.	631,92	41,0	12,6	68,2
Juglans regia L.	413,15	26,5	11,4	46,8
Malus sylvestris Mill.	1.229,47	26,0	10,2	32,4
Prunus avium L.	3360,26	29,0	14,2	73,8
Pyrus communys L.	3554,62	22,8	11,0	42,9
Pyrus amigdaliformis Vill	115,17	15,0	8,8	44,1

Table 2.2 Distribution at different altitudes (%)

Altitude (m a.s.l.)	*Castanea sativa* Mill.	*Corylus colurna* L.	*Juglans regia* L.	*Malus sylvestris* Mill.	*Prunus avium* L.	*Pyrus communys* L.	*Pyrus amigdaliformis* Vill
0–200	43,5	2,1	34,5	10,1	0,1	0	0
201–400	6,2	4,0	22,9	1,8	0,5	5,2	0
401–800	49,7	16,1	30,7	8,1	15,2	11,0	56,7
801–1000	0,6	40,3	10,6	30,8	26,6	34,4	26,4
1001–1200	0	36,4	1,4	46,6	29,1	31,4	16,9
˃1200	0	1,0	0	2,5	28,4	18,0	0

change depends on their surface, the degree of naturalness, of biodiversity and structure. To enable easier and better adaptation of forests to climate change, management should be directed to save natural composition of the given stands and to save diversity (Curovic and Spalevic 2010).

The resistance of the walnut to low temperatures depends on the characteristics of the genotypes. The damage from the late spring frosts is quite frequent, especially at higher altitudes in the continental part of Montenegro. In the continental part, early autumn frosts are also causing significant damage. However, natural selection significantly eliminates such genotypes from natural populations. Therefore, in the continental part walnuts with shorter vegetation are represented.

Fruit size (from 4.0 to 19.5 g) in natural populations of walnut in Montenegro is very variable. The quality of the fruit, beside to the properties of the genotype, also depends on the altitude of the site. At higher altitudes fruits are smaller. Genotypes of walnut in Montenegro are characterized by small and medium weight fruits, high seed germination and vitality of seedlings (Jovančević and Balijagic 2009).

Distribution at different altitudes is important also due to the dynamics of phenological phases. The average flowering of Wild pears genotypes for example, at an altitude of 750–850 m is April 30, and at the altitude of 1250–1350 m wild pear trees bloom on May 22. The size of the vertical phenological gradient, in the conditions of Montenegro is 3.66 days (Jovančević et al. 1999).

Wild pears genotypes that have the ability to use autochthonous varieties and cultivars (at lower altitudes) as pollinators give seedlings of higher variability than genotypes that can be used mainly wild pearls as pollinators at higher altitudes.

The differences in the genotype of wild pear have caused differences in pollen germination, which is pear in the interval from 56.6% to 99.9% (Jovančević et Vujanić-Varga 1997).

2.4 Conclusions

Due to its generally scattered and rare occurrence, the genetic diversity of populations of wild fruit tree species of Montenegrin forests can be considered to be under threat from the various factors such as:

- The felling of trees and destructions of habitat
- Transfer of seedlings from areas with different ecological conditions
- Non-organized collection of seeds
- Pest and disease
- Low natural regeneration and competition with other species.

The progress in wild fruit trees gene conservation in Montenegro depends on financial support and co-ordination at the state level. Definition of plus trees or seed stands followed by establishment of seed orchards should start immediately for wild fruit tree species. This will cover gene resource conservation and an increase in the utilization of the species' genetic potential. Distribution of wild fruit tree species as a first step in identifying potential collecting sites areas for in situ conservation shows that there is still time and place for such intervention.

The creation of a realistic information base of structural characteristics is a first step in wild fruit tree species protection and needed condition in order to be able to adequately monitor the situation of these species. Limited spread of this species gives even greater importance of the results presented in this paper.

References

Curovic, M., & Spalevic, V. (2010, March 9–10). Climate changes impacts on forests in Montenegro mitigation and adaptation. In *International conference – Forest ecosystems and climate changes* (pp. 283–287). Belgrade: IUFRO, EFI, Institute of Forestry Belgrade.

Ducci, F., & Vannuccini M. (2006). Definition of provenance regions for chestnut (*Castanea sativa* Mill.) and flowering ash (*Fraxinus ornus* L.) in Tuscany. In Bozzano, et al. (Eds.), *Noble hardwoods network* (pp. 43–51). Report of the sixth (9–11 June 2002, Alter do Chão, Portugal) and seventh meetings (22–24 April 2004, Arezzo, Italy).

Ducci, F., Germani, A., Janin, G., Proietti, R., & Signorini, G. (2006). Clone selection for wild cherry (*Prunus avium* L.) with special reference to some traits used. In Bozzano, et al. (Eds.), *Noble hardwoods network* (pp. 53–61). Report of the sixth (9–11 June 2002, Alter do Chão, Portugal) and seventh meetings (22–24 April 2004, Arezzo, Italy).

Eichhorn, M. P., Paris, P., Herzog, F., Incoll, L. D., Liagre, F., Mantzanas, K., Mayus, M., Moreno, G., Papanastasis, V. P., Pilbeam, D. J., Pisanelli, A., & Dupraz, C. (2006). Silvoarable systems in Europe – Past, present and future prospects. *Agroforestry Systems, 67*, 29–50.

Fernández-Martínez, M., Vicca, S., Janssens, I. A., Espelta, J. M., & Peñuelas, J. (2017). The role of nutrients, productivity and climate in determining tree fruit production in European forests. *New Phytologist, 213*(2), 669–679, 11p. https://doi.org/10.1111/nph.14193.

Gixhari, B., Dias, S., Hodaj, B., Ismaili, H., & Vrapi, H. (2014). Geo-information analysis of fruit trees species in Albania. *Agriculture & Forestry, 60*(1), 63–74.

Huntley, B., & Birks, H. J. B. (1983). *An atlas of past and present pollen maps for Europe: 0–13000 years ago* (pp. 238–242). New York: Cambridge University Press.

Jolivet, C., Höltken, A. M., Liesebach, H., Steiner, W., & Degen, B. (2011). Spatial genetic structure in wild cherry (*Prunus avium* L.): I. variation among natural populations of different density. *Tree Genetics & Genomes, 7*(2), 271–283.

Jovančević, M., & Vujanić-Varga, D. (1997). Karakteristike populacije Pyrus sp. u području Bijelog Polja. *Contemporary Agriculture*, Novi Sad, *3–4*, 101–108.

Jovančević, M., & Balijagic, J. (2009). Characteristics of walnut population (*Juglans regia* L.) in the area of Andrijevica, Berane, Bijelo Polje and Rozaje. *Contemporary Agriculture, 58*(3–4), 129–135.

Jovančević, M., Šebek, G., & Balijagić, J. (1999). Biological gharacteristic of pear (*Pyrus comuunis* L.) in dependence of the height above the sea level. *Contemporary Agriculture, 48*, 5–6 str 75–78. Novi Sad.

Kleinschmit, J., Stephan, B. R., Liesebach, M., Schiite, G., & Steinhoff, S. (1996, March 24–27). Noble hardwood species in Germany: Occurrence and gene conservation measures. In Turok, et al. (Eds.), *Noble hardwoods network. Report of the first meeting, Escherode, Germany* (pp. 101–111).

Kobliha, J. (2002). Wild cherry (*Prunus avium* L.) breeding program aimed at the use of this tree in the Czech forestry. *Journal of Forest Science, 48*(5), 202–218.

Korać, M., Cerović, S., Gološin, B., & Miletić R. (1990) Population variability of domestic Walnut and selection results. *Jugoslovensko Voćarstvo*, Čačak, *94*, 3–10.

Lazarevic, J. (2012). Genetic resources in the forestry sector of Montenegro. *Agriculture & Forestry, 57*(1), 51–71.

MARD (Ministry of Agriculture and Rural Development Montenegro). (2013). *The first national forest inventory of Montenegro* (pp. 1–347). Podgorica.

Miranda, R. A. (1996, March 24–27). Noble hardwoods in Spain: Conservation of genetic resources. In Turok, et al. (Eds.), *Noble hardwoods network. Report of the first meeting, Escherode, Germany* (pp. 23–25).

Odalovic, A., Prenkic, R., Dubak, D., Jovancevic, M., Cizmovic, M., & Radunovic, M. (2013). Effect of ecological conditions on expression of biopomological characteristics of chestnut (*Castanea sativa* Mill.) in natural populations of Montenegro. *Genetika, 45*(1), 251–260.

Orlovic, S., Matovic, N., & Pilipovic, A. (2006). Noble hardwoods in Serbia and Montenegro. In Bozzano, et al. (Eds.), *Noble Hardwoods Network. Report of the sixth (9–11 June 2002, Alter do Chão, Portugal) and seventh meetings (22–24 April 2004, Arezzo, Italy)* (pp. 21–24).

Šebek, G. (2013). Morphological characteristics of fruits of selected types of wild apples (*Malus silvestris* L.) in the area of Bijelo Polje. *Agriculture & Forestry, 59*(2), 167–173.

Šijačić-Nikolić, M., & Milovanović, J. (2007). Conservation and sustainable use of forest genetic resources. *Bulletin of the Faculty of Forestry*, Belgrade, *95*, 7–21.

Villani, F., Eriksson, G., Bucci, G., Aravanoupoulos, P., Botta, R., Diamandis, S., Fernandez Lopez, J., Garrod, G., Robin, C., Romane, F., Russell, K., Vannini, A., Akkak, A., Alizoti, E., Barreneche, T., Blanco Silva, R., Buck, E., Casasoli, M., Cherubini, M., Drouzas, A., Grandjanny, M., Grossman, A., Kremer, A., Lauteri, M., Marinoni, D., Mattioni, C., Mavrogiannis, M., Monteverdi, C., Perlerou, H., Pliura, A., Sansotta, A., Scarpa, R., Spalato, F., Vettraino, A., & Zas, R. (2006). Complexity of gene resources conservation and utilisation of a differently managed multipurpose tree species (*Castanea sativa* Mill.): what, where and how to conserve. *Advances in Horticultural Science, 20*(1, Special issue on Chesnut (2006)), 117–129.

Chapter 3
Forests, Forestry and the Slovenian Forest Genetic Resources Programme

Hojka Kraigher, Marko Bajc, Gregor Božič, Robert Brus, Kristjan Jarni, and Marjana Westergren

Abstract The chapter gives an overview of the development of forestry in Slovenia in the light of managing and conserving forest genetic resources. Special attention is given to the description of the legislative framework, the international collaboration, the extent and functioning of the Slovenian forest gene bank, forest genetic monitoring as the necessary precondition of managing and conserving forest genetic resources in the changing climates, and the Slovenian forest genetic resources programme.

Keywords Forest genetic resources · SIFORGEN · Slovenian Forest Gene Bank · Sustainable forest management · Climate change · Slovenia

3.1 Introduction

Slovenia is characterized by its heterogeneity in topography, diversity in climate and ecological conditions, its biogene origin, resulting in high biodiversity, and large areas of well preserved forests, conserving the natural biodiversity at different levels (Fig. 3.1). The climate is from mild submediterranean to continental and alpine, and country reaches from the Adriatic Sea to the highest peak at 2864 m a.s.l. The phytogeographic territories, which correspond approximately also to the provenance regions, include Alpine, Prepannonian, Dinaric and Submediterranean regions with a mosaic of intermediate ones, determined as Predinaric and Prealpine. The ground rock material is predominately calcareous limestone, dolomite, flisch

H. Kraigher (✉) · M. Bajc · G. Božič · M. Westergren
Slovenian Forestry Institute, Ljubljana, Slovenia
e-mail: hojka.kraigher@gozdis.si; marko.bajc@gozdis.si; gregor.bozic@gozdis.si; marjana.westergren@gozdis.si

R. Brus · K. Jarni
Department for Forestry and Renewable Forest Resources, Biotechnical Faculty, University of Ljubljana, Ljubljana, Slovenia
e-mail: robert.brus@bf.uni-lj.si; kristjan.jarni@bf.uni-lj.si

M. Šijačić-Nikolić et al. (eds.), *Forests of Southeast Europe Under a Changing Climate*, Advances in Global Change Research 65,
https://doi.org/10.1007/978-3-319-95267-3_3

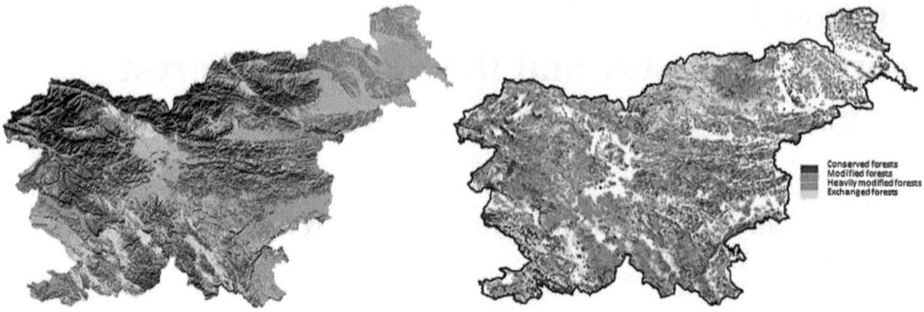

Fig. 3.1 (**a**) Slovenian topography (http://www.hervardi.com/zemljevidi_slovenije.php). (**b**) Forests cover around 60% of the Slovenian territory, and are well preserved (green: species composition equal to (dark green) or very similar to the natural one (light green), only in 13% of forest tree species composition is different form the natural one (yellow colours; Westergren et al. 2006)

and gneiss, with some metamorphic and magmatic areas (Vrščaj et al. 2017). Only one region is characterized by magmatic material, the Pohorje region. The average inclination of the territory is 40%, and the same is the share of the areas with specific karst phenomena, as described by J.V. Valvasor in his The Duchy of Carniola (1689), with the terms from Slovenian language from the Karst region in SW Slovenia defining them. Namely, when water reacts with limestone ground rock material it forms a number of characteristic structures (such as škraplje), valleys with soil piled in the center (dolina, vrtača, koliševka), karst fields (polje) which are flooded as lakes a part of the year, a number of caves with different underground formations, while water is scarce as it disappears into the ground forming streams inaccessible to plants and humans, and forming underground habitats for some highly specialized animals (*ibid.*). The soils on limestone are strongly dependent on the vegetation, and if the forest is cut, the soil disappears along with it, and the ground rock material is exposed for centuries or millennia before the soils are reformed (Vrščaj et al. 2017).

From approximately 3300 plant species, around 350 are woody plants, around 75 are forest tree species (Mala Flora Slovenije, Martinčič et al. 2007; Brus 2004) and 22 are endemic to Slovenia. Including the animal and fungal component, Slovenia belongs to the European 'biodiversity hotbelt' (Mršić 1997). The only national park, the Triglav national park, has been established in 1924, and at present 50% or 41616 ha within it are forests; if also *Pinus mugo* were considered as forest, then 63% or 52965 ha of the national park are forests (Klopčič et al. 2015). In addition to the national park, there is a number of other nature conservation areas of different categories; while in total as much as 37,16% of Slovenian territory is included into NATURA 2000 areas, from which about 29% is in different nature conservation categories and 71% is forests (ZGS 2016).

The present forest area is close to 60% of the total country area with 1183433 ha of forests (1074985 ha managed forests, 98947 protective forests and 9501 forest reserves), which is the third to fourth position regarding the share of forests among the European countries. Forests are, besides water, the most important renewable natural resource in Slovenia. Most are located within the area of beech, fir-beech and beech-oak sites (70%), with high production capacity (Perko 2007). Forests are the key to biodiversity conservation, carbon storage and the provision of renewable raw materials. In Slovenia, they also significantly contribute to air and ground water quality, deterioration of which would lead to an increase in human and animal health issues and would have a negative economic impact. The total growing stock is 337 Mm^3, the average annual increment is 285 m^3 and the yearly possible cut is 5,7 Mm^3 (ZGS 2016). Around 70% of all forests are dominated by beech associations, however the growing stock of beech is only around 32% of the total, while the share of the growing stock is 31% for Norway spruce, 7,5% for silver fir, 5,6% for pines, 1,5% for larch and other conifers, 7% for oaks, 5,2% for noble hardwoods, 8% for other hardwoods, and 1,7% of soft broadleaves (*ibid*).

Forests in Slovenia provide 0,3% GDP, with an increasing tendency, and provide more than 6000 full day work equivalent jobs. Forests and the forest based industry in Slovenia has reached its minimum in 2012 of providing for the whole sector (including forestry, woodworking and pulp and paper industry) merely 1% GDP and 24000 employees, however, the sector has regained its growth in the last 5 years (MKGP 2017).

As the main renewable resource, the need for research and professional based support for future forests is growing, considering especially the large-scale disturbances and climate change related outbreaks of pests and diseases decimating a number of tree species, and largely affecting their distribution areas. A great responsibility for forestry and forest science is to provide support to management, planning and protection of forests and the conservation of the exceptional biodiversity found in Slovenian forests.

3.2 The 'Slovenian Forestry School'

3.2.1 History: Ordnances and Recommendations by Maria Teresia 1774

Historically the first forest ordnances were linked to mining ordnances (Mihelič 2008), aiming at sustainable wood production for mining purposes. However, these were early on followed by forest ordnances, as overviewed by Boštjan Anko (cit. by Zupančič 2013), who organized reprinting of facsimile of a number of forest ordnances: The forest ordnance from Ortenburg 1406, 1985; The Forest Ordnance of Maria Teresia for Carniola 1771, 1985; The Bamberg forest ordnances for the Kanal valley and Bleiberg 1584, 1987; The Temporary forest ordnance for Stiria 1539, 1987; The forest ordnance for Illyrian provinces, 1810, 1989; The forest ordnance for Istria, Furlania and Karst, 1541, 1989; The forest ordnance for Lower Austria, 1813, 1989; and the five-language manuscript by Josef Ressl from 1842 Planning of reforestation of community lands in Istria, 1993. These ordnances, followed by forest management plans, recommended that no clearcutting is done in forests on sensitive karst terrains, in which any plant cover removal would lead to soil erosion, resulting in exposure of bare ground limestone rock material.

The sustainable forest management persisted through centuries, yet the German forestry school had an important influence by actively favorizing Norway spruce at the cost of European beech distribution in actual vegetation cover in Slovenian forests. In addition, the land-use change into agricultural lands led the area of forests to reach only about 38% of the total area of Slovenia by mid twentieth century.

3.2.2 Silviculture Based on Forest Genetics

After the Second World War and a brief episode of large exploitation of forests (of about 5 years), clearcutting was again forbidden by the law (from 1949 onwards), while abandoned agricultural lands underwent natural reforestation. At establishment of the Slovenian Forestry Institute (in 1947) and the Forestry Department (in 1948) of the Biotechnical Faculty (in 1947) in Ljubljana, the first five scientists have also established the basis for future development of forests and forestry in Slovenia (Kraigher and Žitnik 1999). The phytocoenologist dr. Maks Wraber divided Slovenia into phytogeographical regions, and founded silvicultural measures on genetic basis (1950). This scientific background was incorporated into forest gene conservation practice by dr. Miran Brinar (1961), the founder of the first national Register of forest seed objects, and certification of forest reproductive material (FRM) (Kraigher et al. 1996; Westergren et al. 2006). Therefore, the practice and legislation in Slovenia preceded the two EC directives of forest seeds and seedlings (EC/404/66 (1966) and EC/161/71 (1971)) in which the origin of FRM was stipulated to be of primary importance for a successful reafforestation. The principles by M. Wraber and M. Brinar were that biology is the theoretical and practical basis for contemporary forestry, in which the success is based on (Wraber 1951):

- Enlargement of forest areas
- Improvement of yield regarding quantity and quality
- Improvement of wood quality
- Use of site-adapted high value species' FRM being the first priority in professional silviculture and forest management.

For practice Brinar (1961) defined principles and methods for approval of seed stands, and delineated seven Slovenian forest seed regions (Fig. 3.2), based on ecological, phytocenologial, technological, silvicultural criteria, and defined criteria for approval of seed stands, based on source (autochtonity), homogeneity, size, site, adaptation, age, mixture, silvicultural state, density, isolation, and technical characteristics of wood. He also established the Register of seed stands (the first revision made available in 1971 (Semenski objekti 1971)), which was later revised by Pavle (1987, 1997), before the new Act on Forest Reproductive Material (ZGRM 2002)

Fig. 3.2 The map of Slovenian Seed regions. (Semenski objekti 1971)

formed the basis for the present National list of forest seed objects, published yearly in the Official gazette (the last revision in January 2018), and in the EU database FOREMATIS.

3.2.3 The 'Slovenian Forestry School'

The 'Free silvicultural technique' as defined by Dušan Mlinšek (1968) is based on learning from processes in natural forests, aiming at sustainable management and active support of all functions and roles of forests. The 'Slovenian Forestry School' is based on the following principles (as reported in IPGRI/FAO and EUFORGEN country reports by Kraigher et al. 1996; Smolej et al. 1998):

- Small-scale flexible forest management, adapted easily to site characteristics and natural development of forests;
- Active protection of natural populations of forest trees;
- Protection and conservation of biological diversity in forests;
- Support of the bio-ecological and economic stability of forests by improving the growing stock;

- Tending of all developmental stages and all forest forms for supporting of vital and high-quality forest trees, which could fulfil optimally all functions of forests;
- Natural regeneration is supported in all forests;
- If seedlings are used, they should derive from adequate seed sources/provenances, and only adequate species can be used.

These principles form the basis of the current Forestry act (ZOG 1993) and the resolution on the National Forest Programme (NFP 2007).

3.3 The Slovenian Forest Genetic Resources Programme (SIFORGEN)

3.3.1 The Legislative Basis for Conservation of Forest Genetic Resources

The principles for conservation of forest genetic resources in Slovenian legislation are formalized within the Nature conservation act (ZON 1999), the Forest Act (ZOG 1993), the Slovenian Biodiversity Conservation Strategy (2001), and the National Forest Programme (NFP 2007), while detailed rules are defined in the Act on Forest Reproductive Material (ZGRM 2002) and its subordinate regulations (available at http://www.mkgp.gov.si/zakonodaja_in_dokumenti/veljavni_predpisi/).

The Slovenian Forest Act (ZOG 1993) is unique in Europe in that it defines the use of forest reproductive material according to the written decree based on detailed silvicultural plans, while the Slovenian Act on Forest Reproductive Material (ZGRM 2002) is unique in that it clearly states the need for conservation of forest genetic resources, and that it is valid for all forests, as well as all permanent forest tree plantations outside the forest.

The Forest Act (ZOG 1993) defines the tasks of the Slovenian public forest service, to be performed in all forests, dedicated to the Slovenia Forest Service and the Slovenian Forestry Institute, as follows:

- collecting data on the state and development of forests,
- keeping records as a database for its work and for statistical generalisations,
- monitoring the biological balance in forests,
- monitoring degradation and damage to forests,
- ensuring implementation measures for the protection of forests,
- producing plans for protection of forests against fire,
- carrying out the tasks of reporting, prognostic and diagnostic service,
- producing the professional basis for the programme of development of the forests in Slovenia,
- …
- providing seedlings of forest tree and shrub varieties,….

and for the Slovenian Forestry Institute, in addition to research activities, and other tasks, the following tasks of a public forestry service:

- Orienting and professionally heading the monitoring of the level of degradation and damages to forests.
- Orienting and professionally heading a reporting, prognostic and diagnostic service for forestry,
- Professionally orienting forest seed and nursery activities,
- …

The Slovenian Forestry Institute shall have the following public authorities:

- To issue certificates for forest seed and seedlings in accordance with regulations on seed and seedlings,
- To perform professional and health supervision of forest seed and seedling activities.

Therefore professional-advisory services and development of forest seed husbandry and nurseries, together with the Slovenian Forest Gene Bank (see Sect. 3.3.3), as defined within the Act on Forest Reproductive Material (ZGRM 2002), are an integrative part of the Slovenian public forest service. An important part of the public forest service is evaluation, approval and registration of forest seed objects, the yearly revision of the National list of seed objects, certification of forest reproductive material, testing of seed quality from the Seed storage facility lead by the Slovenia Forest Service, and from the Slovenian seed bank (lead by the Institute), testing and control of the source of FRM using molecular methods and databases, and preparation of professional background documents, expert opinions and proposals for modification and harmonization of regulations on FRM. Furthermore, the public forest service provides basis for development of national and international strategies and action plans (such as SIFORGEN (Slovenian programme on forest genetic resources (FGR)) and EUFORGEN (European programme on FGR)), contributions to the EUFGIS information system (on dynamic conservation units – forest gene reserves), collaboration in dedicated European projects, and development of the system for forest genetic monitoring (such as LIFEGENMON).

Through harmonization process with the Directive EC/105/1999 (1999) the provenance regions had to be adapted in order to simplify the fate of FRM from the stand to the end-user. The current Regulation on delineation of regions of provenances in Slovenia (PO 2003/2012/2017) combines the ecological regions, soil types, and main topographical and administrative borders, including the 14 Slovenian forest regions, while each region is divided into 4 elevation zones (Fig. 3.3). For majority tree species Slovenia is divided into seven provenance regions, while for the minority species it would not be feasible to produce and use FRM in the same way, therefore Slovenia as a whole represents a single provenance region for these species, still divided into the four elevation zones. The Regulation includes recommendations for use of FRM, while the decisive role is on the forest rangers formalizing the use in their Decree on detailed silvicultural plan.

Fig. 3.3 Regions of provenances based on ecological regions and subregions: 1. Alpine, 2. Pohorje, 3. Prepannonian, 4. Prealpine, 5. Predinaric, 6. Dinaric, 7. Submediterranean (based on Kutnar et al. 2002)

3.3.2 The European Dimension with Special Emphasis on EUFORGEN and EUFGIS

EUFORGEN – European forest genetic resources programme (www.euforgen.org) – has been established in 1994 based on the resolutions of the Ministerial conferences for the protection of forests (MCPFE, now Forest Europe, FE) in Strassbourg (in 1990) and Helsinki (in 1993) (Kraigher 2010). It aims at developing strategies for conservation of forest genetic resources, overviews policies and strategies contributing to it, provides access to research results to forestry professionals and policy makers and organizes the European information system on dynamic conservation units of forest trees (EUFGIS). Currently it is in its fifth 5-year phase, combines efforts of 27 countries, and contributes to the development of the indicator 4.6 of the pan-European Criteria and Indicators for sustainable forest management, on production and use of forest reproductive material, on development of a decision support tool for monitoring and managing gene conservation units, and provides the coordination for bridging information systems and approaches of Gene Banks across domains of animal, plant/crop and forest genetic resources.

Slovenia has been actively participating in EUFORGEN since 1995, was a partner in the EUFGIS (Agri Gen Res) project, and is currently involved in the GenRes Bridge project. Within the EUFGIS information system, 41 forest gene reserves (gene conservation units) for 21 forest tree species are from Slovenia, all also a part of the National list of the seed objects and the *in situ* part of the Slovenian Forest Gene Bank. Furthermore, all harmonization processes in development of the new Slovenian legislation on FRM have been largely enabled through participation at EUFORGEN events. Among important outputs of the EUFORGEN programme was also the organization of the Steering Committee meeting in Novo mesto in 2007, the organization of the Conifers network meeting in Brdo/Kranj in 2010, and the initiation of the current LIFEGENMON project (LIFE13/ENV/SI/000148), a 6-years (2014–2020) international LIFE+ project aiming at development of a system for forest genetic monitoring.

Furthermore, Slovenian scientists have been partners in a number of bilateral and international projects. Among the bilateral projects, the projects FRANGUSAVA with Croatia and with Serbia have contributed to population genetic studies of ash species (*Fraxinus excelsior* and *F. angustifolia*) along the Sava river, and the feasibility for division of the area of Slovenia on different provenance regions for the major and selected minority species. Among COST actions, EUROSILVA E6 was the first, also deriving from the resolutions of the MCPFE process in Strassbourg, contributing to belowground physiological and ecophysiological studies for protection of forests (followed by actions E38, FP803, FP903, FP1305 on belowground processes and climate change impacts on forest ecosystems), E28 to genetic population studies, E52 to development of the international beech provenance trial and analysis of its results, FP1202 to identification of common problems and development of common strategies for marginal populations of forest trees, E42 to enhance the growing of valuable brodleaved tree species, and the current COST FP1405 NNEXT contributes to development of common overviews, risks and applications of non-native forest tree species in Europe.

Within IUFRO (International union of forest research organizations) the contributions of Slovenian scientists were maximal during the organization of the XVIII World IUFRO Congress in Ljubljana under the presidency of prof. Dušan Mlinšek, when also The Slovenian Forestry School was presented to the world forestry scientists and professionals. At present, the activities within the forest gene conservation scopes in Slovenia are linked to the IUFRO sessions on Norway spruce genetics, Air pollution effects on forest trees and forest ecosystems, as well as others. Furthermore, the Slovenian Forestry School and conservation of hidden biodiversity were presented as a plenary keynote presentation at the 125th Anniversary IUFRO congress in 2017, and at the session on Biodiversity and policy interface, presenting predominantly the LIFEGENMON aims and outputs.

The excellent international networking and personal collaboration has enabled also several PhD Thesis to be finalized partly abroad (here the special recognition is on the Forestry Faculty Zvolen, Slovakia, and dr. Ladislav Paule and dr. Dušan

Gömöry, and to the ASP in Teisendorf, Germany, with the past directors Albrecht Behm and dr. Monika Konnert) as well as participation and coordination of several international and European projects. Among these, the 7FW project EUFORINNO, and the LIFE environment fund project LIFEGENMON are of special importance.

3.3.3 *The Slovenian Forest Gene Bank and the Register of Forest Seed Objects*

Forest management and silvicultural measures depend on the long life span of forest trees, well adapted to the site conditions, while seeds of a number of tree species are impossible to store over longer periods in seed storage and seed banks. Therefore, the Slovenian Forest Gene Bank (SFGB; Kraigher and Žitnik 1999) is a *sensu lato* Gene Bank, including *in situ* and *ex situ* components of the gene bank, as defined in the Nature conservation act (ZON 1999):

> Gene banks shall be the controlled or cultured populations or parts of animals and plants, in particular seeds, spores, reproductive cells and other biological materials which are managed for the purposes of conserving species or their gene pools.

The *in situ* components of the SFGB *sensu latissimo* consider as its parts all natural populations of autochtonous forest tree species, managed according to the principles of co-natural sustainable forest management, and particularly forest stands in protection forests and in different nature conservation areas (Table 3.1). SFGB *sensu lato* comprise all *in situ* and *ex situ* forest seed objects – approved basic material for production of forest reproductive material, while SFGB *sensu stricto* comprises *in situ* dynamic gene conservation units – forest gene reserves, *ex situ* living archives, provenance tests, the Forest seed bank, the DNA library and molecular databases.

The National list of forest seed objects is part of the Register, organized and lead by the Slovenian Forestry Institute (SFI), who is obliged to lead the approval, the database and information system, and to report – publish the National list yearly in the Official gazette and submit the list into the FOREMATIS database. Currently, the list includes 219 seed objects "for forestry purposes", or "with a limited wood-production purpose", and additional seed objects "not for forestry purposes" for the needs of certification of FRM, not intended for use in forestry and not produced under any official control.

The approval starts upon the application submitted by the forest owner (or his legal representative) to the SFI. The seed object is then visited and evaluated by a committee, lead by representatives from the SFI, and participated by the regional head of Department for silviculture and forest protection at the Slovenia Forest Service (SFS), and the representative of the forest owner, which is usually the local forest ranger. The SFS provides the description of the forest stand (when the seed object is *in situ* population of forest trees, either for production of FRM of the

Table 3.1 Components of the Slovenian Forest Gene Bank

Component of SFGB	Area (ha)/number/established in	Reference
Protection forests	98.759,80	UVG (2005)
Forest reserves (173)	9.508,16	UVG (2005)
Forests in TNP	52.965	Klopčič et al. (2015)
Forests in 36 regional parks	30.045	Kraigher (1996)
National list of seed objects	3.304,75 ha/219 SO	Off. g. 2017/4
Forest gene reserves – dynamic conservation units	1131,94/41 FGR	Off. g. 2017/4 and EUFGIS database
Forest Seed Bank	154 accessions/26 spp	SFI report 2017 (unpublished)
DNA library (2014–2016)	6163 extractions	SFI report 2017 (unpublished)
Provenance trials/species:		
Fagus sylvatica	Est. 1998, 38 provenances, Kamenski hrib – Straža	Božič et al. (2000) and Westergren et al. (2010)
Picea abies	Est. 2017, two sites, 6 provenances	SFI report 2017 (unpublished)
Pseudotsuga menziesii	Est. 1971 and 1972, 27 provenances	Breznikar (1991)
Living archives/species:		
Metasequoia glybtostroboides	Est. 1993, 350 trees, Zadobrova	Kraigher (1996)
Populus ssp. and their artificial hybrids	43 clones, Zadobrova	Božič and Krajnc (2012) and Božič (2016)
	31 clones, Pince-Marof	
	13 clones, Ižakovci	
Populus nigra	Est. 2013, 14, 17, Ižakovci, Mura river	Božič (2016), SFI report 2017 (unpublished)

category 'source identified' or 'selected', while for the categories 'qualified' and 'tested', the SFI is responsible for overseeing the establishment and planning for any measures undertaken in it. The *in situ* stand or group of seed trees are further evaluated on ten criteria, corresponding to the criteria set-up in the Directive EC/105/1999 (1999), but with more stringent requirements regarding the category 'source identified':

1. Type of the seed object: forest stand or group of seed trees
2. Origin: autochtonous, non-autochtonous, origin if non-autochtonous
3. Isolation: distance from another stand of the same species with different origin or characteristics
4. Effective Population Size: minimum area and number of trees/groups of trees (for majority species over 5 ha and over 70 trees)
5. Age and Development: sufficient for the different characteristics to be evaluated

6. Adaptation to the ecological conditions: whether the population can sustain itself on the site, based on regeneration (mainly generative: flowering, fructification, germination, survival of natural regeneration)
7. Health and Resistance: presence of any pests and diseases, mechanical damages, and resistance to biotic/abiotic stressors
8. Volume production: in comparison to other stands in the region
9. Wood Quality: in comparison to the other stands in the region
10. Form or Growth Habit: share of trees in the population with unfavorable (presumable inherited) growth habit

For 'source identified' the criteria under 8 and 9 are not important, while in no. 10 a higher share of unfavorable phenotypes can be present.

At the time of the evaluation, the decision is taken whether the seed object is to be registered, and under which category. Furthermore, the criteria for dynamic gene conservation units – forest gene reserves are discussed and if conforming to the minimum criteria (Lefèvre et al. 2013) the owner is consulted whether the seed stand can also be considered for a forest gene reserve, or for a forest genetic monitoring plot. Finally, the directives for tending and production of FRM are discussed and written into the Decree on approval of the seed object by SFI, and these directives become an integral part of the forest management plans. The directives mainly contribute to conservation of forest genetic diversity in produced seeds from an appropriate and still feasible number of forest trees, and on supporting the stability of the stand and supporting the species in concern, while at the same time eliminating the negative phenotypes.

At FRM production the operator is obliged to announce the timing to the local forest office, and providing the proof about the consent of the forest owner. During production the local forest ranger oversees the production on a daily basis, while the whole production is signed at the end by the head for silviculture at the Regional SFS Unit. At the time of production, a specified sample is collected from each tree from which FRM has been collected and sent immediately to the SFI for inspection of the origin of the material and of the diversity (number of trees from which FRM has been produced). After processing, the SFI issues a master certificate to the operator, which has to accompany the seed/FRM unit at all time. When needed, the SFI analyses the identity and genetic diversity of the collected FRM with molecular methods, an example of such traceability was published for beech seeds recently (Westergren et al. 2017). The production of FRM is reported yearly to the ministry by the FRM operator, and the issuing of master certificates by the SFI.

Within the only still active seed orchard, established in 1990 for black alder (*Alnus glutinosa*), at 55 clones and 256 trees on 1,05 ha, the fructification per clone needs to be assessed by the SFI, and seeds/cones are obtained from every tree in the plantation prior to issuing the master certificate.

Additionally, the Register of Plus trees is under development, containing 107 categorized trees of wild cherry (see also Jarni et al. 2017, and the contribution by Brus et al., this issue), and of autochtonous European black poplars. The formal approval of individual trees is under development.

Among the provenance tests two, which are consistent with international standards on establishment in three blocks and with sufficient number of replicates per provenance, have been established in 1998 (The international beech provenance trial, established in 15 countries, in Slovenia at the location Kamenski hrib), and the recent small spruce provenance trial, established at two sites with 6 provenances. The IUFRO international provenence trial with 27 provenances of Douglas fir was established in 1971 and 1972 in Brkini. Several other provenance trials have been established between 1949 and 1990, however, they are mostly not supported, or are cut, and the database is under reconstruction for very few remnants.

The living archives have also undergone a few changes in the past 10 years, since several were cut, while some new ones have been established. The main old living archive is that of of *Metasequioia glyptostroboides* established in 1993, originating from seeds from Sechuan. A new living archive of autochtonous black poplar trees from the floodplain forests alongside the Mura river, established in the years 2013, 2014 and 2017 is a conservation archive. The poplar clones are entered into the live archives in order to preserve the selected material ex situ, and for the identification of particular clones.

3.3.4 Development of a System for Monitoring of Forest Genetic Diversity

Conservation and management of FGR is an essential part of sustainable forestry, however not an easy task. To recognize the state of and changes in the composition of the genetic variation and track the undisturbed transfer of genetic information to subsequent generations, forest genetic monitoring is needed. Genetic monitoring can track changes of FGR adaptive and neutral genetic variation through time caused by climate change, forest management and conservation measures through well defined indicators and their verifiers (Namkoong et al. 1996; Aravanopoulos et al. 2015; Fussi et al. 2016).

Forest genetic monitoring is an essential prerequisite for maintenance and control of sustainable forest management aiming to conserve genetic and consequently biological diversity at species, ecosystem and landscape levels, especially in marginal and peripheral populations (Fady et al. 2016). The need for monitoring genetic diversity has been recognised by the United Nation's Convention on Biological Diversity, which in Article 7 calls for action to "monitor through sampling and other techniques the components of biological diversity" (CBD 1992). Further, genetic monitoring has been recognised to be an integral part of managing gene conservation units in forests (Koskela et al. 2013).

Genetic monitoring was first proposed by experts from FAO (Namkoong et al. 1996), and later on simplified for practical use by the German programme for conservation of forest genetic resources (Konnert et al. 2011) and by the EUFORGEN working group on forest genetic monitoring (Aravanopoulos et al. 2015). The

German concept of forest genetic monitoring was put into practice and the baseline data for selected indicators collected (Konnert et al. 2011). Recognising the importance of forest genetic monitoring, Slovenia is together with Germany and Greece under the framework of the LIFEGENMON project, and within the SIFORGEN programme, developing and implementing the first internationally coordinated system for forest genetic monitoring for two ecologically and economically important species in the region, *Abies alba* (silver fir) and *Fagus sylvatica* (European beech). Two genetic monitoring plots have been established in 2015 where 250 adult trees and 200 saplings have been marked, measured and sampled for genetic analysis, in addition to yearly phenology and seed set observations. Additionally, seed from 20 adult trees has been sampled to undergo seed quality (germination, vitality) and genetic analysis. Next sampling of the natural regeneration is planned in 2019.

In addition to plot selection (representativeness for the area/ecogeographic region/genetic lineage) and establishment, the development of the forest genetic monitoring system entails:

- transfer of the state-of-the art scientific knowledge into an operational form expressed in indicators and their verifiers,
- checking of the feasibility of collecting such verifiers in the field and the laboratory,
- careful cost-benefit analysis of the information value of each verifier with regards to the expenses for its estimation,
- definition of measurement intervals and definition of thresholds for management and conservation of FGR, and
- development of a general support among stakeholders, policy makers and the general public for its implementation, rendering the policy interface, and communication with foresters from practice, NGOs, and others among important goals for long-term sustainability of the system under development.

3.3.5 The Slovenian Forest Genetic Resources Programme (SIFORGEN)

Based on the tradition of forest gene conservation in Slovenia starting with Maks Wraber and Miran Brinar in 1950–1960 (see Sect. 3.2.2) the Slovenian forest genetic resources programme (SIFORGEN) was established alongside with our participation within the EUFORGEN programme. Within different phases of EUFORGEN, we contributed country reports on the state of forest genetic resources within all EUFORGEN networks, contributed to the working group on forest genetic monitoring and are currently actively involved within two working groups, on the decision support tool for gene conservation units and on production and use of FRM.

The most visible past activity of SIFORGEN was publication of the translations of 21 issues of Technical guidelines for conservation of forest genetic resources with Slovenian additions, some of which combine more species than the originals (available at http://www.euforgen.org/member-countries/slovenia/). At present the SIFORGEN overview, strategy and action plan are being revised and prepared for a country publication, taking into account the discussion line with the Slovenia Forest Service, on problems and measures per tree species, the current situation in forest seed husbandry and nurseries, and the needs for further development of the Slovenian forestry school, development of a system for forest genetic monitoring, and for better supporting forest genetic diversity in development of silvicultural practices. Considering the anticipated fast climate change and other stressful events which are diminishing the chances of forests to remain in their current distribution area and structure as we know today (see for example Schüler et al. 2014), SIFORGEN is under continuous development, concentrating especially on measures which would enhance genetic diversity of all components in the forest ecosystem, and help maintain the adaptability potential of forest trees to the future environments.

The "Measures for genetic protection of forests" comprise and consider:

- Every silvicultural/forest management measure to be considered with respect to its impact on genetic diversity of the stand/population(s)
- Support of natural regeneration,
- Assist regeneration by co-planting and co-sawing of a high number of tree species based on site-matching (enrichment planting),
- Use adequate forest reproductive material (FRM) of high genetic diversity, through:
  - Defining the minimum number of seed trees for FRM production,
  - Collection of FRM in full mast years,
  - Controlled and prescribed mixing of seed units,
- Use advanced seed and seedling production systems,
- Test provenances for transfer and mixing of FRM.

Among forestry practices that have shown to help maintain processes in genetic diversity among the adult stand and young regeneration centers the irregular shelterwood system was shown to be especially appropriate (Westergren et al. 2015).

However, since all anticipated measures for genetic protection of forests can only be accepted if forestry practice, decision makers and the general public recognize the role of forest genetic diversity and its monitoring, communication is of an utmost importance for the future of our forests. The communication strategy 'for future forests' demands to:

- Communicate to general and target audience the needs for conservation of FGR, FGM and measures for genetic protection of forests through a well developed narrative

- Develop the positive attitude towards forests and forestry through the whole national education system
- Forestry should get to the hearts of people by helping people in need
- Formalize the "science on communicating science" in forestry
- Establish a national and international science – policy interface.

3.4 Conclusions

Forests provide numerous environmental and social functions, and currently their economic role is increasing in Slovenia. Forest genetic resources are their basic building block, enabling not only their adaptation to novel environments but also to more diverse human needs and the volume of these needs. The role of forest genetic resources, both economic and environmental, especially on the karst terrain, has been traditionally recognized in Slovenia, and incorporated into the country's forestry and nature conservation programmes and legislation as early as the 1950s. This enabled Slovenia to contribute to the development and quick implementation of the European gene conservation initiatives through SIFORGEN, while providing a remarkable example of the role of close-to-nature forest management in conservation of forest genetic resources. The Slovenian Forest Gene Bank and the SIFORGEN programme are currently gaining in recognition and importance especially due to the large-scale disturbances Slovenian forests have been exposed to in the last 4 years. The Slovenian Forestry School has reached the point at which further development of the co-natural doctrine is envisaged in order to adapt to and better resist current speed of the changing environmental conditions.

Acknowledgements The programme SIFORGEN, the Slovenian Forest Gene Bank and professional advisory services of the Slovenian Forestry Institute are financed through the Civil forest service by the Ministry of agriculture, forestry and food (MAFF). The development of the system for forest genetic monitoring is part of the project LIFEGENMON (LIFE13/ENV/SI/000148), co-financed in Slovenia by the Ministry for the environment and spatial planning, MAFF, and the Slovenian Forestry Institute through its Research programme P4-0107, financed by the Slovenian Research Agency (SRA). The development of measures for genetic protection of forests, development of the proposal for forest genetic monitoring, and categorization of plus trees of wild cherry, were financed from 2011 through 3-year target developmental projects V4-1140, V4-1428 and V4-1616, co-financed by MAFF and SRA, and the 7FW project EUFORINNO (RegPot. 315982).

References

Aravanopoulos, F. A., Tollefsrud, M. M., Graudal, L., Koskela, J., Kätzel, R., Soto, A., et al. (2015). *Genetic monitoring methods for genetic conservation units of forest trees in Europe. European Forest Genetic Resources Programme (EUFORGEN).* Rome: Bioversity International xvi+46 pp. http://www.euforgen.org/publications/publication/geneticmonitoring-methods-for-genetic-conservation/.

Bozic, G. (2016). *Poplars and willows in Slovenia: Report of the National Poplar Commission: time period: 2012–2015*. Ljubljana: Poplar Commission of the Republic of Slovenia 5 p. Available at: http://www.fao.org/forestry/44764-0b9ae9686b8739401695396d139afc14b.pdf.

Bozic, G., & Krajnc, N. (2012). Wood biomass production with fast growing trees on arable land in Slovenia: Current state, past experience, and future prospects. *Folia Biologica et Geologica, 53*(1/2), 129–140.

Bozic, G., Brus, R., Breznikar, A., & Kump, B. (2000). A review on research methods and studies in forest population genetics. *GozdV, 58*(9), 361–369.

Breznikar, A. (1991). *Mednarodno provenienčno preučevanje duglazije (Pseudotsuga menziesii (Mirb) Franco) v Sloveniji*. Diploma thesis, Ljubljana, Biotechnical Faculty, Department of Forestry (62 pp.).

Brinar, M. (1961). Načela in metode za izbiro semenskih sestojev. *GozdV, 19*, 1–20.

Brus R. (2004). *Drevesne vrste na Slovenskem* (399 pp). Mladinska knjiga, Ljubljana.

CBD. (1992). The convention on biological diversity. UN Conference on Environment and Development, Rio de Janeiro. https://www.cbd.int/history/.

EC/105/1999. (1999). Council Directive 1999/105/EC of 22 December 1999 on the marketing of forest reproductive material.

EC/161/71. (1971). Council Directive 71/161/EEC of 30 March 1971 on external quality standards for forest reproductive material marketed within the Community.

EC/404/66. (1966). Council Directive 66/404/EEC of 14 June 1966 on the marketing of forest reproductive material.

Fady, B., Aravanopoulos, F. A., Alizoti, P., Mátyás, C., von Wühlisch, G., Westergren, M., Belletti, P., Cvjetkovic, B., Ducci, F., Huber, G., Kelleher, C. T., Khaldi, A., Kharrat, M. B. D., Kraigher, H., Kramer, K., Mühlethaler, U., Peric, S., Perry, A., Rousi, M., Sbay, H., Stojnic, S., Tijardovic, M., Tsvetkov, I., Verela, M. C., Vendramin, G. G., & Zlatanov, T. (2016). Evolution-based approach needed for the conservation and silviculture of peripheral forest tree populations. *Forest Ecology and Management, 375*, 66–75.

Fussi, B., Westergren, M., Aravanopoulos, F., Baier, R., Kavaliauskas, D., Finzgar, D., Alizoti, P., Bozic, G., Avramidou, E., Konnert, M., & Kraigher, H. (2016). Forest genetic monitoring: An overview of concepts and definitions. *Environmental Monitoring and Assessment, 188*, 493. https://doi.org/10.1007/s10661-016-5489-7.

Jarni, K., Gajšek, D., Božič, G., Kraigher, H., & Brus, R. (2017). Izdelava registra plus dreves divje češnje (*Prunus avium* L.) v Sloveniji. *Gozdarski Vestnik, 75*(5–6), 264–272.

Klopčič, M., Pisek, R., & Poljanec, A. (2015). Razširjenost, zgradba in razvojne spremembe gozdov. In A. Poljanec (Ed.), *Gozdovi v Triglavskem narodnem parku* (pp. 27–44). Ljubljana. Available online: https://issuu.com/tnp-publikacije/docs/triglavski_final.

Konnert, M., Maurer, W., Degen, B., & Kätzel, R. (2011). Genetic monitoring in forests – Early warning and controlling system for ecosystemic changes. *iForest, 4*, 77–81. https://doi.org/10.3832/ifor0571-004.

Koskela, J., Lefèvre, F., Schueler, S., Kraigher, H., Olrik, D. C., Hubert, J., Longauer, R., Bozzano, M., Yrjänä, L., Alizoti, P., Rotach, P., Vietto, L., Bordács, S., Myking, T., Eysteinsson, T., Souvannavong, O., Fady, B., de Cuyper, B., Heinze, B., von Wühlisch, G., Ducousso, A., & Ditlevsen, B. (2013). Translating conservation genetics into management: Pan-European minimum requirements for dynamic conservation units of forest tree genetic diversity. *Biological Conservation, 157*, 39–49. https://doi.org/10.1016/j.biocon.2012.07.023.

Kraigher, H. (1996). Quality categories of forest reproductive material, seed orchards and measures to enhance fructification. *Zbornik Gozdarstva in Lesarstva, 51*, 199–215.

Kraigher, H. (2010). Ohranjanje genskih virov – kratek pregled aktivnosti v Sloveniji: tehnične smernice za ohranjanje in rabo genskih virov: Slovenija. *GozdV, 68*(2), 95–96.

Kraigher, H., & Žitnik, S. (1999). Slovenian forest gene bank. *Sodobno Kmetijstvo, 32*(1), 46–50.

Kraigher, H., Božič, G., Brus, R., Veselič, Ž., & Pavle, M. (1996). Forest genetic resources. In M. Černe & H. Kraigher (Eds.), *The Republic of Slovenia: Country report for International*

Conference and Programme for Plant Genetic Resources – ICPPGR (pp. 14–27). Ljubljana: Ministry of Agriculture, Forestry and Food.
Kutnar, L., Zupančič, M., Robič, D., Zupančič, N., Žitnik, S., Kralj, T., Tavčar, I., Dolinar, M., Zrnec, C., & Kraigher, H. (2002). The delimitation of the regions of provenance of forest tree species in Slovenia based on ecological regions. *Zbornik Gozdarstva in Lesarstva: Research Reports – Forest and Wood Science & Technology, 67*, 73–117.
Lefèvre, F., Koskela, J., Hubert, J., Kraigher, H., Longauer, R., Olrik, D. C., Schueler, S., Bozzano, M., Alizoti, P., Bakys, R., Baldwin, C., Ballian, D., Black-Samuelsson, S., Bednarova, D., Bordacs, S., Collin, E., De Cuyper, B., De Vries, S. M. D., Eysteinsson, T., Frydl, J., Haverkamp, M., Ivankovic, M., Konrad, H., Koziol, C., Maaten, T., Notivol Paino, E., Ozturk, H., Pandeva, D., Parnuta, G., Pilipovic, A., Postolache, D., Ryan, C., Steffenrem, A., Varela, M. C., Vessella, F., Volosyanchuk, R. T., Westergren, M., Wolter, F., Yrjana, L., & Zariŋa, I. (2013). Dynamic conservation of forest genetic resources in 33 European countries. *Conservation Biology, 27*(2), 373–384.
Martinčič, A., Wraber, T., Jogan, N., Podobnik, A., Turk, B., & Vreš, B.(2007). *Mala flora Slovenije* (967 pp.). Ljubljana: Tehniška založba Slovenije.
Mihelič, D. (2008). Usoda gozdov na slovenskem do 16. stoletja. *Ekonomska I Ekohistorija, 4*(4), 27–51.
MKGP. (2017). Brief country profile: Slovenia. In *International conference on China-CEECs Forestry Research and Education Cooperation on October 30–31, 2017* (21 pp).
Mlinšek, D. (1968). *Sproščena tehnika gojenja gozdov na osnovi nege* (117 pp). Ljubljana: Poslovno združenje gozdnogospodarskih organizacij
Mršić, N. (1997). *Biotska raznovrstnost v Sloveniji: Slovenija – vroča točka Evrope*. Ljubljana: Ministrstvo za okolje in prostor, Uprava RS za varstvo narave pp.
Namkoong G, Boyle T, Gregorious HR, Joly H, Savolainen O, Ratman W, et al. (1996). *Testing criteria and indicators for assessing the sustainability of forest management: Genetic criteria and indicators*. (Centre for International Forestry Research (CIFOR) Working Paper No. 10). Bogor.
NFP. (2007). Nacionalni gozdni program. Uradni list RS 111/07.
Pavle, M. (1987). *Semenski sestoji v Sloveniji – Register* (145 pp). Ljubljana: IGLG.
Pavle, M. (1997). *Semenski sestoji v Sloveniji – Register. 3. Revizija*. Ljubljana: IGLG.
Perko, F. (2007). *Gozd in gozdarstvo Slovenije*. Ljubljana: Zveza gozdarskih društev Slovenije 39 pp.
PO. (2003/2012/2017). PRAVILNIKo določitvi provenienčnih območij. Ur.l./Official gazette RS 72/03, 58/12, 69/17.
SBCS (2001) Biodiversity conservation strategy of Slovenia. Ministry of the Environment and Spatial Planning RS, 2002
Schüler, S., Falk, W., Koskela, J., Lefèvre, F., Bozzano, M., Hubert, J., Kraigher, H., Longauer, R., & Olrik, D. C. (2014). Vulnerability of dynamic genetic conservation units of forest trees in Europe to climate change. *Global Change Biology, 20*(5), 1498–1511.
Semenski objekti. (1971). *Semenski objekti*. Ljubljana: BF in IGLG 37 pp.
Smolej, I., Brus, R., Pavle, M., Žitnik, S., Grecs, Z., Bogataj, N., Ferlin, F., & Kraigher, H. (1998). Beech and oak genetic resources in Slovenia. In J. Turok, A. Kremer, & S. de Vries (Eds.), *First EUFORGEN meeting on social broadleaves: 23–25 October 1997, Bordeaux, France* (pp. 64–74). Rome: International Plant Genetic Resources Institute.
UVG. (2005). Uredba o varovalnih gozdovih in gozdovih s posebnim namenom. Uradni list/ Official gazette RS, 88/05, 56/07, 29/09, 91/10, 1/13, 39/15.
Valvasor, J. V. (1689). *The Duchy of Carniola*. Ljubljana: Facsimile.
Vrščaj, B., Repe, B., & Simončič, P. (2017). *The soils of Slovenia* (p. 133). Dordrecht: Springer.
Westergren, M., Božič, G., & Kraigher, H. (2006). Silviculture in the light of genetics (M. Wraber 1950) – Principles, development and realisation until 2005. *Razprave, 47*(1), 231–245.
Westergren, M., Božič, G., & Kraigher, H. (2010). Tehnične smernice za ohranjanje in rabo genskih virov: bukev (*Fagus sylvatica*) Slovenija. *GozdV, 68*(2), 103–106.

Westergren, M., Božič, G., Ferreira, A., & Kraigher, H. (2015). Insignificant effect of management using irregular shelterwood system on the genetic diversity of European beech (*Faguss ylvatica* L.): A case study of managed stand and old growth forest in Slovenia. *Forest Ecology and Management, 335*, 51–59.
Westergren, M., Bajc, M., Finžgar, D., Božič, G., & Kraigher, H. (2017). Identification of forest reproductive material origin of European beech using molecular methods. *GozdV, 75*(7/8), 328–343.
Wraber, M. (1950). *Gojenje gozdov v luči genetike. Strokovna in znanstvena dela*. Ljubljana: GIS 67 pp.
Wraber, M. (1951). Nova pota gozdne semenarske službe. *GozdV, 9*, 3–14.
ZGRM. (2002). Zakon o gozdnem reprodukcijskem material, Uradni list/Official.gazette RS 58/02, 85/02, 45/04, 77/11.
ZGS. (2016). Poročilo o gozdovih. Zavod za gozdove Slovenije. Available online at: http://www.zgs.si/slo/zavod/publikacije/letna_porocila/index.html.
ZOG. (1993). Zakon o gozdovih, Uradni list/Official.gazette RS, 30/93, 56/99, 67/02, 110/02, 115/06, 110/07, 106/10, 63/13, 101/13, 17/14, 24/15, 9/16, 77/16.
ZON. (1999). Nature conservation act. Uradni list/Official.gazette RS 56/99.
Zupančič, M. (2013). Boštjan Anko (1939–2013), Novi Slovenski biografski leksikon.

Chapter 4
Conservation and Management of Forest Genetic Resources in Macedonia

Vlatko Andonovski and Nikolco Velkovski

Abstract Conservation of forest biological diversity, which includes forest genetic resources, is essential for sustaining the productive value of forests, and for maintaining the health and vitality of forest ecosystems and thereby maintaining their protective, environmental and cultural roles. The purpose of conserving of forest genetic resources is to maintain hereditary variation in species and local populations far into the future so that their viability and adaptability would be sufficient to cope with changing environmental conditions. Since forest genetic resources are essential part of the adaptation processes of populations and tree species in a changing environment, their conservation and appropriate management have a crucial importance for sustainable forest management. Thus, forests need to be managed in a way that ensures their sustainable utilization, safeguarding the ecological process and genetic diversity essential for the maintenance of the resource base. In this paper a general introduction is given to the state of forest genetic resources conservation in FYR Macedonia with comprehensive overview of policies with respect to their conservation and management. Based on the current situation of conservation of forest genetic resources, some strategic suggestions concerning the future development of genetic conservation is given, taking into consideration the conservation objectives and future trends of great impact on existing forest genetic resources. National policies and regulatory frameworks that target forest genetic resources in Macedonia are partial and inadequate. Most relevant policies neglect the importance of forest genetic resources, but rather use it in the context of biological diversity. Sound social and economic policies are needed at national level to ensure the integration of forest genetic resources into wider national forest policy frameworks and thereby promote the sustainable management of forest genetic resources.

Keywords In situ · Ex situ · Conservation · Forest genetic resources · National policies · Macedonia

V. Andonovski (✉) · N. Velkovski
Faculty of Forestry in Skopje, University Ss. Cyril and Methodius, Skopje, Macedonia
e-mail: vandonovski@sf.ukim.edu.mk; nvelkovski@sf.ukim.edu.mk

M. Šijačić-Nikolić et al. (eds.), *Forests of Southeast Europe Under a Changing Climate*, Advances in Global Change Research 65,
https://doi.org/10.1007/978-3-319-95267-3_4

4.1 Introduction

The need to conserve forest genetic resources better has been widely recognized in recent times because of the risks associated with global changes in the environment, including climate change (Koskela et al. 2013).

The unique characteristics of forest trees and their continued ability to provide goods and services depends on the maintenance and management of forest genetic resources. Critical information on the status, distribution, extent and threats to genetic diversity are prerequisite for planning effective conservation strategies (Young and Clarke 2000).

The aim of conservation and sustainable management of forest genetic resources is to secure the adaptability of forest tree species in a changing environment by maintaining a sufficient level of genetic variability (Namkoong 1998).

The genetic variation of most agricultural and horticultural crops as well as of farm animals can be collected and conserved in so-called gene banks. However, forest genetic resources are usually conserved as living trees in growing forests (Rajora and Mosseler 2001).

Considering the objective of preserving the broadest genetic diversity, not only the most representative trees, populations or important single trees should be subject to gene conservation. Populations from marginal localities also need to be conserved, despite their lower economic importance as such populations and trees may carry genes of importance for breeding (adaptability, resistance). In situ and ex situ measures are necessary to complement each other (Andonovski and Velkovski 2011).

Conservation and use of forest genetic resources is one of the topics that have remained at the cross-roads of many global, European and national policies. This means that various policies often influence the conservation and use of forest genetic resources in one way or another, even if the policies themselves do not specifically target forests or genetic resources. Issues related to genetic resources also offer an illustrative example of a situation in which the forest sector has often found itself: reacting and adapting to policies developed by other sectors (de Vries et al. 2015).

In 1990, European countries initiated a pan-European forest policy process as a response to concerns on the impacts of environmental pollution and forest degradation in Europe. This process, now called FOREST EUROPE (earlier the Ministerial Conference on the Protection of Forests in Europe – MCPFE) has served well as a high-level forum for addressing forest policy issues and it has also contributed significantly to the implementation of sustainable forest management in Europe (Mayer and Buck 2005). In 1992, the United Nations Conference on Environment and Development (UNCED), held in Rio de Janeiro, adopted two documents that have been highly relevant for global efforts to improve the management of forests. The first document was a non-legally binding statement called 'Forest Principles' in which countries affirmed their commitment for conservation, management and sustainable development of all types of forests. The second document was a legally binding agreement for the conservation of all levels of biological diversity, i.e. the management of forests.

As a result of the UNFF work, the UN General Assembly adopted a Non-Legally Binding Instrument on All Types of Forests in 2007. The Rio Conference in 1992 also adopted two other international agreements, namely the UN Framework Convention on Climate Change (UNFCCC) and the UN Convention to Combat Desertification (UNCCD). These conventions did not focus on forests but they have played a major role in promoting the implementation of sustainable forest management around the world.

However, of all the above mentioned policy initiatives, only FOREST EUROPE and the CBD have specifically addressed forest genetic resources (FGR) and recognized their important role.Under the CBD, the discussions on forest biological diversity started in 1995. Later, in 2000, the topic was identified as one of the thematic areas to be addressed within the Convention. In 2002, the CBD adopted its Expanded Work Programme on Forest Biological Diversity, which called for development of information systems and strategies on forest genetic resources, implementation of these strategies and sustainable use of forest genetic diversity, and promotion of fair and equitable sharing of benefits resulting from the utilization of forest genetic resources (Plamberg-Lerche et al. 2004).

After the Strasbourg and Rio Conferences, many European countries started to develop specific national programmes or strategies for managing their forest genetic resources (e.g. Graudal et al. 1997; Tessier du Cros 2001) or to strengthen the efforts they had already started in the 1980s (Behm et al. 1997). However, the implementation of these national programmes and strategies has been hindered by political problems, such as complexities of national legal and administrative structures that have made practical conservation work difficult (Geburek and Konrad 2008). Furthermore, development of other forest related national policies and their implementation, rarely recognized the importance of forest genetic resources.

Several European policies, such as FOREST EUROPE (earlier the Ministerial Conference on the Protection of Forests in Europe – MCPFE), Convention on Biological Diversity (CBD), EU Forest Strategy and Rural Development Programme (RDP) have recognized the importance of forest genetic resources and specifically mention it. Furthermore, Global Plan of Action for the Conservation, Sustainable Use and Development of Forest Genetic Resources (GPA-FGR) was adopted by FAO in 2013 in order to support the implementation of CBD commitments, and to strengthen the conservation and use of FGR as part of sustainable forest management. Additionally, European Union (EU) policies on invasive alien species (Regulation 1143/2014) and protective measures against the introduction and spreading of organisms harmful to plants health (Council Directive 2000/29/EC) may also have an influence on FGR by minimizing the genetic erosion that might be potentially caused by these treats.

4.2 Overview of the National Policies

In the Republic of Macedonia, national resources, flora and fauna are defined as goods of general interest and enjoy special protection under the Constitution of the Republic of Macedonia.

The **Law on Environment** (Official Gazette of the Republic of Macedonia no. 53/05, 81/05, 24/07, 159/08, 83/09, 48/10, 124/10, 51/11, 123/12, 13/13, 163/13, 41/14) regulates the rights and the responsibilities of the Republic of Macedonia in providing conditions for environment protection and improvement, for the purpose of exercising of the citizens' right to a healthy environment.

General principles and measures for biodiversity conservation are described in the Law on Nature Protection with more amendments, mainly due to the ongoing process of transposition of EU Acqui into national legislation. Additionally, legal framework includes a number of legal acts relevant for biodiversity conservation and the sustainable use of natural resources (e.g. legislation on forestry).

The **Law on Nature Protection** (Official Gazette of the Republic of Macedonia No. 67 from 12/10/2004) among other things regulates the conservation, protection, restoration and improvement of the biological and landscape diversity. The main goals of the Law on Nature Protection are defined in Article 4 and these are:

1. Determination and monitoring of the state of nature;
2. Conservation and restoration of the existing biological and landscape diversity in a state of natural balance;
3. Establishment of a network of protected areas for the purpose of sustainable protection of the features on the basis of which they have acquired the status of natural heritage;
4. Providing for sustainable use of natural wealth in the interest of the present and future development, without significant damage of parts of the nature and with the least possible disturbance of natural balance;
5. Prevention of harmful activities of individuals and legal entities and disturbance in nature as a result of technological development and performance of activities, i.e. providing for the best possible conditions for protection and development of the nature;
6. Providing for the citizen to fulfill their right to healthy environment.

The law on nature protection through its definitions of "biological" and "genetic diversity" gets in touch with forest genetic resources. The Article 45 of this law defines Gene banks that shall be established for the purposes of preservation of the species or of their genetic material of this law defines the categories of protected areas, while the Article 54 defines the forest habitats and ecosystems conservation and that biological diversity of the forest ecosystems shall be carried out by way of protecting the forests within the frames of the protected areas, as well as through the general and specific plans for management and protection of forests, in accordance with the provisions of this law and other laws.

The **First Biodiversity Strategy with Action Plan** was adopted in 2004, with Action Plan validity by 2008. In the course of 2012, information was prepared on the need to elaborate the Second Action Plan for biological diversity protection and it was considered by the Government of the Republic of Macedonia as information material. The process of revision of the National Biodiversity Strategy with Action Plan started in January 2013.

The Biodiversity Strategy with Action Plan sets out a series of measures to strengthen the awareness of the need for environmental protection in Republic of Macedonia.

As regarding to the conservation of forest genetic resources in Macedonia, the national legislation has its base in the **Law on Forests** (Official Gazette of the Republic of Macedonia No. 64 from 22/05/2009) pursuant to this law under the Article 19, the conservation and the development of forest genetic resources is defined in terms of the public interest. Moreover the same article regulates the protection of rare and endangered tree species.

Conservation and management of forest genetic resources and the production of forest seeds and seedlings are defined in Articles 10, 13, 19, 30, 41 and 42 of the Law on forests. Additionally, the Law on forests in the Article 95 anticipates the possibility of using allocated funds in order to preserve, enhance and direct the biodiversity utilization in forest ecosystems.

Strategy for Sustainable Development of Forestry (adopted in 2006) also defines and highlights the need for conservation of forest genetic resources in the Republic of Macedonia.

However, the greatest emphasis on forest genetic resources is put by the **Law on Reproductive material of Forest Trees** (Official Gazette of the Republic of Macedonia No.55/07 and 148/11). The law defines and regulates the basic features of the initial and the reproductive material of forest trees, conditions, methods to assess and recognize the initial material, production, inspection and treatment of reproductive material of forest trees, as well as its quality, sale and use, and the act of keeping a register of forest sorts and other registers.

The law also defines the necessity to establish a genetic bank of forest plants in the Republic of Macedonia in order to preserve genetic diversity and its sources. Additionally, the law by its Article 9 in paragraph 3 defines the manner of establishing provenance regions for the production of source material of recognized origins and breeding materials. The law defines the ways to recognize the source material, keep the register, as well as how the production, supervision, refinement, quality, trade and the use of reproductive material will be carried out.

4.3 In Situ Conservation of Forest Genetic Resources

In situ conservation of forest trees comprises the conservation of viable populations in their natural environment. The aim of in situ conservation is often to conserve the function of an ecosystem and evolutionary processes rather than just species (Isajev et al. 2004). In Macedonia, in situ conservation of forest genetic resources is mainly achieved through the establishment of protected areas and nature reserve forests. In addition to these there are also long-term genetic studies and breeding programs (Andonovski 1995).

Protected areas of different categories are of great importance for maintaining or improving the forest genetic resources. These areas in Macedonia are classified as follows presented in Table 4.1 (Orlović et al. 2014).

Table 4.1 National parks and nature protected areas (coniferous species)

Name	Area (ha)	Description
1. National park "Pelister"	12,500	The best preserved natural stand of Macedonian pine (*Pinus peuce*)
2. National park "Mavrovo"	73,088	Natural stand of fir and spruce (*Abies borisii-regis, Picea abies*)
3. Nature reserve "Rozden"	3,5	Crimean pine (*Pinus nigra ssp. pallasiana*)
4. Nature reserve "Tumba"	5,0	Fir (*Abies borisii-regis*)
5. Nature reserve "Golem Kozjak"	4,0	Scots pine (*Pinus sylvestris*)
6. Nature reserve "Popova sapka"	5,2	Norway spruce (*Picea abies*)
7. Nature reserve "Rupa"	7,6	Fir (*Abies borisii-regis*)
8. Natural reserve "Tsam Tsiflik"	490,0	Crimean pine (*Pinus nigra ssp. pallasiana*)
9. Natural reserve "Rutsica"	1785,0	Dwarf mugo pine (*Pinus mugo var. mughus*)

Table 4.2 Selected seed stands (conifers)

Species	Area (ha) total	Area (ha) reduced	Age (years)	Provenance
Abies borisii-regis	182,9	84,4	81	Native
Pinus nigra ssp. pallasiana	258,9	137,5	73	Native
Pinus sylverstris	45,5	32,8	80	Native
Pinus peuce	5,0	3,4	95	Native
Pseudotsuga menziesii	2,8	2,1	35	Exotic

A comprehensive resource inventory on the nature protected areas was set up and management regulations were established for the nature reserves. It is important that the genetic resources of selected forest tree species should be conserved in the protected areas and nature reserves and that appropriate management is needed to fulfill thisobjective. Such management should be an integrated part of the management plan of proected areas and nature reserves.

According to the Law on Forests, selected natural seed stands for production of seeds belong to category of forests with special purposes and they are under special management regime. During the latest period, an increased effort to conserve and enhance the forest genetic resources has been undertaken on the basis of present knowledge about variability and heritability.

The first mass and individual selection in Macedonia was performed in 1962–1965 and the following coniferous seed stands presented oin Table 4.2, were selected and registered (Andonoski 1994).

In 2008, new program for conservation of forest genetic resources in Macedonia started with the revision of the current seed stands of various economically important native and exotic tree species. This program includes preregistration of the

Table 4.3 Registered seed stands under the latest Law on forest reproductive material

Species	Number of seed stands	Area (ha)
Pinus nigra ssp. Pallasiana (native)	8	218,7
Pinus sylvestris (native)	7	131,8
Abies borisii regis (native)	16	375,9
Pinus peuce (native)	3	68,3
Larix decidua (exotic)	3	34,9
Pseudotsuga menziesii (exotic)	9	59,2
Sequoiadendron giganteum (exotic)	1	4,4
Robinia pseudoacacia (exotic)	1	1,2
Fagus moesiaca (native)	7	326.1
Quercus petraea (native)	2	35,2
Total	57	**1255,7**

current seed stands and seed orchards and registration of new, including those of broadleaved species. Following is the table of registered seed stands under the latest Law on forest reproductive material (Table 4.3).

In situ forest genetic conservation in Macedonia includes the "dynamic" approach which encourages the adaptation of forest trees to the changing environment through naturally occurring evolutionary processes. This can maximize adaptability with the sufficient number of trees in the genetic resource population (Andonoski 1974).

4.4 Ex Situ Conservation of Forest Genetic Resources

Ex situ conservation of forest genetic resources in Macedonia includes establishment of arboreta, clonal archives, and ex situ gene conservation stands or seed orchards as a part of the breeding and tree improvement programs. Conservation of individual coniferous tree species was carried out using "plus" trees selected for the development of tree improvement programs. The following "plus" trees were selected (Andonoski 1988):

Abies borisii-regis	42 "plus" trees
Pinus sylvestris	62 "plus" trees
Pinus nigra ssp. pallasiana	82 "plus" trees
Pinus peuce	20 "plus" trees

On the basis of these selected "plus" trees the following Seed orchards, presented in Table 4.4, were established.

Table 4.4 Seed orchards (conifers)

Species	Type	Year of rising	Area (ha)	Fructification
Pinus peuce	Clonal	1963	1,1	Full
Abies borisii-regis	Clonal	1963	0,5	Full
Pinus sylvestris	Clonal	1978–1980	2,5	Full
Pinus sylvestris	Generative	1978	5,0	Full
Pinus nigra ssp. pallasiana	Clonal	1988	1,5	Started

4.5 Conclusions

Genetic diversity is of significant importance for the resilience in an ecosystem and to maintain viable populations of species. However, the knowledge about genetic diversity in general is very low and has received far less attention than biodiversity on the species and ecosystem levels.

Biological diversity is high on the agenda at national level in Macedonia. Many strategies and programs have already been, or are being, made in order to ensure the persistence of biological diversity. However, these strategies and programs often take only into consideration diversity at the more general level. Thus, forest genetic resources are usually mentioned in strategies and programs, but very seldom directly addressed specifically. In combination with a primary focus on conservation of species, habitats and ecosystems, there is a considerable risk of neglecting and forgetting forest genetic resources. Thus, it is important to make sure that diversity is secured all the way from habitats down to the level of genes and genetics.

There is another important incentive to make sure that forest genetic resources are specifically incorporated in the national strategies and not just mentioned indirectly as part of the overall biodiversity. While in situ and ex situ conservation networks are in place as a part of the selection and tree improvement programs, there is no specific program in place for management and conservation of forest genetic resources in Macedonia. Forest genetic resources management can be effective only if treated as an integral element of overall sustainable forest management. Conservation concerns should be integrated into broader national and local development programs, which promote inter sectorial cooperation.

It is obvious that implementation of forest genetic conservation cannot be effective unless there is a national programme with clear objectives and adequate long-term resources. Since solid legal background for genetic conservation is missing in Macedonia, the implementation of the Nagoya Protocol may help to change this situation. One aspect to be considered with all funding mechanisms and policies is that they normally need clear and measurable indicators for follow up and evaluation. There is a need to develop measurable indicators to serve both conservation programmes and wider policies. Another problem is that genetic conservation, particularly forest genetic conservation, is planned and implemented on a long term basis, whereas most policies need targets that can be achieved within 5–10 years.

In order to promote awareness of forest genetic resources in Macedonia, a separate genetic conservation programme should be established that includes a number of genetic conservation units for several tree species. Besides their importance in conserving forest genetic resources, these units are also a way of showing policy makers how management of forest genetic resources is transformed into action.

In the past in Macedonia more emphasis was placed on the conservation and study of "plus" trees, so now it is necessary to focus on study and conservation of the most valuable populations. The majority of gene reserves were selected in the early 1960, thus it needs repeated inventory with biochemical, cytological and molecular genetic methods.

For most of the national tree species genetic is scarce and getting better information about genetic structure and differentiation of the tree species should have high priority. This is an particular important in view of the changing climate conditions, both for the development of adapted reproductive material for the commercial tree species and for the management of forest genetic resources.

The major challenges for the conservation of forest genetic resources in Macedonia include population decline and population structure changes due to forest removal and conversion of forest land to other uses, forest fragmentation, forestry practices, climate change, disease conditions, introduced pests, atmospheric pollution, and introgressive hybridization. Developing scientifically sound conservation strategies, maintaining minimum viable population size, and deployment of genetically engineered organisms represent other important challenges in conservation. Both in situ and ex situ forest genetic resource conservation strategies must include the use of various biochemical and molecular genetic markers, adaptive traits, and genetic diversity measures. So, major opportunities for conservation of forest genetic resources in Macedonia include: use of molecular genetic markers and adaptive traits for developing conservation strategies; in situ conservation through nature reserves, protected areas, and sustainable forest management practices; ex situ conservation through germplasm banks, common garden archives, seed banks, DNA banks, and tissue culture and cryopreservation; incorporation of disease, pest, and stress tolerance traits through genetic transformation; plantation forestry; and ecological restoration of rare or declining tree species and populations.

References

Alexandrov, A. H., & Andonovski, V. (2011). *EUFORGEN technical guidelines for genetic conservation and use of Macedonian pine (Pinus peuce)*. Rome: Bioversity International.

Andonoski, A. (1974). *Application of genetical principles on the production of forest seeds* (pp. 43–67). Forest review 5 and 6 Skopje, Macedonia.

Andonoski, A. (1988). *The state of Macedonian forest genofund and measures for its conservation* (pp. 144–152). X Congress of dendrologists, Sofia.

Andonoski, A. (1994). *Genetics and tree improvement* (pp. 226–261). Skopje.

Andonoski, A. (1998). *Multy-country forestry program (National report for the Republic of Macedonia)*, Skopje, Macedonia (pp. 18–34).

Andonovski, V. (1995). *Adaptive capability of some introduced forest tree species* (pp. 82–94). Master thesis, University Ss. Cyril and Methodius, Faculty of Forestry – Skopje, Macedonia.

Andonovski, V., & Velkovski, N. (2011). Conservation of Macedonian pine (*Pinus peuce* Griseb.) genetic resources in Pelister National Park. *eco.mont, 3*(1), 49–53.

Behm, A., Becker, A., Dorlinger, H., Franke, A., Kleinschmit, J., Melchior, G. H., Muhs, H. J., Schmitt, H. P., Stephan, B. R., Tabel, U., Weisberger, H., & Widmaier, T. (1997). Concept for the conservation of forest genetic resources in the Federal republic of Germany. *Silvae Genetica, 46*, 24–34.

CBD. (2003). *Handbook of the convention on biological diversity* (2nd ed.). Montreal: Secretariat of the CBD.

CBD [Convention on Biological Diversity]. (2011). *Nagoya protocol on access to genetic resources and the fair and equitable sharing of benefits arising from their utilization to the convention on biological diversity. Text and annex*. Secretariat of the Convention on Biological Diversity, Montreal. EU [European Union]. 2014. Regulation 1143/2014 on invasive alien species.

CBD FIFTH NATIONAL REPORT to the Convention on Biological Diversity of the Republic of Macedonia. (2014). *Ministry of environment and physical planning*. Skopje.

de Vries, S. M. G., Alan, M., Bozzano, M., Burianek, V., Collin, E., Cottrell, J., Ivankovic, M., Kelleher, C. T., Koskela, J., Rotach, P., Vietto, L., & Yrjänä, L. (2015). *Pan-European strategy for genetic conservation of forest trees and establishment of a core network of dynamic conservation units*. Rome: European Forest Genetic Resources Programme (EUFORGEN), Bioversity International xii + 40 p.

European Commission's Seventh Framework Programme (FP7). (2008). *The European bioeconomy in 2030, BECOTEPS –Bio-economy technology platforms* (pp. 5–22).

FAO [Food and Agriculture Organization of the United Nations]. (2014). *State of the world's forest genetic resources*. Rome: FAO.

Food and Agriculture Organization of the United Nations. (2014). *The state of the worlds forest genetic resources*. Rome: FAO.

FOREST EUROPE. (2015). *State of Europe's forests 2015*. Spain: FOREST EUROPE Liaison Unit Madrid.

FOREST EUROPE/UNECE/FAO. (2011). *State of Europe's forests 2011. Status and trends in sustainable forest management*. Aas: FOREST EUROPE Liaison Unit Oslo.

Geburek, T., & Konrad, H. (2008). Why the conservation of forest genetic resources has not worked. *Conservation Biology, 22*(2), 267–274.

Graudal, L., Kjær, E., Thomsen A., & Larsen, A. B.. (1997). *Planning national programmes for conservation of forest genetic resources* (Technical Note 48). Humlebaek: Danida Forest Seed Centre.

Isajev, V., Fady, B., Semerci, H., & Andonovski, V. (2004). *EUFORGEN technical guidelines for genetic conservation and use for European black pine (Pinus nigra)*. Rome: International Plant Genetic Resources Institute.

Koskela, J., Lefèvre, F., Schueler, S., & and 19 others. (2013). Translating conservation genetics into management: Pan-European minimum requirements for dynamic conservation units of forest tree genetic diversity. *Biological Conservation, 157*, 39–49.

Law on Environment (Official Gazette of the Republic of Macedonia no. 53/05, 81/05, 24/07, 159/08, 83/09, 48/10, 124/10, 51/11, 123/12, 13/13, 163/13, 41/14). Skopje.

Law on Forests (Official Gazette of the Republic of Macedonia No. 64 from 22/05/2009). Skopje.

Law on Nature Protection (Official Gazette of the Republic of Macedonia No. 67 from 12/10/2004). Skopje.

Law on Reproductive material of Forest Trees (Official Gazette of the Republic of Macedonia No.55/07 and 148/11). Skopje.

Mayer, P., & Buck, A. (2005). A look at Pan-European forest policy: The ministerial conference on the protection of forests in Europe and the 'Living Forest Summit'. In T. Geburek & J. Turok (Eds.), *Conservation and management of forest genetic resources in Europe* (pp. 33–43). Zvolen: Arbora Publishers.

Namkoong, G. (1998). Forest genetics and conservation in Europe. In J. Turok, C. Palmberg-Lerche, T. Skrøppa, & A. S. Quédraogo (Eds.), *Conservation of forest genetic resources in Europe, proceedings of the European forest genetic resources workshop, 21 November 1995, Sopron, Hungary* (pp. 3–10). Rome: International Plant Genetic Resources Institute.

Orlović, S., Ivanković, M., Andonovski, V., Stojnić, S., & Isajev, V. (2014). Forest genetic resources to support global bioeconomy. *Annals of Silvicultural Research, 38*(2), 2014: 51–61.

Palmberg-Lerche, C., Turok, J. & Sigaud, P. (2004). Conservation and management of forest genetic resources in Europe. In Th. Geburek, & J. Turok (Eds.), *Contributions of a training course*. Gmunden. 30 April–11 May 2001.

Rajora, O. P., & Mosseler, A. (2001). Challenges and opportunities for conservation of forest genetic resources. *Euphytica, 118*(2), 197–212.

Šijačić-Nikolić, M., & Milovanović, J. (2010). *Konzervacija i usmereno korišćenje šumskih genetičkih resursa*. Beograd: Univerzitet u Beogradu Šumarski fakultet 200p.

Strategy for Sustainable Development of Forestry. (2009). *Ministry of agriculture, forestry and water economy*. Skopje.

Tessier du Cros, E. (2001). *Forest genetic resources management and conservation. France as a case of study. Ministry of agriculture and fisheries*. Paris: Bureau of Genetic Resources and Commission of Forest Genetic Resources.

Young, A. G., & Clarke, G. M. (2000). *Genetics, demography and viability of fragmented populations*. Cambridge: Cambridge University Press.

Chapter 5
The Current State of Forest Genetic Resources in Romania

Flaviu Popescu, Dragoş Postolache, and Alexandru Lucian Curtu

Abstract In Romania, the forest genetic resources are included in two National Catalogues. The first one is *The National Catalogue of Forest Genetic Resources* which includes 698 core units with the main objective of conservation, and having a total area of 19889.6 ha. The second one is *The National Catalogue of Forest Reproductive Material*, including all base units designated to produce forest reproductive material, 2930 units for 45 forest tree species being listed, with a total area of 40209.3 ha. More than 300 source units of the basic material category are also entirely and partly included in the FGR conservation category, due to the importance of using these resources as base reproductive material. There are GIS databases for both catalogues.

Keywords Forest reproductive material · Basic material units · Genepool · Seed sources · Romania

F. Popescu (✉)
National Institute for Research and Development in Forestry "Marin Drăcea", Simeria, Romania
e-mail: flaviu.popescu@icashd.ro

D. Postolache
National Institute for Research and Development in Forestry "Marin Drăcea", Cluj-Napoca, Romania

University of Agricultural Sciences and Veterinary Medicine, Cluj-Napoca, Romania
e-mail: dragospostolache@yahoo.com

A. L. Curtu
Department of Forest Sciences, Transilvania University of Braşov, Braşov, Romania
e-mail: lucian.curtu@unitbv.ro

M. Šijačić-Nikolić et al. (eds.), *Forests of Southeast Europe Under a Changing Climate*, Advances in Global Change Research 65,
https://doi.org/10.1007/978-3-319-95267-3_5

5.1 General Aspects

Romanian forests cover an area of 6.55 million ha, representing 27.5% of the total country surface. About 48.8% is public property, 33.8% is private property and 17.4% belong to the local communities.

Having a diverse and complex structure of landscapes, with a well represented mountain area (28%) – supported on the Carpathian Range, large highlands and hills (42%) and planes (30%), the Romanian forest has a double distribution: zonal distribution, from 44 to 49° northern latitude, and altitudinal one, with several vegetation layers from 200 to 1900 m. Due to its geographic location, Romania belongs to temperate climate, with great variety of climatic particularities, from continental influences in the eastern part, to the submediterranean influences in south-westerns part.

Under climatic condition and specific relief the forest area is uneven distributed on phytoclimatic layers, from the steppe zone to the mountain story. The forests are well represented in the mountain region (59.7% from the total forest area), covered mainly by spruce on high altitude and by mixt or pure forest of beech, fir and sessile oak, in mixture with other species like: European larch, sycamore, common ash, mountain elm, Swiss stone pine etc.

In the highland and hill region the dominant species are beech and sessile oak, mixt with other scattered species or noble hardwoods: Norway maple, wild cherry, elm, wild service tree, hornbeam, common ash. The total surface of this region represents 33.8% of the total forest area (Stănescu et al. 1997; Şofletea 2005).

The forests from the plain region represent only 6.5% of the total forest area. This area is characterized by a large number of vegetation units. The steppe forest is characterized by grayish oak (*Quercus pedunculiflora*), pubescent oaks (*Q. pubescens* and *Q. virgiliana*), field elm, tatarian maple, common hawthorn and connect steppe with the forest zone. The forest zone can be divided in next subzones: mezzo-xerophytic forest of oaks (with Hungarian and Turkey oak), mezophytic oak forest (pedunculate oak, linden) and also riparian forest (narrow-leave ash, poplar, black alder).

According to the Forest State Report (2015) and National Forest Inventory Report (2008–2012), the main characteristics of the Romanian forest resources, can be summarized by:

- Age-structure of stands is unevenly distributed, with age-forests containing predominantly 20 to 80-year old trees;
- The total standing volume in Romania is about 2221 mil. m^3, of which 31% coniferous tree species, 39% beech, 13.7% oak species and 16.3% other species,
- The average stand productivity is 321.9 m^3, much higher than European average of 147 m^3;
- The annual mean increment of forest stands is 7.8 m^3/year/ha, comparative with European forests of 4.4 m^3/year/ha;

- The annual round wood production, allowed to be extracted according to the management plans, is 22.2 mil m^3;
- Natural regeneration surface in the last 10 years was between 12,000 and 17,000 ha and forest plantation of surface between 10,000 and 15,000 ha;
- In the last 10 years the annual logging volume from Romanian forest, ranged between 16.5 and 19.2 mil m^3 (33% beech, 31% coniferous trees, 10% oaks, 13% hardwood, 10% softwood).

The conservation of genetic resources is an old concern in Romanian forestry. Even in the absence of a special program dedicated to ensuring conservation of FGR, there were early concerns into preservation of valuable genotypes and their use in the Romanian forestry. These issues emerged by identification and selection of valuable stands as seed sources and transforming them into seeds reservations, trees identification and the installation of seed orchards. In 1975 there were over 65,000 ha of FGR stands, of which 15,000 ha of spruce, 13,000 ha of silver fir (Scărlatescu et al. 2012), 16,000 ha of sessile oak, 4400 ha pedunculated oak (Budeanu et al. 2016), 4000 ha of scots pine, 3300 ha of beech, 1500 ha of Hungarian oak, 1700 ha of lime and 1100 ha of larch (Enescu 1969, 1975).

The first catalogue of the Romanian seed reserves was issued in 1986 (Lalu 1996), including the most valuable forest stands, mainly selected after phenotypic and productive traits.

The establishment and characterization of forest seeds harvesting areas and their utilization was realized in 1988 (Enescu et al. 1988), and represented a big step towards the conservation and utilization of seminologic materials and implicitly the genetic resources.

Later, between 1996 and 2000, the Forest Research and Management Institute (ICAS) was involved in updating the catalogue released in1986, which were concluded in 2001 by publishing the National Catalogue of Forest Reproductive Material.

In 2010 a new zoning system was developed, based on provenance regions, which replace the older system of harvesting zones, but using the same criteria – the ecological homogeneity of an ecosystem (Pârnuță et al. 2010).

From the information presented above, it can be seen that the first concerns about the identification, utilization and conservation of the valuable forest genetic resources were related to practical forestry purposes. Later, besides the concerns regarding the utilization of valuable genetic resources, the concept of in situ and ex situ conservation was developed.

In situ conservation is not realized only in forests with conservation functions of valuable gene pool, but also in natural protected forests, natural and national parks, protected area, and in other forests with special protection functions (water, soil and clime protection) as a result of the protection regime.

5.2 Forest Genetic Resources in Romania

The establishment and conservation of Romania's forest genetic resources was realized in two separate phases: the first one, which is also the oldest, had objectives to identify, to describe and to establish new seed reserves, with the purpose of using the valuable genetic material in forestry, and the second one consisted in the identification and conservation of valuable populations especially for adaptive and protective purposes, but also taking in consideration the productive aspects.

This is the reason why at the national level, there are two types of genetic resources: (1) genetic resources for use in forestry practice, which are included in "***National Catalogue of Forest Reproductive Material***" – 2012 edition (Pârnuță et al. 2012), and (2) genetic resources for conservation, established in genetic resources cores and included in "***National Catalogue of Forest Genetic Resources***" – 2011 edition (Pârnuță et al. 2011).

5.2.1 Conservation Units of Forest Genetic Resources

Forest genetic resources (FGR) were established for 45 forest tree species and summarized in 698 conservation units. In every FGR unit there is a core area and a buffer zone. As a general rule, the minimum area of each genetic resource core is 10 ha that include one or more management units. Depending on the distribution of targeted species, the core area may have smaller surfaces, but must respect the following criteria: a minimum number of 500 reproductive trees – if the objective is genetic diversity conservation, minimum of 50 trees – if the objective is conservation of marginal populations and conservation of adaptive traits and minimum of 15 trees in populations with rare or endangered species (Pârnuță et al. 2011).

The effective area of the 698 cores of genetic resources included in the catalogue is 11242.5 ha, the total unit's area being 19889.6 ha. The number of created FGR cores for every species varies by their economic importance at a national level, as well as considering their distribution area.

In the establishment of the FGR it was attempted that in every provenance region of each species to exist at least one genetic resource. Therefore, most conservation units were established for the native forest tree species, with a vast area at a national level like: beech, spruce, sessile oak, fir, pedunculated oak, larch etc. Fewer FRG units were established for indigenous species with low or zonal distribution (European black pine, English yew, gray alder, Turkish hazel), also for exotic species (grand fir, Douglas fir, white pine, plane, shagbark hickory).

The overwhelming majority of FGR are composed of native species in natural stands (in situ), with a total surface of 12047.7 ha, which represent more than 94% from all the FGR area. Just 121 FGR units, with a total surface of 875.1 ha are established in artificial valuable stands (ex situ), representing almost 6% for the total FGR area (Pârnuţă et al. 2011).

The average age of the established FGR stands is around 100 years. Generally, the forest stands are composed of forest tree species selected as FGR with an age over 50 years, but the oldest ones can reach an age of over 150 years. The majority of stands included in the catalogue are in the high (62.6%) and middle (34%) productive classes. Just 3.4% of FGR were established in low productive stands, and these ones refer to marginal populations or to the ones situated in extreme vegetation conditions.

Forest management in FGR cores is done according to the national law (Forestry Code, Technical Norms) with the following special conservation activities being allowed:

- Extraction of trees affected by biotic or abiotic factors or dead trees, in order to maintain a corresponding phytosanitary condition;
- Extraction of trees for creating optimal conditions for seedling development, with the purpose to ensure natural regeneration and the continuity of the forest and existing genepool.

In the buffer zone are allowed silvicultural interventions to ensure protection of FGR cores. Generally, intensive treatments are recommended, with a long period of regeneration. Afforestation works, for completing the natural regeneration, should be realized with native species, using of local origin forestry material, with the purpose to avoid genetic pollution and preserving the local genepool.

In the case of FGR cores included in natural protected areas, the interventions are conditioned by the functional zoning, and silvicultural operations should be performed respecting the management plan, approved for each protected area. The status of protected area is not in disagreement with the purpose of established FGR cores and on the contrary it contributes to a total conservation of the whole ecosystems without detaching the genetic population from the complex relationships inside the community (Geburek et al. 2005). The number of FGR cores established in different natural protected areas is 138, with a total surface of 2140.7 ha, which represent 16.57% of the total area (Pârnuţă et al. 2011).

The inclusion/exclusion of forest genetic resources in the National Catalogue can be done once while updating the catalogue. According to the Ministry Order No.945/2012 and as a EUFGIS focal center, the National Institute for Research and Development in Forestry "Marin Drăcea" is responsible for monitoring and updating FGRs database and also to elaborate and publish every 5 years the National Catalogue of Romanian's Forest Resources. Unfortunately the last update scheduled for 2017 was not possible to be realized due to the lack of funds allocated by the national authority for this purpose.

5.2.2 Forest Reproductive Material

The necessity of increasing the productivity and quality of the stands led to the identification and selection of most valuable forest stands as seeds sources and transforming them into seed reserves. This process started over five decades ago, by publishing in 1986 The Catalogue of Reserves and in 2001 The National Catalogue of Resources for Reproduction Forestry Materials.

With the purpose to harmonize the national legislation with the European EU directive 199/105/CE and related regulations, after Romania joined the European Union in 2007, efforts were consistently intensified to revise and update the "***National Catalogue of Forest Reproductive Material***" published in 2001.

The last published Catalogue summarized those efforts when it was realized in 2012 (Pârnuţă et al. 2012) and approved by the Ministry Order No. 1645/2013.

The basic material sources consist of the following categories: "selected", "qualified" and "tested", which are being established in stands, seed orchards, parents of family, clone or clone mixture.

There were identified and listed in the catalogue 2659 "selected" units within stands, with a surface of 37461.06 ha and 98 "tested" units, with a total surface of 2060.1 ha (Table 5.1). Additionally, 113 seed orchards from the "qualified" category, with a surface of 625.2 ha and 2 seed orchards from the "tested" category with a surface of 8.7 ha were selected (Pârnuţă et al. 2012).

The 58 clone cultures are included in the catalogue and refer to species and hybrids of poplar and willow, being listed in the "qualified" category, with a total surface of 54.1 ha.

The National Catalogue of Forest Reproductive Material is structure by grouping categories of materials and species. For each source unit listed in the catalogue there is complex information that includes: source code (unique code), provenance region, geographic localization (in WGS 1984 system), source type, administrative localization, total and effective area.

A total number of 2930 source units were established for a number of 45 forest tree species, the number of units is depending on the area covered by each species at the national level, as well as its economic importance (Fig. 5.1). For the forest tree species the objective was also to establish at least one source unit in every region of provenance, in its area of distribution. The most units were established fort sessile oak (458) and spruce (440) (Table 5.1).

The addition of new materials in the National Catalogue of Forest Reproductive Material is done by the authority, at the request of forest owners, administrators or delegated persons by the authority, with the compliance of the legal means regarding the rules of marketing.

The central public authority which is in charge of forestry has the responsibility, according to law (Law No.107/2011), to evaluate the opportunity of maintaining the base materials in the catalogue, and to approve or withdraw the approval for the inadequate materials.

Table 5.1 Core units for conservation of forest tree species in Romania

		Basic material										Core units of FGR		
		Stands				Seed orchads				Clones				
		Selected		Tested		Tested		Qualified		Qualified				
No.	Forest tree species	No	S (ha)	No	S (ha)	No	S (ha)	No	S (ha)	No	S (ha)	No.	Effective surface area (ha)	Total surface area (ha)
1	*Abies alba* Mill.	240	3344.8	30	607.2			10	84.9			61	1198.5	2693.1
2	*Abies grandis* (Dougl.) Lindl	–	–	–	–	–	–	–	–	–	–	1	0.2	2.1
3	*Acer platanoides* L.	4	5.8									4	13.5	134.8
4	*Acer pseudoplatanus* L.	79	248.1					1	4.6			14	72.3	655.3
5	*Ailanthus altissima* (Mill.) Swingle.	–	–	–	–	–	–	–	–	–	–	1	2.9	4.9
6	*Alnus glutinosa* (L.) Gaertn.	14	85.1									5	23.3	107.4
7	*Alnus incana* (L.) Moench.	1	2.3									1	4.3	4.3
8	*Alnus viridis* (D.C.) Chaix	–	–	–	–	–	–	–	–	–	–	1	33.9	42.4
9	*Carya ovata* (Mill.) K. Koch.	1	2.1									1	2.1	2.6
10	*Castanea sativa* Mill.	9	43.9									2	26.1	26.1
11	*Corylus colurna* L.	–	–	–	–	–	–	–	–	–	–	1	3.0	30.3
12	*Fagus sylvatica* L.	368	7561.1									123	2881.9	3940.4
13	*Fraxinus angustifolia* Vahl.	10	250.9	5	52.8							14	134.3	392.3
14	*Fraxinus excelsior* L.	122	540.6	1	10.9			5	45.3			24	105.5	451.9
15	*Fraxinus pennsylvanica* Marsh.	1	7.7									1	0.6	1.4
16	*Juglans nigra L.*	9	23.2									7	28.7	56.3
17	*Larix decidua Mill.*	60	301.6	5	30.0	1	5.6	20	119.2			30	174.1	555.3
18	*Larix x eurolepis*					1	3.1	1	2.0			–	–	–
19	*Picea abies* (L.) Karst.	440	8564.0	25	747.0			9	72.9			89	2405.8	3707.6

(continued)

Table 5.1 (continued)

		Basic material										Core units of FGR		
		Stands				Seed orchads				Clones				
		Selected		Tested		Tested		Qualified		Qualified				
No.	Forest tree species	No	S (ha)	No	S (ha)	No	S (ha)	No	S (ha)	No	S (ha)	No.	Effective surface area (ha)	Total surface area (ha)
20	*Picea abies* (L.) Karst. "Columnaris"							1	2.7			–	–	–
21	*Picea abies* (L.) Karst. X							4	3.2			–	–	–
22	*Pinus cembra* L.	3	7.4					3	8.0			4	22.3	209.8
23	*Pinus mugo Turra*	1	33.9									1	33.9	42.4
24	*Pinus nigra ssp. banatica* (Borb.) Novak							3	8.1			1	6.8	22.8
25	*Pinus nigra ssp. nigra* Arn.	34	261.2					10	35.5			15	112.5	161.4
26	*Pinus strobus* L.							4	13.2			1	4.3	14.2
27	*Pinus sylvestris* L.	75	757.1					10	55.1			22	208.8	434.8
28	*Platanus hybrida* Brot.	–	–	–	–	–	–	–	–	–	–	1	1.3	2.5
29	*Populus alba* L.	19	94.0									9	45.4	57.2
30	*Populus nigra* L.	7	19.0									3	14.0	22.9
31	*Populus ssp.*									37	44.6	–	–	–
32	*Prunus avium* L.	23	55.3					9	46.1			2	5.0	63.6
33	*Pseudotsuga menziesii var. glauca* (Beissn.) Franco							1	7.5			1	5.2	5.2
34	*Pseudotsuga menziesii var. menziesii* (Mirbel) Franco	29	86.1					5	28.1			9	30.6	54.4
35	*Quercus cerris* L.	91	904.4									26	220.6	465.2
36	*Quercus frainetto* Ten.	54	838.5					1	7.0			20	243.6	417.3
37	*Quercus palustris* Muenchh.	–	–	–	–	–	–	–	–	–	–	1	0.4	0.4

38	*Quercus pedunculiflora* K.Koch	47	520.3									16	248.6	318.4
39	*Quercus pedunculiflora x Q. robur*							1	6.1			–	–	–
40	*Quercus petraea* (Matt.) Liebl. *ssp. petraea* (Liebl.)	458	8341.4	15	285.1							81	1656.5	2331.4
41	*Quercus pubescens* Willd.	21	104.9									2	7.0	60.2
42	*Quercus robur* L.	205	2526.4	17	327.2			4	25.0			54	743.3	1416.3
43	*Quercus rubra* L.	32	112.1					2	11.3			8	22.6	36.6
44	*Robinia pseudacacia* L.	86	722.1					6	27.5			11	140.8	144.1
45	*Salix ssp.*									21	9.5	–	–	–
46	*Sorbus torminalis* L.	1	0.5									–	–	–
47	*Taxodium distichum* (L.) Rich.	3	5.2									2	3.5	3.5
48	*Taxus bacata* L.	–	–	–	–	–	–	–	–	–	–	1	3.7	18.5
49	*Tilia cordata* Mill.	14	114.6					1	5.0			7	78.0	192.3
50	*Tilia platyphyllos* Scop.	7	56.8									3	33.2	67
51	*Tilia tomentosa* Moench.	90	916.0					2	6.9			16	237.9	516.6
52	*Ulmus pumila* L.	1	2.5									1	2.1	2.1
Total	2659	37460.9	98	2060.2	2	8.7	113	625.2	58	54.1	698	11242.9	19889.6	

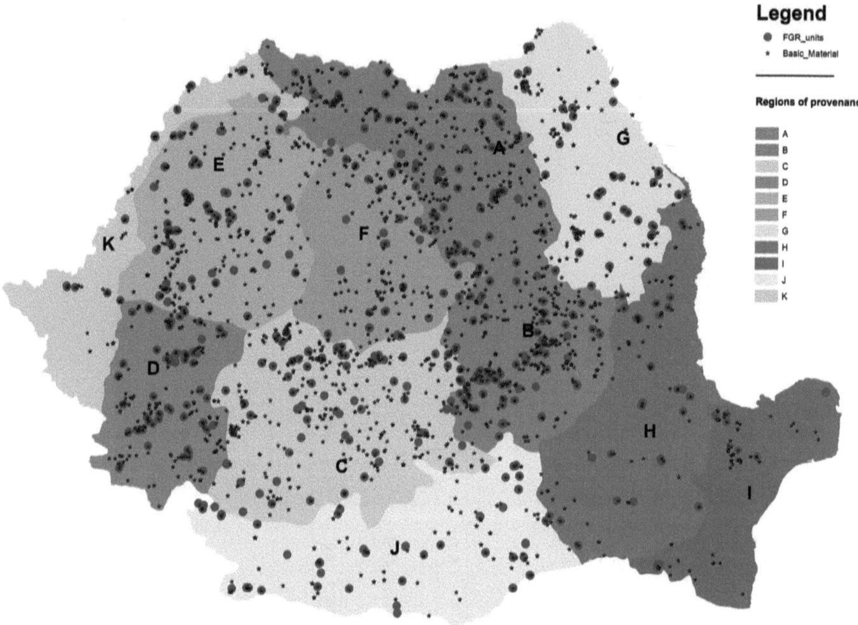

Fig. 5.1 Map of core units for conservation of forest tree species in Romania

5.3 Conclusions

In Romania 698 FGR core units for the conservation of 45 forest tree species with an effective surface of 11242.5 ha have been established (Table 5.1, Fig. 5.1) and are included in the National Catalogue of Forest Genetic Resources (Pârnuţă et al. 2011).

At the same time, activities to update the resources for the production of the basic materials were also initiated with 2930 units for 45 forest species being selected, with a total area of 40209.3 ha, which are listed in the National Catalogue of Forest Reproductive Material – 2012 (Pârnuţă et al. 2012). According to national legislation, it is allowed to use for forest purpose only the reproductive material which has a certain source, from the present catalogue, and it can be transferred only inside the same provenance region or an adjacent border region (Fig. 5.1).

Although it may seem somewhat paradoxical, more than 300 source units of the basic material category are included, entirely or partly, also in the FGR conservation category, due to their importance of use as a base reproductive material (Fig. 5.1).

Both FGR cores database and the units of forest reproductive material database have been created in GIS system and contain all information on: source identification code, geographical and administrative location, general ecosystem conditions, main features of stands, total and effective area, as well as placement in the region of provenance (Popescu et al. 2015).

It can be concluded that:

- The core area of FGR must capture the entire genetic diversity of species, their number being conditioned by the species distribution area and their genetic variability;
- The base material for the production of forest reproductive material must provide the required material for afforestation, for each region of provenance;
- Selection of new sources or exclusion of some selected units must be a dynamic process, connected to the global environmental changing processes, to the ecosystem evolution trends and to the forestry end-users requirements;
- Native species are of first concern in in situ conservation of FGR and valuable gene pool, both with adaptive and productive traits, and should be use into sustainable forest management;
- There are also valuable non-native species and hybrids included in ex situ conservation programs that should be developed together and linked with breeding programs;
- It should be taken into consideration to increase the use of forest reproductive material of ex situ units, in order to ensure the quantity and the quality required in forest practice;
- Genetic studies should be extended in the future, and selection should be based on knowledge of genetic diversity.

Acknowledgements The authors acknowledge to the Department of Forest Genetics and Tree Breeding – National Institute for Research and Development in Forestry "Marin Drăcea" for necessary facilities and to funding agencies (ANCSI, PN09460209, PN16330201, PN16330203, and UEFISCDI, Contract 618/2005) for financial assistance.

References[1]

Budeanu, M., Stuparu, E., & Tănasie, Ş. (2016). Identificarea de noi resurse genetice forestiere de cvercinee cu adaptabilitate ridicată. *Revista de Silvicultură şi Cinegetică, 38*, 21–26.

Enescu, V. (1969). *Arborete-rezervaţii pentru producerea seminţelor selecţionate*. Bucureşti: Editura Agro-Silvică.

Enescu, V. (1975). *Ameliorarea principalelor specii forestiere*. Bucureşti: Editura Ceres.

Enescu, V., Doniţă, N., Bândiu, C., Contescu, L., Chiriţă, C. D., & Roşu, C. (1988). *Zonele de recoltare a seminţelor forestiere în R.S. România*. Ministerul Silviculturii, ICAS, Seria II a, 60pg.

Geburek, T., & Turok, J. (2005). *Conservation and management of forest genetic resources in Europe*. Zvolen: Arbora Publishers 694pg.

Lalu, I. (1996). Catalogul Naţional al Resurselor Genetice Forestiere – Partea I. ICAS Manuscript.

Pârnuţă, G., Lorenţ, A., Tudoroiu, M., & Petrila, M. (2010). *Regiunile de provenienţă pentru materialele de bază din care se obţin materialele forestiere de reproducere din România*. Bucureşti: Editura Silvică.

Pârnuţă, G., Stuparu, E., Budeanu, M., Scărlătescu, V., Marica, F. M., Lalu, I., Tudoroiu, M., Lorenţ, A., Filat, M., Teodosiu, M., Nica, M. S., Cheşnoiu, E. N., Pârnuţă, P., Mirancea, I.,

[1] Asterisk indicates public reports, legislation, etc – no authors.

Marcu, C., Pepelea, D., Dinu, C., Marin, S., Daia, M., Dima, G., Şofletea, N., & Curtu, A. L. (2011). *Catalogul naţional al resurselor genetice forestiere*. Bucureşti: Editura Silvică.
Pârnuţă, G., Budeanu, M., Stuparu, E. L., Scărlătescu, V., Chesnoiu, E. N., Tudoroiu, M., Filat, M., Nica, M. S., Teodosiu, M., Lorent, A., Daia, M., & Dinu, G. C. (2012). *Catalogul Naţional al Materialelor de Bază pentru producerea Materialelor Forestiere de Reproducere*. Bucureşti: Editura Silvică.
Popescu, F., Postolache, D., & Pitar, D. (2015). Aspecte privind conservarea şi managementul resurselor genetice forestiere din România. *Revista de Silvicultură şi Cinegetică, 37*, 13–17.
Scărlătescu, V., Stuparu, E., & Budeanu, M. (2012). Resursele genetice forestiere de Brad (*Abies alba* Mill.) din România. *Revista de Silvicultură şi Cinegetică, 31*, 30–33.
Şofletea, N. (2005). *Genetică şi ameliorarea arborilor*. Braşov: Editura "Pentru Viaţă".
Stănescu, V. (1984). *Aplicaţii ale geneticii în silvicultură*. Bucureşti: Editura Ceres 291pg.
Stănescu, V., & Şofletea, N. (1998). *Silvicultura cu bazele geneticii forestiere*. Bucureşti: Editura "Ceres".
Stănescu, V., Şofletea, N., & Popescu, O. (1997). *Flora forestieră lemnoasă a României*. Bucureşti: Editura Ceres 451pg.
***Inventarul Forestier Naţional. Rezultatele primului ciclu (2008–2012). INCDS "Marin Drăcea". Available online at: http://www.mmediu.ro/articol/rezultate-ifn-ciclul-i-2008-2012/1667.
***Legea 46/2008. Codul Silvic – Monitorul Oficial Partea I, Nr.238/2008.
***Legea 107/2011. Comercializarea materialelor forestiere de reproducere – Monitorul Oficial, Partea I, Nr.430/2011.
***Ordinul Ministrului 945/2012 pentru aprobarea Catalogului Naţional al Resurselor Genetice Forestiere şi a Instrucţiunilor tehnice privind managementul durabil al resurselor genetice forestiere – Monitorul Oficial, Partea I, Nr.419/2012.
***Ordinul Ministrului 1028/2010. Aprobarea delimitării şi descrierii regiunilor de provenienţă pentru materialele de bază din care se obţin materiale forestiere de reproducere din categoria "sursă identificată" şi "selecţionat" pentru speciile de interes forestier din România. Monitorul Oficial, Partea I, Nr.617bis/2010.
***Ordinul Ministrului 1645/2013 pentru aprobarea Catalogului Naţional al materialelor de bază şi a modelului filei Catalogului Naţional al materialelor de bază - Monitorul Oficial, Partea I, Nr.539/2013.
***Raport privind starea pădurilor Românie. Ministerul Mediului, Apelor şi Pădurilor. Available online at: http://www.mmediu.ro/app/webroot/uploads/files/2016-12-16_Raport_Starea_padurilor_2015.pdf.

Chapter 6
State of Biodiversity and Forest Genetic Resources in Greece in Relation to Conservation

Filippos A. Aravanopoulos, Paraskevi G. Alizoti, Anna-Maria Farsakoglou, Ermioni Malliarou, Evangelia V. Avramidou, and Nikolaos Tourvas

Abstract The different levels of biodiversity (genetic diversity, species diversity, ecosystem diversity) present in Greece are introduced, quantified, where possible, and analysed in comparison to the rest of the European continent and internationally. Greece ranks third and fourth regarding key biodiversity areas in Europe and Mediterranean Basin respectively and first in the number of species per unit area. The genetic diversity and differentiation in natural ecosystems and especially in forest tree species is very high. For a number of species (e.g. in chestnut) a significant portion of their total European genetic diversity is present in Greece and some metrics of genetic diversity parameters present high values when compared to European and world-wide average values. Nevertheless, the conservation of biodiversity has insofar focused at the ecosystem level. About one-third of the country land mass is under some form of biodiversity protection (27% included in the Natura 2000 network), however the forest genetic resources protection lags behind as officially only 5 species and 15 populations are part of the European Forest Genetic Resources Network. Overall, the richness for biodiversity in Greece is almost irreversibly proportional to the stage of its protection, especially at the genetic resource level.

F. A. Aravanopoulos (✉) · P. G. Alizoti
School of Forestry and Natural Environment, Aristotle University of Thessaloniki, Thessaloniki, Greece
e-mail: aravanop@for.auth.gr; alizotp@for.auth.gr;

A.-M. Farsakoglou · E. Malliarou · N. Tourvas
Laboratory of Forest Genetics and Tree Breeding, Faculty of Agriculture, Forestry and Natural Environment, Aristotle University of Thessaloniki, Thessaloniki, Greece
e-mail: amfarsakoglou@gmail.com; ermionimalliarou@yahoo.gr; nikostourvas@gmail.com

E. V. Avramidou
Laboratory of Silviculture, Forest Genetics and Biotechnology, Institute of Mediterranean Forest Ecosystems and Forest Products Technology, Athens, Greece
e-mail: aevaggelia@yahoo.com

M. Šijačić-Nikolić et al. (eds.), *Forests of Southeast Europe Under a Changing Climate*, Advances in Global Change Research 65,
https://doi.org/10.1007/978-3-319-95267-3_6

Keywords Biodiversity · Gene diversity · Species diversity · Conservation genetics · Forest trees · Greece

6.1 Introduction

The different levels of biodiversity (genetic diversity, species diversity, ecosystem diversity and, according to some authors, a fourth level of landscape diversity), reflect the natural wealth of a particular geographical area, which can also be the area defined by political boundaries, i.e. the area of a State. The biodiversity wealth, reflects many facets of opportunities: ecosystem services (protection of water resources, soil generation and protection, nutrient storage and recycling, pollution breakdown and absorption, contribution to climate stability, maintenance, resilience and recovery of ecosystems), biological resources (food, breeding stocks, genetic resource reservoirs including diversity in genes, species and ecosystems, medicinal resources and pharmaceuticals, wood products, ornamentals), and social benefits (research, education and monitoring, financial opportunities, recreation and tourism, cultural values).

In this chapter, the different levels of biodiversity present in Greece are introduced, quantified, where possible and analyzed in comparison to the rest of the European continent. Emphasis is given to forest ecosystems that cover a major part of the land mass of the country (49.4%, of which 25.5% high forest, Aravanopoulos (2010)) and host a wealth of biodiversity and especially a wealth of forest tree genetic resources. The State conservation efforts are also presented, both in terms of general biodiversity protection and the conservation of forest tree genetic resources. This paper is attempting to show that the richness for biodiversity in Greece is almost irreversibly proportional to the stage of its protection, especially at the genetic resource level.

6.2 Levels of Biodiversity in Greece

Greece is situated in the south-eastern part of the European continent and is extended over an area of approximately 132,000 km^2, bordering with Asia in the East and Africa in the South (Commission on Genetic Resources for Food and Agriculture 2010). The country's biogeographic position, at the easternmost part of the Mediterranean Basin, and its high topographical and geological diversity explain its biodiversity richness. Greece contains a great percentage of the biodiversity of the Mediterranean Basin (Table 6.1). In total, 22% of the species of the Mediterranean basin are found in Greece and a 10% of the plant species present, are endemic of the country (Commission on Genetic Resources for Food and Agriculture 2010). In Greece, over one third of the mammal and freshwater fish species of the Mediterranean Basin, are found.

Table 6.1 Number of species in Greece and Mediterranean basin, and the contribution of the Greek species in the Mediterranean basin given in percentage per taxonomic group (Commission on Genetic Resources for Food and Agriculture 2010; Derneği 2010)

Taxonomic group	Greece	Mediterranean basin	Percentage
Plants	6308	30,000	21%
Endemic plants	1278	13,000	10%
Mammals	111	330	34%
Birds	436	600	73%
Reptiles	61	357	17%
Amphibians	22	115	19%
Freshwater fish	154	400	39%
Total	7092	31,802	22%

Greece ranks third and fourth with 103 key biodiversity areas in Europe and Mediterranean Basin, respectively (Derneği 2010). Aravanopoulos (2010) has indicated, using a simple statistic metric, that on a per land unit basis Greece ranks first in Europe regarding biodiversity richness. In fact, in Greece which covers 1.29% of the European land mass, 31.60% of the biological species are found (in particular, 59% on the average of the plant, mammal, reptile and bird species). Therefore, in Greece are found 25 times more species than its landmass would "suggest", if we assume a linear relationship between biodiversity and land mass (Aravanopoulos 2010). An interesting comparison is that of the biodiversity levels present in Greece and those of a flagship tropical biodiversity hot-spot, in particular of Borneo island (Farsakoglou and Aravanopoulos 2013), (Box 6.1).

6.3 Genetic Diversity of Forest Tree Species in Greece

Besides species and ecosystem diversity, Greece's position in the eastern Mediterranean basin, contributes to the fact that it encompasses very high levels of genetic diversity (Aravanopoulos 2010). A significantly higher genetic diversity is found in the Eastern Mediterranean Basin compared to the Western and a higher within-species biodiversity, (as for example in four conifer species *Abies, Cedrus, Cupressus, Pinus*) is found in the Mediterranean than other (conifer) species around the world (Fady-Welterlen 2005). The level of genetic diversity can be crucial for detecting areas with high conservation value in the Mediterranean and thus for the conservation of the forest genetic resources (Fady 2012).

In Greece, in particular a number of studies have demonstrated the high levels of genetic diversity found. For instance, Aravanopoulos et al. (2005) studied the genetic diversity of 82 *Castanea sativa* populations from Greece, Italy, France, Spain and Britain, using 89 genetic loci and examining more than 2100 trees. They established by means of a multivariate analysis that about one half of the total chestnut genetic

Box 6.1 A Comparative Biodiversity Analysis Between Borneo and Greece

Conservation strategies are focused on the areas, with the highest biodiversity levels and the greatest percentage of endemic species, called "hotspots" (Fig. 6.1); (Hobohm 2003; Hrdina and Romportl 2017; Myers et al. 2000). Greece (part of the Mediterranean Basin hotspot) and Borneo (part of the Sundaland hotspot) are two of the 36 hotspots meeting such criteria, with Borneo drawing the attention as one of the most important hotspots worldwide (Hrdina and Romportl 2017). Here, we perform a comparative analysis of the biodiversity of Greece and Borneo, using the absolute values of the species on each area and then comparing the number of the species per unit area.

Borneo, the largest island in Asia and the third largest island in the world, is located at the south-eastern part of that continent and it is extended over an area of 743,330 km^2. Biodiversity of Borneo is one of the richest in the world, with 15,000 plants species (5000 of which are endemic), 222 mammal species, 420 bird species, 254 reptile species, 149 amphibian species and 430 fish species (ASEAN Centre for Biodiversity 2010). Borneo's biodiversity and its conservation have been at the centre of attention over the last three decades. Numerous scientific papers have been written on the importance of conserving such tropical areas and their species (Mittermeier et al. 1998), including plants (Ashton et al. 2003; Beaman 2005; De Bruyn et al. 2014; Slik et al. 2009), primates (Meijaard and Nijman 2003), frogs (Grafe and Keller 2009; Kueh et al. 2002), butterflies (Orr and Hauser 1996), bats (Struebig et al. 2012) etc.

Greece is seven times smaller (131,957 km^2) than Borneo (743,330 km^2). As Fig. 6.2 shows Greece has almost half of the total number of species of Borneo and about one-third of the plant species. A percentage of 20.3% and 33.3% of the plant species are endemic in Greece and Borneo, respectively, and almost the same number of bird species can be found in both areas. Greece has about 30% of reptiles and freshwater fish species compared to Borneo. Thus, Borneo generally reflects excessive diversity in absolute species numbers. When species diversity is considered on a per land unit basis (Fig. 6.3), it can be seen that, excluding amphibians, all the other categories of species per unit area in Greece display significantly higher values compared to Borneo. More specifically Greece ranks higher in plant, mammal, and fish species by showing a difference of 237%, 282% and 202% respectively, compared to Borneo and in bird species by an impressive 585%. Those values indicate the importance of conserving such areas that may not have high absolute species numbers but show high diversity when calculated as species per unit area as the cost of protection and conservation is also related to the magnitude of the area being conserved. The assignment of biodiversity hotspots has received some criticism (Brummitt and Lughadha 2003), however, the overall concept is still in use. Taking all the above into consideration, we can conclude that the conservation of such hotspots cannot only depend on absolute species numbers, but should also be evaluated from a different perspective highlighting the biodiversity per unit area as well. This study has also shown the high significance of biodiversity richness in SE Mediterranean and in Greece in particular when the per land unit basis is taken into account.

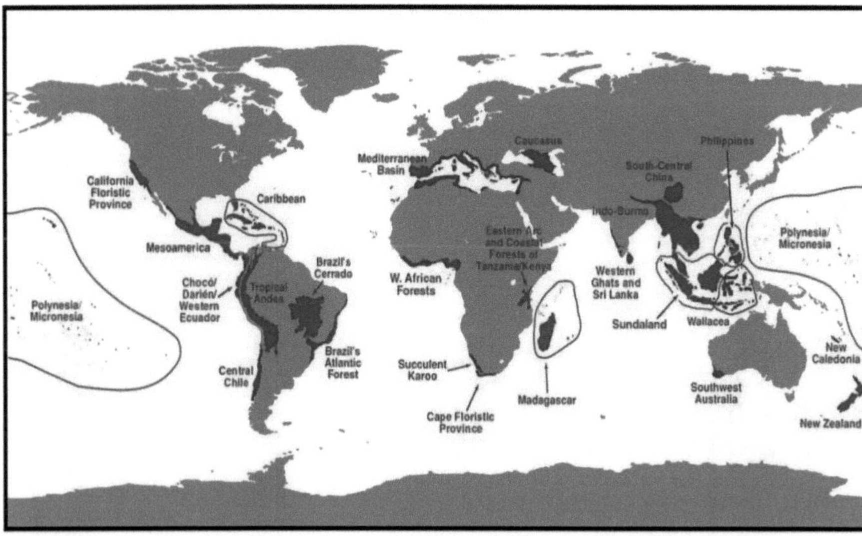

Fig. 6.1 The 25 out of 36 biodiversity hotspots across the world as described by Myers et al. (2000). Among them, Greece that belongs to the Mediterranean Basin and Borneo that belongs to the Sundaland hotspot

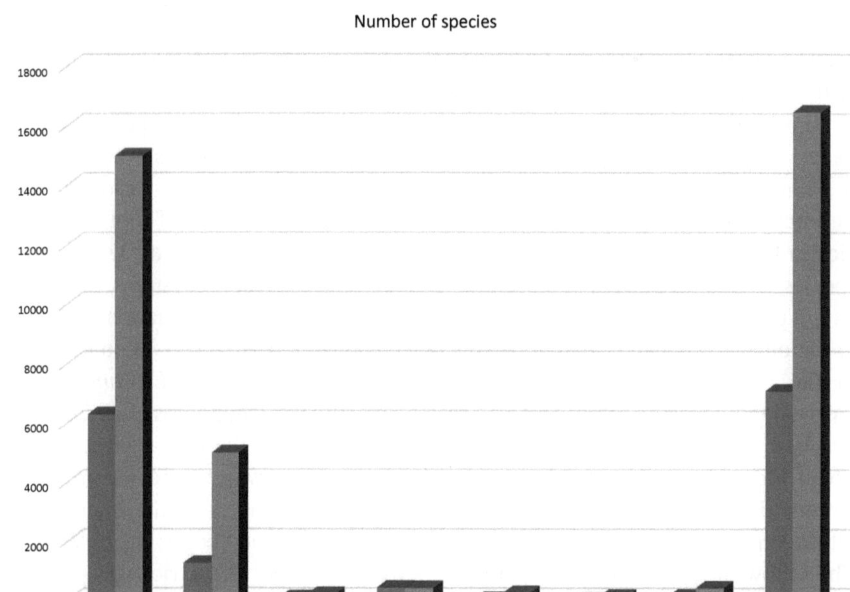

	Plants	Endemic Plants	Mammals	Birds	Reptiles	Amphibians	Fish	Total
GREECE	6308	1278	111	436	61	22	154	7092
BORNEO	15000	5000	222	420	254	149	430	16475

Fig. 6.2 Biodiversity based on the absolute number of the species of Greece and Borneo

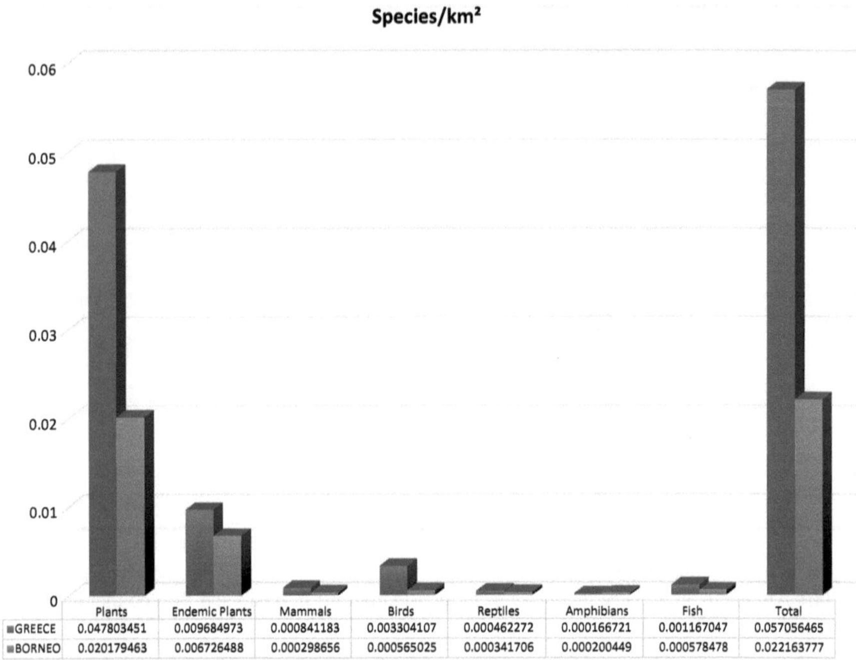

Fig. 6.3 Biodiversity of Greece and Borneo based on the species per unit area (km^2)

diversity in Europe is found in Greece (Fig. 6.4) and the main European gene pool of the species is the Hellenic one (Aravanopoulos et al. 2005).

Similar results have been found in other perennial woody angiosperms, such as in *Fagus sylvatica* (Magri et al. 2006), as well as in Mediterranean conifers, such as in *Pinus halepensis* (Bucci et al. 1998; Fady-Welterlen 2005). In one of the most important studies regarding genetic diversity and differentiation of forest tree species in Europe, Petit et al. (2003) studied the population genetics of 22 forest tree species in 25 mixed stands representative of European forest ecosystems. In Greece, a high genetic diversity and especially a very high genetic differentiation in relation to the other European populations was detected (Petit et al. 2003), in concordance to the results, found regarding the Hellenic populations of *Castanea sativa* (Aravanopoulos et al. 2005) and also in *Prunus avium* (Ganopoulos et al. 2011).

These results led to a more thorough investigation of the peer-review international literature comparing the genetic diversity of forest tree species present in Greece (Figs. 6.5 and 6.6). Data were collected from papers of the international bibliography using the Web of Science, Google Scholar and Scopus search engines. The following genetic diversity parameters were recorded: allelic richness - average number of alleles per locus (A), percentage of polymorphic loci (P), expected heterozygosity (H_E), observed heterozygosity (H_O), fixation index (F_{ST}, G_{ST}) and the

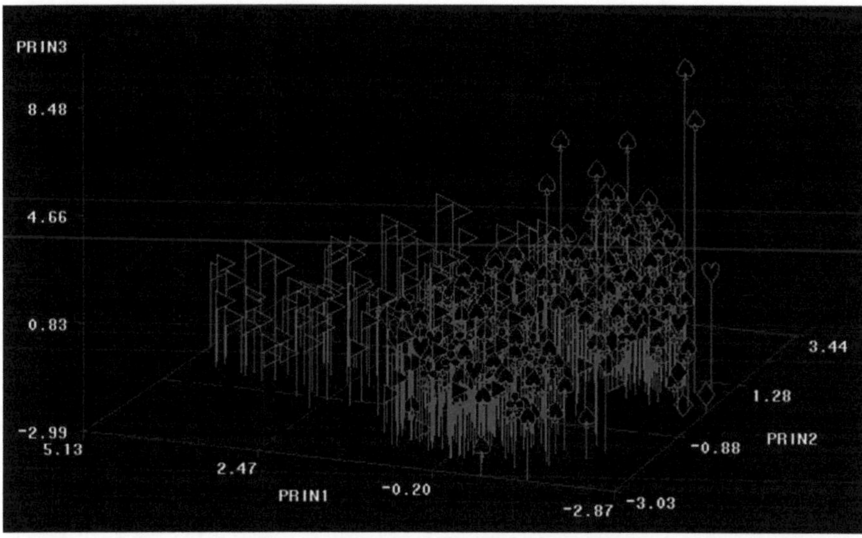

Fig. 6.4 A principal component analysis of chestnut genetic diversity in Europe based on the analysis of 82 populations using 89 loci (France: diamond; Greece: flag; Italy: Spade; Spain: hearts; U.K.: club)

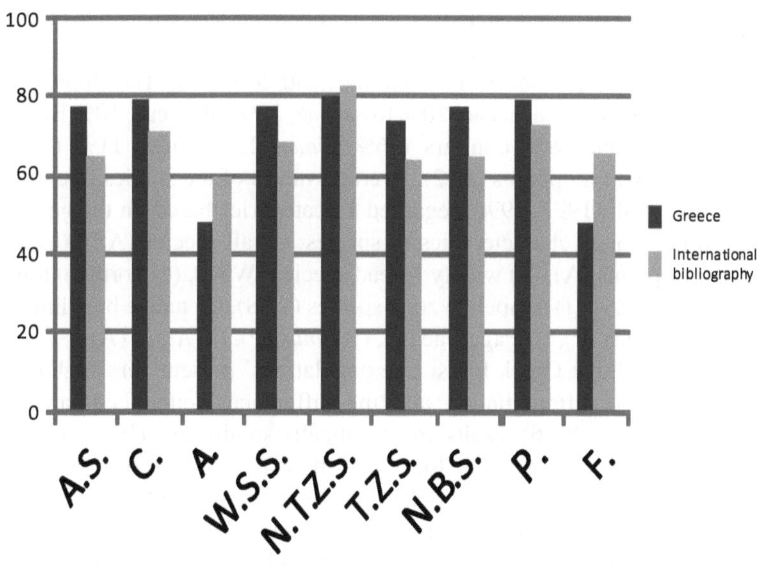

Fig. 6.5 Comparison of the percentage of polymorphic loci values recorded for Greek forest tree populations and those of the international literature (for bar notions see text)

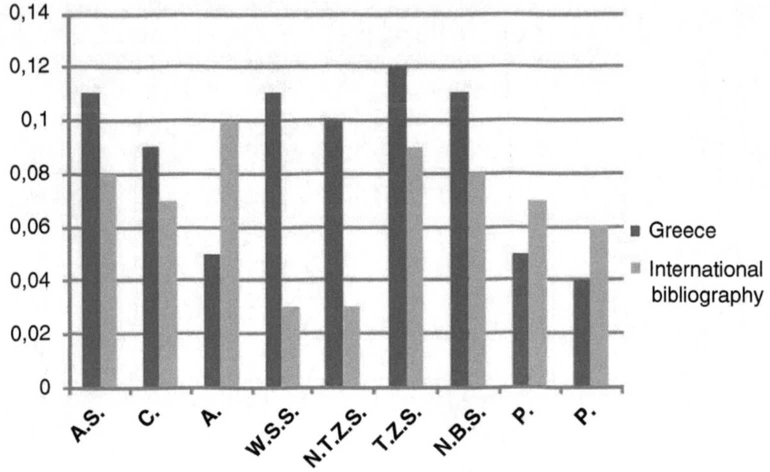

Fig. 6.6 Comparison of the fixation index values recorded for Greek forest tree populations and those of the international literature (for bar notions see text)

Shannon information index (I). The results obtained were compared to those of Hamrick and Godt (1996, 1992) which form probably the most comprehensive presentation of genetic parameters average data for all forest taxa in the international literature. Descriptive statistics were calculated and where feasible results were compared by a Student's t-test.

The results collected concern nine families and 20 genera. The studies concern 31% angiosperms (40% Fagaceae, 20% Rosaceae, 20% Oleaceae, 10% Betulaceae, 10% Salicaceae) and 69% conifers (66% *Pinus*, 23% *Abies*, 11% *Cupressus*) (Malliarou and Aravanopoulos 2012). Average values of the studies in comparison with Hamrick et al. (1992, 1996) regarded 11 categories based on the general biological and taxonomic characteristics of species: (a) all species (AS), (b) conifers (C), (c) angiosperms (A), (d) widely spread species (WSS), (e) northern temperate zone species (NTZS), (f) temperate zone species (TZS), (g) native breeding species (NBS), (h) Pinaceae (P), (i) Fagaceae (F), (j) *Pinus* (P), (k) *Abies* (A).

Results showed that Greek forest tree populations' present very high polymorphism and genetic differentiation and the differences found compared to the Hamrick et al. (1992, 1996) results are statistically significant. The mean values of the average number of alleles per locus and the expected heterozygosity are also higher in Greece compared to Hamrick et al. (1992, 1996), but the difference is not statistically significant (Malliarou and Aravanopoulos 2012). Hence, the Hellenic forest tree populations studied to date, have a broad genetic basis which is characterized by high levels of genetic diversity and differentiation when compared to similar average values of the international literature.

6.4 Conservation of Biodiversity in Greece

The wealth of biodiversity found in Greece has probably received less interest from State agencies and NGOs than it deserves. However, the conservation of species and ecosystem diversity is at rather good levels compared to the international scene. The level of the conservation of forest genetic resources lags behind in comparison. The main means of conservation at the species and ecosystem levels is the Natura 2000 network that includes 419 areas out of which 241 are Sites of Community Importance (SCI) – Special Conservation Zones, 202 are Special Areas of Conservation (SAC), while 24 areas are both SCI and SAC. The land area covered by the Natura 2000 in Greece is about 4,300,000 ha. This corresponds to 27.2% of the State land mass and to 4.5% of the European Natura 2000 network, placing Greece in the 10th position among the 28 EU States. The areas dedicated for biodiversity protection in Greece are presented in Table 6.2. The total area covered by various types of protection amounts to roughly one third of the country land mass.

The specific protection of forest genetic resources in Greece is facilitated by its participation in the European Forest Genetic Resources Network (EUFORGEN). However, thus far only a few species (five species, all conifers) and few populations (only 15) have been included (Table 6.2). This lack of specific direct conservation efforts regarding genetic diversity in Greece is compensated to a certain degree by the seed production areas (SPA) of the Hellenic Forest Service. These are 170 forest stands of 12 forest tree species covering a total area of about 11,500 ha, that are used for seed production (and seed collection) for artificial reforestation purposes. As such, seed production areas are under a special management plan that is directed towards protection. Therefore, these areas and the considerable area that is under different levels of biodiversity protection (Table 6.2) are areas where genetic diversity may be considered to be at least to some level conserved.

Table 6.2 Areas dedicated to biodiversity protection in Greece their type, number and area covered

Protected area type	Number of areas	Total area (ha)	Comments
National Forest Parks	10	211,000	No management/implementation actions are permitted- restrictive law
National Forest Parks	10	107,092	Under protection, management actions are permitted
Natural monuments	51	Not specified	Individual trees, specific natural formations, covering very small areas under full protection
Natura 2000	419	4,294,205	Management under restrictions
Gene conservation units (GCU)	15	2106.77	Genetic diversity conservation units (population within species level)
Seed stands	170	11,500	Seed production and seed collection areas under special forest management

6.5 Conclusions

Greece presents a wealth of biodiversity at all levels: gene, species and ecosystem. This biodiversity richness when quantified (and given all the inherent limitations of any such quantification exercise), ranks Greece among the first countries both in Europe and in the Mediterranean basin and very high in the international domain. While the biodiversity protection at the species and ecosystem level can be seen as adequate (about one-third of the country land mass is under some form of biodiversity protection of which 27% included in the Natura 2000 network), the level of forest genetic diversity protection lags considerably behind and is almost irreversibly proportional to the wealth of the genetic resource.

References

Aravanopoulos, F. (2010). The importance of the biodiversity of Hellenic forest and land-based ecosystems for Europe and the Mediterranean. In M. Ananiadou-Tzimopoulou (Ed.), *Proc. Conf. Environmental Council* (pp. 1–11). Thessaloniki: Aristotle University Press, Aristotle University of Thessaloniki.

Aravanopoulos, F., Bucci, G., & Akkak, A. (2005). Molecular population genetics and dynamics of chestnut (*Castanea sativa*) in Europe: Inferences for gene conservation and tree improvement. *Acta Horticulturae, 693*, 403–412.

ASEAN Centre for Biodiversity. (2010). *ASEAN biodiversity outlook Los Banos, Laguna, Philippines* (pp. 120–121). Los Banos: ASEAN Centre for Biodiversity.

Ashton, P. S., Kamariah, A., & Said, I. M. (2003). *Field guide to the forest trees of Brunei Darussalam and the Northwest Borneo hotspot*. University of Brunei Darussalam in association with Brunei Forestry Department, Brunei Shell Petroleum and Sultan Haji Hassanal Bolkiah Foundation.

Beaman, J. H. (2005). Mount Kinabalu: Hotspot of plant diversity in Borneo. *Biologiske Skrifter, 55*, 103–127.

Brummitt, N., & Lughadha, E. N. (2003). Biodiversity: Where's hot and where's not. *Conservation Biology, 17*, 1442–1448.

Commission on Genetic Resources for Food and Agriculture. (2010). The second report on the state of the world's plant genetic resources for food and agriculture. Food & Agriculture Organization.

De Bruyn, M., et al. (2014). Borneo and Indochina are major evolutionary hotspots for Southeast Asian biodiversity. *Systematic Biology, 63*, 879–901.

Derneği, D. (2010). Ecosystem profile: Mediterranean basin biodiversity hotspot Final Report for submission to the CEPF Donor Council July 27:2010.

Fady-Welterlen, B. (2005). Is there really more biodiversity in Mediterranean forest ecosystems? *Taxon, 54*, 905–910.

Farsakoglou, A. M., & Aravanopoulos, F. A. (2013). Greece-Borneo: A comparative biodiversity analysis. In *Proceeding of the 13th Pan-Hellenic Conference, Hellenic Botanical Society*, Thessaloniki.

Ganopoulos, I., Aravanopoulos, F. A., Argiriou, A., Kalivas, A., & Tsaftaris, A. (2011). Is the genetic diversity of small scattered forest tree populations at the southern limits of their range more prone to stochastic events? A wild cherry case study by microsatellite-based markers. *Tree Genetics & Genomes, 7*, 1299–1313.

Grafe, T. U., & Keller, A. (2009). A Bornean amphibian hotspot: The lowland mixed dipterocarp rainforest at Ulu Temburong National Park, Brunei Darussalam. *Salamandra, 45*, 25–38.
Hamrick, J. L., & Godt, M. (1996). Effects of life history traits on genetic diversity in plant species. *Philosophical Transactions of the Royal Society of London B: Biological Sciences, 351*, 1291–1298.
Hamrick, J. L., Godt, M. J. W., & Sherman-Broyles, S. L. (1992). Factors influencing levels of genetic diversity in woody plant species. *New Forests, 6*, 95–124.
Hobohm, C. (2003). Characterization and ranking of biodiversity hotspots: Centres of species richness and endemism. *Biodiversity and Conservation, 12*, 279–287.
Hrdina, A., & Romportl, D. (2017). Evaluating global biodiversity hotspots – Very rich and even more endangered. *Journal of Landscape Ecology, 10*, 108–115.
Kueh, B. H., Maryati Mohamed, D., & Das, I. (2002). *Application of biogeographical data of frogs to prioritize conservation areas in Borneo*.
Magri, D., et al. (2006). A new scenario for the quaternary history of European beech populations: Palaeobotanical evidence and genetic consequences. *New Phytologist, 171*, 199–221.
Malliarou, E., & Aravanopoulos, F. (2012). A comparative analysis of forest tree species genetic diversity between Greece and the rest of the European continent. In: *Proceeding of the 14th Pan-Hellenic Conference, Hellenic Scientific Society for Plant Genetics and Breeding*, Thessaloniki.
Meijaard, E., & Nijman, V. (2003). Primate hotspots on Borneo: Predictive value for general biodiversity and the effects of taxonomy. *Conservation Biology, 17*, 725–732.
Mittermeier, R. A., Myers, N., Thomsen, J. B., Da Fonseca, G. A., & Olivieri, S. (1998). Biodiversity hotspots and major tropical wilderness areas: Approaches to setting conservation priorities. *Conservation Biology, 12*, 516–520.
Myers, N., Mittermeier, R. A., Mittermeier, C. G., Da Fonseca, G. A., & Kent, J. (2000). Biodiversity hotspots for conservation priorities. *Nature, 403*, 853–858.
Orr, A. G., & Hauser, C. L. (1996). Kuala Belalong, Brunei: A hotspot of old world butterfly diversity. *Tropical Lepidoptrea, 7*, 1–12.
Slik, J., et al. (2009). Environmental correlates for tropical tree diversity and distribution patterns in Borneo. *Diversity and Distributions, 15*, 523–532.
Struebig, M. J., Bożek, M., Hildebrand, J., Rossiter, S. J., & Lane, D. J. (2012). Bat diversity in the lowland forests of the heart of Borneo. *Biodiversity and Conservation, 21*, 3711–3727.

Part II
Legal and Institutional Framework for Forest Genetic Resources Conservation

Chapter 7
Regulatory Framework for Conservation and Sustainable Utilization of Forest Genetic Resources in Serbia

Marina Nonić, Jelena Nedeljković, Dragan Nonić, Jelena Milovanović, and Mirjana Šijačić-Nikolić

Abstract The aim of this chapter is an overall review and analysis of the most important national regulative frameworks related to conservation and sustainable utilization of forest genetic resources in Serbia. A number of legally-binding and non-binding documents (strategies, laws and programs), which have an effect on the conservation and sustainable utilization of forest genetic resources, were adopted after the 2000s in Serbia. This issue is mainly the subject of strategies, laws and programs from the fields of forestry, sustainable development, environmental and nature protection. It could be stated that there is no adopted legally-binding legislation which directly relates to the FGR conservation in Serbia.

Keywords Regulatory frameworks · Genetic resources · Conservation · Forestry

7.1 Introduction

Forest genetic resources (FGR) can be of great importance in reducing, increasingly frequent, negative impacts of climate change both at the individual level and at the level of populations, species and ecosystems. Genetic diversity ensures adaptation of forest trees to changed climate conditions, as the selection of appropriate species and genotypes that are tolerant to altered climatic conditions can reduce the risks of their vulnerability. Due to numerous endangering factors, which frequently lead to genetic erosion of forest ecosystems, there is an increasing need for

M. Nonić (✉) · J. Nedeljković · D. Nonić · M. Šijačić-Nikolić
Faculty of Forestry, University of Belgrade, Belgrade, Serbia
e-mail: marina.nonic@sfb.bg.ac.rs; jelena.nedeljkovic@sfb.bg.ac.rs; dragan.nonic@sfb.bg.ac.rs; mirjana.sijacic-nikolic@sfb.bg.ac.rs

J. Milovanović
Faculty of Applied Ecology FUTURA, Singidunum University, Belgrade, Serbia
e-mail: jelena.milovanovic@futura.edu.rs

M. Šijačić-Nikolić et al. (eds.), *Forests of Southeast Europe Under a Changing Climate*, Advances in Global Change Research 65,
https://doi.org/10.1007/978-3-319-95267-3_7

conservation and targeted use of FGR. Trudić et al. (2015) point out that forests are the most important natural resource of diversity, whose "*...state affects the state of the biosphere as a whole*", so their care and protection "*...are recognized as an imperative in the concepts of sustainable forestry*".

In accordance with global initiatives to mitigate the negative effects of climate change on FGR, their targeted utilization could mitigate the risk and facilitate their adaptation to climate change. Kelleher et al. (2015) state that it would be of great importance if FGR conservation were incorporated into national adaptation strategies to climate change (Kelleher et al. 2015).

Fussi et al. (2016) pointed out that conservation and sustainable use of forest resources, as well as sustainable management of the forest functions are "*...the main goals of monitoring programs in forest ecosystems*" (Fussi et al. 2016).

Regulatory frameworks include all legally binding documents and procedures, such as: laws, regulations, rules, etc., which regulate how certain target groups should behave (Glück 2001). Simple and clear concept of the regulatory framework ensures the conservation and sustainable use of forests and forest lands through certain prohibitions and regulations (Krott 2005). These frameworks "*…have confidence in the authority of the state, official control and threat of coercion (enforcing coercion and punishment) in case of failure*" (Glück 2001).

In UN-FAO publication was stated that national regulatory frameworks for FGR are in many countries "*…partial, ineffective or inexistent, partly because FGR are not well understood*" (FAO 2014). Geburek and Konrad (2008) noted that in Europe "*…any supranational approach to homogenize FGR conservation is overwhelmingly complicated because of significant differences among countries at the legal, administrative, economic, and cultural levels*" (Geburek and Konrad 2008). According to Bouillon et al. (2015), in many countries, a solid legal background for genetic conservation is missing because the forest sector is often caught out by regulatory frameworks of other sectors. Many regulatory documents have already been made in order to ensure the persistence of biodiversity, but often take only into consideration diversity at the more general level and FGR are mentioned very seldom directly addressed specifically. One of the problems could also be that FGR conservation is planned on a long-term basis, whereas most regulatory frameworks need targets that can be achieved within a decade (Bouillon et al. 2015).

Šijačić-Nikolić and Milovanovic (2010) stated that forest genetic resources deserve "*… considerable attention in terms of regulating their use, modification and integration in the commercial production of seed and planting material*", as well as that the conservation of biodiversity and FGR in Serbia "*…is planned, organized and implemented in accordance with the guidelines defined within the ratified international contracts which are of importance for Serbia, and national laws and by-laws*".

This chapter provides a synthesis review and analysis of the most important legally binding and non-binding national frameworks, which regulate the field of conservation and sustainable utilization of FGR in Serbia.

7.2 Legally Binding and Non-binding Documents That Regulates FGR Conservation in Serbia

A number of legally-binding and non-binding documents (strategies, laws and programs), which have an effect on the conservation and sustainable utilization of FGR, were adopted after the 2000s in Serbia. Timeline of adopting these documents is presented in Scheme 7.1.

According to the timeline (Scheme 7.1), it can be noticed that the creation and adoption of laws in certain areas have not always been preceded by strategies, but also that they are often not followed by an appropriate program, as is customary.

In the period of 12 years, Serbia has adopted four strategies (red color), five laws (green) and one program (blue) which in some parts refer to the FGR and their conservation (Scheme 7.1). The National program of conservation and sustainable utilization of FGR for the period 2016–2025 (draft), which is expected to be adopted and is directly related to conservation and sustainable utilization of FGR,[1] is also presented on the timeline.

Table 7.1 shows an overview of 11 documents related to the conservation and sustainable utilization of FGR in Serbia according to the area of regulation.

Based on the data presented in Table 7.1, it could be stated that there is no adopted legally binding legislation which directly relates to the FGR conservation in Serbia.

This issue is mainly the subject of strategies, laws and programs from the fields of forestry, sustainable development, environmental and nature protection, whose review will be shown in the following subheadings.

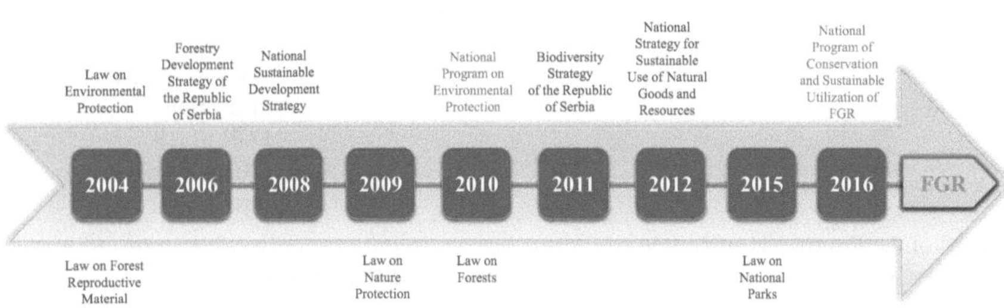

Scheme 7.1 Timeline of adopting the documents related to FGR in Serbia

[1] National Program of conservation and sustainable utilization of FGR of the Republic of Serbia for the period 2016–2025 (draft) is officially posted on the website of the Forest directorate: https://www.upravazasume.gov.rs/wp-content/uploads/2016/10/Nacrt-nacionalne-strategije-konzervacije-sumskih-genetickih-resursa.pdf.

Table 7.1 Overview of the documents related to conservation and directed utilization of FGR in Serbia according to area of regulation

Regulation area	Strategies: Document title	Strategies: Year of adoption	Laws: Document title	Laws: Year of adoption	Programs: Document title	Programs: Year of adoption
Sustainable development	National sustainable development strategy	2008	–	–	–	–
Environmental protection	–	–	Law on environmental protection	2004	National program on environmental protection	2010
Nature protection	National strategy for sustainable use of natural goods and resources	2012	Law on nature protection	2009	–	–
Biodiversity conservation	Biodiversity strategy of the Republic of Serbia	2011	–	–	–	–
Forestry	Forestry development strategy of the Republic of Serbia	2006	Law on forests	2010	–	–
			Law on forest reproductive material	2004		
National parks	–	–	Law on national parks	2015	–	–
Conservation of FGR	–	–	–	–	National program of conservation and sustainable utilization of FGR of RS (2016–2025) (draft)	2016[a]

[a]A draft of the Program is submitted
https://www.upravazasume.gov.rs/wp-content/uploads/2016/10/Nacrt-nacionalne-strategije-konzervacije-sumskih-genetickih-resursa.pdf

7.2.1 Strategies

Strategies that are directly or indirectly related to FGR and its conservation in Serbia are:

- National sustainable development strategy (OG 2008),
- National strategy for sustainable use of natural goods and resources (OG 2012),

- Biodiversity strategy of the Republic of Serbia for the period 2011–2018 (OG 2011),
- Forestry development strategy of the Republic of Serbia (OG 2006).

The **National sustainable development strategy** (OG 2008), in addition to the richness of the genetic, species and ecosystem diversity in Serbia, states the basic problems in relation to the protection of national biodiversity.

These are different strategic goals, important for Conservation and sustainable utilization of genetic resources (2008): establish an efficient system of bio-monitoring components of biodiversity and the system of the living world and other natural values; inventory of biodiversity; implementation of effective control measures of genetically modified organisms. From the aspect of conservation of FGR, the most important goals in this strategy could be: improvement of the methods for sustainable use of genepool and establishment of the bank for conservation of genetic material with increasing support for the conservation of genetic resources, as well as the increase in the number of subjects and areas involved in conservation work (2008).

The **National strategy for sustainable use of natural goods and resources** (OG 2012) as one of the general goals states "*...ensuring conservation, improvement of the state and sustainable use of populations of autochthonous species and communities at the level that will ensure their long-term viability*" (OG 2012). The specific goals which are related to the FGR are: maintaining genetic variability within populations and preserving their overall genepool, especially autochthonous plants; implementation of ex-situ and in-situ conservation measures; improving the capacity of the National gene bank; improving the management system for biological resources in protected natural assets, in accordance with the prescribed protection regimes (OG 2012).

The **Biodiversity strategy of the Republic of Serbia**[2] **for the period 2011–2018** (OG 2011) is a national strategic document in the field of nature protection, which is directly related to FGR. Objectives related to the conservation and sustainable utilization of FGR are: promoting the conservation of forest biodiversity, including genetic diversity, through the development of forest certification programs and best practices for sustainable forestry based on an ecosystem approach; development of forest management measures and guidelines for the prevention of adverse impacts of genetically modified tree species, allochthonous and invasive species on forests and biodiversity (OG 2011). According to the Strategy "*...within the indigenous forest genetic resources, of the highest values are endemic and endemo-relict taxa*", as well as wild fruit species (OG 2011).

The **Forestry development strategy of the Republic of Serbia** (OG 2006) represents the "*basic political and strategic development document of the forestry*

[2] In 2001, Serbia ratified the Convention on Biological Diversity, whereby certain international obligations were undertaken, one of which relates to the development of national strategies, plans or programs for the protection and sustainable use of biological diversity, and in accordance with that, the Biodiversity strategy of the Republic of Serbia was adopted (Nonić et al. 2017a).

sector of Serbia which, based on the guiding principles of the sector, determines the development objectives, measures and methods of support for their implementation" (Nonić 2015).

The Strategy was adopted in 2006 for the purpose of "*...defining the general development goals of Serbia's forestry sector and determining measures for achieving these goals*", emphasizing that "*...forests are the most important source of biodiversity and play a key role in maintaining ecological processes and ecological balance at the local, national, regional and global level...*", so "*...it is necessary to take measures to preserve biodiversity in the forests and the sustainable use of its components*" (OG 2006).

According to these allegations, as one of the strategic goals is defined: "*...conservation, realistic improvement, sustainable utilization and valorization of forest biodiversity*". In order to achieve this goal, among other things, various measures are important from the aspect of conservation of the FGR: improvement of the method for directed utilization of forest tree species genepools through in situ and ex situ conservation and improvement of production of quality forest seed and planting material of controlled origin, as well as active participation in the European forest genetic resources programme (EUFORGEN); support the implementation of international obligations in the conservation of biodiversity in forest ecosystems (OG 2006).

7.2.2 Laws

As laws that, in some of their parts, relate to the FGR and their conservation in Serbia, the following are distinguished:

- Law on environmental protection (OG 2004a) which emphasizes the protection and conservation of forests, biodiversity and genetic resources;
- Law on nature protection (OG 2009b) which recognizes the importance of protecting, improving and preserving the biodiversity of forest ecosystems and conservation of genetic diversity;
- Law on forests (OG 2010a) which recognizes the conservation of forest trees genepool and diversity of forest ecosystems as a general-purpose forest functions;
- Law on forest reproductive material (OG 2004b) which is indirectly significant due to control of production, recognition, traffic, etc. of forest reproductive material;
- Law on national parks (OG 2015) which, among other things, refers to the protection and promotion of flora, vegetation and forests.

In addition to the aforementioned laws, as a law that "*...tackles the problem of conservation and directed utilization of FGR*" (Milovanović et al. 2012), can be specified and the Law on genetically modified organisms, which regulates the procedure for issuing approval and conditions for "*…deliberately introduction into the environment of genetically modified organisms and products of genetically modified organisms*" (OG 2009a).

The **Law on environmental protection** defines the different protection principles, among which are stated (OG 2004a):

- The principle of conservation of natural values, which are used under conditions and in a way that preserves the values of biodiversity, protected natural resources and landscapes (§9);
- The principle of sustainable development, which talks about the economical and reasonable use of natural and created values of the Republic of Serbia, with the aim of preserving and improving the quality of the environment for present and future generations (§9).

Based on this Law, the protection of biodiversity, use of biological resources and biotechnology, as well as the control of the input and growing of allochthous species and regulation of the traffic of endangered and protected species of wild flora and fauna is carried out (OG 2004a, §26).

In §25 of this Law, it is stated that "*…in order to protect and improve forest ecosystems, the forests are managed in such a way as to ensure rational forest management, conservation of genepool, improvement of the structure and achievement of priority forest functions*" (OG 2004a).

The **Law on nature protection** (OG 2009b) regulates the protection, preservation and improvement of biological, landscape and geological diversity. The protection of biodiversity is achieved by implementing measures for the protection and improvement of species, their populations, natural habitats and ecosystems, as well as the corridors that link them (OG 2009b, §15). The protection of forest ecosystems, according to this Law, is achieved by implementing appropriate measures and activities for their protection, improvement and sustainable utilization (OG 2009b, §18). According to this Law, wild species that are endangered or can become, are protected as strictly protected or protected wild species, while prohibiting the introduction of allochthonous wild species and their hybrids in nature (except for scientific purposes, which do not endanger the autochthonous populations), reintroduction of wild species is regulated, as well as traffic, trade and cultivation of protected wild species (OG 2009b, §36, §82, §94).

In §4 of this Law, the term "*in situ protection*" is defined as "*…an active protection measure that includes the protection of populations of species on a natural habitat, preservation of natural ecosystems, preservation and recovery of populations in their natural habitats, as well as the preservation of geodiversity at the site or site of the rock, ore, minerals, crystals and fossils*" (OG 2009b). The term "*ex situ preservation*" is defined as "*…the preservation of components of biological and geological diversity outside their natural habitat…*" (OG 2009b, §4), It can be noticed that these terms relate to several different components of diversity, compared to the in-situ and ex-situ FGR conservation methods.[3] In the same article, the Law defines the concept of genetic diversity as "...*the total number and overall diversity of genes, or genetic information contained in all individual species of plants, animals, mushrooms and microorganisms*" (OG 2009b). According to §7, nature protection is in particular carried out by "... *sustainable use of natural resources and protected natural assets and control of their use by establishing a natural resource management system and protected natural assets*" (OG 2009b).

In accordance with §18 of the Law on nature protection, conservation of the biodiversity of forest ecosystems is done in order to strengthen the general-purpose functions of forests, whereby forest management must "...*be based on the principles of sustainable development and preservation of biodiversity, preservation of the natural composition, structure and function of forest ecosystems, in accordance with the conditions of nature protection, which are an integral part of forests*" (OG 2009b). According to §20 "...*taking genetic material from nature for the purpose of use must not endanger the survival of the ecosystem or the population of wild plants, animals and fungi in their habitats...*", whereas in §21 it is stated that on genetic material created from wild plant species, cannot acquire property (OG 2009b).

The **Law on forests** (OG 2010a) has made significant changes, both in terms of sustainable development of forests and forestry as an economic branch, as well as operationalization of the goals set by the Forest development strategy (Nonić 2015). The Law, as a general-purpose function of forests, prescribes the conservation of genepool of forest trees and other species within the forest community, as well as the biodiversity conservation (§ 6). As forests with a special purpose, the Law defines forests with the purpose of the conservation and utilization of forest tree species genepool and forests for the conservation of biodiversity of genes, species, ecosystems and lanscapes (§6). This Law regulates, among other things, the preservation, protection, planning, cultivation and use of forests, the disposal of forests and forest land (§1). In §9 of this Law, different forest preservation measures are prescribed, but the in-situ and ex-situ conservation of FGR has not been emphasized (OG 2010a).

According to §13, in order to conservation and sustainable utilization of genepool of protected species of forest trees, forest management plans should include measures for the protection, use and extension of distribution of these species. It also states that it is prohibited "...*cutting of trees, destruction of offspring and seed collection from the protected forest tree species by special regulation governing the protection of nature, unless they are source of infection from diseases and pests, or if they endanger people and objects*" (OG 2010a).

The **Law on forest reproductive material** regulates "...*the conditions and method of recognizing the starting material for the production of reproductive*

[3] Šijačić-Nikolic et al. (2014a) state that in-situ conservation implies "*conservation of forest genetic resources in natural populations and the protected nature areas*",whereas the ex-situ conservation is a "*form of conservation of forest genetic resources outside their natural habitat, the establishment of seed orchards, clonal archives, progeny tests, botanical gardens, arboreta and live archives*".

material of forest trees, production, control of production, processing, quality, traffic and use of forestry reproductive material" (OG 2004b, §1).

Law, in §6, defines four categories of reproductive material: reproductive material of known origin, selected reproductive material, qualified reproductive material and tested (varietal) reproductive material (OG 2004b). In accordance with §7, reproductive material of known origin can only be used within the same provenance region if there are insufficient reproductive material in the remaining three category or due to fire and natural disasters, while the remaining three categories can be used for raising and restoring forests (OG 2004b).

The **Law on national parks** (OG 2015) aims to preserve and enhance the values of biodiversity, ecologically important areas, habitats, and populations of wild flora and fauna (§2). National parks represent a significant form of *in-situ* conservation, so this Law can also be considered important from the aspect of FGR conservation. In accordance with this Law (§20) "*…the protection and improvement of flora, fauna (especially hunting and fishing) and vegetation, forests and water, planning of use and use of natural resources and space, regulation of construction and agricultural land in the National Park*" is being implemented (OG 2015).

Based on the review of national legislative frameworks, it can be noticed that there is no legally binding regulation that directly relates to the conservation of the FGR, but it is partly regulated in the area of environmental protection, nature protection and forestry.

7.2.3 Programs

From the aspect of the FGR conservation, programs related to this area are also very significant. The analysis included two programs. The National program on environmental protection is adopted in 2010, while the adoption and implementation of the National program for conservation and sustainable utilization of FGR of the Republic of Serbia for the period 2016–2025 (draft), is still expected. Although it has not been formally adopted, the draft of this Program has been analyzed, which provides concrete activities and measures that need to be applied for the purpose of conservation and sustainable utilization of FGR.

In the **National program on environmental protection** negative anthropogenic impacts on forest ecosystems are particularly pronounced, while emphasizing that "*…an excessive and uncontrolled exploitation of natural resources that have limited capacity…*" done "*…the greatest pressure on biodiversity and geological diversity*" (OG 2010b).

Goals that are indirectly related to FGR (OG 2010b):

- Improving the protection and sustainable use of wild plant species;
- Implementing effective measures for controlling the introduction of genetically modified organisms;

- Harmonization of national regulations in the field of nature, biodiversity and forest protection with EU legislation and international conventions;
- Establishing monitoring of biodiversity components;
- Preserving, improving and extending existing forests;
- Establishing synergies of sectoral policies and strategies;
- Improving the protection of indigenous species and preventing the introduction of invasive species.

The **National program for conservation and sustainable utilization of FGR of the Republic of Serbia for the period 2016–2025 (draft)** provides an overview of the state in this area, defines the objectives, priorities and concrete conservation measures of FGR (Šijačić-Nikolić et al. 2016), and its adoption and implementation are expected. The program was created as a result of the same project funded by the Ministry of Agriculture and Environmental Protection – Forestry Directorate, in its preparation were participated: University of Belgrade – Faculty of Forestry, Institute of Forestry in Belgrade and Institute of Lowland Forestry and Environment from Novi Sad.

The program consists of various chapters, among which are: factors that threaten FGR, legal framework for conservation and sustainable utilization of FGR, the state of FGR in Serbia and previous activities for FGR conservation, etc. whereby the chapter "*Climate change and conservation of forest genetic resources*", in which priorities and measures for conservation of forest genetic resources in line with climate change are defined, is particularly significant (Šijačić-Nikolić et al. 2016).

The Program predicts the activities, which are expected to have an impact on:

- Development of the forestry sector;
- Improvement of forest management;
- Informing the general public about the need to conserve FGR;
- Compliance with international obligations and the possibility of joining activities within the FAO organization related to FGR (Šijačić-Nikolić et al. 2016; Nonić et al. 2017a).

Šijačić-Nikolić and Sovilj (2017) for this Program state that "*...it is necessary to ensure its implementation in a way that all relevant entities in the Republic of Serbia, within their competences, commit to its implementation*", since it is not prescribed as a binding legal act in accordance with the current Law on Forests (OG 2010a), and is important for preserving Serbia's FGR. The same authors point out that the highest strategic planning document in the field of forestry is the Forest Development Program in the Republic of Serbia, which defines the development directions of forests and forestry with an action plan for its implementation, which is prescribed by the Law on Forests, stating: "*Forest Development Program in the Republic of Serbia will include the National Program for Conservation and Sustainable Utilization of Forest Genetic Resources of the RS as its integral part, which will ensure its full implementation*" (Šijačić-Nikolić and Sovilj 2017). The need for implementation of documents aimed the conservation and enhancement of natural resources (including FGR) undertake both domestic and international processes and agreements.

7.2.4 Decrees and Rulebooks (Sub-legal Acts)

In addition to laws, strategies and programs, in Serbia in the period after 2000, various decrees and rulebooks were adopted, which are in some parts related to the conservation of genetic resources. Below are listed some of the more significant sub-legal acts, from the aspect of conservation of FGR.

Decree on putting under control the use and trade of wild flora and fauna determines "...*wild species of flora, fauna and fungi whose collection from natural habitats, use and trade are placed under control*" (OG 2005, §1).

By the **Rulebook on declaration and protection of strictly protected and protected wild species of plants, animals and fungi** are declared wild species of plants, animals and fungi, in order to "...*conserve biological diversity, natural genepool*" as strictly protected wild species or protected wild species and are determined the measures of their protection and protection of their habitats (OG 2010d, §1).

Rulebook on the transboundary movement and trade in protected species prescribes, inter alia, the conditions under which are performed "...*import, export, entry, amount or transit, trade and cultivation of endangered and protected plant and animal wild species, their parts and derivatives*" (OG 2009c, §1).

In **Rulebook on conditions for establishing a gene bank of wild plants, animals and fungi, the manner of work of gene bank, manner of handling biological material, content of applications and documentation to be submitted with the application for the issuance of permit for establishing a gene bank**, in §3, state that research and preservation of biological material for the needs of gene bank may be carried out "in situ – *which implies the research and preservation of genetic material in the ecosystem or natural habitat and the maintenance and restoration of a sustainable population of species in their natural environment; and ex situ – research and preservation of genetic material outside its natural habitat*" (OG 2010c). According to the Rulebook, the use of biological material must not endanger the survival of wild plant populations in their habitats (OG 2010c, §3).

7.2.5 Analysis of Regulatory Frameworks

By analyzing the content of the documents, goals that are important for the conservation of the FGR were selected and grouped into thematic areas (Table 7.2).

It can be noted that, in terms of the strategic documents presented in Table 7.2, the highest number of goals which are important for the conservation of FGR is covered by the Biodiversity Strategy. In other analyzed strategies, more than half of the selected thematic areas are covered individually.

Based on the review of the aforementioned strategic documents, it can be stated that the problem of conservation of FGR is recognized in national strategies, wherein are determined "...*certain guidelines for actions in the field of forestry and environmental protection*" (Milovanović et al. 2012).

Table 7.2 Overview of legally binding and non-binding documents and aims of importance for conservation of FGR in Serbia

Thematic area (goals significant from the aspect of FGR conservation)	Strategies	Laws	Programs
Conservation and sustainable utilization of FGR	National sustainable development strategy	Law on Nature Protection	National program of conservation and sustainable utilization of FGR RS
	National strategy for sustainable use of natural goods and resources		
	Biodiversity strategy of RS		
	Forestry development strategy of RS		
Improving methods for sustainable use of genepool	National sustainable development strategy	–	National program of conservation and sustainable utilization of FGR RS
	National strategy for sustainable use of natural goods and resources		
	Biodiversity strategy of RS		
Conservation, improvement and sustainable use of biodiversity	National sustainable development strategy	Law on environmental protection	National program on environmental protection
	National strategy for sustainable use of natural goods and resources	Law on nature protection	National program of conservation and sustainable utilization of FGR RS
	Biodiversity strategy of RS	Law on forests	
	Forestry development strategy of RS	Law on national parks	
Improving governance and sustainable forest management	National sustainable development strategy	Law on forests	National program on environmental protection
	Biodiversity strategy of RS		
	Forestry development strategy of RS		National program of conservation and sustainable utilization of FGR RS
Improving international activities related to conservation and sustainable use of FGR	Forestry development strategy of RS	–	National program on environmental protection
			National program of conservation and sustainable utilization of FGR RS

(continued)

Table 7.2 (continued)

Thematic area (goals significant from the aspect of FGR conservation)	Strategies	Laws	Programs
Harmonization of national regulations among themselves and with international legislation	National sustainable development strategy	–	National program on environmental protection
	Biodiversity strategy of RS		National program of conservation and sustainable utilization of FGR RS
Promotion of conservation and sustainable utilization of FGR	Biodiversity strategy of RS	–	National program of conservation and sustainable utilization of FGR RS
Establishing a gene bank	National sustainable development strategy	–	National program of conservation and sustainable utilization of FGR RS
	National strategy for sustainable use of natural goods and resources		
Conservation and sustainable use of indigenous species	National strategy for sustainable use of natural goods and resources	Law on national parks	National program on environmental protection
	Biodiversity strategy of RS		National program of conservation and sustainable utilization of FGR RS
	Forestry development strategy of RS		
Improving the production of planting material	Forestry development strategy of RS	Law on forest reproductive material	National program of conservation and sustainable utilization of FGR RS

Guidelines are related to:

- Improvement and conservation of forest biodiversity whith maintainance of genetic variability within populations;
- Sustainable use of forest biodiversity and available genepools;
- Establishing a gene bank;
- Inventory of biodiversity and establishment of monitoring of biodiversity components;
- Participation in international processes and initiatives;
- Fulfillment of international obligations in the field of FGR conservation;
- Improvement of forest management system;
- Public education on the importance of conservation and sustainable utilization of FGR (Nonić et al. 2017a).

The possibility of influence on conservation and sustainable utilization of forest genetic resources is not always clearly recognizable and visible within the various laws of forestry and other related sectors. Numerous laws also apply to other sectors and are mainly indirectly related to the forestry sector (with exception of the Law on Forests and the Law on Forestry Reproductive Material, which are directly related to the forestry sector), and consequently their impact on conservation and sustainable utilization of FGR is more indirect.

Conservation of biological diversity is common to most of the aforementioned laws, however, the terms "*forest genetic resources*" and "*conservation of forest genetic resources*" are not clearly defined nor directly mentioned in them.

Indirectly, they can be recognized within certain terms, such as:

- "*Conservation of genepool*", "*protection and conservation of forests*", "*biodiversity protection*", "*protection and use of flora and fauna*" – Law on environmental protection (OG 2004a);
- "*Conservation of biological diversity of forest ecosystems*", "*conservation of genetic diversity*", "*conservation of nature*" – Law on nature protection (OG 2009b);
- "*Conservation of forest trees and other species within the forest community*" – Law on forests (OG 2010a);
- "*Protection and improvement of flora and fauna, vegetation, forests and water*" – Law on national parks (OG 2015).

The Law on nature protection (OG 2009b, § 22) states the establishment of a gene bank, which is one of the important aspects of *ex-situ* conservation, preservation of genetic diversity of wild plants, or "...*their genetic fund and potentials*" (OG 2009b). Considering that national parks represent a significant form of in-situ conservation, the Law on national parks (OG 2015) can also be considered significant from aspect of FGR conservation.

The National program for conservation and sustainable utilization of FGR of the Republic of Serbia for the period 2016–2025 (draft) covers all goals of importance for FGR conservation which are listed in Table 7.2.

Looking at the number of regulatory documents whose goals include some of the thematic areas (Table 7.2), it can be noted that *conservation, improvement and sustainable use of biodiversity* is recognized in almost all documents (except the Law on forest reproductive material).

7.3 Concluding Remarks

In the previous research, it was also stated that there is no legally binding regulation in Serbia that is directly related to the conservation of FGR (Milovanović et al. 2012; Šijačić-Nikolić et al. 2014b; Lalović 2016; Nonić et al. 2017a, b).

Lalović (2016) analyzed the attitudes of the representatives of experts toward legal regulations and concluded that the majority of respondents (94%) were familiar with the regulations related to this area, whereby the most commonly listed regulations were: Law on forests (83%), Law on forest reproductive material (70%), Law on nature protection (60%), Law on environmental protection (56%), Forest development strategy (36%), Biodiversity strategy (23%) and National strategy for sustainable use of natural goods and resources (20%), which indicates that "…*the experts are largely familiar with the applicable regulations related to the conservation and sustainable utilization of FGR, which is understandable, since they are obliged in their work to comply with the provisions of the aforementioned laws*" (Lalović 2016).

In the same study were also analyzed the attitudes of experts toward the validity of national legal regulations related to the conservation of FGR, wherein it was concluded that 71% respondents are not satisfied,[4] or have only been partially satisfied with the legal regulations related to the FGR conservation in Serbia. According to the opinion of the majority of respondents (90%) from the mentioned research, the programs of FGR conservation "…*should also apply to rare and endangered species of forest trees and to economically important species for forestry, indicating that there is a developed awareness among the experts about the importance of conserving rare and endangered species, and gives hope that in the future more work will be done on the conservation of these species both in their natural habitat and the use of rare and endangered species in reforestation*" (Lalović 2016).

Koskela et al. (2013) stated that many countries have developed specific national programs and strategies for managing forest genetic diversity based on the dynamic conservation approach. The progress in their implementing has been slower than expected and FGR conservation "…*is still riddled with methodological and political problems, and complexities of national legal and administrative structures*" and "…*these problems are not uniquely related to forest genetic diversity but to genetic conservation in general*", so many international and national actions on biodiversity conservation have "…*largely neglected genetic diversity*" (Koskela et al. 2013).

Geburek and Konrad (2008) pointed out that if "…*forests are to be managed sustainably and biodiversity is to be conserved, then the importance of FGR conservation needs to be better recognized and appropriate measures have to be implemented*", so they suggest that forest geneticists "…*take advantage of all opportunities to discuss and better integrate good science with responsible decision making*" (Geburek and Konrad 2008).

Recognizing the problem and importance of conserving biodiversity and genetic diversity in the analyzed documents in the field of forestry, environmental protection and nature protection in Serbia, indicates the existence of intentions for the prevention of various threatening factors. However, there is insufficient commitment

[4]As an explanation of their attitude, respondents indicated: "*lack of binding regulations, globally and in the domestic legislation, which are directly related to the conservation of FGR*", therefore is necessary to "*the adopt strategy of FGR conservation, which will clearly define the goals of the FGR conservation*", and emphasized that "*existing laws are satisfactory, but lacks consistency in applying punitive measures*" as well as "*the jurisdiction of environmental protection and forestry are not well coordinated (forestry needs are in the second plan)*" (Lalović 2016).

to the category of FGR, since in Serbia, there are still no adequate, legally binding, legislative frameworks directly related to this field.

If FGR are not directly mentioned in the national regulatory frameworks, there is the risk that they may be overlooked, so there is a need to define the terms "*forest genetic resources*" and "*conservation of forest genetic resources*" in the existing laws in the field of forestry, nature protection and environmental protection.

Although there is no single legally-binding document that deals exclusively with conservation of FGR, there are a number of strategic documents that are directly or indirectly related to the FGR conservation in Serbia. These documents indicate the existence of possible overlapping of the defined objectives and proposed measures, and their unification in one document and its adoption would be of great importance for the regulation of this area in Serbia.

The draft of such a document was prepared and its adoption and implementation could make a step forward in resolving the problems determined in the lack of appropriate regulatory frameworks related to the conservation and sustainable utilization of forest genetic resources in Serbia.

Acknowledgements The research was financed by the project "*Establishment of forest plantations to increase the afforested areas in Serbia*" (TR 31041), project "*Studies of climate changes and their impact on the environment – monitoring impacts, adaptation and mitigation*" (no. 43007), sub-project "*Socio-economic development, mitigation and adaptation to climate change*" (no. 43007/16-III) and project TR31078, funded by the Ministry of education, science and technological development of the Republic of Serbia.

References

Bouillon, P., Hubert, J., Bakkebø Fjellstad, K., Rusanen, M., Zavrl Bogataj, A., Olrik, D. C., Bordács, S., Longauer, R., Paitaridou, D., Kõiv, K., Koskela, J., Orlovic, S., Black-Samuelsson, S., & Wolter, F. (2015). *The implications of global, European and national policies for the conservation and use of forest genetic resources in Europe*. Rome: European Forest Genetic Resources Programme (EUFORGEN), Bioversity International xii and 42 p.

FAO. (2014). Political and institutional recommendations. In *The State of the World's Forest Genetic Resources, Commission on Genetic Resources for Food and Agriculture – Food and Agriculture Organization of the United Nations* (pp. 227–236). Rome.

Fussi, B., Westergren, M., Aravanopoulos, F., Baier, R., Kavaliauskas, D., Finzgar, D., Alizoti, P., Bozic, G., Avramidou, E., Konnert, M., & Kraigher, H. (2016). Forest genetic monitoring: An overview of concepts and definitions. *Environmental Monitoring and Assessment, 493*(1–12), 188.

Geburek, T., & Konrad, H. (2008, April). Why the conservation of forest genetic resources has not worked. *Conservation Biology, 22*(2), 267–274.

Glück, P. (2001). *Forst- und Holzwirtschaftspolitik*. Institut für Sozioökonomik der Forst- und Holzwirtschaft, BOKU, Wien.

Kelleher, C. T., de Vries, S. M. G., Baliuckas, V., Bozzano, M., Frýdl, J., Gonzalez Goicoechea, P., Ivankovic, M., Kandemir, G., Koskela, J., Kozioł, C., Liesebach, M., Rudow, A., Vietto, L., & Zhelev Stoyanov, P. (2015). *Approaches to the conservation of forest genetic resources in Europe in the context of climate change*. Rome: European Forest Genetic Resources Programme (EUFORGEN), Bioversity International xiv+46 pp.

Koskela, J., François Lefèvre, F., Schueler, S., Kraigher, H., Olrik, D., Hubert, J., Longauer, R., Bozzano, M., Yrjänä, L., Alizoti, P., Rotach, P., Vietto, L., Bordács, S., Myking, T., Eysteinsson, T., Souvannavong, O., Fady, B., Cuyper, B., Heinze, B., von Wühlisch, G., Ducousso, A., & Ditlevsen, B. (2013). Translating conservation genetics into management: Pan-European minimum requirements for dynamic conservation units of forest tree genetic diversity. *Biological Conservation, 157*(2013), 39–49.

Krott, M. (2005). *Forest policy analysis* (Vol. 323). Dordrecht: Springer.

Lalović, V. (2016). *Basic elements of conservation of forest genetic resources and their institutional-legal aspects* (pp. 1–78). Master thesis. University of Belgrade – Faculty of Forestry, Belgrade.

Milovanović, J., Šijačić-Nikolić, M., Nonić, M., & Radojević, U. (2012). Šumski genetički resursi u međunarodnim procesima i zakonskoj regulativi. *Šumarstvo, 3–4*, 111–131.

Nonić, D. (2015). *Organisation and business administration in forestry – book, Electronic resource*. Belgrade: University of Belgrade – Faculty of forestry 398 p.

Nonić, M., Nedeljković, J., Lalović, V., Šijačić-Nikolić, M., & Nonić, D. (2017a). *Legal framework and stakeholders' attitudes toward the conservation of forest genetic resources.* IUFRO 125th Anniversary Congress, 18–22 September 2017, Freiburg, Book of abstracts: 484.

Nonić, M., Nedeljković, J., & Nonić, D. (2017b). Strategic and legal framework of forest genetic resources conservation in Serbia. *Selekcija i Semenarstvo, XXIII*(1), 11–26.

OG. (2004a). Law on environmental protection. "Official Gazette of the Republic of Serbia", no. 135/2004, 36/2009, 36/2009, 72/2009, 43/2011, 14/2016).

OG. (2004b). Law on forest reproductive material. "Official Gazette of the Republic of Serbia", no. 135/2004, 8/2005, 41/2009.

OG. (2005). Decree on putting under control the use and trade of wild flora and fauna. "Official Gazette of the Republic of Serbia", no. 31 /2005, 45/2005 22/2007 38 /2008 9/2010 and 69 /2011.

OG. (2006). Forestry development strategy of the Republic of Serbia. "Official Gazette of the Republic of Serbia", no. 59/2006.

OG. (2008). National sustainable development strategy. "Official Gazette of the Republic of Serbia", no. 57/2008.

OG. (2009a). Law on genetically modified organisms. "Official Gazette of the Republic of Serbia", no. 41/2009.

OG. (2009b). Law on nature protection. "Official Gazette of the Republic of Serbia", no. 36/2009, 88/2010, 91/2010, 14/2016.

OG. (2009c). Rulebook on the transboundary movement and trade in protected species. "Official Gazette of the Republic of Serbia", no. 99/2009 from 1 Dec 2009.

OG. (2010a). Law on forests. "Official Gazette of the Republic of Serbia", no. 30/2010; 93/2012 i 89/2015.

OG. (2010b). National program on environmental protection. "Official Gazette of the Republic of Serbia", no. 12/2010.

OG. (2010c). Rulebook on conditions for establishing a gene bank of wild plants, animals and fungi, the manner of work of gene bank, manner of handling biological material, content of applications and documentation to be submitted with the application for the issuance of permit for establishing a gene bank. "Official Gazette of the Republic of Serbia", no. 65/10 from 14 Sept 2010.

OG. (2010d). Rulebook on declaration and protection of strictly protected and protected wild species of plants, animals and fungi. "Official Gazette of the Republic of Serbia", no. 5/2010 from 5 Feb 2010.

OG. (2011). Biodiversity strategy of the Republic of Serbia for the period 2011–2018. "Official Gazette of the Republic of Serbia", no. 13/2011.

OG. (2012). National strategy for sustainable use of natural goods and resources. "Official Gazette of the Republic of Serbia", no. 33/2012.

OG. (2015). Law on national parks, "Official Gazette of the Republic of Serbia", no. 84/15.

Šijačić-Nikolić, M., & Milovanović, J. (2010). *Conservation and sustainable use of forest genetic resources*. Faculty of Forestry University of Belgrade, Planeta Print, 200 p. (in Serbian).

Šijačić-Nikolić, M., & Sovilj, L. (2017). Implementation of the program for conservation and sustainable utilization of forest genetic resources in Republic of Serbia. *Selekcija i Semenarstvo, XXIII*(2), 35–44.

Šijačić-Nikolić, M., Milovanović, J., & Nonić, M. (2014a). Conservation of forest genetic resources. In: Ahuja, M.R., Ramawat, K.G. (Eds.): *Biotechnology and biodiversity* (pp. 103–128). Springer.

Šijačić-Nikolić, M., Milovanović, J., Nonić, M. (2014b). Šumski genetički resursi u Srbiji – stanje i predlozi za unapređenje ove oblasti, Glasnik Šumarskog fakulteta, Specijalno izdanje povodom naučnog skupa "Šume Srbije i održivi razvoj": 51–70.

Šijačić-Nikolić, M., Vilotić, D., Ivetić, V., Milovanović, J., Stanković, D., Nonić, M., Devetaković, J., Jokanović, D., Maksimović, Z., Popović, V., Rakonjac, L., Lučić, A., Orlović, S., Galović, V., Pilipović, A., Stojnić, S., Kovačević, B., & Trudić, B. (2016). *National program for conservation and sustainable utilization of FGR of the Republic of Serbia for the period 2016–2025*. Novi Sad: Faculty of Forestry, Belgrade, Institute of forestry, Belgrade, Institute of lawland forestry and environment.

Trudić, B., Orlović, S., Stojnić, S., Pilipović, A., Matović, B., & Novčić, Z. (2015). Forest ecosystem services in the context of climate change – A new concept for forestry in the Republic of Serbia? *Topola/Poplar, 195/196*(2015), 55–83.

Chapter 8
Institutional Framework for Conservation of Forest Genetic Resources in Serbia

Dragan Nonić, Jelena Nedeljković, and Marina Nonić

Abstract In this chapter is analysed the institutional framework for conservation of forest genetic resources (FGR) in Serbia, i.e. institutions and organizations responsible for this issue at the republic level. The results show that a number of institutions and organizations are involved in conservation work. The top institutions of public administration are the Ministry of Agriculture, Forestry and Water Management (through the work of the Directorate for Forests) and the Ministry of Environmental Protection (through the work of the Sector for Nature Protection and Climate Change). The public service for nature conservation is responsible for the support and control in the performance of professional activities. The public enterprise for state forests management and public enterprises for management of national parks, as well as other organizations, including non-governmental organizations, are responsible for the implementation of professional activities on FGR conservation. The work of educational and research institutions in relation to the conservation of FGR is done through different projects and programs, at the national and regional level. In most cases, the competencies of institutions and organizations do not relate exclusively to the FGR conservation but also include other activities. Conservation of FGR is a primary objective within the competence of public administration in forestry and nature conservation (competent ministries); public service in nature conservation; public enterprises in forestry and nature conservation; educational and research organizations. Bearing in mind a wide range of the institutional framework, it is necessary to build and implement a policy of cross-sectoral dialogue for successful implementation of FGR conservation activities. Although the conducted analysis did not show any no overlapping of the competencies of institutions and organizations, continuous cooperation and information exchange is necessary.

Keywords Institutions · Organizations · Forestry · Genetic resources · Serbia

D. Nonić (✉) · J. Nedeljković · M. Nonić
Faculty of Forestry, University of Belgrade, Belgrade, Serbia
e-mail: dragan.nonic@sfb.bg.ac.rs; jelena.nedeljkovic@sfb.bg.ac.rs; marina.nonic@sfb.bg.ac.rs

M. Šijačić-Nikolić et al. (eds.), *Forests of Southeast Europe Under a Changing Climate*, Advances in Global Change Research 65,
https://doi.org/10.1007/978-3-319-95267-3_8

8.1 Introduction

The **institutional framework** in the forestry sector includes "*organizational and administrative settings of forestry policy and its implementation*" and defines "*the responsibilities and competencies of various bodies, public and private, at different levels*" (FOREST EUROPE 2011). In the context of forestry, "*the development of a sector depends on an institutional framework that includes "rules of the game", as well as a large number of organizations, public and private, with different mandates, roles and functions*" (FAO 2018).

The most important goal of international forestry policy is to forestall additional forest degradation, thereby playing a part in the implementation of Millennium Development Goals to "*protect climate, conserve biodiversity, combat desertification and alleviate poverty, especially in rural regions*". In the last four decades, numerous internationall and national institutions and mechanisms dealing with forests and biodiversity of forests has significantly increased (FAO 2014).

When it comes to the institutional framework in forestry, consideration should be given to "*a wide range of organizations, incentives, mandate systems and regulations*", which "*affects the results of the sector*" (World Bank 2005). Institutions in the forestry sector have been "*adapted to the changed circumstances, especially those related to the increased number of stakeholders and the complexity of the sector itself*" (UNECE 2005) and "*have a decisive role in achieving forest policy objectives and implementation of forest development strategies and programs*" (Nonić et al. 2006).

Organizations[1] include "*structures or networks that cannot function without rules, communication, membership, and sovereignty*" (Hodgson 2006). According to North (1994), organizations consist of "*groups of individuals bound together by some common purpose in order to achieve certain objectives*".

Organizations in the forestry sector include "*public agencies, private companies, non-governmental organizations, trade unions, cooperatives*" (Gane 2007). Their common characteristic is direct or indirect dependence on forest resources. Some of these organizations are active in several subsectors within the forestry sector, and some outside the sector, or in other economic sectors (Gane 2007).

In that sense, the role of organizations as a particular type of institution is to implement the rules and create opportunities for achieving goals set within a particular institutional framework (Nonić 2015).

[1] They include: "*political bodies (e.g. political parties, the Senat, a city council, regulatory bodies); economic bodies (e.g. firms, trade unions, family farms, cooperatives); social bodies (e.g. churches, clubs, athletic associations); educational bodies (e.g. schools, universities, vocational training centers)*" (North 1994).

According to the UN-FAO publication "*The State of the World's Forest Genetic Resources*", the importance of **forest genetic resources** (FGR) for society is reflected in their contribution to solving important issues, such as climate change mitigation and adaptation, supply of food (edible forest products), energy and wood, as well as ecosystem services (primarily in terms of environmental protection) (FAO 2014). In order to fully exploit the potential of FGR in a sustainable way, political commitment is needed at all levels, which implies, above all, the appropriate regulative framework, as well as the effective institutions that would implement such a policy (FAO 2014).

Many different public and private institutions and organizations play a role in FGR management. In different countries, different government levels are responsible for FGR and policy related to them. Conservation and directed utilization of FGR can be incorporated into national forest programs and managed by institutions such as ministries in charge of agriculture, forestry or environment. Still, most often the public administration body responsible for the forests is the institution involved in FGR conservation and directed utilization (FAO 2014).

Forest governance effectiveness is essential for the sustainable use and conservation of FGR. Ineffective governance may lead to loss of forests and its genetic diversity (Geburek and Konrad 2008). However, the importance of managing intra-specific variation in tree species variations to support rural life has received little attention from policy makers (Dawson et al. 2014).

In that sense, the important issue to consider when planning the FGR conservation is the national organization. For example, in different countries, there are different administrative bodies in charge for this issue (e.g. Forest Service, Center for Tree Seeds, Forest Research Institute) (Graudal et al. 1997). However, the conservation of FGR for both public and private forests is the responsibility of national institutions (Paul et al. 2010). Although usually there is some collaboration among different institutions, often there is a lack of coordination arrangement for FGR activities (FAO 2014). It should be highlighted that "*established networks make possible to increase the extent and effectiveness of the knowledge about this important pool of genetic material and engage in systematic morphological characterization and biochemical evaluation*" (Barata et al. 2016). Yet, in order to ensure quality coordination and cooperation of all participants in the process of FGR conservation, "*it is necessary to analyze the common interests of forestry and nature conservation sectors, and work on their harmonization*" (Šijačić-Nikolić et al. 2017).

In developed countries, numerous institutions and organizations can be actively engaged in FGR conservation, including national and provincial administration and services, universities, research institutes, non-governmental organizations (NGOs), etc. Their work may include in situ and ex situ conservation, forest seed testing, sustainable use of forest resources, registration and control of forest reproductive materials, applied research, etc. (FAO 2014).

8.2 Institutional Framework for Conservation of Forest Genetic Resources in Serbia

This chapter presents and analyzes the institutional framework (Scheme 8.1), which consists of institutions and organizations responsible for the conservation of FGR at the republic level[2] as well as organizations that perform their activities at the central Serbia level (without autonomous provinces).

Numerous institutions and organizations are involved in work on FGR conservation. The top institutions of public administration are the Ministry of agriculture, forestry and water management (through the work of the Forest Directorate) and the Ministry of environmental protection (through the work of the Sector for nature conservation and climate change). The public service for nature conservation is responsible for the support and control in the performance of professional activities. The public enterprise (PE) for state forests management and public enterprises for management of national parks (PE NP)[3] as well as other organizations, including

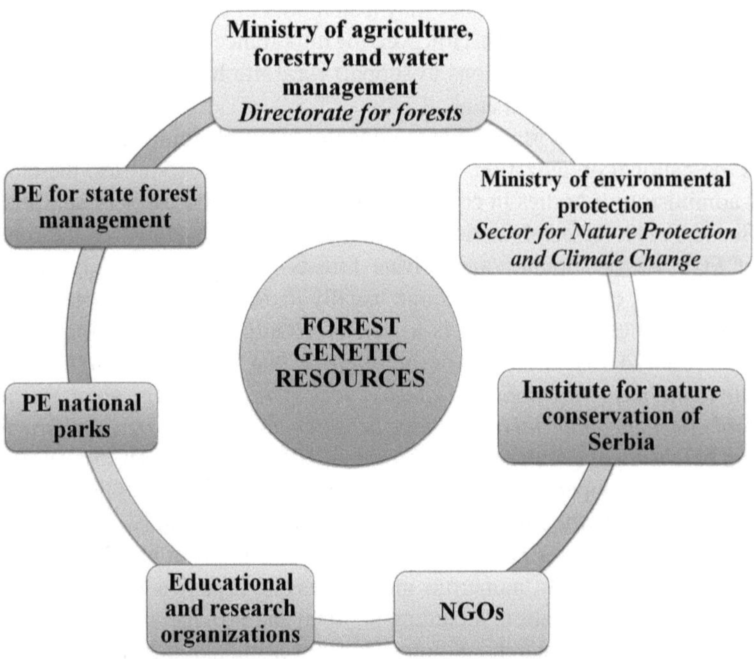

Scheme 8.1 Institutional framework for the conservation of FGR in Serbia

[2] The institutional framework for conservation of the FGR in Serbia includes institutions and organizations responsible for the conservation of FGR at the republic, provincial and local level.

[3] Within this research, the competencies of three PE NP were presented and analyzed: PE "NP Djerdap", PE "NP Tara" and PE "NP Kopaonik".

NGOs, are responsible for the implementation of professional activities on FGR conservation. The work of educational and research institutions in relation to the conservation of FGR is done through different projects and programs, at the national and regional level.

8.2.1 Overview of the Institutional Framework

The **Ministry of agriculture, forestry and water management**, through the work of the Forest Directorate, is responsible for the formulation and implementation of the state forestry policy and relevant legislation at the national level.

The Forest Directorate, as an administrative body within the Ministry, was established with the aim of carrying out tasks related to (FD 2016):

- Forest and forest conservation policy;
- Promotion and use of forests and wildlife;
- Implementation of measures for protection of forests and wildlife;
- Control of seeds and planting material in forestry, etc.

When it comes to tasks related to the conservation of the FGR, the Forest Directorate issues (prepares) the strategic frameworks and regulations that the forest users (PE for state forest management, PE NP, etc.) are obliged to implement (e.g. Law on reproductive material of forest trees, Law on forests, with accompanying regulations) and controls the implementation of laws (inspections), controls the performance of tasks entrusted to the institutes and the Faculty of Forestry, finances projects involving faculties, institutes, public companies. In addition, it keeps registers of stoolbeds, forest nurseries and seed sources (Lalović 2016).

Within the **Ministry of environmental protection**, the Sector for nature conservation and climate change carries out state administration affairs, which inter alia relate to: protection and improvement of the environment; inspection in the field of environmental protection; nature conservation; approving cross-border traffic of protected plant and animal species, etc. (MEP 2017). The Sector for nature conservation and climate change participates directly in the process of FGR conservation by preparing regulations and controls its implementation (e.g. Law on environmental protection, Law on nature conservation, Law on national parks), approving management plans for certain protected areas (for NPs the approval is given by the Government of Republic of Serbia) (Lalović 2016).

The **Institute for nature conservation of Serbia**, Belgrade, is a public professional service responsible for the protection and promotion of natural heritage in Serbia. Activities are mainly related to professional and research work on valorisation and monitoring of natural values, monitoring of the status and level of vulnerability of particular ecosystems and species, defining the system of measures for their protection, activities on the preparation of studies and proposals for the protection of natural areas, etc. (INC 2017).

The Institute plays an important role in the conservation of FGR, because it works on biodiversity conservation (as a starting point for the preservation and improvement of rare and endangered plant species), proposes the designation of protected areas, develops protection studies, which determine the values of the areas proposed for protection and the way management of these areas, issues decisions on the conditions of nature conservation in the process of drafting project documentation, spatial and urban plans, strategies, etc. in all activities that affect nature (INC 2017).

PE for state forest management "Srbijašume", Belgrade, was established by the Republic of Serbia with the aim of managing forests in state ownership and performing professional, technical and advisory activities in private forests, in accordance with the principles of sustainable management. In addition, PE "Srbijašume" is a manager of numerous protected areas.

PE "Srbijašume" participates in the improvement of the state of FGR through the management measures, provides information from the field on the state and current problems, and participates in the preparation of regulations, the allocation of seed sources and the production of planting material in the forest nurseries. Within the General Directorate of PE "Srbijašume", the work of two divisions (within the Sector for forestry and environmental protection) is of special importance for the conservation of the FGR: the Division for genepool, seed science and nursery production and the Division for forest protection and protected areas.

The Division for genepool, seed science and nursery production is engaged in the FGR conservation in a direct way by defining the business policy of PE "Srbijašume" within the scope of activity of genepool, seed science and nursery production, creating guidelines and monitoring the preparation and implementation of the Plan of production of reproductive material of forest and horticultural species trees at the level of PE "Srbijašume". Part of the Division's activities is regular control of the production and realization of forest and horticultural planting material as well as the monitoring of the status of registered seed sources within the areas managed by PE "Srbijašume" is carried out. The Division also deals with the provision of technical support to other parts of the PE "Srbijašume" in order to allocate and register new ones, as well as the revision of existing seed sources, the establishment of new and the implementation of care measures in existing seed plantations (Maksimović et al. 2015).

The Division for forest protection and protected areas is in charge of coordinating the management of protected areas. The Division is in charge of coordinating the adoption of protected areas' management plans, as well as supervision of its implementation. In addition, the Division's activities include improvement, promotion, sustainable use and preservation of protected areas, as well as keeping records of natural values and human activities. In addition to activities related to nature conservation, Division is also responsible for activities on protection against diseases, harmful insects, rodents, fire, illegal logging, etc. (Srbijašume 2018).

This company cooperates with the Forest Directorate, because it manages seed sources and nurseries of forest reproductive material, and they are obliged to register in the registers kept by the Forest Directorate, to ask for certificate on the recognition of planting material, decisions on designated seed sources, etc. In addition, the Forest Directorate finances establishing of seed sources, and collection of seed. There is also cooperation with the Institute of forestry and the Faculty of Forestry, because these organizations inspect nurseries and seeds sources. Cooperation with the above mentioned educational and research organizations is advisory, as well. Research is often carried out on the territory managed by PE (Lalović 2016).

National parks (NP) are protected areas managed by public enterprises,[4] in accordance with the nature conservation and sustainable development principles. They are a significant factor in the FGR conservation, as in situ conservation objects.

Additionally to nature conservation activities, PE NPs are managing state forests, and cooperating with public administration institutions in forestry, when it comes to different approvals, for example, consent to the management plans. They are cooperating with the Forest Directorate because they are obliged to register seed objects (nurseries) that are in their territory. Occasionally institutes and faculties are carrying out research in the territory of NP (Lalović 2016).

Faculty of Forestry, University of Belgrade, is a scientific-educational institution that, through its curriculum, directly contributes to the improvement of knowledge about conservation and directed utilization of the FGR. At the Chair of seed science, nursery production and afforestation, as part of Master Academic Studies, there is a special module entitled "Plant Production and Conservation of Forest Genetic Resources". Within this module, students gain knowledge, among other things, on the significance of the FGR and their conservation within the subject "Conservation and directed utilization of forest genetic resources", for which the the same titled textbook has been written. Also, at the Chair have been conducted numerous studies on genetic diversity, intra- and inter-population variability of some of the most important forest tree species, etc. Various projects have been implemented in relation to the conservation of rare and endangered species, as well as concrete activities related to the conservation of FGR, through the various in situ and ex situ methods, with research within master theses and doctoral dissertations related to this field. The Faculty of Forestry also performs tasks entrusted by the Ministry of agriculture, forestry and water management related to certificate on the recognition of planting material, and control of seed sources and nurseries. The Faculty of Forestry has also played a significant role in the preparation of the National program for conservation and sustainable utilization of forest genetic resources of the Republic of Serbia for the period 2016–2025 (further information avalilable in Chap. 7). Through concrete activities on the project with the same name, a draft of this Program was prepared, and its adoption and implementation are expected. In addition to aforementioned, the Faculty of Forestry is the manager of two protected areas: the monument of nature "Arboretum of the Faculty of Forestry" and the special nature reserve "Goč-Gvozdac".

Faculty of Biology, University of Belgrade, is an educational and scientific institution in the field of biological sciences. The Faculty of Biology deals with education in the area of FGR conservation in an indirect way, through three organizational units: Chair of ecology and geography of plants, Biodiversity Information Center and

[4] According to the Law on national parks (OG 2015).

Institute of botany and Botanical Garden "Jevremovac". Chair of ecology and geography of plants is engaged in education and scientific work in the field of flora of vascular plants of the Balkan Peninsula, vegetation diversity as a very important aspect of plant biodiversity of Serbia and the Balkan Peninsula, and biodiversity protection. A part of the activities is directed towards ex situ conservation and reintroduction of endangered species to natural habitats (FB 2017b). The overall goal of the Center's work is to collect, organize, verify, analyze and distribute data on biodiversity in Serbia in a modern IT environment, based on dynamic online databases and GIS processing and presentation (FB 2017a). In addition to to aforementioned, the Faculty of Biology is engaged in the conservation of the FGR through the management of the protected area (the monument of nature "Botanical Garden Jevremovac").

Institute for biological research "Siniša Stanković", University of Belgrade, is the leading multidisciplinary scientific institution in Serbia in the field of biological sciences. Institute is engaged in conservation of FGR in an indirect way, through the work of the Ecology Division, which combines the most important knowledge acquired through research of an ecosystem character. The Division carries out research focusing on the development, preservation and functioning of biodiversity in relation to ecological processes and interactions with abiotic factors on the habitat. These studies are specifically aimed at preserving the biodiversity of economically important (or potentially important), endemic-relict and endangered plant species in Serbia (IBISS 2017).

Institute of molecular genetics and genetic engineering, University of Belgrade, is one of the leading scientific research organizations in Serbia and the region in the field of fundamental and applied research in the field of molecular genetics, biomedicine and biotechnology. Teaching activity is realized mostly by the implementation of basic academic, master and doctoral programs within the Faculty of Biology. The Institute's associates also participate in different levels of study at other faculties and universities in the country (IMGG 2016). The Institute works on FGR conservation in an indirect way, through the work of the Laboratory for molecular biology of plants, which carries out research on various aspects of molecular biology of plants (examination of mechanisms of plant response to various factors of abiotic stress, research in the field of microevolution – population genetics, etc.) (IMGG 2017).

Institute of forestry, Belgrade is a research organization, which through scientific and professional work contributes to the improvement of the status of FGR. Department of genetics, plant breeding, seed and nursery production carries out research on the genetic potential of forest trees, as well as the diversity of forest and ornamental trees and shrubs (IF 2017a). Division for forest protection carries out activities of public interest in the field of diagnostics of harmful organisms and protection of the health of forest plants on the territory of the Republic of Serbia[5] (IF 2017b). Also, the Forest Directorate entrusted the Institute of forestry with the tasks of controlling seeds and production and health status of nurseries in the territory of Serbia (Lalović 2016). In addition, the Institute has been engaged in the preparation

[5] Without the territory of the Autonomous Province of Vojvodina.

of the National Program for conservation and sustainable utilization of forest genetic resources of the Republic of Serbia for the period 2016–2025.

NGOs in Serbia appear only after 2000 as notorious participants in public life. There are several[6] NGOs in Serbia whose activities are partly related to the issues of the FGR, primarily through the topics of nature and biodiversity conservation, sustainable forest management, climate change, etc. However, there are still no specialized NGOs involved in the conservation of the FGR.

The impact of NGOs on the conservation of FGR is mainly reflected in informative instruments, participation and organization of public awareness campaigns, financing of projects, publication of information materials, etc. (Lalović 2016).

Based on the review, it can be concluded that in most cases the competencies of institutions and organizations are not exclusively related to the FGR conservation, but also include other activities.[7]

8.2.2 *Analysis of the Institutional Framework*

A review of the competencies of institutions and organizations dealing with FGR conservation issues is presented in Table 8.1.

Based on these data, it can be concluded that the FGR conservation is the primary objective within the competence of the following institutions and organizations:

- Public administration in forestry and nature conservation (competent ministries);
- Public service in nature conservation (Institute for nature conservation of Serbia);
- PE in forestry and nature conservation (PE "Srbijašume" and PE NPs);
- Educational and research organizations (Faculty of Forestry, Institute of botany and Botanical Garden "Jevremovac" within the Faculty of Biology, Institute of forestry).

The following can be stated regarding the internal organization of the competent institutions and organizations:

- Within the public administration, the departments are competent;
- Within the public service, these jobs are performed by divisions;
- At the level of PE in forestry, the divisions (as organizational units of the sector within the General Directorate) are responsible;

[6] International Union for Conservation of Nature (IUCN), World Wildlife Fund (WWF), Regional Environmental Center Serbia (REC).

[7] Besides the Division for genepool, seed science and nursery production within the Sector for forestry and environmental protection at the General Directorate of PE "Srbijašume", whose activities are exclusively focused on the conservation of FGR.

Table 8.1 Competencies of institutions and organizations regarding the conservation of FGR in Serbia

Name of the institution/organization		Conservation of FGR	
Institutions and organizations	Organizational unit	Primary objective	Secondary objective
Public administration			
Ministry of agriculture, forestry and water management	Forest Directorate (Department for forestry policy and implementation of measures to improve forestry)	Monitoring of the state of production of forest seed and planting material and preservation of genetic variability of forest species and proposing measures for improvement of this activity	
		Give opinions to annual and medium-term management programs for protected areas	
Ministry of environmental protection	Sector for nature conservation and climate change (Division for protected areas, geodiversity and ecological network – Department for protected areas)	Development of strategic documents as well as plans and programs in field of nature and biodiversity conservation, sustainable use of natural resources and goods	
	Sector for nature conservation and climate change (Division for biodiversity – Department for protection and conservation of wild species)	Implementation of international conventions, agreements, laws, and other regulations related to the nature conservation and environment	
		Protection of nature and preservation of biodiversity	
		Protection, improvement and management of protected areas and ecological networks	

Public service			
Institute for nature conservation of Serbia	Division for natural goods, landscape values and spatial planning documents	Launch of the protection procedure	
		Professional supervision (monitoring of the state of the protected area)	
	Division for biodiversity, ecological networks and sustainable development (Department for biodiversity)	Research	
		International cooperation	
PE for state forest management			
PE "Srbijašume"	Sector for forestry and environmental protection (Division for genepool, seed science and nursery production)	Coordination of the development of the Plan of the production and realization of reproductive material of forest and horticultural tree species at the level of PE "Srbijašume"	
		Definition of business policy of PE "Srbijašume" within the scope of activities on genepools, seed science and nursery production	
		Monitoring of the status of registered seed sources within the areas managed by PE "Srbijašume"	
		Cooperation with scientific research organizations and public administration	

(continued)

Table 8.1 (continued)

Name of the institution/organization		Conservation of FGR	
Institutions and organizations	Organizational unit	Primary objective	Secondary objective
	Sector for forestry and environmental protection (Division for forest protection and protected areas)	Improvement, promotion, sustainable use and preservation of protected areas	
		Adoption of a management plan and control of its implementation	
		Control over the implementation of nature protection measures	
		Keeping records	
PE national parks			
PE "NP Đerdap"	Sector for protection and development	Protection, preservation and improvement of natural values	Organization of scientific-research activities
PE "NP Tara"	Sector for planning, protection and development of the area	Biodiversity protection	
PE "NP Kopaonik"	Sector for protection and development	Sustainable forest management	
Educational and research organizations			
Faculty of Forestry	Department of forestry (Chair of seed science, nursery production and afforestation)	Education	Tasks entrusted by the Ministry
		Research	
Faculty of Biology	Institute of botany and botanical garden "Jevremovac"	Implementation of programs for the protection of endangered plants	
		Education	

	Biodiversity Information Center		Provide scientific and IT support to activities in the field of biodiversity protection, especially in the implementation of international conventions, strategies and directives
			Inventory and ongoing analysis and monitoring of the state of biodiversity components of Serbia
			Support the launch of interdisciplinary projects and the involvement of national experts and institutions in international projects in the field of biodiversity
			Improving communication and cooperation between researchers, institutions and organizations in the field of nature protection, users and/or managers of natural resource, public administration and general public, in national and international frameworks
			Popularization of scientific knowledge about biodiversity and its significance
	Chair of ecology and geography of plants		Education
			Research
Institute for biological research "Siniša Stanković"	Division for ecology		Research
Institute of molecular genetics and genetic engineering	Laboratory for molecular biology of plants		Research

(continued)

Table 8.1 (continued)

Name of the institution/organization		Conservation of FGR	
Institutions and organizations	Organizational unit	Primary objective	Secondary objective
Institute of forestry	Department of genetics, plant breeding, seed and nursery production	Tasks entrusted by the Ministry	
		Research	
Other organizations			
NGOs (IUCN, WWF, REC)			Assistance in the shaping of the regional protection policy
			Introducing innovative approaches to efficient conservation of biodiversity and natural resources management
			Support to institutional strengthening and capacity building
			WWF program "Protected areas for nature and people"
			Expansion of support at the local level for the proclamation and proper management of new protected areas through the promotion of REC brand "Education for Natura 2000"
			Strengthening the role of rural communities in the management of areas with rich biodiversity, including protected areas
			Strengthening cooperation and knowledge sharing on issues relevant to biodiversity between national REC offices

- At the level of PE in nature conservation, the relevant sectors (in all three analyzed PE NPs, these jobs are performed by the sectors in charge of protection and development).

There are differences in the number of organizational units involved in the conservation work of the FGR within the institutions and organizations of forestry and nature conservation. Thus, e.g. within the ministry responsible for nature conservation, the FGR conservation is represented in the work of two departments, while, on the other hand, only one department deals with these activities within the ministry responsible for forestry. However, when it comes to PE in forestry and nature conservation, the situation is reversed. Within PE "Srbijašume", FGR conservation is the competence of two departments, while these activities within the PE NPs are represented only within one sector in each enterprise.

On the other hand, when analyzing the competencies of educational and research organizations, a large number of organizational units, such as departments, chairs, centers, laboratories, etc., can be identified. They are all engaged in education and research in the field of FGR conservation. However, it is very important to emphasize that these activities are the main goal within the jurisdiction of the Faculty of Forestry, Institute of forestry, and Institute of botany and Botanical Garden "Jevremovac" within the Faculty of Biology.

The tasks that are the primary objective within the competence of the analyzed institutions and organizations are as follows:

- Development of strategic documents and preparation of legislative framework in forestry, environmental protection and nature conservation (public administration);
- Implementation of international obligations (public administration);
- Control of the implementation of the laws and other legal regulations (public administration);
- Giving opinions on planning documents in forestry and nature conservation (public administration);
- Formulation of measures for improving FGR conservation (public administration);
- Conservation of biodiversity (public service, PE in forestry and nature conservation, educational organizations, as managers of protected areas);
- Sustainable forest and protected areas management (PE in forestry and nature conservation);
- International cooperation in conservation FGR (public service);
- Education and research activities (educational and research organizations).

Conservation of the FGR is a secondary objective within the competence of the following institutions and organizations:

- PE in nature conservation (all three observed PE NPs);
- Scientific and educational organizations (Chair of Ecology and Geography of Plants and Biodiversity Information Center within the Faculty of Biology,

Institute for biological research "Siniša Stankovic" and Institute of molecular genetics and genetic engineering);
- Other organizations (NGOs).

The tasks that are the secondary objective within the competence of the analyzed institutions and organizations include:

- Education and scientific-research activities (PE in nature conservation and Chair of ecology and geography of plants within the Faculty of Biology);
- Management of the protected area (Faculty of Forestry);
- Control of seeds and nurseries (Faculty of Forestry);
- Participation in the preparation of strategic documents (Faculty of Forestry);
- Strengthening cooperation between all stakeholders involved in nature and biodiversity conservation, and sustainable protected areas management (Biodiversity Information Center within the Faculty of Biology).

Based on the analysis, one can notice the common tendency of numerous institutions and organizations in forestry, environmental protection and nature conservation at the national level is to preserve the FGR, both from the aspect of forming strategic and legislative frameworks, as well as from the aspect of their implementation, education of professional staff and providing scientific knowledge.

8.3 Concluding Remarks

Conservation goals are usually related to the welfare of present and future generations. Areas with FGR can be owned and used by different individuals, communities, or public institutions and organizations. Thus, the conservation of FRG may be complex, due to a generally large number of interests (Graudal et al. 1997).

The distribution of FGR does not respect national boundaries. Thus, issues related to FGR management should be coordinated and addressed in a cooperation between different countries and regions (FAO 2014; Graudal et al. 1997). Furthermore, many factors affecting FGR, such as e.g. climate change also act beyond political borders. For this reason, there is a worldwide development of international cooperation in forestry, i.e., a trend has been observed that many countries are developing cooperation with other countries and international organizations in order to preserve their FGR. According to FAO (2014), the main forms of international and regional cooperation are "*international networks, bilateral and multilateral cooperation, and international conventions*". However, many countries luck financial means dedicted to "*international gene conservation purposes*" (FAO 2014).

According to Graudal et al. (1997), responsibilities and activities on FGR conservation should be distributed among existing units of newly established units at the different administrative levels. In that sense, it is necessary to have information on different stakeholders, their competences and interests (Graudal et al. 1997).

The previous (e.g. Graudal et al. 1997; FAO 2014) and this research confirmed that, in FGR conservation, stakeholders can be: government authorities (public administration, public service), state and private companies, educational and research organizations, NGOs and individuals (private forest owners, farmers, etc.). All of them, most often, have different competencies and interests. Thus, e.g. public administration (usually ministry responsible for the environment, forest service, etc.) generally has long-term interests, such as the conservation of FGR and biodiversity. State and private companies usually have economic interests, such as improved wood production. NGOs in this area most often have environmental and nature conservation interests (Graudal et al. 1997). However, the government has to be very careful when distributing tasks on FGR conservation, because every stakeholder is not nor should be competent, despite the interest. For example, private forest owners should not be obliged to implement FGR conservation measures, but they should be enabled, if interested and willing, to participate and support such activities. Research activities should be conducted by institutes and universities, and government institutions can support them professionally and financially (Paul et al. 2010). This means that "*in order to implement the activities on FGR conservation, it is necessary to provide initiatives from scientists, financial support from national and international funds and donors, as well as logistical support in the field*" (Šijačić-Nikolić et al. 2017). National research organizations have a pivotal role in this process, but they need assistance from relevant national authorities and international programmes and initiatives (FAO 2014).

For the well-functioning of all FGR conservation activities, it is crucial that all tasks are organized in an integrated way. Only in the case of effective integration, it is possible to avoid overlapping of competencies and tasks, to ensure the exchange of information and knowledge, thus optimizing the use of human, physical and financial resources (Graudal et al. 1997).

Previous research in Serbia found that "*cooperation between participants in the process of conservation of FGR can be improved by the following measures: better flow of information between related sectors, continuous financing of conservation activities by institutions, regular involvement of scientific institutions in the conservation work, responsible forests management, more active involvement of PE NP in research projects and more active cooperation of NP with other participants in the process of FGR conservation*" (Lalović 2016).

Asked about proposals of measures to improve FGR in Serbia, the key stakeholders said that the "*multidisciplinary approach and better communication between institutions*" would improve the current situation (Šijačić-Nikolić et al. 2017). The majority of respondents also believe that in addition to the professional activities and support from the institutions, it is necessary to intensify activities to promote conservation through: "*further education of forestry professionals, and afterwards the general public about the importance of conservation and directed utilization of FGR, through debates, workshops, public meetings and lectures; and training of forest owners and local population*" (Šijačić-Nikolić et al. 2017).

The results of previous research on institutional framework and the impact of environmental protection organizations indicate that (Stamatović et al. 2013):

- Individually, the impacts within an organization can be contradictory (at the same time there is a stimulating and restrictive influence);
- The impacts of the sector, and individual organizations within the sector, are of different character and intensity compared to the different aspects observed.

Concerning the conservation of FGR, no contradictory influence was perceived within an organization. However, it can be concluded that all analyzed institutions and organizations do not have the same character and intensity of influence, which primarily depends on their primary competence.

The organizational structure of all forestry institutions and organizations should effectively support the planning and implementation of measures to improve the state of forests, as well as climate change strategies and policies. Also, the responsibility of key organizations in the forestry and nature conservation, as well as other relevant sectors, should be clearly defined (Nonić et al. 2014).

For the purpose of strengthening institutional capacity to address major issues related to sustainable FGR management, the following should be done (FAO 2014):

- *develop national strategies for in situ and ex situ conservation and sustainable use of FGR;*
  - *integrate FGR conservation and management into wider policies, programmes and frameworks of action at the national, regional and global levels;*
  - *develop collaboration and promote coordination of national institutions and programmes related to FGR;*
  - *reinforce regional and international cooperation, including networking, to support education, knowledge dissemination, research, and conservation and sustainable management of FGR;*
  - *promote public and international awareness of the roles and value of FGR;*
  - *strengthen efforts to mobilize the necessary resources, including financing, for the conservation, sustainable use and development of FGR.*

The following is crucial for countries be able to respond to needs and pressures in FGR conservation and management: institutional strengthening, training and support to research (FAO 2014).

Conservation of the FGR requires inter-sectoral approach and cooperation between institutions and organizations of forestry, environmental protection and nature conservation and all other relevant sectors. In that sense, for the successful implementation of the conservation activities of the FGR, it is necessary to build and implement a policy of cross-sectoral dialogue. Although the conducted analysis did not show any overlapping of the competencies of institutions and organizations, continuous cooperation and information exchange is necessary.

Acknowledgements The research was financed by the project "*Establishment of forest plantations to increase the afforested areas in Serbia*" (TP 31041) and project "*Studies of climate changes and their impact on the environment – monitoring impacts, adaptation and mitigation*" (no. 43007), sub-project "*Socio-economic development, mitigation and adaptation to climate change*" (no. 43007/16-III), funded by the Ministry of education, science and technological development of the Republic of Serbia.

References

Barata, A. M., Rocha, F., Lopes, V., & Carvalho, A. M. (2016). Conservation and sustainable uses of medicinal and aromatic plants genetic resources on the worldwide for human welfare. *Industrial Crops and Products, 88*, 8–11.

Dawson, I., Leakey, R., Clement, C., Weber, J., Cornelius, J., Roshetko, J., Vinceti, B., Kalinganire, A., Tchoundjeu, Z., Masters, E., & Jamnadass, R. (2014). The management of tree genetic resources and the livelihoods of rural communities in the tropics: Non-timber forest products, smallholder agroforestry practices and tree commodity crops. *Forest Ecology and Management, 333*, 9–21.

FAO. (2014). Institutional framework for conservation and management of forest genetic resources. In *The state of the world's forest genetic resources* (pp. 207–211). Commission on Genetic Resources for Food and Agriculture – Food and Agriculture Organization of the United Nations, Rome.

FAO. (2018). Organizational development of forestry institutions, FAO. Available online at: www.fao.org/forestry/institutions/en/.

FB. (2017a). Biodiversity Information Center, Faculty of Biology, Belgrade. Available at: http://www.bio.bg.ac.rs/01_centar_osnovno.php?id_centra=14.

FB. (2017b). Chair of Ecology and Geography of Plants, Faculty of Biology, Belgrade. Available at: http://www.bio.bg.ac.rs/01_katedra_osnovno.php?id_katedre=3.

FD. (2016). *Information about work scope* (52 pp.). Ministry of agriculture and environmental protection-Forest Directorate, Belgrade.

FOREST EUROPE. (2011). State of Europe's Forests 2011 – Status and trends in sustainable forest management in Europe. Ministerial Conference on the Protection of Forests in Europe FOREST EUROPE, Liaison Unit Oslo, Oslo (341 pp.).

Gane, M. (2007). *Forest strategy-strategic management and sustainable development for the forest sector* (414 pp.). Dordrecht: Springer

Geburek, T., & Konrad, H. (2008). Why the conservation of forest genetic resources has not worked. *Conservation Biology, 22*(2), 267–274.

Graudal, L.O.V., Kjær, E.D., Thomsen, A., & Larsen, B. (1997). *Planning national programmes for conservation of forest genetic resources* (Technical Note no. 48, 54 pp.). Danida Forest Seed Centre, Humlebaek.

Hodgson, G. M. (2006). What are institutions? *Journal of Economic Issues, 40*(1), 1–25.

IBISS. (2017). Department for Ecology, Institute for Biological Research "Siniša Stanković", Beograd. Available at: http://www.ibiss.bg.ac.rs/index.php/sr-yu/ekologija/odeljenje.

IF. (2017a). Department of Genetics, Plant Breeding, Seed and Nursery Production – Research, Institute of Forestry, Belgrade. Available at: http://www.forest.org.rs/?sr_%D0%98%D1%81%D1%82%D1%80%D0%B0%D0%B6%D0%B8%D0%B2%D0%B0%D1%9A%D0%B0,51.

IF. (2017b). Forest Protection Division – Research, Institute of Forestry, Belgrade. Available at: http://www.forest.org.rs/?sr_%D0%98%D1%81%D1%82%D1%80%D0%B0%D0%B6%D0%B8%D0%B2%D0%B0%D1%9A%D0%B0,61.

IMGG. (2016). *Development strategy of the Institute of Molecular Genetics and Genetic Engineering 2016–2026* (17 pp.). Institute of Molecular Genetics and Genetic Engineering, Belgrade.

IMGG. (2017). Laboratory for Molecular Biology of Plants, Institute of Molecular Genetics and Genetic Engineering, Belgrade. Available at: http://www.imgge.bg.ac.rs/index.php/yu/istrazivanje/oblasti/molekularna-genetika-biljaka/laboratorija-za-molekularnu-biologiju-biljaka.

INC. (2017). *Information about the work scope* (40 pp.). Belgrade: Institute for nature conservation of Serbia.

Lalović, V. (2016). Basic elements of the conservation of forest genetic resources and their institutional-legal aspects, master thesis, University of Belgrade-Faculty of Forestry, Belgrade, (78 pp.).

Maksimović, Z., Šijačić-Nikolić, M., Aleksić, P., & Sovilj, L. J. (2015). Seed sources in SE "Srbijašume" as a basis for conservation and directed utilization gene pool. *Bulletin of the Faculty of Forestry, 111*, 63–82.

MEP. (2017). *Sector for nature conservation and climate change.* Ministry of Environmental Protection. Available at: http://www.ekologija.gov.rs/organizacija/sektori/sektor-za-zastitu-prirode-i-klimatske-promene/.

Nonić, D. (2015). *Organization and business administration in forestry, electronic source* (380 pp.). Belgrade: University of Belgrade-Faculty of Forestry.

Nonić, D., Avdibegović, M., & Krilašević, E. (2006). Institutional organization of forestry in Serbia. *Naše šume, 6–7*, 78–81.

Nonić, D., Avdibegović, M., Nedeljković, J., Radosavljević, A., & Ranković, N. (2014). *Sustainable governance in forestry and nature protection* (pp. 113–140). Bulletin of the Faculty of Forestry, Special issue on the occasion of the conference: "Forests of Serbia and sustainable development".

North, D. (1994). Economic performance through time. *The American Economic Review, 84*(3), 359–368.

OG. (2015). Law on National Parks, "Official Gazette of the Republic of Serbia", no. 84/15.

Paul, M., Hinrichs, T., Janßen, A., Schmitt, H. P., Soppa, B., Stephan, B. R., & Dörflinger, H. (2010). *Forest genetic resources in Germany: Concept for the conservation and sustainable utilization of forest genetic resources in the Federal Republic of Germany* (76 pp.). Bonn: Federal Ministry of Food, Agriculture and Consumer Protection.

Šijačić-Nikolić, M., Nonić, M., Lalović, V., Milovanović, J., Nedeljković, J., & Nonić, D. (2017). Conservation of forest genetic resources: Key stakeholders' attitudes in forestry and nature protection. *Genetika, 49*(3), 875–890.

Srbijašume. (2018). *Internal documentation about work tasks, public enterprise for state forest management.* Belgrade: Srbijašume.

Stamatović, S., Vukotić, M., & Nedeljković, J. (2013). Institutional framework and impact of environmental organizations on issues of woody biomass utilization for energy. *Bulletin of the Faculty of Forestry, 108*, 191–210.

UNECE. (2005). *European forest sector outlook study 1960–2000-2020 main report* (Geneva Timber and Forest Study Paper 20, 235 pp). Geneva: UNECE.

World Bank. (2005). *Forest institutions in transitions: Experiences and lessons from Eastern Europe* (PROFOR book number 4, 74 pp). Washington, DC: The World Bank.

Chapter 9
International and National Legislative in Light of Climate Changes and Its Impact on Forest Genetic Resources in Bonsia and Herzegovina

Dalibor Ballian, Milan Mataruga, Rifet Terzić, Vanja Daničić, and Branislav Cvjetković

Abstract Bosnia and Herzegovina is a consignee of numerous acts relating to protection and sustainable use of biodiversity. The international adopted legislative has been followed by the state laws and rulebooks. The protection of nature is regulated at an entity level in Bosnia and Herzegovina (the Republic of Srpska, the Federation of Bosnia and Herzegovina and Brcko District as separate districts). Many of laws and documents have been adopted in National assemblies. The main problem is enforcing of adopted documents due to lack of financing. The future guidelines are presented in order to improve the state of forest genetic resources.

Keywords Legislative · Forest genetic resources · Bosnia and Herzegovina

9.1 Introduction

In the world, a general consensus regarding to resolution of problems relating to climate changes and the central problem of sustained growth and lowering of poverty, exists. Forest cutting and its degradation are second in line causes of global warming. Although there are varying opinions on how to include the cutting and

---

D. Ballian (✉)
Faculty of Forestry, University of Sarajevo, Sarajevo, Bosnia and Herzegovina
e-mail: balliandalibor9@gmail.com

M. Mataruga · V. Daničić · B. Cvjetković
Faculty of Forestry, University of Banja Luka, Banja Luka, Bosnia and Herzegovina
e-mail: milan.mataruga@sf.unibl.org; vanja.danicic@sf.unibl.org; branislav.cvjetkovic@sf.unibl.org

R. Terzić
Faculty of Natural Sciences, University of Tuzla, Tuzla, Bosnia and Herzegovina
e-mail: rifet.terzic@untz.ba

M. Šijačić-Nikolić et al. (eds.), *Forests of Southeast Europe Under a Changing Climate*, Advances in Global Change Research 65,
https://doi.org/10.1007/978-3-319-95267-3_9

degradation in any future regimes relating to climate changes it is a widespread opinion that this problem needs to be solved effectively and permanently.

Better forest management plays a key role in facing the climate changes as these changes intrinsically connected in functionality of forests. On the other side of the climate changes, pressures are already being made on forests over the past years by the actual climate changes such as high temperatures, changes to the rainfall and extreme weather conditions. At the same time the forests and wood itself present a product that can effectively oppose the climate changes as they convert carbon dioxide and prevent the extreme weather events. Also, when the forests are destroyed by uncontrolled cutting and burning, there is a potential for harmful gas emissions that can create the glass house effect extremes which are dominated by the carbon dioxide. It is noted that at a global level, approximately 18% of global glass house gas (GHG) emissions happens as a direct result of deforestation and degradation of forests. To prevent the negative trend of habitat destruction and within it the destruction of forest resources, several international legislations and acts have been adopted and which aim to preserve, improve and protect the current resources. Hence Bosnia and Herzegovina is a consignee of numerous acts relating to protection and sustainable use of biodiversity:

- The United Nations Framework Convention on Climate Change – 2000;
- International Plant Protection Convention – 2003;
- Convention (UN) on Biological Diversity – 2002;
- Convention on Wetlands of International Importance, in particular as a Wetland Habitat – 2001;
- Convention on the Establishment of the Organization for the Protection of European and Mediterranean Plants – 2005;
- *United Nations Convention* to Combat *Desertification* in Those *Countries* Experiencing Serious *Drought and/or Desertification, Particularly in Africa* (UNCCD) – 2002;
- Kyoto Protocol – 2007;
- Convention on International Trade in Endangered Species of Wild Animals and Plants (CITES) – ratified on 05.12.2008 year ("Official Gazette BiH", No: 11/08).

There are further numerous international agreements and accords which are currently under process of being ratified, however there are also several other equally important agreements that Bosnia and Herzegovina hasn't adopted or accepted yet and which are being worked on at the moment in view of being allowed to join the European Union, these being:

- Protocol on Biosafety to the United Nations Convention on Biological Diversity;
- Convention on International Trade in Endangered Species of Wild Fauna and Fauna;
- Protocol on Biosafety within the United Nations Convention on Biological Diversity;
- Convention on the Protection of European Wildlife and Natural Habitats (Bern Convention);
- Convention on the Protection of the World Cultural and Natural Heritage.

As Bosnia and Herzegovina is not a member/participant of the OECD Scheme it cannot export forest reproductive materials outside of its borders. We will provide an overview of the most important acts adopted by Bosnia and Herzegovina in the following text.

9.2 Bosnia and Herzegovina Legislations and Genetic Resources of the Forests

The protection of nature is regulated at an entity level in Bosnia and Herzegovina (the Republic of Srpska, the Federation of Bosnia and Herzegovina and Brčko District as separate districts). Part of the responsibility in the area of protection is done at a state level and relates to application of international and responsibilities of Bosnia and Herzegovina (monitoring and coordination of preparations for the conclusion and implementation of international agreements, co-ordination of projects in cooperation with international organizations, programs and funds, co-ordination of cooperation with institutional structures in Bosnia and Herzegovina, entities and the Brčko District).

State level responsibilities are mainly the province of Ministry of Foreign Trade and Economic Relations of Bosnia and Herzegovina, and to a lesser extent to the Ministry of Civil Affairs (UNESCO Convention). The Inter-Entity Environmental Protection Agency is dealing with issues of environmental protection that require a harmonized approach to the entity. This includes, inter alia, issues of international agreements and programs, participation in cooperation with international organizations, co-ordination of application and adoption of legal acts.

Forest management, delineation and establishment of seed facilities, collection of seed material, processing, marketing of seed and planting material, as well as production control are defined by entity laws on forests and laws on seed and planting material of forest and horticultural trees and shrubs.

According to earlier laws adopted in the the twentieth century (especialy in the second part of the century) in Bosnia and Herzegovina, for every cut cubic meter of trees it was necessary to replant an area of 14 m^2 which in practical terms was rarely realized in full and complimented by absence of forest-breeding work after the establishment of cultures. This led to the failure of afforestation (Butulija and Nenad 1980).

Bosnia and Herzegovina endeavors to align legislation with EU legal acts, but although it has harmonized laws with EU legislation, e.g. with Directive 1999/105/EC, due to the lack of regional provenance and other professional and legal proceedings, Bosnia and Herzegovina is not a member of the OECD Forest Reproductive Transfer Scheme material although a methodology was used to extract the seed zone (Ballian 2011). Bosnia and Herzegovina, as mentioned above, is not even an EUFORGEN member.

Legislation regulating forestry in Bosnia and Herzegovina is made at the level of two entities. According to the Forest Law of the Republic of Srpska, forests and forest land are natural goods of general interest and enjoy special care and protection of the Republic of Srpska (Official Gazette RS 75/08). The law provides for measures for sustainable forest management, stating that the obligation of the user and forest owners to maintain, preserve and improve the biodiversity in forest ecosystems. Forest law in Federation of Bosnia and Herzegovina is at the state of adopting.

Protected zones are defined in forests

1. Protection of biodiversity and other natural values of forests,
2. Protection of genpool, production of forest seed and planting material,
3. Educational, scientific research, cultural - historical and aesthetic function,
4. Health-recreational and tourist function.

Law on Nature Protection in the Federation of Bosnia and Herzegovina (Official Gazette of the Federation of Bosnia and Herzegovina No. 66/13 of 28.8.2013) and The Nature Protection Law of the Republic of Srpska (Official Gazette of Republic of Srpska 20/14) regulate the area of nature conservation, biological, ecological and landscape diversity as a part of the environment.

Laws define the term "genetic diversity" as the total number and total diversity of genes, i.e. genetic information contained in plants species, animals, fungi and microorganisms speceie as well as the sustainability of natural resources. Allochtonous and autochotonous types have been defined, as well as in-situ and ex-situ conservation measures. Also, a favorable condition is defined as a condition that permits species survival, genetic stability of populations, and gene exchange among populations. Laws prescribe a sustainable management of biological and landscape diversity and monitoring of the state of nature in which they fall and jeopardizing factors such as climate change. The Ministry of Agriculture of Forestry and Water Management can restrict temporary or permanent discontinuation of the use of forest resources hijacked the relevant institutions of endorsement of natural resources by the beneficiaries. This Article of the Law may have implications for the use of forests in the light of climate change, whereby climate change is destabilized and inadequate utilization measures can endanger forest resources. The introduction of allochtonous species is also regulated by this law.

The Federation of Bosnia and Herzegovina Ministry of Environment and Tourism performs nature protection activities in co-operation with other relevant ministries responsible for individual resources, while the Ministry of Spatial Planning, Civil Engineering and Ecology is in the Republic of Srpska. The ministries cooperate with relevant ministries for certain branches of industry of importance for nature protection, including genetic resources and the implementation of documents:

- Strategy of Bosnia and Herzegovina with Action Plan for the Protection of Biological and Landscape Diversity (National Strategy);
- Federal Nature Protection Strategy.

Based on the Federation of Bosnia and Herzegovina Environmental Protection Strategy, the following characteristics of diversity can generally be distinguished:

- A high degree of ecosystem diversity;
- A high degree of preservation of the whole of landscape diversity of European and global importance;

- A significant degree of change in relation to distribution and composition of climatogenic ecosystems;
- A marked trend of loss of diversity, conditioned by anthropogenic influences;

The Republic of Srpska Nature Protection Strategy has been adopted by the National Assembly of the Republic of Srpska for a period of 10 years. All these documents at the level of both entities aim to protect and improve the state of forest genetic resources as one of the components of the overall genetic resources available to Bosnia and Herzegovina.

Law on Seed and Planting Material of Forest and Horticultural Trees and Shrubs in the Federation of BiH ("Official Gazette of Federation of B&H", No: 71/05 and 01-02-77/10) and the Law on Forest Reproductive Material of the Republic of Srpska (Official Gazette of Republic of Srpska, No. 60/09) define the species, purpose, use and transfer of reproductive material in BiH.

Both laws state that the use of reproductive material must be approved by the competent ministries in the Federation of Bosnia and Herzegovina and the Republic Srpska, i.e. if it meets the quality standards and the health status for internal marketing, and will in the future not be a threat to autochthonous genpool. The emphasis is placed on the protection of the native genpool, its optimum exploitation in order to preserve genetic variability, as well as the ability to import qualitative reproductive material from abroad that can respond to various challenges such as climate change.

The laws on national parks (Official Gazette of the Republic of Srpska, No. 75/10), as well as specific laws on individual national parks: the Law on NP "Sutjeska" and the Law on NP "Kozara" concern the issues of conservation of forest genetic resources and their preservation. Laws define the preservation of the genotype of the species in situ and protect some of the most important genetic resources in BiH, such as the largest prosciutto in Europe - the Perucica primeforest. The Federation of BiH has the Law on NP "Una" ("Official Gazette of Federation of B&H", No. 44/08).

In addition to the aforementioned laws, several laws apply to forest genetic resources, such as the Hunting Act. Natura 2000 (Milanović et al. 2015) has a number of touch points to preserve forest genetic resources through conservation programs of specific habitats and forests.

9.3 Documents in Bosnia and Herzegovina Related to the Problems of Climate Changes and FGR

In the Federation of Bosnia and Herzegovina, the Forestry Program of the Federation of Bosnia and Herzegovina is being drafted, which will, while respecting international agreements and obligations, define the general forestry policy and game management policy in the Federation of Bosnia and Herzegovina, oriented towards the conservation and sustainability of forests management, including maintenance and improvement of biodiversity in forests forest land.

Strategic documents in the field of forestry may include the Strategy for Development of the Forestry of the Republic of Srpska 2012–2020 which provides for guidelines for the conservation of the biodiversity in the forests as well as the improvement of the existing Forest Law of the Republic of Srpska, as well as its amendments (Official Gazette RS No. 75/08, Official Gazette RS No. 60/13), which in some segments indicates the importance of climate change. The adopted Strategy defines forest genetic resources, cites potential forest fires and defines directions for action to prevent threats as well as improve the status of forest genetic resources.

The basic principles of forest management and management are based on international documents, notably conventions (Rio de Janeiro, 1992, Helsinki, 1993). These are:

1. The principle of continuity (durability), strives for the permanent provision of the benefits of forests;
2. Principle of polyfunctionality, striving to optimize the overall benefits of forests;
3. The principle of economics, striving to achieve the most favorable relationship between the overall benefits of the forests, on the one hand, and on labor and resources consumption, on the other hand, or the principle that strives to achieve the greatest degree of realization of the overall benefits of the forests with the smallest (minimum) means.

In 2013, the Program for the Conservation of Forest Genetic Resources of the Republic of Srpska was adopted in 2013–2025 (Mataruga et al. 2013). This program, adopted by the Government of the Republic of Srpska, defines the importance of climate change from the point of view of conservation of genetic resources (biodiversity) in forest ecosystems. Among the partial measures, the project estimates the impact of climate change on forest genetic resources as well as clearer (more specifically) defining the significance of conservation of genetic resources in terms of adaptation of forest ecosystems according to predicted climate change.

The Program defines some of the endangered species, some of which are considered as widespread and forestry public of known species such as *Arctostaphylos uva urus* (L.), *Picea omorika* (Panc./Purk.), *Taxus baccata* L., *Betula pubescens* Ehrh., *Corylus colurnoides* C.K. Schneid., *Rhus coriaria* L., *Sorbus* x *halfpipe* (Roth) Hedl., *Acer helderichii* Oprh., Ex. Boiss. subsp. *vissianii* K. Maly., *Acer percussion* auct. balcan. not Fish. & C.A.Mey., *Salix pentandra* L. and others. The most important species of wild fruit and herbs are also mentioned. The Federation of Bosnia and Herzegovina has issued the Red List of vegetation vegetation of Bosnia and Herzegovina that defines the endangered species and extent of their endangerment (Federal Ministry of Environment and Tourism, 2013). It is also noted that in Bosnia and Herzegovina 42% of species are "considered vulnerable or vulnerable" and 42.3% of rare and potentially endangered species compared to 678 species analyzed.

Forest noise and fragmentation, damage to forests as a result of aerosolization, water regime, biotic and abiotic damage, the effects of historical forest management, the use of unproductive reproductive material, the genetic "pollution" of the forest, the effects of inadequate practices and particularly climate change are listed in the Program as the basic factors endangering the genetic diversity of forests. Provide detailed information on in situ conservation and relevant procedures for separating objects, separation criteria, conservation strategies, conservation areas sizes, populations, plans for each population individually. Ex situ conservation has been elaborated in detail with answers to relevant ex situ conservation issues from planning and sampling through establishment to management of ex situ plantations. Also mentioned are other ex situ conservation strategies such as: storage of seeds, cryopreservation of seeds and germs, in vitro conservation, storage of pollen and DNA storage.

The Republic of Srpska forest genetic resources conservation program identified the objectives, measures and organization of the implementation of the foreseen activities, the interest groups as well as the international organization on which the documents relied on the Program.

For the purpose of the Forestry Program of the Federation of Bosnia and Herzegovina, a study on Forest and Climate Change was carried out in 2011. This document, among other things, provides an overview of relevant international conventions, agreements, programs, resolutions and declarations, then Bosnia and Herzegovina's Adaptation Plan to Combat Climate Change with INC through the Climate Change Mitigation Plan and the Assessment of Potential for Forestry Development as well as the Strategy Proposal and a plan for possibly addressing the future/expected EU obligations.

However, it should be beared in mind that in the previous period (through the development of the First and Second National Report and part of the Third Report) there have been no significant changes in the forestry sector regarding to climate change, directly through the change of the management system, greater reforestation. It can be noted that the sectoral strategy in this area is very slow and the events in forestry do not attach importance to climate change in terms of the importance of existing forests in Bosnia and Herzegovina. Missing capacities and strategic documents that would recognize forests in Bosnia and Herzegovina are main reason weak combat against climate changes.

In addition to legal and other planning documents, Bosnia and Herzegovina has certified FSC-certified forests owned by the entity, which has favored the more significant exports of wood and forest products, and this represents a new impact on forests and genetic resources that are stored in forests.

Taking into account the limitations of the existing financial and institutional mechanisms regarding technology transfer in the Bosnia and Herzegovina forestry sector, there is a need for new policies and mechanisms and their institutional implementation. When it comes to climate change and forestry, it is possible to implement a large number of different practices such as improvement in forestry practices (and in practice sustainable management), protection of genetic superior planting material, strengthening of protected area management systems, bioenergy replacement of fossil fuels, efficient processing and utilization of forest products, monitoring of vegetation status, especially in the area of afforestation of quarrying areas. The mitigation technologies in the forestry sector are minimal and mostly aim at biodiversity aspects.

9.4 Measures for Lowering the Effects of Climate Changes on FGR

In essence, several basic measures can be identified and applied in order to raise the existing potential of reducing the forest complex in Bosnia and Herzegovina to a higher level. The importance of all these measures is mainly related to the improvement of the forest management system through a series of different activities as well as through the reduction of the recent negative trend in the surface change of the forest cover.

In this regard, the following mitigation measures may be most important:

1. Maintenance of the existing and increase of the future increase of carbon density per unit of area (tons of C/ha) on the basis of the application of certain forestry methods. At the same time, the advancement of the management system is influenced by the prevention of all processes that negatively reflect the existing growth. Performing these activities would certainly have resulted in increased forest productivity, i.e. by linking larger quantities of carbon wood to wood biomass.
2. Improvement of seed and nursery production through the separation of the provenance region, the separation of new and revision of old seedlings, the work on breeding of species through selection, hybridization and establishment of generative and vegetative seed plantations and inclusion in the OECD scheme.
3. Establishing short-term research findings and making recommendations for the targeted transfer of reproductive material through stimulation measures in the upstream based on the recommendations of scientific and state institutions dealing with forestry issues.
4. Continuous reforestation of boulders as well as lands with another purpose dedicated to reproductive material, translation of degraded and spotted forests into a higher breeding form, all for the purpose of maintaining and preserving existing and increasing forest areas in the forthcoming period.
5. Demining of existing mined forest areas that have additional potential for increasing carbon storage potential.
6. Improvement of the existing fire protection system for forest fire protection, including permanent monitoring and observation mechanisms and rapid and efficient interventions in cases of their occurrence.
7. Establish effective mechanisms for preventing all illegal activities in the forestry sector in Bosnia and Herzegovina, which have very significant negative

implications in the last period. This is primarily related to the illegal felling of forests, which is quite intensive in B&H both from the perspective of the cut biomass and from the aspect of the number of actors it is realizing.

8. Certification of the entire forestry fund in Bosnia and Herzegovina for the purpose of improving the sustainable management of forest complexes.
9. Continuous growth of energy obtained using wood biomass, for the purpose of adequate replacement for fuels with high reproductive greenhouse effect.
10. Increase in the area of protected forest areas. This measure is based on the percentage increase of the forests areas under different types of forests in order to maintain and strengthen the general forest functions, the conservation of biodiversity, the extension of protected zones, the rehabilitation of degraded forests, the protection of forests and wildlife, monitoring, scientific research and the education of cadres in forestry.

References

Anonymous. (1999). Council Directive 1999/105/EC of 22 December 1999 on the marketing of forest reproductive material. *Official Journal L 011* (15/01/2000), 17–40.

Ballian, D. (2011). Osnovni principi rajonizacije sjemenskih objekata za proizvodnju sjemena na genetičkim principima. *Radovi Hrvatskog društva za znanost i umjetnost, 12*(13), 18–41.

Butulija, S., & Nenad, B. (1980). Neke karakteristike sistemskih rješenja za pošumljvanje u Bosni i Hercegovini. Šumarstvo i prerada drveta, Vol. 7–9, XXXIV, 217–225.

CITES convention on international trade in endangered species. – Official gazette FB&H No. 11/08.

Forest program of Federation of Bosnia and Herzegovina (draft)., Federal Department of agriculture, water management and environment.

Kyoto protocol to the UN framework convention on climate change. https://unfccc.int/resource/docs/convkp/kpeng.pdf (date of accessing: October 17, 2017).

Law on forest reproductive material of the Republic of Srpska (Official Gazete of the Republic of Srpska No. 60/09).

Law on forests, Official gazette the Republic of Srpska No. 75/08.

Law on national park "Una"., Official gazette FB&H No. 44/08.

Law on national parks., Official gazette the Republic of Srpska No. 75/10.

Law on nature protection., Official gazette FB&H No. 66/13.

Law on nature protection of the Republic of Srpska (Official Gazette of the Republic of Srpska No. 20/14).

Low on forest and horticultural reproductive material of trees and shrubs. Official gazette FB&H No. 71/05.

Mataruga, M., Isajev, V., Orlović, S., Đurić, G., Daničić, V., Cvjetković, B., Balotić, P., & Ćopić, M. (2013). The Program of Forest Genetic Resources conservation in Repulic of Srpska, 2013–2025. Department of agriculture, forestry and water management, Government of Republic of Srpska (pp. 36–145).

Milanović, Đ., Brujić, J., Đug, S., Muratović, E., & Lukić Bilela, L. (2015). Vodič kroz tipove staništa BiH prema Direktivi o staništima EU, Prospect C&S s.a. Rue du Prince Royal 83, 1050 Brussels, Belgium (pp. 18–180).

Strategy of Forest Development of the Republic of Srpska 2011. – 2021. (2012). Government of Republic of Srpska, Department for agriculture, forestry and water management.

Zakon o izmjenama Zakona o šumama Republike Srpske, Official gazette RS No. 60/13.

Chapter 10
Legal and Institutional Aspects of the Conservation and Management of FGR in Romania

Alexandru Lucian Curtu, Flaviu Popescu, and Dragoş Postolache

Abstract Romania has started to identify and characterize its forest genetic resources shortly after the adoption of Strasbourg S2 and Helsinki H2 Resolutions of the Ministerial Conference on the Protection of Forests in Europe. At the same time, the country has experienced major changes (transition from a centralized to a market economy, property restitution process, and privatization) which have strongly affected the forest sector. In 2011, the National Catalogue of Forest Genetic Resources was finalized by the National Institute for Research and Development in Forestry "Marin Drăcea" (former ICAS). One year later, the catalogue and the corresponding set of management measures were endorsed by the central authority for forestry under the Ministry of Environment and Forests. According to the ministerial order, forest owners and managers have to comply with the requirements in order to manage genetic conservation units included in the catalogue. The management measures allow low-intensity silvicultural interventions and promote natural regeneration. There is a need to evaluate the existing conservation units and to assess the quality of the management measures taken so far.

Keywords Forest genetic resources · Legal and institutional framework · National catalogue · Genetic conservation units

A. L. Curtu (✉)
Department of Forest Sciences, Transilvania University of Braşov, Braşov, Romania
e-mail: lucian.curtu@unitbv.ro

F. Popescu
National Institute for Research and Development in Forestry "Marin Drăcea", Simeria, Romania
e-mail: flaviu.popescu@icashd.ro

D. Postolache
National Institute for Research and Development in Forestry "Marin Drăcea", Cluj-Napoca, Romania

University of Agricultural Sciences and Veterinary Medicine, Cluj-Napoca, Romania
e-mail: dragospostolache@yahoo.com

M. Šijačić-Nikolić et al. (eds.), *Forests of Southeast Europe Under a Changing Climate*, Advances in Global Change Research 65,
https://doi.org/10.1007/978-3-319-95267-3_10

10.1 Introduction

In the mid-1990s, Romania has started to implement Strasbourg Resolution S2 “Conservation of Forest Genetic Resources” (Anonymous 1990) and Helsinki Resolution H2 “General Guidelines for the Conservation of the Biodiversity of European Forests” (Anonymous 1993) of the Ministerial Conference on the Protection of Forests in Europe (MCPFE). Thus, a first national survey for identification and characterization of forest genetic resources was conducted between 1993 and 1996 using a set of technical guidelines elaborated by the Forest Research and Management Institute (ICAS) (Enescu 1993). In addition, Romania has established collaborations in the conservation and sustainable use of forest genetic resources through the European Forest Genetic Resources Programme (EUFORGEN) that was established in October 1994 as a pan-European implementation mechanism for Resolution S2. However, although Romania has signed the resolutions of MCPFE, the legislative framework in the field of conservation of forest genetic resources was consolidated much later, in a more general context, with the adoption of the Law on environmental protection in 2006. Furthermore, measures for biodiversity conservation of forest ecosystems were stated in the Forest Act adopted by the Romanian Parliament in 2008. The central public authority for forestry (ministerial department for forests) is in charge for the conservation of forest genetic resources in Romania.

The implementation of concrete measures for the conservation and management of forest genetic resources in Romania was influenced by several factors. The property restitution process was one of the major changes that affected the entire forest sector in post-socialist Romania. In 1990, all forests belonged to the Romanian state, many of them being nationalized in 1948. The situation gradually changed as three restitution laws came into power (18/1991, 1/2000 and 247/2005, respectively). As a result, the ownership structure has changed and the Romanian state is currently the owner of about 49% of the forest land. Private persons, associations and companies hold about one third of the forests, and approximately 16% is owned by municipalities, towns and villages.

Romania’s accession to the European Union in 2007 has triggered a series of legislative changes such as the implementation of EU directive on forest reproductive materials. Furthermore, a new system of regions of provenance for forest species was adopted in 2010 (Ministerial order 1028/2010) following the European Council Directive (EEC) 1999/105/CE and the scheme of the Organization for Economic Co-operation and Development (OECD).

10.2 National Catalogues of Forest Genetic Resources and of Approved Basic Material for Forest Reproductive Material

A first part of the National Catalogue of Forest Genetic Resources focusing on a set of main forest tree species was issued in 1996 (Lalu 1996). The work on identification and characterization of genetic resources continued for an additional set of forest tree species (Lalu 1999). However, due to major changes affecting the forest sector in the early 2000 no further action was taken in this respect.

The current version of the National Catalogue of Forest Genetic Resources (Pârnuţă et al. 2011) has been developed by the National Institute for Research and Development in Forestry "*Marin Drăcea*" (former ICAS) in cooperation with Forest State Service (Regia Nationala a Padurilor – Romsilva) and Transilvania University of Brasov. The catalogue was conceived between 2005 and 2008 in the framework of the national project Conservation and sustainable management of forest genetic resources in Romania (COREGE- FOREST). The catalogue and the technical instructions for sustainable management of forest genetic resources were endorsed by the Ministry of environment and forests in 2012 (Ministerial Order No. 945/2012).

The National Catalogue of Forest Genetic Resources comprises 698 genetic conservation units for 45 forest tree species both native and non-native. The core zone of the genetic conservation units for conifer and broadleaved species occupies an area of approximately 4.700 ha and 8.200 ha, respectively. The genetic conservation units represent less than 0.2% of the Romanian forest cover. The vast majority (about 93%) of the genetic conservation units are in situ units. Most of conservation units are located in state owned forests (76%) and forests that belong to municipalities and villages (12%). Less than 12% are located on private land. Databases and GIS-based maps were produced for the target tree species.

In addition, a National Catalogue of Approved Basic Material for Forest Reproductive Material (Pârnuţă et al. 2012) under the OECD scheme was elaborated and endorsed (Ministerial Order No. 1645/2013). The most recent version of the catalogue complies with the requirements of the national law on forest reproductive material (no. 107/2011) and European Council Directive (EEC) 1999/105/CE. In the catalogue, there are 2659 and 98 registered sources (units of approval) in the category "Selected" and "Tested", respectively; 115 seed orchards, of which 113 in the category "Qualified" and two in the category "Tested"; a total of 37 poplar and 21 willow clonal collections.

10.3 Management of Forest Genetic Resources

The current forest legislation states the obligation of forest owners and forest managers to follow the technical instructions for the management of forest genetic resources. According to the functional zoning of forests in Romania, the core and buffer zone of the conservation units are included in the category I-5H and I-5L, respectively. Category I and subcategory 5 are assigned to forests with special protective function and for safeguarding of forest genepool, respectively. Only silvicultural interventions of low-intensity (so called conservation works in the Romanian classification system) are allowed in the core zone of the genetic conservation units. The wood volume which can be extracted in a decade is usually less than 10% of the total standing volume. Silvicultural treatments like group shelterwood systems can be used in the buffer zone of the genetic conservation units. The type of silvicultural intervention for every stand is mentioned in the forest management plans which are approved by the central public authority for forestry (department of forests within the ministry). Natural regeneration is promoted in both core and buffer zone; plantations with local material are done only in very rare cases (absence of natural regeneration).

The genetic resources included in the National Catalogue of Approved Basic Material are used for increasing the productivity and adaptability of forest ecosystems. Only registered sources are accepted for the production of forest reproductive material. The main management purpose of these genetic resources is to provide high quality forest reproductive material. Management measures consist of thinnings, crown formation and fertilization.

The implementation of the forest management plans is done by state and non-state forest companies. The control of the application of forest management plans is done by regional public institutions (called Forest Guards) that represent the central authority for forestry in the territory. The National Institute for Research and Development in Forestry "*Marin Drăcea*" is responsible for a regular update of the national catalogues but this activity has to be done with the financial support of central public authority for forestry.

10.4 Conclusions

The legal and institutional framework concerning the conservation and management of forest genetic resources at national level has been developed and is well in place. However, it is essential to continue to raise awareness of the importance of forest genetic resources among forest owners, managers, practitioners and central/regional forest authorities. There is a need to evaluate the existing conservation units and to assess the quality of the management measures which have been taken so far.

References

Anonymous. (1990). Ministerial conference on the protection of forests in Europe. 18 Dec 1990, Strasbourg. Ministère de l'Agriculture et des Forêt, Paris, France.

Anonymous. (1993). Ministerial conference on the protection of forests in Europe. 16–17 June 1993 in Helsinki. Sound forestry – Sustainable development. In *Conference proceedings*. Ministry of Agriculture and Forestry.

Enescu, V. (1993). Identificarea, eşantionarea, prospectarea, descrierea şi clasificarea resurselor genetice forestiere. Îndrumări metododologice. Manuscris ICAS.

Lalu, I. (1996). *Catalogul National al Resurselor Genetice Forestiere. Partea I. Ministerul Apelor, Pădurilor şi Protecţiei Mediului, RNP-Romsilva* (p. 187). Bucureşti: ICAS.

Lalu, I. (1999). *Identificarea, eşantionarea şi prospectarea resurselor genetice forestire din fondul forestier (Etapa a II-a). referat final de etapa, Tema nr. A41/1999.* ICAS.

Pârnuţă, G., Stuparu, E., Budeanu, M., et al. (2011). *Catalogul naţional al resurselor genetice forestiere*. Bucureşti: Editura Silvică.

Pârnuţă, G., Budeanu, M., Stuparu, E., et al. (2012). *Catalogul naţional al materialelor de bază pentru producerea materialelor forestiere de reproducere*. Bucureşti: Editura Silvică.

Chapter 11
Conserving Biodiversity and Plant Genetic Resources: The Hellenic Legal Framework

Paraskevi G. Alizoti and Filippos A. Aravanopoulos

Abstract The Mediterranean Basin is among the top biodiversity hotspots globally due to the high endemism and the vast number of flora and fauna species harbored in its ecosystems. Hellenic ecosystems, being also Mediterranean ecosystems, are biodiversity hotspots that harbor 22% of the species found across the Mediterranean basin, while 10% of the plants occurring in them are endemic. The value of conserving biodiversity in natural ecosystems was recognized at the national level as early as 1937 and ever since a great part of the Hellenic ecosystems is conserved in situ under national and European legislation, as well under International Conventions. A solid, but also large and complicated legal framework is in effect in the country for the protection of the biodiversity and the genetic resources. The legal framework includes Laws that have incorporated Regional (EU) and International legislation into the national legal framework, as well as the national legislation on the conservation of nature, biodiversity and genetic resources. In the current chapter the above legal framework, under which a notable part of the surface of the country is currently protected, will be presented in short.

Keywords Forest ecosystems · Protected areas · National laws · Regional laws · International conventions for biodiversity

11.1 Introduction

Biodiversity, as defined in Article 2 of the Convention on Biological Diversity (CBD) (Secretariat of the Convention on Biological Diversity 2005) is "*the variability among living organisms from all sources including, inter alia, terrestrial, marine and other aquatic ecosystems and the ecological complexes of which they are part; this includes diversity within species, between species and of ecosystems*".

P. G. Alizoti (✉) · F. A. Aravanopoulos
School of Forestry and Natural Environment, Aristotle University of Thessaloniki, Thessaloniki, Greece
e-mail: alizotp@for.auth.gr; aravanop@for.auth.gr

M. Šijačić-Nikolić et al. (eds.), *Forests of Southeast Europe Under a Changing Climate*, Advances in Global Change Research 65,
https://doi.org/10.1007/978-3-319-95267-3_11

In the same CBD article it is defined that the term 'Biological Resources' includes "*genetic resources, organisms or parts thereof, populations, or any other biotic component of ecosystems with actual or potential use or value for humanity*", as well as that the term 'Genetic Resources' means "*the genetic material of actual or potential value*". The CBD treaty, signed by 196 nations, in its first article defines its objectives which are "*the conservation of biological diversity, the sustainable use of its components and the fair and equitable sharing of the benefits arising out of the utilization of genetic resources, including by appropriate access to genetic resources and by appropriate transfer of relevant technologies, taking into account all rights over those resources and to technologies, and by appropriate funding*" (Secretariat of the Convention on Biological Diversity 2005). This was first time that the significance of genetic resources was recognized and was clearly stated in the very first article of the Convention on Biological Diversity treaty (Glowka et al. 1994).

Biodiversity distribution is not even across the globe, as areas with high and low species richness can be identified. Attempts to understand the distribution of biodiversity have focused particularly on the identification of species richness spatial patterns, and on the mechanisms that have given rise to these patterns (Gaston and Spicer 2004). The value of biodiversity harbored in natural ecosystems, and especially in forest ecosystems is extremely high, as they provide numerous goods and services (i.e. wood and secondary non-wood products, carbon sequestration, water resources, recreation, food products, climate regulation, soil protection, pharmaceutical substances). Wild species have been, since antiquity, the source of important pharmaceutical products, and it is still questionable if the synthetic chemistry can match the complexity of many of the natural compounds that have been proved effective as drugs, and are found in wild flora and fauna species (US National Research Council 1999). Biodiversity though suffers losses across the globe mainly due to its direct exploitation, the loss of habitats, degradation and fragmentation, the impacts of introduced species, and the extinction cascades (Gaston and Spicer 2004). Following the above, it can be easily recognized that our present and future rely on the biodiversity of natural ecosystems, rendering its protection of vital importance.

The Mediterranean Basin is ranking among the top biodiversity hotspots globally, as its ecosystems contain a vast number of species and exceptional concentrations of endemic ones that may face threats due to habitat loss (Myers et al. 2000; Myers 2001). The high biodiversity and endemism can be due to historic (i.e. glacial and post glacial history, past connection of the mainland to the islands, barriers that prohibited migration and vegetation changes especially during the last 20,000 years) and ecological factors (i.e. temperature, humidity, soil types, topography, mosaic of different types of ecosystems, human activity since prehistoric times) (Legakis and Maragkou 2009). Hellenic forest ecosystems are significant biodiversity hotspots of the Mediterranean that include more than 400 forest tree species, which serve as keystone or umbrella species. Their role and significance was recognized quite early and their protection was launched at a national level in 1937. Since then a notable part of the surface of the country is protected under the national, regional and international legislation; only the Natura 2000 sites cover 27.2% of land (Biodiversity Information System for Europe 2015).

In the current chapter, the legal framework, at the national, regional and international level, regulating the conservation of biodiversity and genetic resources in Greece, will be covered.

11.2 Legislation at the National Level

The Law 856/1937 on 'National Forest Parks', published in 1937 in the Official Gazette of the Government, was the first law regulating the protection of natural forest ecosystems and their biodiversity. Following this Law the first two National Forest Parks of Olympus and Parnassos were established in 1938.

In 1950 the Law 1465/1950 on 'Landscapes of natural beauty' was published, following which more than 300 areas have been designated as protected ones. Despite their designation status, these areas were not protected accordingly and following the provisions of the law.

Later on, in 1969 the Legal Decree 86/1969 (Official Government Gazette) on 'The Forest Code' was published and was regulating, among others, issues related to protected areas. The above Legal Decree was amended in 1971 with the Legal Decree 996/1971 on 'National Forest Parks, Aesthetic Forests and Preserved Monuments of Nature'. Since 1938, ten National Forest Parks, 19 Aesthetic Forests and 51 Preserved Monuments of Nature have been designated.

Milestone for the protection of natural environment in Greece was the Constitution of 1975, in the articles 24 and 117 of which the constitutional protection of the environment has been provisioned for the first time. According to article 24, paragraph 1, subparagraph a' it is stated that *"the protection of natural and cultural environment is an obligation of the State"*. This obligation is further clarified in subparagraph b' of the 1st paragraph of article 24, following which the State is obliged to take particular preventive and repressive measures for the preservation of natural and cultural environment. Following the Constitutional revision of 2001, in article 24 the right to a healthy and ecologically balanced environment is enshrined as a prerequisite for human life, health and human dignity, which is expressed as a requirement for quality of life and as an asset of self-worth for the mankind. Thus, with the 2001 revision of the Constitution, the environment has been recognized as everybody's right and the principle of sustainability of the environment was established; namely the protection of the environment for the sake of the future generations (Koutoupa-Regkakou 2005). The article 117 of the Constitution in paragraph 3 states that public or private forests or forest areas destroyed, burned down, being currently or in the past deforestated in any way, they do not lose the status they had before being destroyed, they are declared areas for obligatory reforestation and they cannot be used for any other purpose. In paragraph 4 of the same article it is provisioned that the obligatory expropriation of forests or forest lands, belonging to individuals or legal entities under private or public law, is only allowed in favor of the

State, following the provisions of article 17, for reasons of public benefit and their status as forests or forest lands remains unchanged.

The Law 1650/1986 'For the protection of the environment' in the articles 18 and 19 defines the environmental zones of different protection level; namely areas of absolute nature protection, National parks, protected natural formations, landscapes and landscape features, and areas for eco-development. Terrestrial or aquatic areas or mixed ones (including terrestrial and aquatic parts), elements or the whole of nature and the landscape can be protected and conserved due to their ecological, geomorphological, biological, scientific or aesthetic significance. In article 19 of the same Law the following categories of protected areas are defined:

1. *Areas of absolute nature protection*: Delicate ecosystems, biotopes or ecotopes of rare or threatened with extinction flora and fauna species or areas playing a decisive role in the life cycle of rare or threatened with extinction wildlife species. In these areas any human activity is prohibited. In special cases, scientific research or works aiming to the preservation of the characteristics of those areas may be allowed, but only if the higher conservation level is secured.
2. *Areas of nature protection*: Extensive areas of high ecologic and biological value. In those areas the natural environment is protected from every activity or intervention that could potentially change or alter its condition, composition or evolution. Scientific research, works and mainly traditional activities may be allowed, only in case they are not contradictory to the objectives of the protection.
3. *National Parks*: Extensive terrestrial or aquatic or mixed (terrestrial and aquatic) areas, which are non or slightly affected by human activities and harbor a large number and variety of significant biological, ecological, geomorphological and aesthetic elements. When the whole or a big part of the Park is a forest area or sea area then it can be characterized as National Forest Park or National Marine Park. For the protection of Natural Parks specific measures are taken, so that they are protected from natural causes resulting in degradation or from human actions, interventions and activities. Under the existing operation and management regulations, scientific research and traditional activities may be allowed.
4. *Protected natural formations, protected landscapes and landscape elements*: Protected natural formations are considered the functional parts of nature or isolated natural formations of particular scientific, ecological or aesthetic value. They may also be areas that are significant for the conservation of natural processes or natural resources. All activities that may cause destruction, deterioration or alteration of the above areas are prohibited.
5. *Areas of eco-development*: Extensive areas, that may include villages or settlements, which are of specific interest and value, due to their natural and cultural characteristics and their potential for the development of activities being in harmony with the protection of nature and landscape.

The Law 2742/99 on 'Spatial Planning and Sustainable Development and other provisions' clarifies issues regarding the operation and management of the protected areas and is complementary to the Law 1650/1986. The Law regulates the administration and management of areas, elements and the whole of nature and landscapes, based on the categories defined in the Law 1650/1986. Provisions are also included for the Natura 2000 'Special Areas of Conservation', following the Council Directive 93/43/EEC. In article 15, the establishment of Management Bodies for protected areas, as legal entities under private law, is provisioned. Following the Law 3044/2002, twenty-five Management Bodies have been established, while two more were established following two recent Presidential Decrees (Presidential Decree-Official Government Gazette 395/2000 for the National Park of Marathonas-Schoinias, Presidential Decree- Official Government Gazette 906A/1999 for the Zakynthos National Marine Park) (Source: Hellenic Ministry of Environment and Energy).

The main Law that regulates issues on biodiversity and genetic resources is the Law 3937/2011. The Law regulates the sustainable management and effective maintenance of biodiversity, as a valuable, irreplaceable and important national capital. It amends the categories of protected areas as following: *Areas of absolute nature protection*; the highly sensitive ecosystems, biotopes or habitats of rare or threatened species of wildlife – no human activity is allowed in these areas, except for scientific research, *Areas of nature protection*; the areas of significant ecological or biological value – any activity or intervention that may alter their condition, composition or evolution is prohibited, *Natural National or Regional Parks*; areas of high ecological importance – the demarcation of zones within the Natural Park is allowed, as for instance the 'core' area, which is of absolute protection and the 'buffer zone', where specific interventions can be allowed, *Areas of habitat and species protection;* they are distinguished in Special Conservation Areas (EZZ), Wildlife Refugia (KAZZ) and Special Protection Areas (SPAs) and they include protected landscapes and landscape elements or protected natural formations. In the above Law all the Natura 2000 sites are explicitly described and the most significant categories of flora and fauna species are defined. There are also provisions for the conservation of the autochthonous biodiversity, the conservation of natural ecosystems and the prevention of their degradation, specific provisions for the invasive species and the environmental biosafety. It is defined that the State is responsible for taking the necessary measures against all the potential threats or risks that may harm the structure of the ecosystems, and to prevent, reduce and restore every environmental damage. It is also clearly stated that *all the genetic resources of Greece* are *considered as protected national capital*. The use of genetic resources is subjected to the conditions and restrictions regarding the access to genetic resources, as well as to the Law concerning the equal sharing of benefits accruing from their use, in accordance with the Nagoya Protocol, which is expected to be ratified by Greece in the near future, in the framework of the Convention on Biological Diversity, and to the Treaty of the Food and Agriculture Organization (FAO) on plant genetic resources for nutrition and agriculture, as ratified by Law 2014/1992 (Government Gazette 29/A/1992) (Alizoti 2014). The use of genetic resources is regulated by the Joint Decision of the Ministers of Rural Development and Environment & Energy. Besides the above legal framework, under the Presidential Decree 67/1981, 775 autochthonous species of flora and 225 wildlife species are protected.

A number of additional Laws, Presidential and Ministerial Decrees regulate at the national level issues related to nature conservation, and they are in detail: the Law 177/1975 and the Law 2637/1998 on 'Refugia of wildlife', the Law 998/1979 on the 'Protection of forests and forest areas of the country', the Presidential Decree 67/1981 on the 'Protection of autochthonous and wild flora', the Law 1564/1985 on 'Organization of production and trading of plant reproductive material', the Presidential Decree 126/1986 on 'Concession of the exploitation, conservation and improvement of publicly owned forests to forest cooperatives', the Presidential Decree 80/1990 and Law 3165/2003 on the 'Conservation and protection of the plant genetic resources of the country', the Ministerial Decision 303206/1997 on 'Requirements for granting plant reproductive material trading licenses', the Law 3208/2003 on the 'Protection of forest ecosystems, establishment of forest cadastre, regulation of legal rights on forests and forest areas and other provisions', and the Presidential Decree 17/2003 on 'Trading of forest reproductive material'.

11.3 Legislation at the Regional and International Level

At the Regional Level, the Natura 2000 network is a pan-European nature protection program for species and their habitats. Its goal is the long-term preservation of the most valuable and most endangered species and habitats at a satisfactory level. It consists of two categories of areas: the Special Protection Areas (SPAs) for the Birds, as defined in Directive 79/409/EC "on the conservation of wild birds", and the Sites of Community Importance (SCI), as defined in Directive 92/43/EEC. Greece has designated 202 Special Protection Areas (SPAs); published in the Official Government Gazette 1945/B/2010, as an annex in the law that incorporated the European Directive 79/4009/EEC in the national legislation, and 241 Sites of Community Importance (SCI); as defined in the Law 3937/2011 (Official Government Gazette 60/A/2011). The two types of Natura 2000 areas may overlap with respect to the land they cover.

At the international level, the Ramsar Convention was ratified in Greece by the Law 191/1974, (Official Government Gazette 350/A/1974). The most important eleven wetlands in the country were characterized as protected ones under the above mentioned Convention. The first goal of the Ramsar Convention is the protection of wetlands of high priority and the second one is the protection of the wetlands as ecosystems in conjunction with the aquatic birds, but also the rest of the fauna and flora species.

The Ramsar Convention initiated a period of high global activity regarding the protection of the wildlife. This whole activity resulted in:

1. The UNESCO Convention on the Protection of the World Cultural and Natural Heritage (1972) that was ratified by Greece in the Law 1126/1981 (Official Government Gazette 32/A/1981). Two areas were recognized as sites of World

Cultural and Natural Heritage in Greece; namely the Meteora area in Trikala region, Central Greece, and Mount Athos in Chalkidiki area, Northern Greece.

2. The Convention on International Trade in Endangered Species of Wild Fauna and Flora – CITES (1973). The Convention was ratified by the country in the Law 2055/1992 (Official Government Gazette 105/A/1992), while the law enforcement mechanism was defined in the Law 2637/1998. The CITES Convention regulates the international trade of wild fauna and flora species, which applies for all countries when importing or exporting wild species.
3. The Bonn Convention (1979) on the Conservation of Migratory Species of Wild Animals that protects the wild life and includes also the main provisions for ecotopes. In Greece, it was ratified by the Law 2719/1999, (Official Government Gazette 106/A/1999).

In 1979 the European Ministerial Conference on the Environment signed in Bern the Convention for the Conservation of the Wildlife and Natural Habitats of Europe. The Convention of Bern is quite similar to the provisions of the EU Directive 79/409/EEC on the conservation of wild birds, but it is much broader as it concerns the protection of species in general and habitats. The purpose of the Convention, as specified in Article 1, is to protect the wild flora and fauna species and their habitats; in particular, species and ecosystems that require an international concerted effort. In Greece the Bern Convention was ratified by the Law 1335/1983 (Official Government Gazette 32/A/1983).

The Convention on Biological Diversity (CBD), that was based on the globally raised awareness on the need for sustainable development, came into force on the 29th of December 1993, after being signed by 196 nations. It was first opened for signature on the 5th of June 1992, during the Rio de Janeiro "Earth Summit", which became a historical United Nations Conference on Environment and Development due to its scope, size and the decisions taken. The Convention is the main tool to tackle the issues of biodiversity conservation and sustainable use of its components, as well as the equal and fair sharing of benefits arising from the use of genetic resources. In Greece the Convention was ratified by the Law 2204/1994 (Official Government Gazette 59/A/1994).

The Cartagena Protocol on Biosafety was adopted in 2000 as a complementary agreement to the Convention on Biological Diversity (CBD). It was signed for the first time in 2000 in Montreal, Canada and ever since it has been signed by more than 50 countries. The Cartagena Protocol, that entered into force on the 11th of September 2003, is an international treaty ruling the movement of genetically modified organisms (LMOs) among countries. The importance of the Protocol lies in the recognition of the right, of the sovereign countries that have signed it, not to accept Genetically Modified Organisms (GMOs), on the basis of the "precautionary principle". The Cartagena Protocol, which is above the national and the regional (EU) law, was ratified in Greece by the Law 3233/2004.

The Protocols and Acts relating to the UNEP "Convention for the Protection of the Marine Environment and the Coastal Region of the Mediterranean" were ratified by the Law 1634/1986 (Official Government Gazette 104/A/1986) in Greece.

11.4 Conclusions

The value of biodiversity harbored in natural ecosystems, and especially in forest ecosystems is extremely high, as they provide numerous goods and services. The Mediterranean Basin is among the top biodiversity hotspots globally due to the high endemism and the vast amount of flora and fauna species harbored in the Mediterranean ecosystems. Hellenic forest ecosystems, as part of the Mediterranean ones, are also characterized by endemism and great diversity among and within species, and for this reason they are considered to be among the most important ones at a European level. The value of conserving biodiversity in natural ecosystems was recognized at the national level as early as 1937 and ever since a great part of the Hellenic ecosystems is conserved in situ under national and European legislation, as well under International Conventions. A solid, but also complicated legal framework is in effect in the country for the protection of the biodiversity and the genetic resources, and consists of Laws that have incorporated the Regional (EU) and International legislation into the national legal framework, as well as the national legislation on the conservation of nature, biodiversity and genetic resources, that traces its history back to 1937. Result of this extensive and solid legal framework is the protection of a big part of the surface of the country under the national, regional and international Law.

References

Alizoti, P. G. (2014). Securing equal sharing of benefits coming from utilization of native genetic resources in European Union – The role of genetic improvement. In *Proceeding of the 15th conference of Hellenic Scientific Society for Plant Genetics and Breeding*, Larisa (pp. 24–30).

Biodiversity Information System for Europe. (2015). *Mid- term report.* Member states contribution to the mid-term review of the EU biodiversity strategy to 2020 – Country. Greece. https://biodiversity.europa.eu/mtr/countries/greece.

Gaston, K., & Spicer, J. (2004). Biodiversity: An introduction. Blackwell Publ.

Glowka, L., Burhenne-Guilmin, F., Synge, H., McNeely, J. A., & Gündling, L. (1994). *A guide to the convention on biological diversity*. Gland: IUCN.

Koutoupa-Regkakou, E. (2005). *Environmental law* (p. 211). Thessaloniki: Sakkoula Publ.

Legakis, A., & Maragkou, P. (2009). *The red book of endangered fauna species* (p. 521). Athens: Hellenic Zoological Society.

Myers, N. (2001). Hotspots. In S. A. Levin (Ed.), *Encyclopedia of biodiversity* (Vol. 3, pp. 371–381). San Diego: Academic.

Myers, N., Mittermeier, R. A., Mittermeier, C. G., da Fonseca, G. A. B., & Kent, J. (2000). Biodiversity hotspots for conservation priorities. *Nature, 403*, 853–858.

National Research Council (US). (1999). *Perspectives on biodiversity* (p. 153). Washington, DC: National Academic Press.

Secretariat of the Convention on Biological Diversity. (2005). *Handbook of the convention on biological. Diversity including its cartagena protocol on biosafety* (3rd ed., pp. 5–6), (Montreal, Canada).

Chapter 12
Legal and Institutional Framework for Forest Genetic Resources Conservation in Bulgaria

Mariya Belovarska, Denitsa Pandeva, Svilena Bozhinova, Emil Popov, and Petar Zhelev

Abstract The chapter presents information about the legal instruments and institutions concerning the conservation of genetic resources of forest tree species in Bulgaria. Due to the specific circumstances, the vast majority of the Bulgarian forests are state-owned. Therefore, the main regulation documents concerning management of state forests are relevant in the context of tree genetic resources sustainable use and conservation. A brief survey of the most important regulation documents determining the legislation framework for conservation of genetic resources of forest trees is presented.

Keywords Regulations · State forest administration · Seed control stations

12.1 Introduction

Forests and forest territories occupy roughly 38% of the Bulgarian territory. Total forest area in the country is 4,222,874 ha (hectares), and it could be further subdivided as follows: State-owned forests are 73.23%, including 2,906,508 ha (68.83%) forests managed by state forestry authorities and the enterprises under and

M. Belovarska (✉) · D. Pandeva
Executive Forest Agency, Ministry of Agriculture, Food and Forestry, Sofia, Bulgaria
e-mail: belovarska@iag.bg; denica@iag.bg

S. Bozhinova
Forest Seed Legislation Station, Sofia, Bulgaria
e-mail: GSSSofia@nug.bg

E. Popov
Forest Research Institute, Bulgarian Academy of Sciences, Sofia, Bulgaria
e-mail: emilpopov99@hotmail.com

P. Zhelev
University of Forestry, Sofia, Bulgaria
e-mail: zhelev@ltu.bg

M. Šijačić-Nikolić et al. (eds.), *Forests of Southeast Europe Under a Changing Climate*, Advances in Global Change Research 65,
https://doi.org/10.1007/978-3-319-95267-3_12

174,463 ha (4.13%) protected territories – national parks and reserves – managed by Ministry of Environment and Water (MoEW). The private forest amount 24.68% of the forest territory, but they include 551,334 ha (13.06%) forest area managed by municipalities, 427,573 ha (10.13%) owned by private owners, 42,849 ha (1.01%) owned by legal entities, and 20,345 ha (0.48%) owned by religious communities (Executive Forest Agency, EFA 2016). In fact, the forests managed by municipalities were formerly state-owned and were transferred to municipalities for the need of local people still at the beginning of twentieth century.

The low share of privately owned forests is mostly due to the fact that until the end of nineteenth century Bulgaria was a part of the Ottoman Empire and Bulgarians were not allowed to own forest territories (Branchev 1918; Dimitrov 1934). It should be pointed out also that the share of the forests of artificial origin (forest plantations) is substantial. More than 680,000 ha new forests were established during the second half of twentieth century and the total area where afforestations took place is more than 2 million ha, including reconstructions of low-productivity forests, assistance of regeneration and some other categories of afforestation (Milev et al. 2017). All these plantations, although important from economic and environmental points of view, were not primary target of gene conservation activities.

Many studies published in the recent decades concern the sustainable use and conservation of genetic resources of forest trees in Bulgaria (see Alexandrov and Dobrev 2012, for review). Also a number of targeted activities have been undertaken in the country to conserve genetic diversity as a constituent part of biological diversity. These activities were carried out both internationally and nationally in the process of harmonizing the Bulgarian legislation with the European one. This process is related to the understanding that the destruction of forests on a global, European and national scale is associated with substantial loss of biodiversity. The objective of the present paper is to summarize the achievements and to present national and international legal and regulatory documents adopted by different authorities regulating and contributing to the conservation of forest genetic resources.

12.2 Legal Base

Conservation of forest genetic resources is done by considering legislation and regulations at national and international level, and especially, after the synchronization of Bulgarian legislation with the European one, done before Bulgaria joined EU in 2007.

12.2.1 National Legislation and Regulations

Conservation and utilization of forest genetic resources in Bulgaria is covered by four acts – one concerning forests ant three concerning nature conservation:

- Forest Act (2011),
- Protected Areas Act (1998),
- Biological Diversity Act (2002),
- Act of Environment Protection (2002).

The last National Forest Sector Development Strategy (Anonymous 2013), Strategic Plan for the Development of the Forest Sector (Anonymous 2014) and number of regulations have been developed:

(a) Regulations:

- Regulation No. 4 of 8 July 2003 issued by MoEW on the conditions and procedures for issuing permits for introduction of non-native or reintroduction of native plant and animal species in the wild.
- Regulation No. 14 (2005) issued by MAFF (2005) Concerns the conditions and procedures for issuing permits for the introduction of non-native or reintroduction of native trees, shrubs and game species in nature and consideration of the public opinion in the region of the reintroduction.
- Regulation № 2 of 07.02.2013 issued by MAFF (2013) on the terms and procedures for afforestation of forest areas and agricultural land used for creating special, protective and commercial forests and forests in protected areas, inventory of created plantations, their accounting and registration.

(b) Instructions, schemes, rules, and maps:

- Instruction for Establishing and Mapping Forest Habitat Types and Determining the Composition of Dendrocoenoses (EFA 2011a);
- Classification scheme of types of forest habitats in Bulgaria (2011b);
- Scheme on utilization of Forest Reproductive Material produced in Bulgaria from sources of categories "identified" and "selected", depending on their provenance (EFA 2010a),;
- Rules and specific exceptions concerning transfer of forest reproductive material (FRM) within and between regions of origin (EFA 2010b); see also Konnert et al. (2015).
- Map of the regions of origin in the Republic of Bulgaria (EFA 2010c).

According to the Forest Act, Art. 9. Forest planning is done on three levels and the first is: National Strategy for Forest Development and Strategic Plan for the development of the forest sector. The next levels of forest planning 2. Regional Development Plan of forest areas and 3. Forestry plans are the key legal and institutional framework for forest genetic resources conservation on forest territories in Bulgaria. The choice of tree species for afforestation and reforestation in Bulgaria is carried out according to regulations as follows: Regulation № 2 of 07.02.2013 on the terms and procedures for afforestation of forest areas and agricultural land used for creating special, protective and commercial forests and forests in protected areas, inventory of created plantations, their accounting and registration (MAFF 2013), Instruction for Establishing and Mapping Forest Habitat Types and Determining the Composition of Dendrocoenoses (EFA 2011a), Classification scheme of types of forest sites in Bulgaria (EFA 2011b). Everywhere a general recommendation for use of native species with the priority has been made. Classification scheme of types of forest sites in Bulgaria describe precisely 155 forest sites (it should be clarified that these sites are different from Natura 2000 habitats) including: (1) number; (2) region, altitudinal belt and sub belt; (3) sub region; (4) site name, (5) site code; (6) soil type, depth, stoniness; (7) exposure; (8) slope and shape of the terrain; (9) current species composition, percentage of area occupied by and average productivity class for the main tree species; (10) suitable tree species for the site and expected potential productivity class, where applicable, are listed. Licensed experts elaborate Forest management plans according to those regulations.

For Natura 2000 protected zones batch of texts referred to the regulations issued by the Ministry of Environment and Water. Ministry of Agriculture and Food (MAF) and Executive Forest Agency (EFA) published "Regimes for sustainable management of forests in Natura 2000 (MAFF and EFA 2011) were texts not allowing non-native tree species for afforestation and recommending reduction of participation share of existing non-native trees in species composition for 35 Natura 2000 forest habitats.

Three Nature Conservation Acts exist in Bulgaria under which the protected areas are managed.

12.2.2 International Legislation and Regulations

International provisions governing the conservation of biodiversity and forest genetic resources can be found in worldwide and/or pan-European documents. Here we list the most important of them:

- 1990: First Ministerial Conference on the Protection of Forests in Europe (MCPFE), Strasbourg: Resolution S2 "Conservation of Forest Genetic Resources",
- 1992: UN Conference on Environment and Development (UNCED), Rio de Janeiro: "Convention on Biological Diversity" (CBD),
- 1993: Second Ministerial Conference on the Protection of Forests in Europe (MCPFE), Helsinki: Resolution H2 "General Guidelines for the Conservation of the Biodiversity of European Forests",
- 1994: Ordinance of the EU Council on the conservation, description, collection and utilization of the genetic resources of agriculture, Brussels (forest resources included),
- 1995: "European Forest Genetic Resources Programme" (EUFORGEN) of IPGRI (today Bioversity International) affiliated with the Food and Agriculture Organization (FAO), Rome: European network on existing resources concerning defined tree species
- 1998 – Third Ministerial Conference on the Protection of Forests in Europe (MCPFE), Lisbon: Resolution L2 "Pan-European Criteria, Indicators and Operational Guidelines for a Sustainable Forestry" including Appendices 1 and 2,

- 2002: Sixth Conference of the Parties of the Convention on Biological Diversity (CBD) adopted a programme of work on Forest Biological Diversity (COP 6. Decision VI/22),
- 2003: Fourth Ministerial Conference on the Protection of Forests in Europe (MCPFE), Vienna: Resolution V4: "Conserving and Enhancing Forest Biological Diversity in Europe",
- 2007: Seventh United Nations Forum on Forests (UNFF) adopted the "Non-legally binding instrument on all types of forests",
- 2007: Fifth Ministerial Conference on the Protection of Forests in Europe (MCPFE), Warsaw: Warsaw Declaration "Benefiting Quality of Life" (...maintains, conserve, restore and enhance the biological diversity of forests, including their genetic resources through sustainable forest management...)
- 2008: Ninth Conference of the Parties of CBD; decisions related to forest biological diversity – continuation of the programme adopted in 2002 (COP 9 Decision IX/5).

12.3 Institutional Framework

The institutions responsible for implementing the policy for forest genetic resources conservation are the two ministries managing the forest territory of the country: Ministry of Agriculture, Food and Forestry (MAFF) with the agency affiliated with the ministry, Executive Forest Agency (EFA), and the Ministry of Environment and Waters (MoEW). EFA is responsible for management and conservation of all state-owned forests, including nature parks, and for control of the management in non-state owned forests. MoEW is responsible for conservation (and management, where applicable) of the forests in the three national parks and in all nature reserves, including maintained ones.

There are two seed control stations (SCS) in Bulgaria, situated in Sofia and in Plovdiv. These stations, together with forest management bodies, are responsible for identification conservation and use of forest genetic resources. They manage and control the permanent and temporary seed production stands and also clonal and generative seed orchards. SCS in Plovdiv has a special construction with controlled temperature, humidity and light conditions, where different seed lots with known origin are stored. The station has also a seed testing laboratory and the seed quality of each seed lot is tested periodically. The forestry authorities work in close cooperation with research and educational institutions – University of Forestry, and Forest Research Institute – Bulgarian Academy of Sciences, both situated in Sofia.

Acknowledgements The support of the Governing Bodies of the Ministry of Agriculture, Food and Forestry, Executive Forest Agency, and Ministry of Environment and Waters of Bulgaria is very much appreciated.

References

Act of Environment Protection. (2002). Published. SG. 99 of 25.09.2002, last amend. and suppl., SG. 12 of 03.02.2017.

Alexandrov, A., & Dobrev, R. (2012). State of forest genetic resources in Bulgaria. *Nauka za Gorata, 1–2*, 1–22.

Anonymous. (2013). *National strategy for development of forestry sector* (Adopted by the Government of Bulgaria). Sofia. 207 pp.

Anonymous. (2014). *Strategic plan for development of forest sector (2014–2023)*. Sofia: MAF 123 pp.

Biological Diversity Act. (BDA 2002, 2013). Published. SG. 77 of 9.08.2002, last amend. and suppl., SG. 58 of 26.07.2016.

Branchev S. (1918). The forests and forestry in Bulgaria. In *Proceedings of Bulgarian Academy of Sciences, Branch Mathematics and Natural Sciences*, book X. Royal Courtier Publishing House, .Sofia, 244 pp (in Bulgarian).

Dimitrov, T. (1934). Ascertainment on the status of our forests. *Journal of Bulgarian Economic Society, XXXII*(1), 27–39 (in Bulgarian).

EFA. (2010a). Scheme on utilization of forest reproductive material produced in Bulgaria from sources of categories "identified" and "selected", depending on their provenance.

EFA. (2010b). Rules and specific exceptions to the movement of forest reproductive material (FRM) within and between regions of origin.

EFA. (2010c). Map of the regions of origin in the Republic of Bulgaria.

EFA. (2011a). Instruction for establishing and mapping forest habitat types and determining the composition of dendrocoenoses.

EFA. (2011b). Classification scheme of types of forest sites in Bulgaria.

EFA. (2016). *Annual report of the Executive Forest Agency for 2015*. Sofia: MAFF.

Forest Act. State Gazette, No 19/08.03.2011, last amend. and suppl., State Gazette No 13/07.02.2017.

Konnert, M., Fady, B., Gömöry, D., A'Hara, S., Wolter, F., Ducci, F., Koskela, J., Bozzano, M., Maaten, T., & Kowalczyk, J. (2015). *European Forest Genetic Resources Programme (EUFORGEN). Use and transfer of forest reproductive material in Europe in the context of climate change*. Rome: European Forest Genetic Resources Programme (EUFORGEN), Bioversity International xvi and 75 pp.

Milev, M., Petkova, K., & Iliev, N. (2017). Afforestation in Bulgaria. *Reforesta, 3*, 143–154.

MoEW. (2003). Regulation no. 4 of 8 July 2003 on the conditions and procedures for issuing permits for introduction of non-native or reintroduction of native plant and animal species in the wild.

MAFF. (2013). Regulation № 2 of 07.02.2013 on the terms and procedures for afforestation of forest areas and agricultural land used for creating special, protective and commercial forests and forests in protected areas, inventory of created plantations, their accounting and registration.

MAFF, 2005 Regulation No. 14. (2005). Concerns the conditions and procedures for issuing permits for the introduction of non-native or reintroduction of native trees, shrubs and game species in nature and consideration of the public opinion in the region of the reintroduction.

MAFF, EFA. (2011). Regimes for sustainable management of forests in Natura 2000 2011.

National Strategy for Forestry Sector Development (2013).

Protected Areas Act, State Gazette No 133/11.11.1998, updated and amended State Gazette No 66/26.07.2013.

Part III
In Situ Conservation Case Studies

Chapter 13
Preview of the Activities Related to *In Situ* Conservation of Forest Genetic Resources in Serbia

Vladan Popović and Aleksandar Lučić

Abstract The *in situ* (on-site) conservation implies the continuous maintenance of populations in the environment where they have naturally developed. The positive effect of the *in situ* conservation implies the preservation of the ecosystem as a whole and not only the individual species or genes that are the subject of the protection. The preservation of other plant and animal species inhabiting the preserved area can also be achieved by establishing the *in situ* conservation. Establishing *in situ* conservation in an area does not imply strict nature protection systems. The law permits the implementation of activities that will ensure durability and sustainability of genetic resources. The *in situ* is a desirable form of the conservation of the forest genetic resources because it allows the evolutionary processes to take place in the involved populations. The selection and registration of seed stands is the most commonly used form of the *in situ* conservation of the forest genetic resources in Serbia. This form of the conservation of the forest genetic resources is the most common and the only sustainable, because the funds allocated for conservation are modest. According to the data from the current Register of Seed Facilities of the Ministry of Agriculture, Forestry and Water Management of the Republic of Serbia – Forest Administration, 126 seed stands have been selected in Serbia on the total area of 1.842,89 ha, with 44 seed stands of coniferous, 81 stands of herbaceous species and 1 mixed seed stand. A total of 198 seed facilities for the production of the selected reproductive material and 312 seed facilities for the production of the reproductive material of known origin have been registered. The existing number of the seed facilities and the areas they cover are insufficient and do not reflect the richness of the gene pool of our country's forest tree species to which a special attention should be given in the following period. Protected natural resources represent a significant form of the *in situ* conservation of forest genetic resources, because they contain large areas of forest complexes. The protected natural areas currently cover 575.310 ha, or 6.51% of the territory of Serbia. There are 461 protected natural areas (November 2015) which include: 5 national parks, 17 nature parks, 20 land-

V. Popović (✉) · A. Lučić
Institute of Forestry, Belgrade, Serbia
e-mail: vladanpop79@gmail.com; aleksandar.lucic@gmail.com

M. Šijačić-Nikolić et al. (eds.), *Forests of Southeast Europe Under a Changing Climate*, Advances in Global Change Research 65,
https://doi.org/10.1007/978-3-319-95267-3_13

scapes of extraordinary characteristics, 68 nature reserves, 3 protected habitats, 310 nature monuments, 38 areas of cultural and historical importance.

Keywords *In situ* · Conservation · Genetic resources · Serbia

13.1 Introduction

The *in situ* (on-site) conservation implies the continuous maintenance of populations in the environment where they have naturally developed. This type of conservation in Serbia is the most often used in natural populations that are naturally regenerated in the protected natural areas (national parks, nature parks, landscapes of extraordinary characteristics, nature reserves, protected habitats, nature monuments, and areas of cultural and historical significance) and in forests regularly managed. The conservation of the forest genetic resources in forests with the regular management is mainly achieved through the selection of the seed facilities (seed stands, group of trees or individual trees). This form of the conservation of the forest genetic resources is the most common and the only sustainable, because the funds allocated for conservation are modest.

The positive effect of the *in situ* conservation implies the preservation of the ecosystem as a whole and not only the individual species or genes that are the subject of the protection. The preservation of other plant and animal species inhabiting the preserved area can also be achieved by establishing the *in situ* conservation. Establishing the *in situ* conservation in an area does not imply strict nature protection systems. The law permits the implementation of activities that will ensure durability and sustainability of the genetic resources. The *in situ* is a desirable form of the conservation of the forest genetic resources because it allows the evolutionary processes to take place in the involved populations.

The selection of seed stands is the most commonly used form of the *in situ* conservation of the forest genetic resources in Serbia. The seed stand implies a stand of a high level of tree quality, good health status and phenotypic characteristics (Šijačić-Nikolić et al. 2014). The seed stands represent the form of the *in situ* conservation in a place and in conditions where a species naturally occurs. They consist of one or more groups of trees properly arranged and in sufficient number. The seed stands are parts of a forest complex of sufficient uniformity that are selected based on the phenotypic characteristics of trees whose primary purpose is the production of the reproductive material (OECD 2014). To fulfill the basic purpose of the selected seed stands it is necessary to conduct the genetic amelioration including the selection of seed source, thinning and other activities that will increase productivity in them (Mataruga et al. 2010). The quality of seeds and seedlings is improved by removing phenotypic inferior trees from the seed stands (Sivakumar et al. 2011), but as a consequence, the genetic diversity in subsequent generations can be reduced (Lyngdoh et al. 2013).

In order to ensure the proper use of the seed from the seed stands, pursuant to the Article 9 of the Law on Forest Reproductive Material, the Minister of Agriculture,

Forestry and Water Management made the Decision on delineation of regions of provenance for the following forest tree species: beech, Sessile oak, common oak, Austrian pine, Scots pine, Narrow-leaved ash, common spruce and silver fir. For all other forest tree species, the whole territory of the Republic of Serbia is one region of provenance (OG 2004). The regions of provenance are the legal precondition for the trading and the use of the forest reproductive material of the "source-identified" category because this reproductive material can be used for afforestation only in the region of provenance in which it is produced, in order to perform afforestation with the reproductive material adapted to the ecological conditions of the area in which it is used. Also, in this way the preservation and the targeted use of the gene pool on-site (*in situ*) are ensured within the region of provenance.

13.2 *In Situ* Conservation

The existing number of the seed facilities and the areas they cover are insufficient and do not reflect the richness of the gene pool of our country's forest tree species to which a special attention should be given in the following period. According to the data from the current Register of Seed Facilities of the Ministry of Agriculture, Forestry and Water Management of the Republic of Serbia – Forest Administration, 126 seed stands have been selected in Serbia on the total area of 1.842,89 ha, with 44 seed stands of coniferous, 81 stands of herbaceous species and one mixed seed stand. A total of 198 seed facilities for the production of the selected reproductive material and 312 seed facilities for the production of the reproductive material of known origin (FD 2016) have been registered.

The seed stands represent a dynamic form of the *in situ* conservation in which natural selection processes run smoothly as a basis for the species adaptation, which is very important in the context of the climate change. The emphasis should be on the preserving species diversity and ensuring high genetic diversity within species and seed lots, thereby increasing the adaptive potential of the forest reproductive material for future afforestation and forest regeneration (OECD 2014). Increasing the gene flow and strengthening capacity of the tree populations to adapt to the future environmental changes are achieved by increasing the areas covered by forests and planting well-adapted autochthonous species (Maksimović et al. 2015).

The existing number of seed facilities, the representation of tree species and the spatial distribution in the north-south direction are not at satisfactory level, hence it is necessary to select a large number of seed facilities on the sites where they have not been selected yet. The spatial arrangement of sites where the seed facilities are selected should fully include the coeno-ecological, ecological and population diversity of forest species in order to cover the whole genetic potential. The priority in the selection have the seed facilities of the economically and ecologically important tree species that are in a minority or not represented at all, those in areas without registered seed facilities (Toplica, South Morava and Nišava forest areas), as well as the seed facilities of rare and endangered species (Maksimović et al. 2015).

The protected natural areas are selected for the preservation of the best parts of nature. They currently cover 575.310 ha, or 6.51% of the territory of Serbia. There are 461 protected natural areas (November 2015) which include: 5 national parks, 17 nature parks, 20 landscapes of extraordinary characteristics, 68 nature reserves, 3 protected habitats, 310 nature monuments, 38 areas of cultural and historical importance (2016a). The Spatial Plan of the Republic of Serbia (OG 2010) stipulates that around 10% of the territory of Serbia will be protected by 2015, and by 2021 around 12% of the territory of Serbia. The protected natural resources represent a significant form of the *in situ* conservation of the forest genetic resources, because they contain large areas of forest complexes. The breeding measures are conducted within these forest complexes which are under different protection regimes in accordance with the following existing laws: Law on Nature Protection, Law on Environmental Protection and Law on Forests. The significance of the protected natural areas in the conservation and the directed use of the forest genetic resources depends on the share of forest ecosystems in their total area. According to Šijačić-Nikolić et al.: "*Having in mind the share of the areas covered by forest in the total area of the active protection of national parks in Serbia, the diversity of forest ecosystems in them and a large number of identified endemic and relict species, it is clear that national parks represent an important basis for the preservation and targeted use of the forest tree gene pool in situ, on habitats where species naturally occur*" (Šijačić-Nikolić et al. 2006).

There are five National Parks in Serbia, including: NP Fruška gora, NP Djerdap, NP Tara, NP Kopaonik and NP Šar planina. The total area of the National parks is 158,986.36 ha or 1.75% of the territory of Serbia. Their total natural values exceed the borders of our country, hence they are included in the EUROPARC Federation. The oldest national park in Serbia is the NP Fruška Gora, designated in 1960, the largest is the NP Djerdap and the smallest is the NP Kopaonik (INC 2016).

The National Park Fruška gora rises like a green island from the heart of the Pannonia Plain in gentle Vojvodina. It covers the active protected area of 25.525 ha. The flora of NP Fruška gora consists of about 1400 plant species, among which the species of relict and endemic character stand out (for example, Daphne-laurel, pheasant's eye, pygmy iris and more than 30 species of orchids). The forests are mixed or pure and cover 90% of the total area. Monodominant forests consist of Sessile oak, common beech and sometimes linden or hornbeam, while in the structure of the two-dominant forest types most often are represented linden and beech, and rarely hornbeam and Sessile oak. Pure beech or hornbeam forests are rare. The most common forest type in NP Fruška Gora is the forest of Sessile oak and hornbeam with butcher's-broom. Also, the significant place in the vegetation of the deciduous forests has the community of common beech and linden (INC 2016a).

The National Park Kopaonik is located in the central part of the Southern Serbia. With the area of 11.810 ha encompasses the highest parts of this mountain massif surrounded by river valleys of the Ibar, Jošanica, Toplica and Brzećka river. The flora of Kopaonik consists of 1500 plant species, out of which 91 are of endemic and 82 of sub-endemic character which makes this mountain one of the most important centres of biological diversity and endemism in Serbia and Balkans. A special place in this natural wealth belongs to three local endemics: *Sempervivum kopaonicense Pancic, Viola kopaonicensis* and *Cardamine pancicii*. The most widespread forest-pasture dismembered zone of central Serbia is located in Kopaonik. On the higher parts of this mountain can be found spruce and fir forest, and on its slopes beech and oak forest (INC 2016b).

The National Park Tara with the area of 19.200 ha covers the greatest part of the Tara Mountain situated in the westernmost part of Serbia, in an elbow shaped meander of the river Drina between cities of Visegrad and Bajina Basta. Forest ecosystems occupy about 70% of its total territory hence its importance for preserving forest genetic resources is immeasurable. This forest area with its diversity and preservation represents one of the richest and most valuable forest areas of Europe, with more than 1000 plant species. Numerous forest phytocoenoses of a quite complex structure, very rich from the floristic point of view, with a large number of relict and endemic plant species, hide true natural rarities such as Serbian spruce (*Picea omorika*), European yew (*Taxus baccata*), European holly (*Ilex aquifolium*), *Daphne blagayana, Dafne blagayana Freyer*, Rustyback (*Asplenium ceterach*), cyclamen, gentiana etc. The habitats of these species indicate that many parts of Tara, especially the gorges and canyons are unique mountain habitats where many communities of relict character have preserved completely or to a greater extent their original, ancient characteristics. The mixed forests of silver fir, common beech and common spruce prevail in Tara with individual trees or smaller groups of other conifers such as pine and spruce and broadleaves such as maple, aspen, birch etc. Due to richness of species, the complex Tara represents a real living archive of the flora which is characteristic for the great part of the Balkan Peninsula and the reserve of the gene pool of the European and planetary importance (INC 2016c).

The National Park Djerdap covers the area of 63.680 ha, of which 44.851 ha are covered by forests (Medarević 2005). It is situated on the right bank of the Danube river in the Djerdap Gorge about 130 km long, between the small settlements Golubac and Kladovo. The flora of NP Djerdap is not distinguished only by its great diversity and richness, but also by its distinctive relic character. Over 11.000 plant species have been identified in the area of the National park, among which are particularly significant the tertiary relict species that survived the Ice Age and still exist: Turkish hazel, walnut tree, European bladdernut (*Staphylea pinnata*), *Acer monspessulanum*, lilac (*Syringa*), European nettle tree (*Celtis australis*), *llex aquifolium L., Daphne blagayana* and European yew (*Taxus baccata*). Ancient relict and expansive postglacial species of the following genera can also be found nowadays in the Djerdap Gorge: oak, maple, ash, elm, linden, hawthorn, hornbeam etc. They make a unique and unrepeatable arboretum of the Djerdap Gorge and record the history of the flora from the tertiary period to nowadays. In this area a total of 50 forest and shrub communities have been identified, of which 35 have a relict character (INC 2016d).

The National Park Šar planina (Sar Mountain) spreads at southmost of Serbia, along the border with FYR Macedonia, on the area of 39.000 ha. It represents one of the most important centres of biodiversity on the Balkan Peninsula and covers about 56% of Serbia's flora with about 2000 species of vascular plants and 20 local endemic species among which the best known are Serbian phoenix flower (*Ramonda serbica*), Macedonian Pine (*Pinus peuce*), Bosnian pine (*Pinus heldreichii*) and Macedonian oak. Forests of oak, beech as well as mixed forests of beech and fir, spruce and fir spread to the belt of the forests of Macedonian Pine (*Pinus peuce*) and Bosnian pine (*Pinus heldreichii*) that form the upper forest border (INC 2016e).

The most important local and regional centres of the ecosystem diversity in Serbia undoubtedly are large mountain massifs (Kopaonik, Sar Mountain, Stara Planina (meaning "The Old Mountain"), Tara, Suva planina (meaning "The Dry mountain") etc.), where are represented almost all types of altitudinal zonation of ecosystems in the Southeast Europe. Also, the refugial canyon spaces such as the Djerdap Gorge and the Drina canyon are the areas of exceptional beauty, ecosystem richness and diversity in which the largest number of endemic, relict or endemorelict communities are distributed.

A detailed study of phytocenoses has been performed in the Kopaonik mountain massif and the results showed the presence of 119 plant associations and 39 subassociations classified in 62 alliances, 41 lines and 24 classes within 6 altitudinal vegetation belts. Hence, a total of 285 different syntaxonomic vegetation units were recorded. The gorges and canyons of the eastern and the western Serbia represent the most important refugiums of polydominant ecosystems on the Balkan Peninsula. For example, 45 relict forest and bush phytocenoses of polidominant or oligodominant type have been found just in the area of the Djerdap Gorge. Relict or endemorelict representatives of dendroflora can be found in these communities, including: walnut (*Juglans regia*), Turkish hazel (*Corylus colurna*), European nettle tree (*Celtis australis*), *Acer intermedium*, *Acer monspessulanum*, lilac (*Syringa vulgaris*), European yew (*Taxus baccata*), Oriental hornbeam (*Carpinus orientalis*), silver linden (*Tilia tomentosa*), manna ash (*Fraxinus ornus*), mahaleb cherry (*Prunus mahaleb*), European smoketree (*Cotinus coggygria*), *Quercus virgiliana* etc. (Lakušić 2001).

References

FD. (2016). *Directorate of forestry*. http://www.mpzzs.gov.rs/registri-pumskih-semenskih-objekata/. Visited: January 25, 2016.

INC. (2016). *Institute for nature conservation of Serbia*. http://www.zzps.rs/novo/index.php?jezik=sr&strana=zastita_prirode_osnovni_podaci. Visited: January 11, 2016.

INC. (2016a). *Institute for nature conservation of Serbia*. http://www.zzps.rs/novo/index.php?jezik=sr&strana=zastita_prirode_zp_np_fruska_gora. Visited: January 13, 2016.

INC. (2016b). *Institute for nature conservation of Serbia*. http://www.zzps.rs/novo/index.php?jezik=sr&strana=zastita_prirode_zp_np_kopaonik. Visited: January 13, 2016.

INC. (2016c). *Institute for nature conservation of Serbia*. http://www.zzps.rs/novo/index.php?jezik=sr&strana=zastita_prirode_zp_np_tara. Visited:January 13, 2016.

INC. (2016d). *Institute for nature conservation of Serbia*. http://www.zzps.rs/novo/index.php?jezik=sr&strana=zastita_prirode_zp_np_djerdap. Visited: January 13, 2016.

INC. (2016e). *Institute for nature conservation of Serbia*. http://www.zzps.rs/novo/index.php?jezik=sr&strana=zastita_prirode_zp_np_sar_planina. Visited: January 13, 2016.

Lakušić, D. (2001). Diversity of the ecosystem of Yugoslavia. Biodiversity and the new millennium (pp. 15–28). Society of Ecologists of Serbia, Institute for Nature Protection. (Original title: Diverzitet ekosistema Jugoslavije. Biodiverzitet i novi milenijum (pp. 15–28). Društvo ekologa Srbije, Zavod za zaštitu prirode).

Lyngdoh, N., Joshi, G., Ravikanth, G., Vasudeva, R., & Shaanker, R. U. (2013). Changes in genetic diversity parameters in unimproved and improved populations of teak (*Tectona grandis* L.f.) in Karnataka state, India. *Journal of Genetics, 92*(1), 141–145.

Maksimović, Z., Šijačić-Nikolić, M., Aleksić, P., & Sovilj, Lj. (2015). Seed facilities in PE "Srbijašume" as a basis for conservation and directed use of gene pool (Bulletin of the Faculty of Forestry 111). University of Belgrade–Faculty of Forestry, Belgrade, 63–82 (Original title: Semenski objekti u JP „Srbijašume" kao osnova za konzervaciju i usmereno korišćenje genofonda, Glasnik Šumarskog fakulteta 111, Univerzitet u Beogradu-Šumarski fakultet, Beograd, 63–82).

Mataruga, M., Daničić, V., & Cvjetković, B. (2010). Theoretical basis of the importance of implementing genetic ameliorations in the function of seed stand arrangements. *Journal of Forestry, 62*(3–4), 101–110 (Original title: Teorijske osnove značaja sprovođenja genetskih melioracija u funkciji uređenja sjemenskih sastojina. Šumarstvo, 62, 3–4, 101– 110).

Medarević, M. (2005). Forest types in the National Park "Djerdap", Faculty of Forestry, University of Belgrade, Ministry of Science and Environmental Protection of the Republic of Serbia and NP "Djerdap", Belgrade, pp. 1–158. (Original title: Tipovi šuma nacionalnog parka "Đerdap", Šumarski fakultet Univerziteta u Beogradu, Ministarstvo za nauku i zaštitu Životne sredine Republike Srbije i NP "Đerdap", Beograd, 1–158).

OECD. (2014). Rules and regulations of the OECD forest seed and plant scheme.

OG. (2004). Law on forest reproductive material, "Official Gazette of RS" Nos. 135/04, 8/05, 41/09.

OG. (2010). Spatial plan of the Republic of Serbia, "Official Gazette of RS", No. 88/10.

Šijačić-Nikolić, M., Isajev, V., & Ivetić, V. (2006). National parks of Serbia–preservation and directed use of the forest tree gene pool. In Scientific conference: Management of forest ecosystems of National Parks and other protected areas, proceedings, Jahorina-NP Sutjeska, 5–8. July, 131–136. (Original title: Nacionalni parkovi Srbije – oblik očuvanja i usmerenog korišćenja genefonda šumskog drveća. Naučna konferencija: Gazdovanje šumskim ekosistemima nacionalnih parkova i drugih zaštićenih područja, Zbornik radova, Jahorina-NP Sutjeska, 5–8. jul, 131–136).

Šijačić-Nikolić, M., Milovanović, J., & Nonić, M. (2014). Forest genetic resources in Serbia–condition and proposals for improvement of this field (Bulletin of the Faculty of Forestry special edition, pp. 51–70). Belgrade: University of Belgrade–Faculty of Forestry. (Original title: Šumski genetički resursi u Srbiji – stanje i predlozi za unapređenje ove oblasti, Glasnik Šumarskog fakulteta specijalno izdanje, Univerzitet u Beogradu–Šumarski fakultet, Beograd).

Sivakumar, V., Gurudevsingh, B., Anandalakshmi, R., Warrier, R. R., Sekaran, S., Tigabu, M., & Odén, P. C. (2011). Culling phenotypically inferior trees in seed production area enhances seed and seedling quality of *Acacia auriculiformis*. *Journal of Forestry Research, 22*(1), 21–26.

[illegible]

Chapter 14
Conservation of European White Elm and Black Poplar Forest Genetic Resource: Case Study in Serbia

Jovana Devetaković, Dijana Čortan, and Zoran Maksimović

Abstract European white elm (*Ulmus laevis* Pall.) and black poplar (*Populus nigra* L.) are species which occur in riparian forests and present the most significant types of forest trees in the alluvial habitats. Both species are characterized by their broad range and some type of vulnerability. Primarily due to the changes in their natural habitats caused disappearance of this species across Europe, also as Dutch elm disease at European white elm and hybridization of black poplar with poplar clones. Conservation is recognized as necessary and there is more activity in Europe in the aim. We give the review of activity on the conservation both species in Serbia, and especially present population of European white elm and black poplar at Veliko ratno ostrvo Island as very good candidates for in situ conservation.

Keywords *Ulmus laevis* · *Populu nigra* · Conservation

14.1 Introduction

European white elm (*Ulmus laevis* Pall.) and black poplar (*Populus nigra* L.) are species which occur in riparian forests, plains and river valleys (Jovanović 2007). The elm grows alongside poplar, but it can be found in communities with common oak and narrow-leafed ash, which constitute the group of riparian hardwoods (Jović et al. 2009; Härdtle et al. 1996). Black poplar is one of the most significant types of forest trees in the alluvial habitats (Cagelli and Lefèvre 1995; Smulders et al. 2001;

J. Devetaković (✉)
Faculty of Forestry, University of Belgrade, Belgrade, Serbia
e-mail: jovana.devetakovic@sfb.bg.ac.rs

D. Čortan
Faculty of Education, University of Novi Sad, Novi Sad, Sombor, Serbia
e-mail: dijanacortan@yahoo.com

Z. Maksimović
SE "Srbijašume", Belgrade, Serbia
e-mail: zoran.maksimovic@srbijasume.rs

M. Šijačić-Nikolić et al. (eds.), *Forests of Southeast Europe Under a Changing Climate*, Advances in Global Change Research 65,
https://doi.org/10.1007/978-3-319-95267-3_14

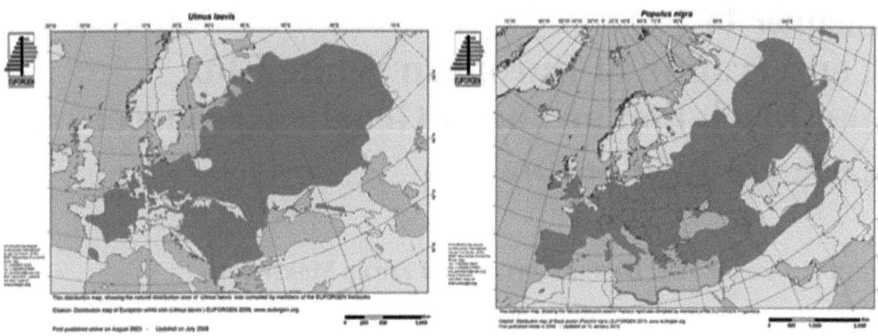

Fig. 14.1 European white elm (left) and black polar (right) distribution (EUFORGEN – http://www.euforgen.org)

Fossati et al. 2003; Storme et al. 2004; Pospíšková and Šálková 2006), which is reflected in the broad ecological valence (Kajba et al. 2005). In floodplain ecosystems it forms meta-populations with the planting of the seeds in the fresh alluvial deposits (Herpka 1986; Vanden Broeck et al. 2003). The black poplar is characterized by a great diversity of population types, from isolated trees to large, pure or mixed stands which are distributed over a vast area (Lefèvre et al. 1998; Toplu 2005). Both species are characterized by their long life span with over 100 years in some habitats (Imbert and Lefèvre 2003; Jovanović 2007) while individual trees may live up to 400 years (Popivshchy et al. 1997; Toplu 2005; Jovanović 2007).

Both species are characterized by their broad range, which stretches from Central and East Europe, through Asia and also parts of Siberia. The black poplar range is slightly broader, covering the entire West and South Europe, North Africa and Asia Minor. The black poplar range is quite vast and disjunctive (Romanić 2000). The natural range of black poplar encompasses the larger part of West Europe (with the exception of Scandinavia, Ireland and Scotland), Ukraine, the Russian Federation, part of North Africa, Asia Minor and northern part of Iran (Toplu 2005). The elm range covers the eastern part of France, through Central Europe to central part of Russia in the east. The northern border is the Baltic Sea and Siberia, and the southern borders are the Iberian and Balkan Peninsula (Fig. 14.1).

14.2 European White Elm and Black Poplar Populations and the Need for Conservation

European white elm and black poplar as species are characteristic to riparian forests in Europe and fall into the category of endangered species primarily due to the changes in their natural habitats. Their natural habitats are decreasing under the pressure of agricultural and forestry activities as well as the pressure of urbanization.

The elm is most commonly found in small fragmented populations, especially on the outskirts of its range (Collin et al. 2000). This populations structure is to a point dependant on the type of the habitat the elm is in (Whiteley 2004), i.e. the limited area of riparian forests along the river banks. However, there is a noticeable trend of the decrease in the number and size of elm populations in its entire range. In addition to the changes in the humid habitats due to the cultivation of poplar and agricultural production (Collin 2003; Hooke 2006), the reduction in the population of European white elm as well as other elm types was also caused by the Dutch elm disease (DED) (Brasier 2000; Mittempergher and Santini 2004; Karadžić 2010). Some research showed that the European white elm is more resilient to DED than other types of European elms (Sacchetti et al. 1990. Webber 2000; Pajares 2004), as well as that numerous European white elm populations in Central and Easter Europe were affected by this disease in the past (Collin 2003). Karadžić (2010) state that, among elm species growing in Serbia, the European white elm shows more resilience than the field elm but also that it is less resilient than the wych elm. These factors have led to the reduction in the number of populations and trees in these populations. It is thought that only 1% of the older elms remain in Germany (Mackenthun 2004). The situation is similar in Belgium (Flanders) where the European white elm exists as individual trees surrounded by other agricultural species (Vander Mijnsbrugge et al. 2005). This trend in the reduction of the numbers of elm trees and the further defragmentation of the European white elm populations has resulted in the need to conserve this species (Eriksson 2001; Koskela et al. 2003; Collin et al. 2004). The European white elm is on the endangered species list in several European countries (REFORGEN 2003). In Serbia, the European white elm is classified as a rare and endangered species (Banković et al. 2009).

The black poplar is one of the most endangered autochthonous tree species in Europe (Lefèvre et al. 2001; Vanden Broeck et al. 2002; Tabbener and Cottrell 2003; Pospíšková and Šálková 2006; Bordács and Bach 2014). Lefèvre et al. (1998) list three factors as being the basic threat to the existence of the natural populations of black poplar. The first factor pertains to changes in the natural habitats of this species due to human activity. The regulation of the river flow leads to irregularities in the natural regeneration of softwood species and enhances replacement by hardwood species. The second factor pertains to the extensive exploitation of the autochthonous types of poplar at the beginning of the twentieth century and the establishment of fast-growing hybrid poplars plantations with the aim to meet the needs of the human population for softwood species timber. And finally, the introgression of the cultivated clones of other types of poplar into the existing gene pool is seen as the potential threat to the natural populations of black poplar. In conjunction with the exotic poplar species, different varieties which are not autochthonous and are present all over Europe, pose a threat to the gene pool and the sustainability of natural populations of black poplar. In addition to the above-mentioned, it was concluded that the European black poplar (*Populus nigra* L.) is slowly disappearing due to its sensitivity to bark canker (*Dothichiza populea*) and diseases of the leaf (*Melampsora* spp. and *Marssonina brunnea*) (Orlović et al. 2005).

Native populations of autochthonous black poplar can be in danger of extinction due to genetic assimilation if they are found in the immediate surroundings of the widely spread perennial plantations of hybrid poplars. Individual trees in small, isolated populations in contact with other taxa are more prone to hybridization, if for no other reason, than due to the fact that there are no same-species individuals in the immediate surroundings. The black poplar is one of the most endangered species in whole Europe. In Belgium, along the Meuse River valley it is completely extinct (van Slycken 1995) and situation is similar in Netherlands as well (de Vries 1995). In Great Britain with just 7000 recorded black poplar trees situation is alarming situation is alarming (Cooper et al. 2002; Preston et al. 2002). In Austria, along the Danube decrease has been evident since 1960 (Heinze 1997). Black poplar is classified as a "vulnerable" species and is on the National Red List (BfN 2008). In addition, due to urbanization and bad management, the black poplar is now one of the most endangered species in Turkey, where it once had wide geographic distribution (Toplu 2005).

Extensive research was done for the purpose of estimating the genetic diversity of European white elm and black poplar in European populations. First studies on European white elm populations in France were conducted through the application of allozymes and have shown a low level of genetic diversity in relation to other elm species (Machon et al. 1995). The application of allozymes determined high genetic differentiation of examined populations as well as isolation during pollination by other elm species, which most likely caused the decrease in the level of genetic diversity (Machon et al. 1997). Low genetic variability in the population and the high degree of genetic differentiation among populations in Finland were also determined with the application of allozymes (Mattila and Vakkari 1997), while Gehle and Krabel (2002), determined a certain degree of variability of allozymes between and among populations in Germany. Subsequently developed microsatellite markers were used in further research of intrapopulation and interpopulation variability, which provided somewhat different results (Whiteley et al. 2003; Collada et al. 2004; Zalapa et al. 2008). The analysis of the chloroplast DNA determined a low degree of genetic diversity, with the differentiation of only three haplotypes in European white elms (Whiteley 2004), while Petit et al. (2003), in studies which included 25 populations of the field and wych elm in Europe found 41 haplotypes. The variability of the morphological features of the European white elm in Serbia was studied on population at Ada Ciganlija (Belgrade) and Klenak near Šabac. The variability of European white elm seed yield and leaves was the subject of these studies (Isajev 1978). In the second half of the twentieth century, studies on the variability of leaves within the *Ulmus* genus were done all over Europe (Jeffers 1999).

Studies of black poplar populations based on molecular markers have shown that the genetic diversity is most prominent in southern countries (France, Italy, and Spain) and lowest in Britain (Storme et al. 2004). In a genetic study based on isoenzymes, Legionnet and Lefèvre (1996) determined less diversity of black poplar in French populations in relation to the populations in Central Europe, which also indicates less diversity in the marginal parts of its range. The variability of black poplar was studied all over Europe as part of the EUROPOP project (van Dam 2002). This project included 9 countries and 6 river basins (Danube, Drôme, Ebro, Rhein, Ticino and Usk), with a total of 23 populations. Research was conducted on natural populations and within the existing gene banks of each participating country and was based on several genetic markers (morphological, isoenzymes, AFLP, cpDNA and SSR). In terms of these markers, great variability was determined on both the intra-population and the inter-population level as well as between different river basins. The variability of the morphological features of black poplar in Serbia as well as in neighboring countries was studied in the natural populations along the Danube in Serbia (Tucović 1965), along Sava and Drava in Croatia (Krstinić et al. 1997; Kajba et al. 1999; Romanić 2000), along Neretva and Vrbas in Bosnia and Herzegovina (Kajba et al. 2004; Ballian et al. 2006), along Sava and Mura in Slovenia (Brus et al. 2010), along the Danube, Sava and Tisza in AP Vojvodina (Čortan 2015). These studies mainly dealt with the intra-population and inter-population variability of black poplar leaves with the use of morphological markers, while the studies on black poplar done in AP Vojvodina also included the application of molecular markers (Čortan 2015).

14.3 Selected Populations of European White Elm and Black Poplar in Serbia

In the total forest fund of the Republic of Serbia, autochthonous poplar forests account for 0.5% up to 1% and European white elm accounts for less than 1% so these species fall in the category of rare/endangered species (Banković et al. 2009). The European white elm tree does not have significant economic value (Vajda 1952) and in terms of economic aspects in forestry this has long been a neglected species. On the other hand, from the perspective of economics, the poplar represents an important aspect of forestry. In Europe, black poplar plantations cover large areas (Coaloa and Nervo 2011), however; in the course of the establishment of plantations the autochthonous poplar species were neglected in favour of different poplar clones (Servo et al. 2017). From the economic and management aspect, the black poplar was completely marginalized in the last few decades (Ballian and Mekić 2008). With the lack of natural regeneration and the aging and cutting of existing marsh forests, the process of degradation of natural stands. Given the fact that poplar is not regenerated by the seeds under the canopy, selective cutting or even by random planting of seed from another location, the regeneration of poplar forests was done through clean cut, by sprouts from stools and roots. This resulted in the degradation of the cultivated forms of high poplar seed stands to the cultivated low forms, the poplar coppice. After a few generations, the poplar coppice forests lost their vitality and were merely the degradation phase of the stands of the autochthonous poplars. As a result, the coppices were gradually transformed to a completely new cultivation form, the hybrid clones of Euro-American poplars (Romanić 2000). With increased contact of the autochthonous poplars with the cultivated plantations, opportunity arose in terms of production of hybrid seeds (Levin et al. 1996), which in interaction with the fragmentation of the habitat caused an erosion of genetic diversity and the disappearance of pure, natural black poplar population all over

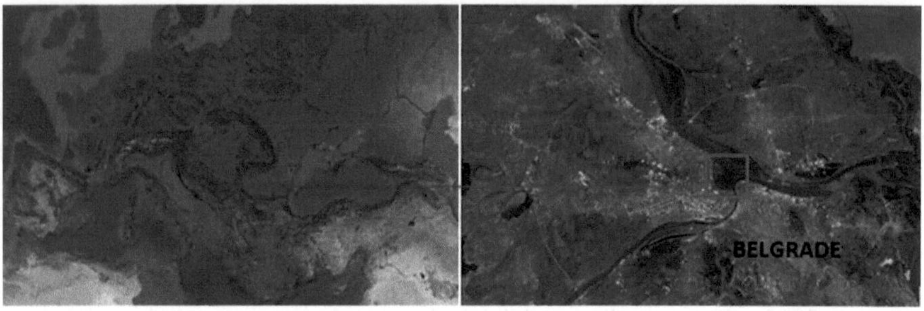

Fig. 14.2 Position of the Veliko ratno ostrvo Island

Europe, as well as in Serbia. In Serbia, only a small part of natural black poplar stands was preserved (Orlović et al. 2005). With continuous human activities within their natural habitats, the majority of the black poplar range in Serbia was downsized to individual, mutually separated areas.

While keeping in mind the status of the European white elm and black poplar in the forest fund of Serbia, an estimate of the status of these populations, in terms of genetics, was performed. The paper presents the overview of activities conducted on the European white elm and black poplar populations on the river island Veliko ratno ostrvo which is located at the river mouth of the Sava and Danube Rivers (20°26′00″ N, 20°27′10″ E), near Belgrade (Fig. 14.2). The central part of the island is permeated by the Galijaš Canal. The island originated as sediment and an alluvial-accumulation deposit, due to the slowing down and complete cessation of the sand deposits at the river mouth of the Sava and Danube. The Veliko ratno ostrvo Island has characteristics of plains, with the average ground level of 72 m and the entire surface of the island is slightly tilted towards the east, in the direction of the Danube River flow. The total land surface of Veliko ratno ostrvo Island is 207.11 ha, out of which 119.41 ha is covered by forest and the island is classified as a protected natural area. In terms of floristic composition, the Veliko ratno ostrvo Island is quite abundant, since the wealth of water and heat enabled lush vegetation in all vertical layers of the vegetation. Since its appearance and then further on, the island remains under the impact of the rivers it is surrounded and occasionally flooded by as well as under the impact of groundwater level. Depending on the moisture regime, a gallery-like distribution of forests can be observed; where rows of forests are distributed and changed, depending on the air humidity percentage. The following types of forests are distinguished in the management unit:

- Willow forest type (*Salicetum albae Issl.26*) on the β –gley;
- Willow forest type (*Salicetum albae Issl.26*) on fresh, moist and multiple-layer alluvial deposit;
- White and Black poplar forest type (*Populetum albo-nigra Slav.52*) on the mosaic of different alluvial soils.

In this area, the European white elm appears in three forest types, while black poplar grows mostly in communities with white poplar.

The presence of black poplar and European white elm in this locality is of utmost importance for the preservation of sensitive hygrophilic forests. The population of black poplar in this area is around 907 trees (Maksimović et al. 2016) and the population of European white elm is only 89 trees (Devetaković 2017), which are irregularly dispersed on the entire surface of the island (Fig. 14.2). The health condition of the population was classified as good. All registered trees are vital; they do not show signs of diseases and have not prominent mechanical damages. The crowns of the large number of trees were classified as good; with the side note that with a large number of trees the crown is to wide or eccentric which is caused by the sparse canopy. The presence of bushes and weed is extensive for the largest portion of the island, so the European white elm and black poplar trees are most commonly surrounded by a very dense layer of indigo bush and blackberry bush which prevent the natural regeneration. Those trees which are located in weed-free areas are mostly the first ones to be hit by flood waves which take away the seeds as well, so the sprouts under them are not recorded. The age structure of these populations points to different ages, which can be considered an asset in conservation terms. Mixing of genetic materials from trees from different generations helps preserve the genetic diversity (Kang et al. 2005) by increasing the adaptation potential (Ivetić et al. 2016) and enabling the transfer of rare alleles to the next generation (Adams et al. 1998).

14.4 Researches Conducted on Populations of European White Elm and Black Poplar

Genetic diversity ensures the basis for adaptation and resilience to stress and change of habitat, making it necessary for long-term forest survival (Booy et al. 2000; Schaberg et al. 2008). The term genetic diversity refers to the appearance of individuals of the same species not having the same genotype, making them different in appearance and behaviour (Koski 2000). Great genetic variability and high probability of selection allow natural selection to result in adaptation (Savolainen et al. 2007). The more variability the higher the chance that at least some individual will have the allele variant which is favorable for new, altered environmental conditions. Low genetic diversity produces stronger and less predictable interaction of the population with the environment, while high genetic diversity ensures greater and safer biological production (Lindgren 2016). Even though the trees exhibit high degrees of genetic diversity, most variability of forest species can be found within the population, while the interpopulation component of variability rarely overpasses 5% (Leonardi and Menozzi 1995; Larsen 1996). Assessment and description of population genetic diversity bears great importance for investing effort into species protection (Lefèvre et al. 2001). Population genetic diversity was examined on the level of morphological and molecular markers, while including testing of posterity.

The populations of European white elm and black poplar were also examined for variability in the following morphological markers: black poplar leaves (Maksimović and Šijačić-Nikolić 2013), European white elm leaves (Devetaković and Šijačić-Nikolić 2013) and variability of fruits with seed (Nonić et al. 2012) among selected test trees. In all cases, a high degree of intrapopulation variability was noted, which confirms with previous research finding a high degree of intrapopulation variability, in most cases higher than interpopulation variability, for poplar (Krstinić et al. 1997; Romanić 2000; Čortan et al. 2013, 2014; Kovačević 2014), elm (Zebec et al. 2010, 2014, 2015) but also for other deciduous species (Franjić 1996; Kajba 1996; Bašić et al. 2007; Ballian et al. 2010, 2014; Poljak et al. 2014; Popović and Kerkez 2016). Morphological characteristics of trees in the population, but also the frequency and abundance of fruit-bearing are under great influence of the environment the tree grows in (Broshtilov 2006; Šrámek and Čermák 2012; Poljak et al. 2014), and they cannot provide the complete picture of the genetic diversity of the population. Testing of genetic progeny for European white elm in the first (Devetaković et al. 2013), and in the following years (Devetaković 2017) have presented a high degree of variability among different half-sib lines, which conforms with previous research for seedlings of various species of conifer and deciduous trees (Mathew and Vasudeva 2005; Jacobs et al. 2006; Šijačić-Nikolić et al. 2008; Woeste et al. 2011; Devetaković et al. 2013; Popović et al. 2013).

Existence of a high level of genetic diversity was confirmed on a molecular level by application of microsatellite molecular markers characteristic of European white elm and black poplar. Population of European white elm on the Veliko ratno ostrvo Island was tested with five microsatellite markers with the obtained results pointing to a high degree of genetic variability (Devetaković 2017). The obtained values of the expected heterozygosity (0.555), point to a higher level of genetic diversity in respect to some of the examined populations in the Netherlands (0.5) and Denmark (0.5), (Nielsen and Kjær 2010a), central Europe (0.513) and the Iberian peninsula (0.49), (Fuentes-Utrilla et al. 2014), but also lower values in respect to some other wind-pollinated species (*Quercus petraea*: He>0.7 in Ireland (Muir et al. 2004), *Ulmus glabra*: He>0.7 in Denmarks (Nielsen and Kjær 2010b). Low genetic diversity by application of allozymes was detected in populations of European white elm in Finland (Vakkari et al. 2009) and in France (Machon et al. 1997). However, it should be kept in mind that the estimates of the level of genetic diversity based on allozymes is always of lower values obtained by application of nuclear microsatellites. The inbreeding coefficient in European white elm population on the Veliko ratno ostrvo Island shows there is no crossing in relation (−0.177).

To obtain information on the degree of European white elm population isolation on the Veliko ratno ostrvo Island when compared to the surrounding populations, the same locus set was used to test a smaller number of trees from the nearest European white elm population. The second population is also located in Belgrade, on the river island of Sava known by the name of Ada Ciganlija numbers around 120 European white elm trees. The obtained results show that the European white elm populations from the Veliko ratno ostrvo Island and Ada Ciganlija are not genetically differentiated and can be considered as one population. This points to high pollen and seed flows between these populations (Devetaković 2017). The research of the degree of differentiation between populations in Europe by applications of microsatellite markers have shown a high degree of differentiation among populations across central Europe and the Iberian Peninsula, as well as inside these areas (Fuentes-Utrilla et al. 2014), while similar results of high differentiation among populations in Finland by application of allozymes were also obtained by Vakkari et al. (2009). The mentioned results are in accordance to the findings of Nielsen and Kjær (2010b) who report that the medial distance for scattering of pollen in conditions of Denmark is very small (<40 m), but also that in certain cases of pollen scattering this distance can be 1000–1200 m and even more. The observed populations are located at greater distances, certain trees even within the populations are located at greater distances, however the absence of physical barriers preventing pollen and seed flow, as well as the direction of winds characteristic of Belgrade, all speak in favour of the obtained results. According to TPM ($p = 0.9375$) model of microsatellite evolution, European white elm population on the Veliko ratno ostrvo Island was not exposed to a sudden decrease during the last 2Ne-4Ne generations, meaning it did not experience a "bottleneck" in the recent past (Devetaković 2017).

The research at the level of chloroplast genome for European white elm in population of the Veliko ratno ostrvo Island has shown a high degree of variability (Devetaković 2017), which is opposite to available data pointing to the existence of only three haplotypes for European white elm (Whiteley 2004; Fuentes-Utrilla et al. 2014). Low detected haplotype levels for European white elm did not conform with numbers of haplotypes of European elm, nor for most deciduous species in Europe (Petit et al. 2003), however this cannot be the consequence of certain genetic and demographic processes which led to low levels of genetic variability for European white elm. The black poplar population on the Veliko ratno ostrvo Island was tested by application of 11 microsatellite loci specific for black poplar. The obtained values of observed (0.702) and expected heterozygosity (0.822) in the population at Veliko ratno ostrvo Island are high (Maksimović et al. 2014) and according to their values are close to results obtained by Čortan (2015) during the research of population diversity in the area of Vojvodina (Northern Serbia).

The research of 12 populations of black poplar along the Danube in Europe (Jelić et al. 2014) established a high level of observed and expected heterozygosity. The mean value of observed heterozygosity for researched populations ranged from 0.66 (Neuburg-Ingolstadt in Germany and the Danube delta in Romania) to 0.74 (Danube Ipoli in Hungary), with the populations of observed heterozygosity being close to each other, as well as the black poplar population in the area of the Veliko ratno ostrvo Island. The expected heterozygosity ranged from 0.76 (Ferto-Hansag population in Hungary) to 0.82 (Danube Ipoli in Hungary), with no greater deviations registered between populations similar to the researched population. A genetic diversity research in black poplar populations along river basins in Europe, Smulders et al. (2008) registered highly observed (0.74, ranging from 0.59 to 0.82) and expected heterozygosity (0.76, ranging from 0.477 to 0.907) which was similar to the researched population. Rathmacher et al. (2010) while studying genetic diversity of black poplar along the Eder river in Germany found high genetic diversity of older black poplar trees (the obtained heterozygosity 0.7, ranging from 0.62

to 0.78; expected 0.73, ranging from 0.58 to 0.80). The observed and expected heterozygosity in the researched population was comparable to the genetic diversity in four populations of black poplar in the Morava river basin in the Czech Republic (Pospíšková and Bartáková 2004) where high average values of obtained (0.79) and expected heterozygosity (0.83) were noted. Similar results were obtained by Imbert and Lefèvre (2003) who studied genetic diversity of black poplar populations along the Drôme River in France and established high genetic diversity of the researched populations, with observed heterozygosity for every locus in each population higher than the researched population (0.78, ranging from 0.71 to 0.85), while the expected heterozygosity was somewhat smaller in comparison with the population of black poplar in the area of the Veliko ratno ostrvo Island (0.73, ranging from 0.65 to 0.80). A little more observed heterozygosity in comparison to the researched population was established in black poplar population along the Morava river in Czech Republic (Pospíšková and Šálková 2006), where observed heterozygosity ranged from 0.70 to 0.92 (medium value 0.80) depending on locus. The expected heterozygosity in this population was in accordance with the researched population and ranged from 0.70 to 0.90 (medium value 0.82).

Comparison of average values of observed and expected heterozygosity shows that the researched population in the area of the Veliko ratno ostrvo Island is the most similar concerning genetic variability to the black poplar populations in the area of Vojvodina, while sharing most similarity with the population in the lower course of the Danube and along Sava, being physically the closest. The researched populations share great similarities concerning genetic variability with populations along the Danube, where the Gornje Podunavlje region in Serbia and Lonjsko polje in Croatia, while the similarity with this population established by application of morphological markers. High heterozygosity levels of black poplar in the researched population and in other populations in Europe conform to expectations, as this is cross species hybridization. High diversity in populations of black poplar Pospíšková and Bartáková (2004) is explained by high levels of polymorphisms of microsatellite markers, as well as the fact that poplar is a dioecious species with present outcrossing with pollen and seed being dispersed by wind and water on great distances.

The medium inbreeding coefficient value calculated for locus bears a positive value slightly higher than zero (0.143) which points to the existence of an addition of homozygotes in the researched population (Maksimović et al. 2014). The positive inbreeding coefficient value shows that inbreeding exists, but is not yet significantly present, making intrapopulation variability still significant. Similar average inbreeding coefficient values were noted in four black poplar populations in the area of Vojvodina (Čortan et al. 2016), as well as 12 black poplar populations along the Danube (Jelić et al. 2014). Low addition of homozygote individuals was established in the black poplar population along the Morava river in the Czech Republic (Pospíšková and Šálková 2006), where 6 out of 12 locus presented additional homozygotes, out of which only three locus showed significant deviation from balance. In opposition, one locus presented a significant addition of heterozygotes.

Rathmacher et al. (2010) establish a positive inbreeding coefficient value for adult and juvenile black poplar population along the Eder river in Germany, with the adult population exhibiting additional heterozygotes in 4 to 6 locus, out of which only two locus presented significant deviation from balance. Similar results were obtained by Smulders et al. (2008) who noted a low addition of homozygous individuals. In opposition to this research, other research based on microsatellite markers show additional heterozygotes (van Dam 2002; Imbert and Lefèvre 2003; Pospíšková and Bartáková 2004). Based on "bottleneck" analysis with the TPM model ($P = 0.0081$), which best reflects the microsatellite assessment in most cases, black poplar population of Veliko ratno ostrvo Island has recently undergone a bottle neck (i.e. in the last 2 Ne–4 Ne generations). This fact points to the need for conservation of the remaining gene pool of black poplar in this locality or the need to define adequate conservation strategy and focus the use of the remaining gene pool (Maksimović and Šijačić-Nikolić 2016).

An overview of all research conducted on populations of European white elm and black poplar on the Veliko ratno ostrvo Island establishes a high degree of genetic variability. Populations are located in range of researched populations in Europe, and in some cases, suggest the existence of a somewhat greater diversity in comparison to European populations.

14.5 Proposed and Implemented Conservation Measures for European White Elm and Black Poplar Populations on the Veliko Ratno Ostrvo Island

Population of European white elm on the Veliko ratno ostrvo Island provides a good basis for species conservation. Results on molecular-genetic analysis show that the European white elm population from the Veliko ratno ostrvo Island presents a high level of genetic diversity on the nuclear genome level. It was concluded that there is no inbreeding or related crossbreeding in the population and that it did not undergo sudden decreases in numbers in the recent past (Devetaković 2017), and based on the number of trees, determines that this population is a good candidate for in situ conservation (Jensen et al. 1999). According to Collin et al. (2004), several in situ and ex situ methods are proposed for the conservation of European white elm, with the advantage being given to in situ method in all cases where it is applicable. Based on the spatial layout of European white elm trees on the Veliko ratno ostrvo Island, a network of conservation stands is defined and marked (Fig. 14.3).

Conservation stand A includes the largest section of European white elm trees (78) and spans on both left and right until the Galijaš canal. The trees are close on the terrain with rich low vegetation without the possibility for natural renewal, making the gene pool time-limited. Conservation stand B is located on an elevated section of the Veliko ratno ostrvo Island which is the least affected by floods and a

Fig. 14.3 Network for in situ European white elm conservation stands (*Ulmus laevis* Pall.) in the area of the Veliko ratno ostrvo Island

Fig. 14.4 Appearance of European white elm seedlings in the nursery in 2012 (left) and the planted European white elm on the Veliko ratno ostrvo Island in 2016

part of the island most often visited by hikers and tourists. This conservation stand includes 11 trees.

Separation of the conservation stands of European white elm on the Veliko ratno ostrvo Island was estimated as inadequate for the survival of this species due to inability of natural renewal, so artificial plantations were formed out of the seedlings produced from the seeds of maternal trees originally from the Veliko ratno ostrvo Island (Fig. 14.4). Progeny was produced for a total of 16 maternal trees, however regular numbers of seedlings during production was maintained for 13 half-sib lines. Three-year-old seedlings (1 + 2) were planted in the fall 2013 on a separate surface in the area of the Veliko ratno ostrvo Island. On this occasion, 55 seedlings from each half-sib line were planted. This method increases the number of European white elm trees and ensures its presence in this area in the following period (Fig. 14.4). Population size or the number of trees taking part in breeding (effective population size) determines in the greatest extent the level of genetic diversity transferred onto the next generation (Ivetić et al. 2016). Small populations

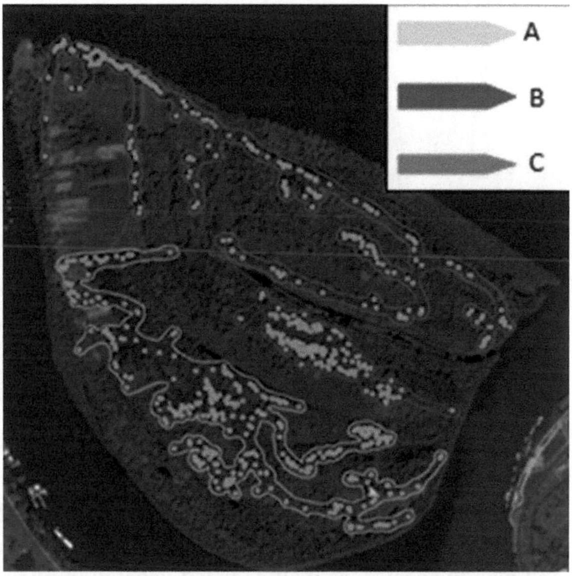

Fig. 14.5 Network for in situ Black poplar conservation stands (*Populus nigra* L.) in the area of the Veliko ratno ostrvo Island

are endangered by genetic drift and inbreeding depression which could lead to a decrease of genetic diversity and vitality (Hensen and Oberprieler 2005; Chanyenga et al. 2011; Del Castillo et al. 2011; Nagamitsu et al. 2014). As genetic diversity can be assessed by effective population size (Wright 1931), this parameter is in use in certain countries to accept reproductive material. For instance, for use in public afforestation in British Columbia, the minimal effective population size in a seed party must be 10 (Stoehr et al. 2004), as it includes 95% of population genetic diversity (Nei 1973; Yanchuk 2001).

Molecular-genetic analysis show that the black poplar population from the Veliko ratno ostrvo Island presents a good candidate for species conservation, as it exhibits high levels of genetic diversity. Based on this, the defining of the strategy for conservation of the existing gene pool has begun. The strategy prescribes adequate measures of in situ conservation harmonized with the existing plan documents and protection measure for this area. Conservation measures are founded on bio-ecological species characteristics, by current conservation status, population state, degree of genetic variability, and the available knowledge and management of these populations (FAO 1989; Šijačić-Nikolić and Milovanović 2010). A network of conservation stands (Fig. 14.5) was established in the area of the Veliko ratno ostrvo Island which includes the best trees and preserved black poplar stands. Conservation stand A include most of black poplar trees in this island (in total 455), which leans onto white willow forest growing on the coast of Danube outlets that surrounds the island. Conservation stand B includes just 192 black poplar trees, which extends on the right side of the canal Galijaš that enters in island, while conservation stand C, that spans between left sides of canal and white willow forest growing on the coast of Danube, includes 260 black poplar trees. The defined conservation stands are

Fig. 14.6 Clone archive (left) and clone test (right) of black poplar in the nursery

spatially separated. In all defined conservation stands, the black poplar trees are located on a field that is thickly covered with indigobush and other invasive shrubs and woody species, as well as weed vegetation of riparian habitats. For these reasons, there is no possibility of natural regeneration; hence the survival of the present black poplar genpool is time limited.

The definition of black poplar conservation stands in the area of the Veliko ratno ostrvo Island is estimated as insufficient for the survival of this species, since it is difficult or almost impossible for black poplar to regenerate under the banner of the old stands, which ais the present conditions in all conservation stands.

Thus, gene pool conservation is ensured by establishment of a clone archive and clone test in the nursery together with planting black poplar seedlings in the area of the Veliko ratno ostrvo Island from the nine trees selected in this population. The clone archive was established in the nursery in April 2012 from cuttings made from 1-year-old shoots collected from selected maternal trees. The primary use of the established clone archive (Fig. 14.6) is a fixation of selected genotypes obtained by tree selection from the natural population. The establishment of the clone archive sets the foundation for preservation of a part of the gene pool and controlled production of qualified reproductive materials.

The clone test was founded in March 2014, in the vicinity of the established clone archive. The cuttings for the establishment of the clone test were obtained by cutting the overground part of the seedlings from the clone archive. The main purpose of establishing the clone test (Fig. 14.6) was to provide information on genetic potential of selected maternal trees (ortet) through their posterity (a collection of vegetative copies makes one clone) obtained through vegetation. Additionally, the basic clone test enabled recognition of genotypes with favourable characteristics in early ontogenetic phases of development and provided information on properties of the tested posterity itself. One-year-old seedlings (1/1) from the clone test we planted in the fall of 2014 on the Veliko ratno ostrvo Island, where they established an artificial black poplar plantation. Eight hundred two 1-year-old seedlings were planted on this

Fig. 14.7 Appearance of planted black poplar seedlings in spring 2015 (left) and in summer 2016 (right)

occasion. Planting of all 9 clones was performed in two blocks with 25 repetitions, with each clone represented in the field gene bank with a minimum of 50 seedlings, while the rest were planted in the remaining rows.

The seedlings were planted in extension of the planted European white elm seedlings, while the appearance of the surface where the black poplar seedlings were planted is shown on the (Fig. 14.7).

The primary purpose of artificial black poplar plantations is preservation of the available section of the gene pool and an increase of surface for genetic resources of this species on the Veliko ratno ostrvo Island. Additionally, balanced habitat conditions for growing seedlings will enable testing of gen-ecological potential of implanted clones. Establishment of black poplar plantations forms a new young population which will contribute the biological stability of the forest ecosystems of the Veliko ratno ostrvo Island.

References

Adams, W. T., Zuo, J., Shimizu, J. Y., & Tappeiner, J. C. (1998). Impact of alternative regeneration methods on genetic diversity in coastal Douglas-fir. *Forest Science, 44*, 390–396.

Ballian, D., & Mekić, F. (2008). Klonski arhiv Bosansko-Hercegovačkih populacija crne topole (*Populus nigra* L.) u Žepču – podizanje i upotreba klonskog materijala. *Naše šume, 12*(13), 16–24.

Ballian, D., Kajba, D., & Idzojtic, M. (2006). Morphological diversity of hairy European Black Poplar (*Populus nigra* subsp. *caudina*) in Bosnia and Herzegovina. *Bulletin Faculty of Forestry University of Banja Luka, 5*, 13–22.

Ballian, D., Memišević, M., Bogunić, F., Bašić, N., Marković, M., & Kajba, D. (2010). Morfološka varijabilnost hrasta lužnjaka (*Quercus robur* L.) na području Hrvatske i zapadnog Balkana. *Šumarski list, 7–8*, 371–386.

Ballian, D., Hajrudinović, A., Franjić, J., & Bogunić, F. (2014). Morfološka varijabilnost lista makedonskog hrasta (*Quercus trojana* Webb.) u Bosni i Hercegovini i Crnoj Gori. *Šumarski list, 3–4*, 135–144.

Banković, S., Medarević, M., Pantić, D., Petrović, N., Šljukić, B., & Obradović, S. (2009). The growing stock of the Republic of Serbia – state and problems. *Bulletin of the Faculty of Forestry, 100*, 7–30.

Bašić, N., Kapić, J., & Ballian, D. (2007). Morfometrijska analiza varijabilnosti svojstva lista hrasta lužnjaka (*Quercus robur* L.) na području sjeverne Bosne. *Radovi šumarskog instituta Jastrebarsko, 42*(1), 5–18.

BfN. (2008). *Nature data 2008*. Federal Agency for Nature Conservation (BfN).

Booy, G., Hendriks, R. J. J., Smulders, M. J. M., Van Groenendael, J. M., & Vosman, B. (2000). Genetic diversity and the survival of populations. *Plant Biology, 2*(4), 379–395.

Bordács, S., & Bach, I. (2014). Restoration and afforestation with *Populus nigra* in Hungary. In M. Bozzano, R. Jalonen, E. Thomas, D. Boshier, L. Gallo, S. Cavers, S. Bordács, P. Smith, & J. Loo (Eds.), *The state of the world's forest genetic resources – Genetic considerations in ecosystem restoration using native tree species* (pp. 233–235). Food and Agriculture Organization of the United Nations.

Brasier, C. M. (2000). Intercontinental spread and continuing evolution of the Dutch elm disease pathogens. In P. D. Christopher (Ed.), *The elms: Breeding, conservation, and disease management* (pp. 61–72). Boston: Kluwer Academic Publishers.

Broshtilov, K. (2006). *Quercus robur* L. leaf variability in Bulgaria. *Plant Genetic Resources Newsletter, 147*, 64–71.

Brus, R., Galien, U., Božič, G., & Jarni, K. (2010). Morfological study of the leaves of two European black poplar (*Populus nigra* L.) population in Slovenia. *Periodicum Biologorum, 112*(3), 317–325.

Cagelli, L., & Lefèvre, F. (1995). The conservation of *Populus nigra* and gene flow with cultivated poplars in Europe. *Forest Genetics, 2*, 135–144.

Chanyenga, T. F., Geldenhuys, C. J., & Sileshi, G. W. (2011). Effect of population size, tree diameter and crown position on viable seed output per cone of the tropical conifer Widdringtonia whytei in Malawi. *Journal of Tropical Ecology, 27*(5), 515–520.

Coaloa, D., & Nervo, G. (2011). Poplar wood production in Europe on account of market criticalities and agricultural, forestry and energy policy. Actas del Tercer Congreso Internacional de las Salicáceas en Argentina 'Los álamos y los sauces junto al paisaje y el desarrollo productivo de la Patagonia' Neuquen, Argentina 16–19 Marzo 2011, 9.

Collada, C., Fuentes-Utrilla, A., Gil, L., & Cervera, M. T. (2004). Characterization of microsatellite loci in *Ulmus minor* Miller and cross-amplification in *U. glabra* Hudson and *U. laevis* Pall. *Molecular Ecology Notes, 4*, 731–732.

Collin, E. (2003). *EUFORGEN: Technical guidelines for genetic conservation and use for European white elm (Ulmus laevis)*. Rome: International Plant Genetic Resources Institute 6.

Collin, E., Bilger, I., Eriksson, G., & Turok, J. (2000). The conservation of elm genetic resources in Europe. In C. P. Dunn (Ed.), *The elms – Breeding, conservation and disease management* (pp. 281–293). New York: J. Chapman & Hall.

Collin, E., Rusanen, M., Ackzell, L., Bohnens, J., de Aguiar, A., Diamandis, S., Franke, A., Gil, L., Harvengt, L., Hollingsworth, P., Jenkins, G., Meier-Dinkel, A., Mittempergher, L., Musch, B., Nagy, L., Pâques, M., Pinon, J., Piou, D., Rotach, P., Santini, A., Vanden Broeck, A., & Wolf, H. (2004). Methods and progress in the conservation of elm genetic resources in Europe. *Investigación agraria: Sistemas y recursos forestales, 13*(1), 261–272.

Cooper, F. M. P., Jones, M., Watkins, C., & Wilson, Z. A. (2002). *Geographic distribution and genetic diversity of black poplar* (R & D Technical Report W1-022/TR). Bristol: Environment Agency.

Čortan, D. (2015). *Procena varijabilnosti prirodnih populacija crne topole (Populus nigra L.) na području Vojvodine primenom genetičkih markera*. Doktorska disertacija. Šumarski fakultet-Univerzitet u Beogradu (pp. 1–213).

Čortan, D., Šijačić-Nikolić, M., & Knežević, R. (2013). Varijabilnost morfometrijskih karakteristika listova crne topole (*Populus nigra* L.) iz dve populacije sa područja Vojvodine. *Šumarstvo, 65*(3–4), 193–202.

Čortan, D., Šijačić-Nikolić, M., & Knežević, R. (2014). Varijabilnost morfometrijskih karakteristika listova crne topole sa područja Vojvodine. *Glasnik Šumarskog fakulteta, 109*, 63–72.

Čortan, D., Schroeder, H., Šijačić-Nikolić, M., Wehenkel, C., & Fladung, M. (2016). Genetic structure of remnant black poplar (*Populus nigra* L.) populations along biggest rivers in Serbia assessed by SSR markers. *Silvae Genetica, 65*(1), 12–19.

de Vries, S. M. G. (1995). *Populus nigra* in the Netherlands. In E. Frison, F. Lefèvre, S. de Vries, & J. Turok (Eds.), *Populus nigra network* (p. 46). Report of the first meeting, Izmit, Turkey, 3–5 October 1994. Rome: IPGRI.

Del Castillo, R. F., Trujillo-Argueta, S., Sánchez-Vargas, N., & Newton, A. C. (2011). Genetic factors associated with population size may increase extinction risks and decrease colonization potential in a keystone tropical pine. *Evolutionary Applications, 4*(4), 574–588.

Devetaković, J. (2017). *Genetički potencijal veza (Ulmus laevis Pall.) za proizvodnju namenskog sadnog materijala* (pp. 1–244). Doktorska disertacija, Univerzitet u Beogradu Šumarski fakultet.

Devetaković, J., & Šijačić-Nikolić, M. (2013). Variability of morphometric characteristics of the leaves of European white elm from the area of Great war island. *Bulletin of the Faculty of Forestry, 107*, 57–70.

Devetaković, J., Šiljačić-Nikolić, M., & Ivetić, V. (2013). Variability of morphometric caracteristics of one-year seedlings of different half-sib European White Elm (*Ulmus effusa* Wild.) from the Great War Island. *Biologica Nyssana, 4*(1–2), 87–92.

Eriksson, G. (2001). Conservation of noble hardwoods in Europe. *Canadian Journal of Forest Research, 31*(4), 577–587.

FAO. (1989). *Plant Genetic Resources: Their conservation in situ for human use*. Rome: FAO of the United Nations.

Fossati, T., Grassi, F., Sala, F., & Castiglione, S. (2003). Molecular analysis of natural populations of *Populus nigra* L. intermingled with cultivated hybrids. *Molecular Ecology, 12*, 2033–2043.

Franjić, J. (1996). Morfometrijska analiza varijabilnosti lista posavskih i podravskih populacija hrasta lužnjaka (*Quercus robur* L., Fagaceae) u Hrvatskoj. *Glasnik za šumske pokuse, 33*, 153–214.

Fuentes-Utrilla, P., Venturas, M., Hollingsworth, P. M., Squirrell, J., Collada, C., Stone, G. N., & Gil, L. (2014). Extending glacial refugia for a European tree: Genetic markers show that Iberian populations of white elm are native relicts and not introductions. *Heredity, 112*, 105–113.

Gehle, T., & Krabel, D. (2002). Genetic differentiation of elm (*Ulmus minor* Mill., *Ulmus laevis* Pallas) in mixed stands from the Elbe flood-plains. *Forest Genetics, 9*(1), 39–46.

Härdtle, W., Bracht, H., & Hobohm, C. (1996). Vegetation und Erhaltungszustand von Hartholzauen (Querco-Ulmetum Issl. 1924) im Mittelelbegebiet zwischen Lauenburg und Havelberg. *Tuexenia, 16*, 25–38.

Heinze, B. (1997). *Populus nigra* in Austria – Rare, endangered, not recognised?. Turok, J., Lefèvre, F., de Vries, S., & Toth, B. (Eds.)., *Populus nigra* network (pp. 34–40). Report of the third meeting, Sarvar, Hungary, 5–7 October 1996, IPGRI, Rome.

Hensen, I., & Oberprieler, C. (2005). Effects of population size on genetic diversity and seed production in the rare Dictamnus albus (Rutaceae) in central Germany. *Conservation Genetics, 6*, 63.

Herpka, I. (1986). A survey of development and possibilities of growing: Natural forests of poplars and willows. In *Poplars and willows in Yugoslavia* (pp. 21–36). Novi Sad: Poplar Research Institute.

Hooke, J. M. (2006). Human impacts on fluvial systems in the Mediterranean region. *Geomorphology, 79*, 311–355.

Imbert, E., & Lefèvre, F. (2003). Dispersal and gene flow of *Populus nigra* (Salicaceae) along a dynamic river system. *Journal of Ecology, 91*, 447–456.

Isajev, V. (1978). *Proučavanje unutarvrsne promenljivosti veza (Ulmus leavis Pallas) i njen značaj za oplemenjivanje ove vrste* (pp. 1–77). Magistarski rad, Beograd: Univerzitet u Beogradu-Šumarski fakultet.

Ivetić, V., Devetaković, J., Nonić, M., Stanković, D., & Šijačić-Nikolić, M. (2016). Genetic diversity and forest reproductive material – from seed source selection to planting. *iForest, 9*, 801–812.

Jacobs, D. F., Woeste, K. E., Wilson, B. C., & McKenna, J. R. (2006). Stock quality of black walnut (*Juglans nigra*) seedlings as affected by half-sib seed source and nursery sowing density. *Acta Horticulturae, 705*, 375–381.

Jeffers, J. R. N. (1999). Leaf variation in the genus Ulmus. *Forestry, 72*(3), 183–190.

Jelić, M., Panteković, A., & Kurbalija Novičić, Z. (2014). Genetic variability of *Populus nigra* L. in the Danube Basin. In Z. Tomović & I. Vasić (Eds.). Variability of European black poplar (*Populus nigra* L.) in the Danube Basin. Procedeengs of the Danubeparks conference held in Novi Sad, 24. April, Serbia, (pp. 86–117).

Jensen, J. S., Rusanen, M., & Pliura, A. (1999). *Gene conservation of Noble Hardwoods in situations where they occur widespread.* Background document for the Fourth meeting of the EUFORGEN Noble Hardwoods Network, Gmunden, Austria.

Jovanović, B. (2007). *Dendrologija* (pp. 1–536). Beograd: Šumarski fakultet Univerziteta u Beogradu.

Jović, N., Jović, D., & Tomić, Z. (2009). *Tipologija šuma* (pp. 87–120). Beograd: Univerzitet u Beogradu, Šumarski fakultet.

Kajba, D. (1996). Međupopulacijska i unutarpopulacijska varijabilnost breze (*Betula pendula* Roth.) u djelu prirodne rasprostranjenosti u Republici Hrvatskoj. *Glasnik za šumske pokuse, 33*, 53–108.

Kajba, D., Idžojtić, M., & Bogdan, S. (1999). Discriminant analysis of leaf morphological characters of the European Black Poplar (*Populus nigra* L.) in natural populations in Croatia. In J. Turok, F. Lefèvre, S. de Vries, N. Alba, B. Heinze, R. Voloyanchul, & E. Lipman, E. (compilers). *Populus nigra* network (pp. 73–76). Report of the fift meating, 5–9 May 1999, Kyiv, Ukraine, International Plant Genetic Resources Institute, Rome.

Kajba, D., Ballian, D., Idžojtić, M., & Bogdan, S. (2004). The differences among hairy and typical European black poplars and the possible role of the hairy type in relations to climatic changes. *Forest Ecology and Management, 197*, 279–284.

Kajba, D., Antić, I., & Pfeifer, D. (2005). Potrajnost i očuvanje genofonda s posebnim osvrtom na evropsku crnu topolu (*Populus nigra* L.). *Šumarski list, 5–6*, 271–278.

Kang, K. S., EL-Kassaby, Y. A., Chung, M. S., Kim, C. S., Kang, Y. J., & Kang, B. S. (2005). Fertility variation and effective number in a clonal seed orchard of *Cryptomeria japonica. Silvae Genetica, 54*, 104–107.

Karadžić, D. (2010). *Šumska fitopatologija. Univerzitetski udžbenik* (pp. 1–774). Univerzitet u Beogradu – Šumarski fakultet, Univerzitet u Banjoj Luci – Šumarski fakultet.

Koskela, J., Vries S. M. G., Gil, L., Mátyás, C., Rusanen, M., & Paule, L. (2003). Conservation of forest genetic resources and sustainable forest management in Europe. In *Proceedings of the Symposium of the North American Forest Commission, Forest Genetic Resources and Silviculture Working Groups and the International Union of Forest Research Organizations (IUFRO)*, Canada (pp. 9–19).

Koski, V. (2000). A note on genetic diversity in natural populations and cultivated stands of Scots pine *Pinus sylvestris* L., *Investigacion Agraria, Sistemas y Recursos Forestales* (Fuera de Serie 1) (pp. 89–95)..

Kovačević, B. (2014). Variability of leaf morphometric characters in *Populus nigra* populations in the Danube Basin. In Z. Tomović & I. Vasić (Eds.). Variability of European Black Poplar (*Populus nigra* L.) in the Danube Basin. Procedeengs of the Danubeparks conference held in Novi Sad, 24. April, Serbia, (pp. 52–81).

Krstinić, A., Trinajstić, I., Kajba, D., & Samardžić, J. (1997). Morphological variability of the leaves of black poplar (*Populus nigra* L.) in natural stands along the Sava River (Croatia). In J. Turok, F. Lefèvre, S. de Vries, N. Alba, B. Heinze, & J. Van Slycken (Eds.), *Populus nigra network* (pp. 71–77). Fourth Report of that meeting, October 3–5, 1997, Geraardesbergen, Belgium. Rome: International Plant Genetic Resources Institute.

Larsen, A. B. (1996). Genetic structure of populations of beech (*Fagus sylvatica* L.) in Denmark. *Scandinavian Journal of Forest Research, 11*(3), 220–232.
Lefèvre, F., Legionnet, A., de Vries, S., & Turok, J. (1998). Strategies for the conservation of a pioneer tree species, *Populus nigra* L., in Europe. *Genetics Selection Evolution, 30*, 181–196.
Lefèvre, F., Kajba, D., Heinze, B., Rotach, P., de Vries, S. M. G., & Turok, J. (2001). Black poplar: A model for gene resource conservation in forest ecosystems. *Forestry Chronicle, 77*, 239–244.
Legionnet, A., & Lefèvre, F. (1996). *Diversité et fonctionnement génétique des populations naturelles de Populus nigra L., espèce pionnière des ripisylves européennes*. Thesis, Universite Montpellier, France.
Leonardi, S., & Menozzi, P. (1995). Genetic variability of *Fagus sylvatica* L. in Italy: The role of postglacial recolonization. *Heredity, 75*, 35–44.
Levin, D. A., Francisco-Ortega, J., & Jansen, R. K. (1996). Hybridization and the extinction of rare plant species. *Conservation Biology, 10*, 10–16.
Lindgren, D. (2016). The role of tree breeding in reforestation. *Reforesta, 1*(1), 221–237.
Machon, N., LeFranc, M., Bilger, I., & Henry, J.-P. (1995). Isozymes as an aid to clarify the taxonomy of French elms. *Heredity, 74*, 39–47.
Machon, N., LeFranc, M., Bilger, I., Mazer, S. J., & Sarr, A. (1997). Allozyme variation in Ulmus species from France: Analysis of differentiation. *Heredity, 78*, 12–20.
Mackenthun, G. L. (2004). The role of *Ulmus laevis* in German floodplain landscapes. *Investigación agraria: Sistemas y recursos forestales, 13*(1), 55–63.
Maksimović, Z., & Šijačić-Nikolić, M. (2013). Morfometrijske karakteristike listova crne topole (*Populus nigra* L.) na području Velikog ratnog ostrva. *Glasnik Šumarskog fakulteta, 108*, 93–108.
Maksimović, Z., & Šijačić-Nikolić, M. (2016). In situ conservation of black poplar (*Populus nigra* L.) gene pool in the protected area "Great War Island". *Reforesta, 2*, 39–49.
Maksimović, Z., Čortan, D., Ivetić, V., Drinić, M. S., & Šijačić-Nikolić, M. (2014). Genetic structure of black poplar (*Populus nigra* L.) population in the area of Great War Island. *Genetika, 46*(3), 963–973.
Maksimović, Z., Šijačić-Nikolić, M., Medarević, M., & Vasić, V. (2016). Stanje populacije crne topole (*Populus nigra* L.) na području Velikog ratnog ostrva kao osnova za konzervaciju i usmereno korišćenje genofonda. *Šumarstvo, 1–2*, 121–136.
Mathew, J., & Vasudeva, R. (2005). Variation of seedling vigour among half-sib families of teak (*Tectona grandis*). *Journal of Tropical Forest Science, 17*(1), 170–172.
Mattila, A., & Vakkari, P. (1997). Genetic variation of *Quercus robur* and *Ulmus laevis* in Finland. Forest studies 28. In *Proceedings of the meeting of the Nordic Group for Tree Breeding*, Estonia, June 3–7 1996, (pp. 63–68).
Mittempergher, L., & Santini, A. (2004). The history of elm breeding. *Investigación agraria: Sistemas y recursos forestales, 13*(1), 161–177.
Muir, G., Lowe, A. J., Fleming, C. C., & Vogl, C. (2004). High nuclear genetic diversity, high levels of outcrossing and low differentiation among remnant populations of *Quercus patraea* at the margin of its range in Ireland. *Annals of Botany, 93*, 691–697.
Nagamitsu, T., Kikuchi, S., Hotta, M., Kenta, T., & Hiura, T. (2014). Effects of population size, forest fragmentation, and urbanization on seed production and gene flow in an endangered maple (*Acer miyabei*). *The American Midland Naturalist, 172*(2), 303–316.
Nei, M. (1973). Analysis of gene diversity in subdivided populations. *Proceedings of the National Academy of Sciences of the United States, 70*, 3321–3323.
Nielsen, L. R., & Kjær, E. D. (2010a). Fine-scale gene flow and genetic structure in a relic *Ulmus laevis* population at its northern range. *Tree Genetics & Genomes, 6*, 643–649.
Nielsen, L. R., & Kjær, E. D. (2010b). Gene flow and mating patterns in individuals of wych elm (*Ulmus glabra*) in forest and open land after the influence of Dutch elm disease. *Conservation Genetics, 11*, 257–268.

Nonić, M., Devetaković, J., Šijačić-Nikolić, M., & Milovanović, J. (2012). Yield variability as a basis for conservation and directed utilization of European White Elm (*Ulmus effusa* Willd.) gene pool at Great war island. *Agriculture & Forestry, 58*(3), 105–113.

Orlović, S., Pilipović, A., Pap, P., Radosavljević, N., & Drekić, D. (2005). Genetički resursi evropske crne topole (*Populus nigra* L.) i topola iz sekcije Leuce Duby u prirodnim populacijama Srbije i Crne Gore. *Topola, 5–8*, 4–8.

Pajares, J. A. (2004). Elm breeding for resistance against bark beetles. *Forest Systems, 13*(1), 207–215.

Petit, R. J., Aguinagalde, I., de Beaulieu, J. L., Bittkau, C., Brewer, S., Cheddadi, R., Ennos, R., Fineschi, S., Grivet, D., Lascoux, M., Mohanty, A., Müller-Starck, G., Demesure-Musch, B., Palmé, A., Pedro Martín, J., Rendell, S., & Vendramin, G. G. (2003). Glacial refugia: Hotspots but not melting pots of genetic diversity. *Science, 300*, 1563–1565.

Poljak, I., Idžojtić, M., Šapić, I., Vukelić, J., & Zebec, M. (2014). Varijabilnost populacija bijele (*Alnus incana*/L./Moench) i crne johe (*A. glutinosa* /L./Moench) na području Mure i Drave prema morfološkim obilježjima listova. *Šumarski List, 1–2*, 7–17.

Popivshchy, I. I., Prokazin, A. E., & Routkovsky, L. V. (1997). Black poplar in the Russian federation. In J. Turok, F. Lefèvre, S. de Vries, & B. Toth (Eds.), *Populus nigra network* (pp. 46–52). Report of the third meeting, Sarvar, Hungary, 5–7 October 1996. Rome: IPGRI.

Popović, V., & Kerkez, I. (2016). Varijabilnost populacija divlje trešnje (*Prunus avium* L.) u Srbiji prema morfološkim svojstvima listova. *Šumarski List, 7–8*, 347–355.

Popović, V., Lučić, A., Šijačić-Nikolić, M., Ćirković-Mitrović, T., Rakonjac, L. J., & Brašanac-Bosanac, L. J. (2013). Analysis of intra-population variability of Bald Cypress (*Taxodium distichum* L. Rich.) in a seed stand near Backa Palanka using morphometric markers. *Archives of Biological Sciences, 65*(3), 1093–1103.

Pospíšková, M., & Bartáková, I. (2004). Genetic diversity of a black poplar in the Morava river basin assessed by microsatellite analysis. *Forest Genetics, 11*(3–4), 257–262.

Pospíšková, M., & Šálková, I. (2006). Population structure and parentage analysis of black 790 poplar along the Morava River. *Canadian Journal of Forest Research, 36*(5), 1067–1076.

Preston, C. D., Pearman, D. A., & Dines, T. D. (2002). *New atlas of the British and Irish flora*. Oxford: Oxford University Press.

Rathmacher, G., Niggemann, M., Kohnen, M., Ziegenhagen, B., & Bialozyt, R. (2010). Short-distance gene flow in *Populus nigra* L. accounts for smallscale spatial genetic structures: Implications for in situ conservation measures. *Conservation Genetics, 11*, 1327–1338.

REFORGEN. (2003). http://foris.fao.org/reforgen/factSheet.jsp

Romanić, B. (2000). *Morfološka varijabilnost lista Evropske crne topole (Populus nigra L.) u prirodnim populacijama na području reke Drave u Hrvatskoj* (pp. 1–129). Magistarski rad. Hrvatska: Šumarski fakultet Sveučilista u Zagrebu.

Sacchetti, P., Tiberi, R., & Mittempergher, L. (1990). Preference of *Scolytus multistriatus* (Marsham) during the maturation feeding for two species of elm. *Redia*, 347–354.

Savolainen, O., Bokma, F., Knurr, T., Karkkainen, K., Pyhajarvi, T., & Wachowiak, W. (2007). Adaptation of forest trees to climate change. In J. Koskela, A. Buck, & E. Teissier du Cros (Eds.), *Climate change and forest genetic diversity: Implications for sustainable forest management in Europe* (pp. 19–30). Rome: Bioversity International.

Schaberg, P. G., Dehayes, D. H., Hawley, G. J., & Nijensohn, S. E. (2008). Anthropogenic alterations of genetic diversity within three populations: Implications for forest ecosystem resilience. *Forest Ecology and Management, 256*(5), 855–862.

Servo, G., Devetaković, J., & Ivetić, V. (2017). Comparison of seedling quality between autochthonous and poplar clones. *Reforesta, 3*, 41–47.

Šijačić-Nikolić, M., & Milovanović, J. (2010). Konzervacija i usmereno korišćenje šumskih genetičkih resursa. *Univerzitet u Beogradu – Šumarski fakultet*, 1–200.

Šijačić-Nikolić, M., Knežević, R., & Milovanović, J. (2008). A contribution to the study of Hackberry (*Celtis occidentalis* L.) juvenile stage of development. *Bulletin of the Faculty of Forestry, 97*, 57–78.

Smulders, M. J. M., van der Schoot, J., Arens, P., & Vosman, B. (2001). Trinucleotide repeat microsatellite markers for black poplar (*Populus nigra* L.). *Molecular Ecology Notes, 1*, 188–190.

Smulders, M. J. M., Cottrell, J. E., Lefèvre, F., van der Schoot, J., Arens, P., Vosman, B., Tabbener, H. E., Grassi, F., Fossati, T., Castiglione, S., Krystufek, V., Fluch, S., Burg, K., Vornam, B., Pohl, A., Gebhardt, K., Alba, N., Agúndez, D., Maestro, C., Novitol, E., Volosyanchuk, R. T., Pospíšková, M., Bordács, S., Bovenschen, J., van Dam, B. C., Koelewijn, H. P., Halfmaerten, D., Ivens, B., van Slycken, J., Vanden Broeck, A., Storme, V., & Boerjan, W. (2008). Structure of the genetic diversity in black poplar (*Populus nigra* L.) populations across European river systems: Consequences for conservation and restoration. *Forest Ecology and Management, 255*, 1388–1399.

Šrámek, M., & Čermák, J. (2012). The vertical leaf distribution of *Ulmus laevis* Pall. *Trees, 26*(6), 1781–1792.

Stoehr, M., Webber, J., & Woods, J. (2004). Protocol for rating seed orchard seedlots in British Columbia: Quantifying genetic gain and diversity. *Forestry, 77*(4), 297–303.

Storme, V., Boejeran, W., Vanden Broeck, A. H., Ivens, B., Halfmaerten, D., Van Slycken, J., Castiglione, S., Grassi, F., Fossati, T., Cottrell, J. E., Tabbener, H. E., Lefèvre, F., Imbert, E., Fluch, S., Krustufek, V., Burg, K., Bordacs, S., Gebhardt, K., Vornam, B., Pohl, A., Alba, N., Bovenschen, J., van Dam, B., van der Schoot, J., Vosman, B., & Smulders, M. J. M. (2004). Ex situ conservation of black poplar in Europe: Genetic diversity in nine gene bank collections and their value for nature development. *Theoretical and Applied Genetics, 108*, 969–981.

Tabbener, H. E., & Cottrell, J. E. (2003). The use of PCR based DNA markers to study the paternity of poplar seedlings. *Forest Ecology and Management, 179*, 363–376.

Toplu, F. (2005). Breeding and conservation of Black poplar (*Populus nigra* L.) gene resources in Turkey. *Unasylva, 221*(56), 26–30.

Tucović, A. (1965). Sistematika i bioekološka istraživanja crne topole (*Populus nigra* L.) u Srbiji. Doktorska disertacija. Šumarski fakultet Univerziteta u Beogradu.

Vajda, Z. (1952). Borba protiv sušenja brestova. *Šumarski List, 9*, 326–335.

Vakkari, P., Rusanen, M., & Kärkkäinen, K. (2009). High genetic differentiation in marginal populations of European white elm (*Ulmus laevis*). *Silvae Fennica, 43*, 185–196.

van Dam, B. C. (2002). EUROPOP: Genetic diversity in river population of european black poplar for evalution of biodiversity, conservation strategies, nature development and genetic improvement. In B.C. van Dam & S. Bordacs (Eds.). *Towards the generic diversity in river populations of european black poplar: Implications for riparian ecosystem management* (pp. 15–32). Proceedings of an international symposium held Csiszàr Nyomda, Hungary.

van Slycken, J. (1995). Short note about *Populus nigra* in Belgium. In E. Frison, F. Lefèvre, S. de Vries, & J. Turok (Eds.), *Populus nigra Network, Report of the first meeting, Izmit, Turkey, 3–5 October 1994*. Rome: IPGRI 1995, 40.

Vanden Broeck, A. H., Storme, V., Cox, K., Quataert, P., Ivens, B., Hostyn, V., Haflmaerten, D., Van Slycken, J., & Boerjan, W. (2002). Mating system of *Populus nigra* in a mixed poplar stand composed of *P. nigra, P. deltoides* and *P.* × *euramericana*. In B. C. van Dam & S. Bordács (Eds.), *Genetic diversity in river populations of European black poplar – implications for riparian eco-system management* (pp. 53–60). Budapest: Csiszár Nyomda.

Vanden Broeck, A., Quataert, P., Roldán-Ruiz, I., Van Bockstaele, E., & Van Slycken, J. (2003). Pollen competition in *P. nigra* females revealed by microsatellite markers. *Forest Genetics, 10*, 219–227.

Vander Mijnsbrugge, K., Vanden Broeck, A., & Van Slycken, J. (2005). A survey of *Ulmus laevis* in Flanders (Northern Belgium). *Belgian Journal of Botany, 138*, 199–204.

Webber, J. F. (2000). Insect vector behaviour and the evolution of Dutch Elm Disease. In C. P. Dunn (Ed.), *The elms – Breeding, conservation and disease management* (pp. 47–60). New York: J. Chapman, & Hall.

Whiteley, R. E. (2004). *Quantitative and molecular genetic variation in Ulmus laevis Pall* (pp. 1–33). Dissertation, Swedish University of Agricultural Sciences.

Whiteley, R. E., Black-Samuelsson, S., & Clapham, D. (2003). Development of microsatellite markers for the European white elm (*Ulmus laevis* Pall.) and cross-species amplification within the genus Ulmus. *Molecular Ecology Notes, 3*, 598–600.

Woeste, K. E., Jacobs, D. F., & McKenna, J. R. (2011). Halfsib seed source and nursery sowing density affect black walnut (*Juglans nigra*) growth after five years. *New Forests, 41*, 235–245.

Wright, S. (1931). Evolution in Mendelian populations. *Genetics, 16*, 0097–0159.

Yanchuk, A. D. (2001). A quantitative framework for breeding and conservation of forest tree genetic resources in British Columbia. *Canadian Journal of Forest Research, 31*, 566–576.

Zalapa, J. E., Brunet, J., & Guries, R. P. (2008). Isolation and characterization of microsatellite markers for red elm (*Ulmus rubra* Muhl.) and cross-species amplification with Siberian elm (*Ulmus pumila* L.). *Molecular Ecology Notes, 8*, 109–112.

Zebec, M., Idžojtić, M., Poljak, I., & Mihaldinec, I. (2010). Varijabilnost nizinskog brijesta (*Ulmus minor* Mill. *sensu latissimo*) na području Hrvatske podravine prema morfološkim svojstvima listova. *Šumarski List, 134*(11–12), 569–579.

Zebec, M., Idžojtić, M., & Poljak, I. (2014). Morfološka varijabilnost nizinskog brijesta (*Ulmus minor* Mill. *sensu latissimo*) na području kontinentalne Hrvatske. *Šumarski List, 138*(11–12), 563–571.

Zebec, M., Idžojtić, M., Poljak, I., & Modrić, I. (2015). Raznolikost gorskog brijesta (*Ulmus glabra* Huds.) na području gorsko-kotlinske Hrvatske prema morfološkim obilježjima listova. *Šumarski List, 139*(9–10), 429–438.

Chapter 15
In Situ Conservation: Case Study Bosnia and Herzegovina

Branislav Cvjetković, Milan Mataruga, Dalibor Ballian, Rifet Terzić, and Vanja Daničić

Abstract The total area under protection on the territory of Bosnia and Herzegovina is around 2% with tendency to increase. The conservation of forest ecosystems "in situ" in Bosnia and Herzegovina is comprises: natural reserves, national parks, groups of trees or individual trees and seed stands. There are 2 Strict Nature Reserves, 4 National Parks, 16 Natural Monuments, 5 Protected Landscape and 2 protected areas with sustainable use of natural resources. The total area of delineated forest seed sources is 2154.93 ha 212 which consist of 121 coniferous and 88 deciduous species. The management in all in-situ areas is prescribed by law and rulebooks.

Keywords In situ conservation · Bosnia and Herzegovina

15.1 Introduction

The conservation of forest ecosystems "in situ" in Bosnia and Herzegovina is realized through:

1. Natural reserves,
2. National parks,
3. Groups of trees or individual trees,
4. Seed stands.

B. Cvjetković (✉) · M. Mataruga · V. Daničić
Faculty of Forestry, University of Banja Luka, Banja Luka, Bosnia and Herzegovina
e-mail: branislav.cvjetkovic@sf.unibl.org; milan.mataruga@sf.unibl.org; vanja.danicic@sf.unibl.org

D. Ballian
Faculty of Forestry, University of Sarajevo, Sarajevo, Bosnia and Herzegovina
e-mail: balliandalibor9@gmail.com

R. Terzić
Faculty of Natural Sciences, University of Tuzla, Tuzla, Bosnia and Herzegovina
e-mail: rifet.terzic@untz.ba

M. Šijačić-Nikolić et al. (eds.), *Forests of Southeast Europe Under a Changing Climate*, Advances in Global Change Research 65,
https://doi.org/10.1007/978-3-319-95267-3_15

15.2 Protected Areas in Bosnia and Herzegovina

Legal solutions have contributed to the significant progress in the protection of nature in the last 10–15 years by establishing new protected areas, where the areas under protection on the territory of Bosnia and Herzegovina (B&H) have increased from 0.5% to 2%. In accordance with the strategic document (NEAP B&H 2013), B&H should raise the percentage under protection up to 3% until 2012, and 10% until 2018. Compared to other surrounding countries, B&H holds the last position on the scale regarding the area under protection.

A collision of economic and ecological interest often occurs when protected areas should be delineated. However, there are forest complexes that are not used. For example, deep river or stream canyons which are not suitable for forest management, then wetlands, mountain ranges as well as the living places of certain forest fauna and the stands of forest trees that are protected by law (Bosnian pine, Serbian spruce, Mountain maple, etc.). Such habitats can be delineated as "in situ" protected areas in B&H.

Regarding the existing protected areas, the largest total area is occupied by national parks (Table 15.1). At this moment in B&H there are four National Parks, "Sutjeska" (Official Gazette of Republic Srpska, No. 124/12), "Kozara" ("Official Gazette of Republic Srpska" number 124/12), "Una" (Official Gazette of the Federation of Bosnia and Herzegovina No. 44/08) and "Drina" (Official Gazette of Republic Srpska, No. 63/17), with the total area of around 47.472 ha. The spatial distribution of national parks in Bosnia and Herzegovina is such that they cover almost all the most important and most preserved ecosystems, from the northwestern flooded forests, wetlands, across the mountainous region of central Bosnia and finally the south-eastern part of Bosnia and Herzegovina (Fig. 15.1). Management plans for the protected areas are regulated by laws as well as by-laws such as the Rulebook on the content and the manner of drafting the Management Plan for protected areas (Official gazette FB & H No. 65/06).

The review of the protected areas is given in Table 15.1 and includes the protected areas in B&H.

Besides the above-mentioned areas, numerous other natural monuments, such as caves, springs, lakes, protected landscapes as well as ramsar areas (Hutovo blato, Livanjsko polje and Bardača).

Two strict nature reserves have been delineated: "Janj" ("Official Gazette of the Republic of Srpska", No: 123/12) and "Lom" ("Official Gazette of the Republic of Srpska", No: 39/13). The reserves are mostly virgin forests which are mainly used for scientific research purposes. Other natural monuments, important for the preservation of forest biodiversity, such as Yellow beech, a unique specimen of the species that is characterized by yellow color of leaves, have been designated by the decision

Table 15.1 Protected areas in B&H

No	Name	Entity[a]	IUCN category	Area [ha]
Strict nature reserve				
1	Virgin forest "Janj"	RS	I a	295.00
2	Virgin forest "Lom"	RS	I a	297.82
Wilderness area				
1	–	–	I b	–
National park				
1	NP Kozara	RS	II	3907.54
2	NP Sutjeska	RS	II	16,052.34
3	NP Una	FBiH	II	19,800.00
Natural monument or feature				
	NM Pećina Orlovača	RS	III	27.01
	NM Pećina Ljubačevo	RS	III	45.45
	NM Žuta bukva	RS	III	0.50
	NM Pećina Rastuša	RS	III	11.39
	NM Prokoško jezero	FB&H	III	2119.00
	NM Skakavac	FB&H	III	1430.70
	NM Tajan	FB&H	III	3591.98
	NM Vrelo Bosne	FB&H	III	603.00
	NM Jama Ledana	RS	III	28.26
	NM Vaganska pećina	RS	III	12.00
	NM Pećina Đatlo	RS	III	43.42
	NM Pavlova pećina	RS	III	13.40
	NM Girska pećina	RS	III	25.37
	NM Pećina pod lipom	RS	III	6.10
	NM Pećina Ledenjača	RS	III	7.40
	NM Velikapećina	RS	III	820.92
Habitat/species management area				
1.	–	–	IV	–
V Protected landscape/seascape				
	PL Blidinje	FB&H	V	35.800,00
	PL Hutovo blato	FB&H	V	11.093,98
	PL Konjuh	FB&H	V	8.016,61
	PL Bijambare	FB&H	V	367,36
	PL Trebević	FB&H	V	400,20
VI Protected area with sustainable use of natural resources				
1.	PA Univerzitetski grad	RS	VI	27,38
2.	PF Slatina	RS	VI	35,73

Source: The Institute for Protection of Cultural and Natural Heritage, the Ministry of Education and Culture of the Republic of Srpska, 2016

[a]Entities: *RS* Republic of Srpska, *FB&H* Federation of Bosnia and Herzegovina

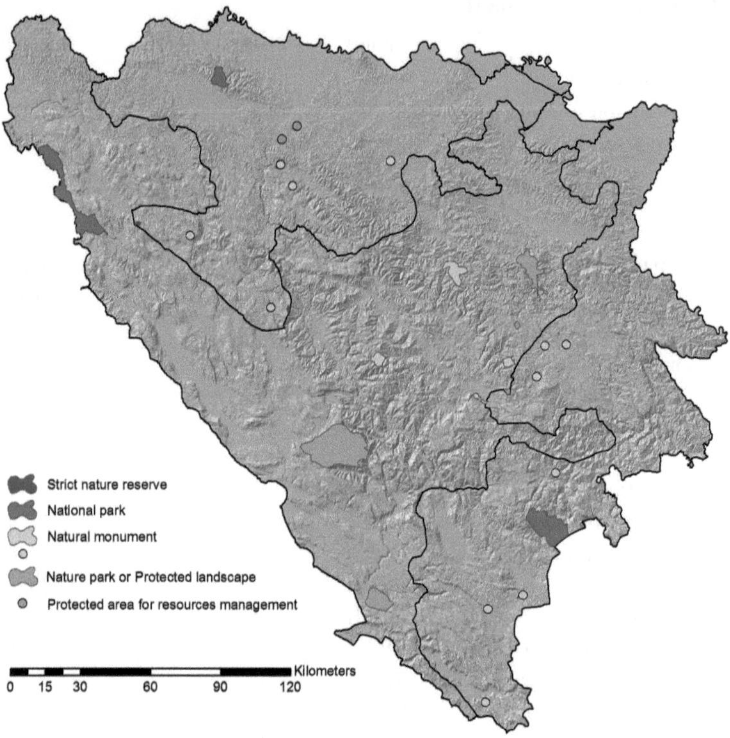

Fig. 15.1 Map of protected areas in B&H

of the Government of the Republic of Srpska ("Official Gazette of the Republic of Srpska", No: 30/12). Yellow beech is a protected, registered variety according to the Rulebook on the Register of Forest Plant Varieties ("Official Gazette of the Republic of Srpska", No. 111/09).

In addition to the protected areas, FSC certification provides the designation of High Conservation Value Forests, which has largely been realized. However, the process of continuous monitoring in the forests imposes the need to continuously monitor these forests alongside with the delineation of new HCVF forests.

15.3 Seed Stands, Groups of Trees and Individual Trees

According to Ballian (2011), the criteria that the basic seed unit should satisfy (seeds stand, seed zone, groups of trees and individual trees), and whose aim is to produce forest reproductive material, are:

- It should contain individual trees of similar genetic properties that will be mutually replicating,

- It should be large enough to guarantee sufficient representation of an autochthonous genepool,
- It should guarantee the permanent production of seed according to the characteristics of species in order to meet the needs of nursery production,
- It should be geographically well positioned.

The same author states that due to the great diversity of habitats in B&H and the combination with anthropogenic factors and climate changes, it is often not possible to achieve genetic homogeneity (Ballian 2011).

The first activities on the designation of seed objects in the form of in situ protection were started in 1953. Since then, continuous actions have been undertaken on the delineation of new seed stands, the revision of old seed stands as well as the work on genetic meliorations. Another important step was made in 1987, when the revision of the existing seed objects and the proposal for the delineation of new ones, based on the needs of forest seed and plant breeding in B&H (Dizdarević et al. 1987), and also based on the Ecological and vegetational regionalization of B&H (Stefanović et al. 1983) were carried out. Strategies for the development of forest seeds and seedlings have been defined and concrete measures on the delineation of new seed facilities, the regulation of the existing seed facilities and the revision of some seed facilities have been proposed.

Unfortunately, the 1992–1995 war interrupted all activities related to forest seed stands. The first steps were made 10 years after 1995 in the Republic of Srpska when, as a result of the multi-year project and the cooperation of the Faculty of Forestry, the University of Banja Luka and PFE "Forests of the Republic of Srpska" published the first Register of forest seed facilities of RS, which included 56 seed objects marked on the field with the detailed data on their bioecological characteristics (Mataruga et al. 2005). The precursor of this was the establishment of a special organizational unit "Center for Forest Seed and Seedlings Production" within the Public Forest Enterprise "Forests of the Republic of Srpska".

After 2012, the activities on the delineation of seed facilities have been carried out by the Public Enterprise "Forests of the Republic of Srpska", "Research and Development Project Center" sector, in cooperation with other organizational units of the company, the Ministry of Agriculture, Forestry and Water Management and the Faculty of Forestry of the University of Banja Luka with the keeping of the Register.

If observed at the level of B&H, a total of 209 seed facilities of domestic and introduced species were delineated, out of which 124 are coniferous and 85 are deciduous species. The total area of seed stands is over 2150 ha (Table 15.2). Reduced areas range from 67% to 82% of the total area of the seedlings.

Table 15.2 Area of seed sources in B&H

Species	Republic of Srpska	Federation of B&H	B&H
Conifers [ha]	640.65	1000.43	1641.08
Deciduous[ha]	309.27	204.58	513.85
Total [ha]	949.92	1205.01	2154.93

Table 15.3 Types of forest seed objects (conifers/deciduous species) in B&H

	RS		FB&H		BIH	
Seed source	Conifers	Decid.	Conifers	Decid.	Conifers	Decid.
Seed stands	47	13	55	9	102	22
Individual trees	12	20	0	0	12	20
Groups of trees	6	14	0	32	6	46
Seed zone	0	0	1	0	1	0
Total	65	47	56	41	121	88

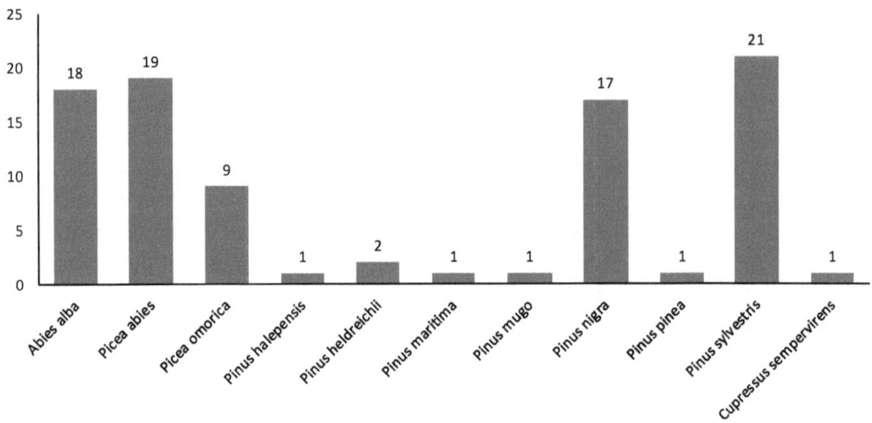

Fig. 15.2 Seed sources of domestic conifer species in B&H

In 2016, 112 seed objects were delineated in the Republic of Srpska. There are 65 seed objects of conifers, out of which 11 are introduced and 11 are autochthonous. There are 47 seed objects of deciduous species, 18 species are native and 9 are introduced. In the Federation of Bosnia and Herzegovina, 97 seeds were registered. Out of this, there are 55 seed objects of conifer species, out of which 6 are domestic and two are introduced. Out of 42 deciduous species, 9 are autochthonous and one is alochtonous species. If observed at the level of B&H, the total number of isolated seeds is 209, out of which 120 are of coniferous species and 88 deciduous, where the dominant categories are seed stands (Table 15.3).

A detailed overview of seed objects by species is given in Figs. 15.2, 15.3, 15.4, and 15.5. Induced species are especially delineated. The reason for this is the increase of the area under introduced species in Europe with the emphases on those which are resistant to climate changes and gives good yield.

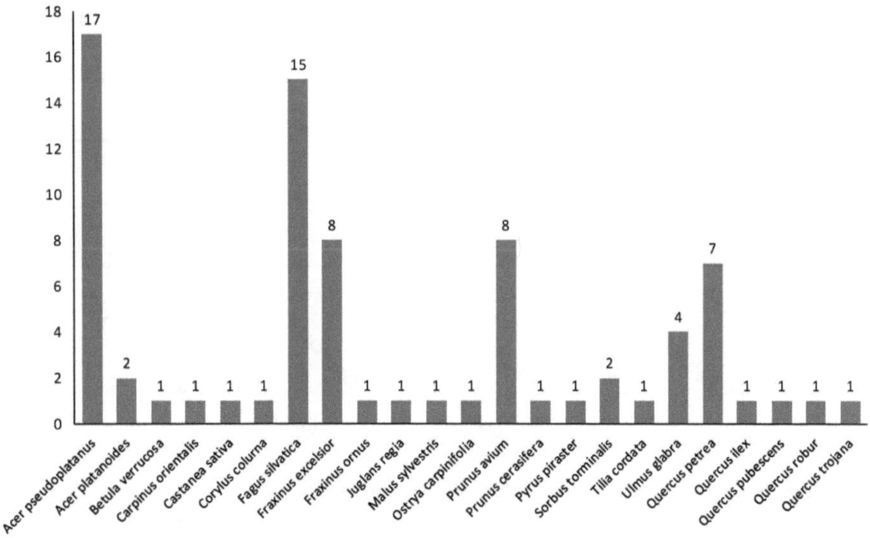

Fig. 15.3 Seed sources of domestic deciduous species in B&H

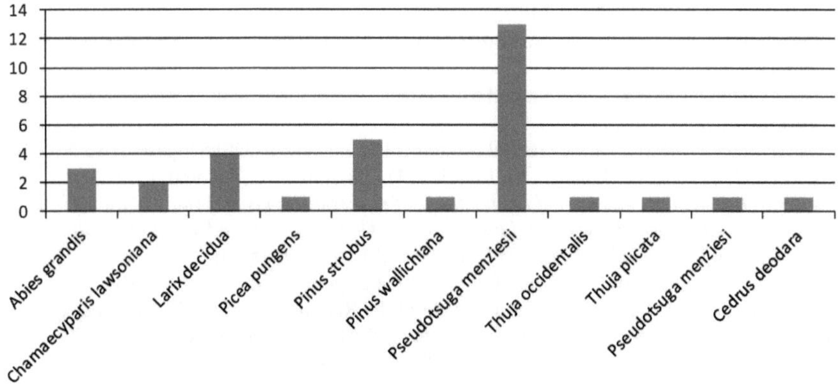

Fig. 15.4 Seed sources of introduced conifers in B&H

It might be expected that seed objects of introduced species which are characteristic for warmer areas such as Arizona cypress, xero-thermophilic pine species, oaks and species from the Far East might be established and delineated.

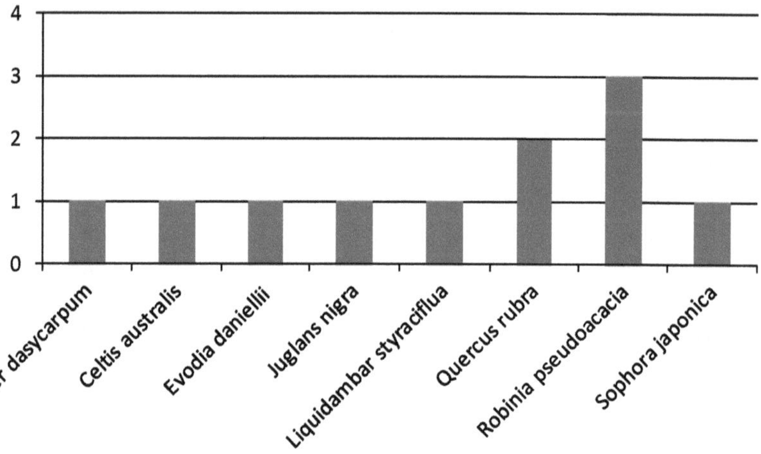

Fig. 15.5 Seed sources of introduced deciduous species in B&H

References

Ballian D. (2011). Osnovni principi rajonizacije sjemenskih objekata za proizvodnju sjemenana genetičkim principima. Radovi Hrvatskog društva za znanost i umjetnost, 12/13: 18-41.

Decision on the Protection of the Natural Monument "Virgin forest Janj" Official gazette RS No. 123/12.

Decision on the Protection of the Natural Monument "Virgin forest Lom". Official gazette RS No. 39/13.

Decision on the Protection of the Natural Monument "Žuta bukva". Official gazette RS No. 30/12.

Dizdarević, H., Prolić, N., Pintarić, K., Luteršek, D., Uščuplić, M., Vukorep, I., & Stefanović, V. (1987). *Revision of existing and delineationd of new seed stands and study of biological characteristics of spruce, fir, white and black pine in the function of production of quality seed for the needs of forestry of BiH* (pp. 2–14). Faculty of Forestry Sarajevo.

Law on Nationa Park "Drina". Official Gazette of the Republic of Srpska, No. 63/17.

Law on Nationa Park "Kozara". Official Gazette of the Republic of Srpska No. 124/12.

Law on Nationa Park "Sutjeska". Official Gazette of the Republic of Srpska No. 124/12.

Law on Nationa Park "Una". Official gazette FB&H No. 44/08.

Mataruga, M., Isajev, V., Lazarev, V., Balotić, P., & Daničić, V. (2005). *Register of forest seeds sources in the Republic of Srpska – basis of improvement of seed production* (pp. 1–222). University of Banja Luka, Faculty of Forestry.

NEAP – Environmental Action Plan. (2013). Federal Ministry of Physical Planning and Environment, Ministry of Urban Planning, Housing and Utilities, Construction and Ecology of the Republic of Srpska.

Rulebook on the content and method of drafting the Protected Area Management Plan. Official gazette FB&H No. 65/06.

Rulebook on the register of varieties of forest plants. Official gazette RS No. 111/09.

Stefanović, V., Beus, V., Burlica, Č., Dizdarević, H., & Vukorep, I. (1983). *Ecological and vegetational dellineation of Bosnia and Herzegovina* (Special edition No 18, pp. 1–49). Faculty of Forestry Sarajevo.

Chapter 16
In Situ Conservation of Forest Genetic Resources in Romania

Marius Budeanu, Flaviu Popescu, and Neculae Şofletea

Abstract In Romania, the forest genetic resources are included in two National Catalogues approved by ministerial orders. First is *The National Catalogue of Forest Genetic Resources* which includes 698 cores with main role of conservation, 577 of them being preserved in situ. Thus, the total area of in situ conservation is 10559.3 ha, being constituted especially for the main species from Romanian forests: European beech, Norway spruce, Silver fir and Oaks. At the same time, a large number of the conservation cores were established for some species with a low participation in the forest area (e.g. European larch, Scots pine) but located in natural populations where the environmental conditions are limited Most of the in situ conservation cores are also included in *The National Catalogue of Forest Reproductive Materials*. A total of 138 cores, accounting for 24% of the total number of the in situ conservation stands, with an area of 2140.7 ha, are the forest genetic resources located in different types of protected areas (national and natural parks, scientific and natural reservations, biosphere reservation, forest park or secular forests).

Keywords Breeding programme · Forest dynamic conservation · Marginal populations · Seed sources

M. Budeanu (✉) · F. Popescu
National Institute for Research and Development in Forestry "Marin Drăcea", Voluntari, Romania
e-mail: budeanumarius@yahoo.com; flaviu.popescu@icashd.ro

N. Şofletea
Faculty of Silviculture and Forest Engineering, Transilvania University of Braşov, Braşov, Romania
e-mail: nic.sofletea@unitbv.ro

M. Šijačić-Nikolić et al. (eds.), *Forests of Southeast Europe Under a Changing Climate*, Advances in Global Change Research 65,
https://doi.org/10.1007/978-3-319-95267-3_16

16.1 Introduction

As a consequence of the fact that Romania has all types of landforms (mountains, hills, tableland, plains) and the forest area is about 6.56 million hectares (INS 2016), the genetic diversity of the forests is high (Popescu et al. 2015). It is known that the adaptability of species to climate change is higher when genetic diversity is higher (Geburek and Turok 2005; White et al. 2007). Over 70 trees and shrubs species that are the object of preservation, especially in situ, exist in Romanian forests. The main species for which the most in situ preservation cores were constituted are: European beech (*Fagus sylvatica*), accounting 31% of the forested area of Romania, Norway spruce (*Picea abies*), covering 23%, and the oaks species (especially *Quercus petraea* and *Quercus robur*, but also *Q. pedunculiflora*, *Q. pubescens*, *Q. cerris*, *Q. frainetto*, etc.), these ones accounting 16% of the national forest fund (Şofletea and Curtu 2007; IFN 2012). Moreover, although having a relatively low share (4% of the Romanian forest area), an important number of preservation cores were constituted also for silver fir (*Abies alba*), one of the most important species for mountainous mixed forests.

The completely different climatic characteristics in certain regions of Romania required the division of the country into 11 regions of provenance: 5 in the mountain area (the Eastern Carpathians, Curvature Carpathians, Southern Carpathians and Western Romanian Carpathians, the latter divided into 2 regions, namely Banatului Mountains and Apuseni Mountains) and 6 in the tableland and plain areas (Transylvania Plain, Moldavia Plain, Siret and Bărăgan Plains, Danube Ponds, Oltenia and Bucharest Plains, Banatului and Crişurilor-Someş Plains). Depending on the general exposure of the mountain, their belonging to tableland or plain, and the specific climatic conditions in certain areas, the regions of provenance have been divided into a total of 26 subregions (Pârnuţă et al. 2010).

The main objective of the latest update of the Romanian National Catalogue of Forest Genetic Resources (FGR), which was conducted from 2005 to 2011, was to establish at least one FGR per sub-region of provenance for the most important species of Romanian forest.

16.2 FGR Conservation History and the Necessary Stages

The first edition of The National Catalogue of Forest Genetic Resources was published in 1996 including only the main species (Lalu 1996). Its revision, carried out during 2005–2011, resulted in the issuing of a new Catalogue (Pârnuţă et al. 2011), by re-analyzing the resources included in the initial catalogue and taking into account the latest data from the literature and the discussions with the forestry specialists, from education, research as well as with forest managers. The necessity of this updating was required not only by the above mentioned reasons but also as a result of the impact of the law changes, as a consequence of which about half of the

Romanian forests moved from state property to private property. Even though the forest owners, private or legal persons, were bound by law to preserve the forest genetic resources (FGR), a research of the stands and discussions with the owners and the administrators of these forests were imperatively necessary. In Romania, FGR preserved in situ include cores constituted in:

- High productivity stands for the respective species;
- Stands situated in marginal populations at altitudinal or latitudinal limits of the distribution range of the species;
- Stands classified under climatic or edaphic ecotypes;
- Stands where valuable biotypes from a scientific and/or economic point of view are present (e.g. Norway spruce stands with resonance wood) (Stănescu and Şofletea 1998; Şofletea 2005).

Stands representative from the point of view of the bio-accumulation capacity, wood quality and resistance to biotic and abiotic disturbing factors were selected and included in The National Catalogue of FGR for each region of provenance. Each FGR is formed of a core and a buffer area. In the case of the main species the requirement was that the core has a minimum area of 10 ha. The buffer area is formed of development units surrounding the core and has the role to protect it. The core represents the strictly protected area and any kind of cuttings (except conservation and hygienically cuttings) is excluded. Only long time regeneration cuttings are accepted in the buffer area (cutting all trees in one intervention is not allowed). A decrease of the core area under 10 ha was accepted for marginal populations, in which case the requirement was the existence of a minimum of 50 mature trees. For the rare or endangered species the condition of constituting the FGR was that at least 15 non-related mature trees exist (Pârnuţă et al. 2011).

The FGR description was done using a template including several details (about 70 blanks to be filled in), beginning with the localization, following with the description of biotope conditions (relief, climate, rock, substrate), herbaceous cover, shrubs, the description of stand and of the buffer area (Pârnuţă et al. 2011). The core limits were marked in the field using yellow paint.

After the completion of the fieldwork, a FGR Catalogue file containing the most important data of the field file has been established. Each FGR is individualized by a unique code composed of: the symbol "RG", followed by the code of the main species for which the genetic resource is constituted, followed by the enumeration of the associated species, then the region of provenance, and finally the FGR sequence number of the basic species in the provenance region, example: RG-LA/BR,MO-B120-1 (genetic resources (RG), main species is *Larix decidua* (code=LA)/*Abies alba* (BR) and *Picea abies* (MO) are associated species, provenance region is B1, Curvature Carpathians, beech-fir-spruce associated phytoclimatic floor, first in the order of *Larix* FGR in the B1 provenance region).

At the same time, in order to increase the resources for the production of the forest reproductive materials, 55 years ago were start a programme for selections of seed stands, which are now listed in the National Catalogue of Forest Reproductive Materials (Pârnuţă et al. 2012).

16.3 Status of the In Situ Conservation of Romanian FGRs

Romanian National Catalogue of Forest Genetic Resources included a total number of 698 cores with an effective area of 10559.3 ha, 577 of them (93% of the total cores area) are preserved in situ (Pârnuţă et al. 2011). The most important species in Romania's forests (European beech, Norway spruce, Silver fir and oaks, which together account for about 75% of the forests area) also benefit from a majority representation in the FGR catalogue, 69% of the FGR cores were constituted for the conservation of these species (Fig. 16.1). The number of cores established for hardwood conservation is almost double in comparison with the number of resinous FGR (Fig. 16.1). In the category of other hardwood are included: wild cherry (*Prunus avium*), alders (*Alnus sp.*), sweet chestnut (*Castanea sativa*), poplars (*Populus sp.*), black locust (*Robinia pseudoacacia*), black walnut (*Juglans nigra*), shagbark hickory (*Carya ovata*), Siberian elm (*Ulmus pumila*), London sycamore (*Platanus* x *hybrida*), tree-of-heaven (*Ailanthus altissima*) and Turkish hazel (*Corylus colurna*). In the category of other resinous (softwood) are included: grand fir (*Abies grandis*), English yew (*Taxus baccata*), Douglas fir (*Pseudotsuga menziesii*) and bald cypress (*Taxodium distichum*).

At the same time, in order to increase the resources for the production of the forest reproductive materials, 2930 units for 45 forest species were selected, with a total area of 40209.3 ha, which are listed in the National Catalogue of Forest Reproductive Materials (Pârnuţă et al. 2012).

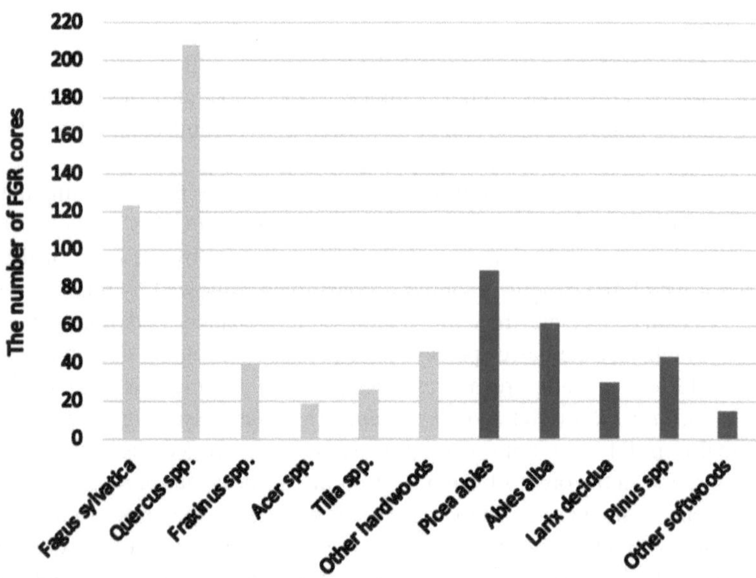

Fig. 16.1 The species distribution of Romanian FGR

In the following, some data about the Romania FGR are presented, by species, within the two large groups (resinous and hardwoods) and in order of the species importance.

16.3.1 Hardwood Tree Species

16.3.1.1 European Beech (*Fagus sylvatica* L.)

A total of 123 FGR were constituted for European beech and the area occupied by the strictly protected area (FGR cores) is 3106 ha, which represents 24% of the total area of FGR cores (Pârnuţă et al. 2011). The distribution of beech FGR by region of provenance (Fig. 16.2) shows a 88% presence in the Carpathian mountain forests, the rest of the preservation cores being located in the hills and plains of the inner Carpathian arch (Transylvania) or outside of them (Moldavia). All beech FGR are preserved in situ, and most of them originate from seed and are of superior productivity, so they relate to the high productivity FGR group. The average age, in 2017, was 115 years, with amplitude of 130 years, from 65 to 190 years. Among these stands are found some of the virgin beech forests that were included in UNESCO World Heritage, in 2017 (July 7).

In the National Catalogue of Forest Reproductive Materials (Pârnuţă et al. 2012), with a leading role of seeds production, are included a total number of 368 seed sources, all of them were selected in natural populations.

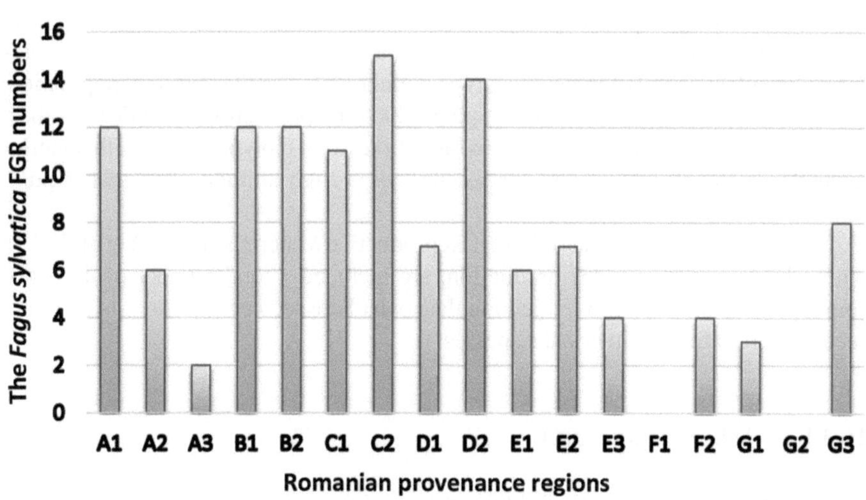

Fig. 16.2 The distribution of the European beech FGR in Romania by regions of provenance (A–E = Carpathians, F = Transylvania plain, G = Moldavia plain)

16.3.1.2 Oaks (*Quercus spp.*)

Forest genetic resources were established for the majority of *Quercus* species, most of them being sampled for sessile oak (*Quercus petraea*) and pedunculate oak (*Quercus robur*): 65% of the 208 oaks FGR. In addition to these two species, FGR was also constituted for: Turkey oak (*Q. cerris*), Hungarian oak (*Q. frainetto*), grayish oak (*Q. pedunculiflora*), pubescent oak (*Q. pubescens*), red oak (*Q. rubra*), pin oak (*Q. palustris*).

For **sessile oak (*Quercus petraea*)** a total number of 81 conservation cores were established, being relatively uniformly distributed in all hill and plateau subregions (Stuparu et al. 2015). The average age of the oak trees in the conservation cores is 110 years (amplitude 60–170 years) and half of them are of superior productivity, the rest being of the medium productivity. FGR of sessile oak are preserved in situ and are located at the altitude between 160 and 800 m.

A total number of 54 FGR where constituted for conservation of **pedunculate oak (*Quercus robur*)**, relatively uniformly distributed in plains and low hills regions, the stands from the conservation cores being, on average, 105 years old (amplitude 60–165 years).

In the forest steppe and forest plains are included in the forest compositions other oaks species for which conservation core units have been established in the most representative areas. Thus, 26 FGR have been established for the Turkey oak, 20 for the Hungarian oak, 16 for the grayish oak and 2 for the pubescent oak, the latter two species being the most suitable for the forest steppe area. Red oak and pine oak have good adaptability in some regions of Romania, so that FGR was also constituted for these species (8 and 1, respectively).

Because the vast majority of oaks' FGR are related to the group of high productivity FGR, the tree breeding specialists from Romania want to identify new FGR for the other four groups. In this regard, in 2016, 10 new FGR of oaks were selected, which will be added to the existing 208 (Budeanu et al. 2016): 4 preservation cores for pedunculate oak (2 of them in ecotype populations, 1 in the marginal population, at low altitude, and another one in a high productivity population), 3 FGR for pubescent oak, 1 FGR for the Turkey oak (white Turkey oak biotype, wood with superior technological qualities), a core for sessile oak at high altitude (marginal population), and a core for the pin oak.

In the National Catalogue of Forest Reproductive Materials (Pârnuţă et al. 2012), with a leading role of seeds production, are included a total number of 473 seed sources for sessile oak, 226 for pedunculate oak (4 are seed orchards) and 157 (3 are seed orchards) for the other oaks species.

16.3.1.3 Ash (*Fraxinus spp.*)

For the species of this genus, 39 FGRs were constituted, of which 33 are for in situ conservation: 20 for common ash (*Fraxinus excelsior*) and 13 for narrow-leaved ash (*Fraxinus angustifolia*). The average age of trees for the preservation cores is

100 years for both species (the amplitude of variation being between 55 and 155 years for the common ash, and between 65 and 145 years for the narrow-leaved ash, respectively). A total number of 144 seed sources (5 seed orchards) were established for ash species, inclusive 1 stand for *Fraxinus pennsylvanica* (Pârnuţă et al. 2012).

16.3.1.4 Maple (*Acer spp.*)

For sycamore maple (*Acer pseudoplatanus*) and Norway maple (*Acer platanoides*) were constituted 14, respectively 4 FGRs, and their distribution is relatively uniform in the natural range of species in Romania. Predominant are the cores identified by the tree productivity criterion. In 2017, trees were, on average, 95 years old, ranging from 70 to 125 years old. A total number of 84 seed sources (one seed orchard) were established for maple species (Pârnuţă et al. 2012).

16.3.1.5 Linden (*Tilia spp.*)

Conservation cores were established for all three existing species in Romania: 16 FGR for silver linden (*Tilia tomentosa*), 7 for small-leaved linden (*Tilia cordata*) and 3 for large-leaved lime (*Tilia platyphyllos*). All originate from natural regeneration, from seed and are preserved in situ. The current average age is 90 years. A total number of 114 seed sources (3 seed orchards) were established for linden species (Pârnuţă et al. 2012).

16.3.1.6 Other Hardwood Species

Forest genetic resources (FGR) were also constituted for wild cherry (*Prunus avium*) – 2 cores, black alder (*Alnus glutinosa*) – 5 cores, sweet chestnut (*Castanea sativa*) – 2 cores, white poplar (*Populus alba*) – 9 cores, black poplar (*Populus nigra*) 3 cores, black locust (*Robinia pseudoacacia*) – 11 cores, black walnut (*Juglans nigra*) – 7 cores, and for other species, like: grey alder (*Alnus incana*), green alder (*Alnus viridis*), shagbark hickory (*Carya ovata*), Siberian elm (*Ulmus pumila*), London sycamore (*Platanus hybrida*), tree of heaven (*Ailanthus altissima*) and Turkish hazel (*Corylus colurna*), were establish one FGR/species (Pârnuţă et al. 2011).

16.3.2 *Softwood Tree Species*

16.3.2.1 Norway Spruce [*Picea abies* (L.) Karst]

From the total number of Romanian FGR, 13% were set up for the conservation of valuable Norway spruce populations, the total area of the spruce preservation cores representing 20% of the total area of the FGR cores. Both spruce and all other basic species appear not only in cores formed in pure stands, but also as associated species in other FGRs. All the spruce conservation core units are located in the mountainous area of Romania, and 97% (86 cores) are constituted for the in situ preservation of the species. In 2017, the mean age of spruce FGR was 110 years, varying between 70 and 165 years. They are naturally regenerated stands, from seeds, and mostly are reported to high productivity FGR category. There are also marginal FGRs, in stands from the upper altitudinal limit of the species (e.g. FGR with Norway spruce and Swiss stone pine at 1700 m altitude, on Bistra Valley in Sibiu County) but also the FGR for intraspecific varieties (the resonant spruce variety in Eastern Carpathian, used for making the violins, but also the narrow-crown spruce ideotype, especially encountered in the Western Romanian Carpathians, much more resistant to wind and snow breaks).

In the National Catalogue of Forest Reproductive Materials (Pârnuţă et al. 2012), with a leading role of seeds production, are included a total number of 454 seed sources for Norway spruce (14 are seed orchards).

16.3.2.2 Silver fir (*Abies alba* Mill.)

Silver fir is a species of high economic and ecological value, for which it is necessary to increase the proportion of participation in the afforestation composition, especially in resinous-beech mixed forests. When the FGR catalogue was developed, it was intended to conserve as many as possible valuable populations, which could also be used as sources of reproductive material. In this context, although the silver fir area in the Romanian forests is almost 6 times smaller than in the case of Norway spruce, the silver fir FGRs' number is only 31% lower (Pârnuţă et al. 2011). The 61 silver fir FGRs are distributed in all regions of provenance from the Romanian Carpathians, being better represented in the Eastern Carpathians than in the Western ones. The stands are of medium (11%) or superior (89%) productivity and are naturally regenerated from seeds. In 2017, the average age of silver fir FGR was 110 years, with a variation amplitude of 125 years (55–180 years), and the stands have an average consistency of 0.75. All silver fir FGRs are preserved in their natural environment (in situ). There was also identified a FGR of grand fir (*Abies grandis*) mixed with Douglas fir, that is located in Brasov County (Scărlătescu et al. 2012). The grand fir trees have, at the age of 65, an average height of 23 m, thus falling into the category of high productivity stands and showing the ability to adapt to the biotope characteristics of that area.

In the National Catalogue of Forest Reproductive Materials (Pârnuţă et al. 2012), with a leading role of seeds production, are included a total number of 280 seed sources for silver fir (10 are seed orchards).

16.3.2.3 European Larch (*Larix decidua* Mill.)

The European larch from Romania relates to the *carpatica* variety (geographical race). Currently, the larch has a proportion of only 0.07% in the Romanian forests, most often being a disseminated species (Şofletea and Curtu 2007). A number of 30 FGRs were established for larch, 77% of them are stands obtained through artificial regeneration which is preserved ex situ (Pârnuţă et al. 2011). In the future, maximum attention should be given to the identification of larch FGR in natural populations. The average age of the seven larch FGRs preserved in natural populations (in situ) is 120, with 95–160 years variation amplitude. Also, a total number of 86 seed sources for European larch (21 are seed orchards) and two seed orchard for hybrid larch (Larix x eurolepis) are included in the catalogue of reproductive materials (Pârnuţă et al. 2012).

16.3.2.4 The Pine (*Pinus spp.*)

In Romania, there were established 44 FGRs for the preservation of 6 species of the genus *Pinus*, most cores being designated for Scots pine (22) and black pine (15). There were also set up four cores for Swiss stone pine (*Pinus cembra*) and one each for mountain pine (*Pinus mugo*), black pine var. Banatica (*Pinus nigra* ssp. *banatica*) and eastern white pine (*Pinus strobus*) (Pârnuţă et al. 2011). The **Scots pine (*Pinus sylvestris*** L.) FGRs are relatively uniformly distributed in all of the provenance regions from the Romanian Carpathians, but also in the plateau regions of Transylvania and Moldavia. Only five of the Scots pine FGRs are preserved in situ. The five natural Scots pine populations are, on average, 105 years old. All the **black pine (*Pinus nigra ssp nigra* Arn.)** FGRs are preserved *ex situ*. Four FGRs are constituted for the conservation of **Swiss stone pine (*Pinus cembra* L.**), three of them are preserved in situ and are located at the upper altitudinal limit of Romanian forests (up to 1860 m), where they form pure stands or mixed stands with Norway spruce and mountain pine. Also, a total number of 85 seed sources for the Scots pine (10 are seed orchards), 47 for black pine (13 are seed orchards) and 11 (seven are seed orchards) for the other pine species, are included in the catalogue of forest reproductive materials (Pârnuţă et al. 2012).

16.3.2.5 Other Softwood Species

In the category of other resinous fall: grand fir (*Abies grandis*), English yew (*Taxus baccata*), Douglas fir (*Pseudotsuga menziesii*) and bald cypress (*Taxodium distichum*), species for which a total number of 14 FGRs (10 for Douglas fir) were constituted. Only English yew is preserved in situ.

A total number of 35 seed sources for Douglas fir (6 are seed orchards) are included in the catalogue of forest reproductive materials (Pârnuţă et al. 2012).

16.4 Conclusions

In Romania, 698 FGR core units were established for the conservation of 45 main species and 31 associated species. The most important species of the Romanian forests (European beech, Norway spruce, silver fir and oaks) are in situ preserved and are, on average, 110 years old. The vast majority of Romanian FGRs is related to the first group of FGRs, namely high productivity genetic resources. The next edition of the National Catalogue will have to include many more stands from the other categories: marginal FGRs, edaphic and climatic varieties, intraspecific varieties with distinct anatomical and morphological traits.

Romanian National Catalogue of FGR was approved by Ministerial order no. 945/1.03.2012, so it has the character of law. All forest managers have the obligation to comply with all recommendations regarding the conservation and sustainable management of stands included in the FGR Catalogue. In Romania exists, at the same time, with the aim of seeds production, a National Catalogue of forest reproductive materials, also approved by Ministerial order (no. 1645/2.07.2013). In the current edition, a total number of 2930 units for 45 forest species are included. In the future, we intend to edit only one catalogue and avoid duplications (in present more than 300 forest reproductive material units are also FGRs).

Acknowledgements Budeanu Marius and Popescu Flaviu acknowledge the Department of Trees Genetic and Breeding of the National Institute for Research and Development in Forestry "Marin Drăcea" for necessary facilities and funding agencies [ANCSI (PN09460209, PN16330203) and UEFISCDI (Contract 618/2005)] for financial assistance.

References

Budeanu, M., Stuparu, E., & Tănasie, Ş. (2016). Identificarea de noi resurse genetice forestiere de cvercinee cu adaptabilitate ridicată (Identification of new forest genetic resources of oaks with highly adaptability). *Revista de Silvicultură şi Cinegetică, 38*, 21–26.

Geburek, T., & Turok, J. (Eds.). (2005). *Conservation and management of forest genetic resources in Europe*. Zvolen: Arbora Publishers.

IFN. (2012). *Inventarul Forestier Naţional, rezultatele primului ciclu (2008–2012) (Romanian forest inventory, results of first edition, from 2008 to 2012)*. INCDS „Marin Drăcea". Available online at: http://www.mmediu.ro/articol/rezultate-ifn-ciclul-i-2008-2012/1667.

Lalu, I. (1996). Catalogul Naţional al Resurselor Genetice Forestiere-Partea I (National catalogue of forest genetic resources, part I). ICAS Manuscript.

Pârnuţă, G., Lorenţ, A., Tudoroiu, M., & Petrila, M. (2010). *Regiunile de provenienţă pentru materialele de bază din care se obţin materialele forestiere de reproducere din România (The provenance regions for the basic materials from which are obtained the forest reproductive materials in Romania)*. Bucureşti: Editura Silvică.

Pârnuţă, G., Stuparu, E., Budeanu, M., Scărlătescu, V., Marica, F. M., Lalu, I., Tudoroiu, M., Lorenţ, A., Filat, M., Teodosiu, M., Nica, M. S., Cheşnoiu, E. N., Pârnuţă, P., Mirancea, I., Marcu, C., Pepelea, D., Dinu, C., Marin, S., Daia, M., Dima, G., Şofletea, N., & Curtu, A. L. (2011). *Catalogul naţional al resurselor genetice forestiere (National catalogue of forest genetic resources)*. Bucureşti: Editura Silvică.

Pârnuţă, G., Budeanu, M., Stuparu, E., Scărlătescu, V., Cheşnoiu, E. N., Tudoroiu, M., Filat, M., Nica, M. S., Teodosiu, M., Lorent, A., Daia, M., & Dinu, G. C. (2012). *Catalogul Naţional al Materialelor de Bază pentru producerea Materialelor Forestiere de Reproducere (National catalogue of forest reproductive materials)*. Bucureşti: Editura Silvică.

Popescu, F., Postolache, D., & Pitar, D. (2015). Aspecte privind conservarea şi managementul resurselor genetice forestiere din România (Aspects concerning the conservation and management of the forest genetic resources in Romania). *Revista de Silvicultură şi Cinegetică, 37*, 13–17.

Scărlătescu, V., Stuparu, E., & Budeanu, M. (2012). Resursele genetice forestiere de Brad (*Abies alba* Mill.) din România (Forest genetic resources of silver fir (*Abies alba*) from Romania). *Revista de Silvicultură şi Cinegetică, 31*, 30–33.

Şofletea, N. (2005). *Genetică şi ameliorarea arborilor (Genetics and tree breeding)*. Braşov: Editura "Pentru Viaţă".

Şofletea, N., & Curtu, A. L. (2007). *Dendrologie (Dendrology)*. Braşov: Editura Universităţii "Transilvania".

Stănescu, V., & Şofletea, N. (1998). *Silvicultura cu bazele geneticii forestiere (Silviculture with the base of forest genetics)*. Bucureşti: Editura "Ceres".

Stuparu, E., Budeanu, M., & Scărlătescu, V. (2015). Consideraţii privind conservarea *in situ* a resurselor genetice forestiere de gorun (*Quercus petraea*) din România (Considerations regarding the *in situ* conservation of sessile oak (*Quercus petraea*) forest genetic resources in Romania). *Revista Pădurilor, 130*(1–2), 3–9.

White, T. W., Adams, W. T., & Neale, D. B. (2007). *Forest genetics*. Cambridge: CABI Publishing.

Chapter 17
From Nature Conservation to Dynamic Genetic Conservation: The Hellenic Case

Paraskevi G. Alizoti and Filippos A. Aravanopoulos

Abstract The importance of the Hellenic ecosystems from a biodiversity point of view at all levels is presented. The value of the existing biodiversity and its significance has been recognized quite early and the first protected areas were designated at a national level in 1937. The protected areas designated ever since under national and regional (EU) legislation, as well as under international Conventions and their protection status is presented in detail. Most of the protected areas though, that cover almost one third of the country, have been designated mainly for nature conservation (ecosystem, species level) and not for conserving the species genetic resources, even if within the general protection framework they were also indirectly conserved. Focused activities aiming towards the dynamic conservation of forest tree species (maintenance of forest tree genetic diversity to ensure the evolutionary continued processes and not just the preservation of existing frequencies of genotypes and alleles) were initiated after the participation of the country in the EUFORGEN program and the EUFGIS project. Dynamic gene conservation units (GCUs) have been designated so far for a number of species, but the process of designation needs to be intensified, given the role of the country as a biodiversity hotspot and an area of high endemism.

Keywords Biodiversity · Forest genetic resources · Dynamic genetic conservation · Protected areas · Gene Conservation Units (GCUs)

P. G. Alizoti (✉) · F. A. Aravanopoulos
School of Forestry and Natural Environment, Aristotle University of Thessaloniki, Thessaloniki, Greece
e-mail: alizotp@for.auth.gr; aravanop@for.auth.gr

M. Šijačić-Nikolić et al. (eds.), *Forests of Southeast Europe Under a Changing Climate*, Advances in Global Change Research 65,
https://doi.org/10.1007/978-3-319-95267-3_17

17.1 Introduction

Biodiversity can be defined as the variation present at the genetic, species (flora and fauna) and ecosystem level, referring to the diversity of genes and genotypes, the richness of species, and the diversity of species communities, habitats and ecological processes (Ramanatha and Hodgkin 2002).

Forests are biologically diverse ecosystems, valuable not only for being habitats for numerous plant and animal species, but also for serving as carbon sinks and key ecosystems for climate change mitigation. Biodiversity is one of the main characteristics of forest ecosystems, due to the ecosystems, communities, species and within species variation that represent diversity at all levels of biological organization (Geburek and Turok 2005). Forest ecosystems are challenged though by threats and risks, such as climate change and human population growth. Human population growth is expected to result in further intensification of the biodiversity and environment deterioration, through emission of greenhouse gases and industrial pollutants, deforestation, overexploitation of natural resources, extensive change of land use, continuously increasing demand for wood and non-wood products, forest services, and water resources.

Intraspecific genetic diversity is the cornerstone of biodiversity, as its existence allows species to evolve over time and space, and secures the maintenance of biodiversity at all levels (species, ecosystems, landscapes). Consequently, the presence and extent of genetic diversity determine the survival of species over time and the stability of forest ecosystems (Ledig 1986; Friedman 1997; Alizoti 2007). Potential loss of genetic diversity from several causes (i.e. human impact, stochastic events, climate change) is expected to have dramatic impacts on species and forest ecosystems and therefore measures are needed to protect and conserve their genetic resources (Alizoti 2010, 2011).

Certain parts of the globe are biodiversity hotspots; biogeographic regions that are significant reservoirs of biodiversity, characterized by high endemism, and are threatened with destruction (Myers 1988; Myers et al. 2000). The Mediterranean Basin, being a cross-road of continents and civilizations, is characterized by the great number of refugia, its great variety of environments and the vast number of flora and fauna species. The great biodiversity of the Mediterranean basin can also be attributed to the postglacial migration and demographic evolution of forest species in different environments, to the isolated populations rescued in refugia that were bearing genes for specific adaptations to adverse or stressful environments, or to the rare and unique alleles characterizing the Mediterranean tree populations (Hobbs et al. 1995). It is thus one of the most significant biodiversity hotspots, including 22,500 endemic vascular plant species, four times the number found in the rest of Europe (Blondel and Aronson 1999, Conservation International). Hellas (Greece) is among the countries of the Mediterranean basin considered as important biodiversity hotspots at a global scale. The country is characterized by the variety of environments and natural ecosystems, the vast number of flora and fauna species harbored in those ecosystems and the high level of endemism. The Hellenic flora includes 575,218 species and 893 subspecies of angiosperms, which represent 6600 taxa that belong to 1072 genera and 185 families (Official Government Gazette 2014). The Hellenic natural ecosystems are among the most important European and Mediterranean centers of endemism, as they harbor 1278 endemic species and 452 endemic subspecies, but also for the vast genetic variation within species and subspecies (Official Government Gazette 2014).

In the Hellenic forests, that cover the 25.4% of the country area, more than 400 forest tree species are growing, most of which serve as keystone species for the forest ecosystems that harbor the majority of the recorded fauna and flora biodiversity. Out of the 400 forest tree species, 42% are evergreen broadleaves, 20% noble broadleaves and 38% conifers. The 65.5% of the forested area is public, managed under the principles of sustainable and close to nature forestry, 22.51% is private, and the rest is owned by enterprises. The high forests cover is 34.7% of the forested area, while the coppice forests cover the 48% and the mixed high coniferous with coppice broadleaved ones cover the 17.3%. Forest tree species and populations are characterized by high genetic variation (Ganopoulos et al. 2011; Aravanopoulos et al. 2001; Morgante et al. 1998), which could be attributed to the variation of the environmental conditions they may grow, as well as to their glacial and post glacial history. Rear edge peripheral and ecologically marginal populations of nearly all forest tree species growing in the country can be identified. These populations may harbor genes for specific adaptations and are of great importance for *in situ* and *ex situ* conservation, given the climate change challenges. Moreover, several species, such as *Pinus halepensis, Pinus brutia, Pinus sylvestris, Fagus sylvatica, Abies alba, Picea abies, Castanea sativa, Prunus avium* meet the limits of their natural range in the country.

For all the above reasons, Hellenic forest ecosystems started to be protected at the national level since 1937 and are currently conserved to a large extent under national and regional legislation, regional programs or international Conventions, focusing on nature conservation, as well as on the dynamic genetic conservation of the forest genetic resources.

17.2 Nature Protection/Conservation

According to IUCN (1994), the protected areas are 'areas especially dedicated to the protection and maintenance of biological diversity and associated cultural resources, and managed through legal or other effective means'. IUCN clarified later on that the above definition should be applied in the context of certain principles and only when the main objective is nature conservation can the areas be considered as protected ones; in those areas additional goals can be set, but if in conflict with nature conservation, the latter one will be of first priority (Dudley 2008). Protected areas may cover various situations, ranging from managed resource areas, protected watersheds, national parks and strictly protected reserves, to sacred forest groves (Thomson and Theilade 2001).

The value of conserving biodiversity in natural ecosystems was recognized quite early (1937) in the country (Law 856/1937 – Official Government Gazette), and the first protected areas (i.e. forests, waterlands) were recognized at a national level, due to their high ecological significance. The first *in situ* protection/conservation approach, followed in 1937 at a national level, was that of the absolute protection of natural ecosystems and restriction of all human activities. Later on, the approach of integrating the protected areas into the surrounding areas and linking of protection to the sustainable use of natural resources has been adopted (Legal Decree 996/71 – Official Government Gazette).

The natural ecosystems of the country are recognized as protected either according to existing national legislation or following international conventions, ratified by the country, as well as international or European initiatives. The Natura 2000 sites are protected areas of habitat types of European Community interest. In several cases there is an overlap among the areas protected at the national, European and international level.

17.2.1 Protected Areas at the National Level

The Protected Areas following the National Legislation (Official Government Gazette – Law 3937/2011) are: Areas of absolute nature protection, Areas of nature protection, National Natural parks or Regional Parks, Habitat and Species Protection Areas (Special Areas of Conservation (EZZ), Special Protection Areas (SPAs) or Wildlife Refuge areas, or a combination thereof, protected landscapes and landscape elements or protected natural formations.

The ten National Forest Parks, that include /forest ecosystems of particular ecological and scientific interest, were designated during the period 1937 (Official Government Gazette-Law 856/1937) to 1971 (Official Government Gazette – Legal Decree 996/71) and are the following: National Park of Parnitha (3812 ha), Sounio (3.500 ha), Pindos (6927 ha), Oete (7210 ha), Prespes (19,470 ha), Vikos-Aoos (12,000 ha), Samaria (4850 ha), Parnassos (3515 ha), Aenos (2862 ha), Olympus (3.513 ha). The National Forest Parks of Prespa, Vikos-Aoos, Pindos, Oete and Sounio include both a core area and a buffer zone, while the rest include only core areas (Hellenic Ministry of Environment – www.ypeka.gr). The core areas are strictly protected and no human activity or intervention is allowed by legislation, while in the buffer zones limited human intervention can be allowed. It is worth mentioning that the first National Park was designated, according to the Law 856/1937, in 1938, and was the Olympus National Forest Park, followed by the Parnassos National Forest Park within the same year.

Besides the 10 National Forest Parks, 17 National Parks that include waterlands, lakes, lagoons, sea parks and forest areas were designated in 2011, under the Law 3937/2011 (Official Government Gazette). Those are the following: Schinias-Marathon National Park, National Park of Lakes Koroneia–Volvi, National Park of North Pindos (National Vikos – Aoos forest and Pindos), Mesolongi Lagoon

National Park, Kerkini Wetland National Park, National Forest Park of Dadia - Lefkimmi – Soufli, Evros River Delta National Park, National Park of Amvrakikos Wetlands, National Park of Eastern Macedonia and Thrace (Delta of Nestos River–Vistonida – Ismarida), Rhodope Mountain Range National Park, National Park of the Axios – Loudias – Aliakmon Rivers Delta, National Forest Park of Prespes, National Park of Kotychi- Strophylia Wetlands, National Park of Tzoumerka, Peristeri and Arachthos Gorge, Helmos-Vouraikos National Park, Zakynthos National Sea Park, Alonissos National Sea Park, National Park of Psalidi coast and wetlands – Kos, Nature Protection Area of Acherontas and Kalamas Rivers estuaries, Nature Protection Area of Kastoria Lake.

Moreover, following the Law 3937/11, four areas, namely Mt. Parnitha, Mt. Ymittos, Milos island biotopes, Mt. Parnon and Moustou wetland, are protected with specific protection measures.

Nineteen Aesthetic Forests, covering an area of 32,506 ha have been designated under the Law Decree 996/71 (Official Government Gazette) and include forest landscapes of particular aesthetic and ecological interest. The protection of those forests aims towards nature protection and recreation activities (Hellenic Ministry of Environment).

The 51 Preserved Monuments of Nature, most of which occupy an area of several square meters, include individual trees or tree clusters of particular botanical, ecological, aesthetic, historical or cultural value. In the same category fall areas of important ecological, palaeontological, geomorphological or other interest (Hellenic Ministry of Environment).

Wildlife Refuge areas (610 in total) have also been designated in natural ecosystems serving as shelters of wild life (Hellenic Ministry of Environment).

17.2.2 Protected Areas at the International, Regional/ European Level

Besides the national legislation, laws ratifying international conventions are significant tools for the protection of nature and biodiversity. Such international conventions ratified by laws in the country are the Convention of Bern for the conservation of wildlife and the natural environment of Europe, the Bonn, the Barcelona, and the Ramsar Conventions.

Following the above mentioned legal framework ten wetlands and lakes of international significance have been designated under the Ramsar Convention. Nine protected areas (out of which six are forest ecosystems) have been designated under the Barcelona Convention.

The Biogenetic Reserves of international or regional interest designated up to now are the following (Source: Hellenic Ministry of Environment): the Virgin forest of Central Rhodope, the core areas of the Olympus, Aenos, Samaria, Oete, Pindos National Forest Parks, the Cedars Forest of Prespes National Park, the Aesthetic

Forest Kouri-Almyros, the natural monuments of Chaidou-Koula beech forest, the Lesinion forest, the Sapienza Island evergreen Platanus Forest, the Almopia-Aridea mixed forest, the Embona-Rhodes cypress forest, the Mt. Grammos mixed forest, the Paranesti-Drama Virgin Forest and the Laganas-Zakynthos Bay. Also, two Biosphere Reserves were designated, namely the Mt. Olympus National Park and the Samaria Gorge, while Mt. Athos has been designated as a Monument of International Heritage. The Samaria Gorge has been also recognized as an Area with Eurodiploma (Eurodiploma is granted to areas of natural heritage and of European interest and are protected accordingly).

The Natura 2000 (European Commission 2008) network is a pan -European nature protection program for species and their habitats and the cornerstone of the European policy for the conservation of nature. It aims to ensure the long-term preservation of the most valuable and most endangered species and habitats at a satisfactory level. It was initiated in 1992 when the Habitats Directive was adopted, complementing the Birds Directive (79/409/EEC). The two EU Directives serve as the legal framework for the network Natura 2000, which includes 241 Sites of Community Importance (SCI) in the country and 202 Special Protection Areas (SPA) for birds, covering a total area of approximately 5.5 million hectares (Hellenic Ministry of Environment). The Natura 2000 network of sites has also included the majority of the protected by national legislation areas, as well as areas that have been internationally classified.

Following the existing legislation (Official Government Gazette – Law 2742/99), protected areas are managed by Management Bodies, public services, special services and legal entities or bodies designated for this specific purpose by management contracts. Protected areas are managed according to 5 year Management Plans that identify the priorities for the implementation of projects, actions and measures aiming towards the effective protection and management of the protected objects. The Management Plans are also accompanied by action plans.

The Hellenic Nature 2000 Commission is the central scientific advisory body of the State for the coordination, monitoring and evaluation of policies and measures to protect Hellenic biodiversity. Its unimpeded operation is the responsibility and duty of the Minister of Environment and Energy (Official Government Gazette-Law 3937/2011). The main responsibility of the Commission is to monitor and enforce the provisions of the Habitats Directive (92/43/EEC), as incorporated into the Hellenic Law, for all the Natura 2000 sites of the country and beyond, covering the whole of the National Area (beyond the Natura 2000 network). The Nature 2000 Commission acts also as the National Committee of Protected Areas, bearing the responsibility to coordinate, monitor and evaluate the plans of action, organize and operate the National System of Protected Areas Administration and Management (Official Government Gazette – Law 2742/1999). It also records, classifies, coordinates, controls and evaluates the work and activities of the Management Bodies. Additionally, it proposes to the appropriate Ministries the general measures and actions that need to be taken for the protection of nature and the sustainable development of the protected areas, and recommends the allocation of resources to

Management Bodies in order to fulfill their goals (Official Government Gazette-Law 3044/2003).

17.3 Dynamic Genetic Conservation

The international efforts towards improving the management of forest genetic diversity and resources were initiated more than 40 years ago (FAO 1975, 1989) and the lack of knowledge regarding the minimum size of a genetic conservation unit for forest trees was recognized as a problem by FAO in 1975. Shaffer (1981) and Ryder (1986) presented the concepts of 'minimum viable population' (MVP – population size ensuring the population persistence for a given period of time) and 'evolutionary significant units' (ESU-populations having independent evolutionary histories), contributing to the inclusion of genetics in conservation that leaded to the dynamic conservation approach that was applied also in forest trees (Namkoong 1997).

At the pan-European level the biodiversity policy has been driven mainly by the Ministerial Conferences on the Protection of Forests (MCPFE) and the 'Environment for Europe' process. The first Ministerial Conference of 'Forest Europe' adopted in 1990 a resolution, following which EUFORGEN – the European Forest Genetic Resources Program – was established. EUFORGEN is an international cooperation program promoting the conservation and sustainable use of forest genetic resources in Europe as an integral part of sustainable forest management. Greece participated as a member in EUFORGEN, as well as in the EUFGIS (European Information System on Forest Genetic Resources) project (http://www.eufgis.org/project) and the EUFGIS expert's group that developed the pan-European minimum requirements and data standards for the dynamic conservation units (Koskela et al. 2012; Lefevre et al. 2012). Thus, EUFORGEN and EUFGIS facilitated the establishment of a pan-European network of dynamic Gene Conservation Units (GCUs), that is fulfilling the minimum requirements and data standards set by the program (Koskela et al. 2012; Lefevre et al. 2012) and can be found in the EUFGIS portal (http://www.eufgis.org/). Lefevre et al. (2012) assessed the network of dynamic genetic conservation units, established across 33 European countries that are managed for the dynamic conservation of forest genetic resources, based on the information available in the EUFGIS portal, the species distribution maps and the environmental stratification of Europe. The gap analysis they performed for eleven species revealed that for almost all species the marginal parts of the species distribution ranges were not covered by the network of GCUs, while multiple GCUs were recorded in core areas. The above authors, based on their results, proposed a combination of species and site oriented strategies to achieve an efficient conservation of genetic resources across the whole continent that could capture, if possible, whole the existing forest genetic resources.

Within the above framework, 15 GCUs were designated in Greece, based on previous results of research on genetic and adaptive variation of five species. In detail, three GCUs have been designated for each one of the species *Pinus halepensis*,

Pinus brutia, *Pinus nigra*, *Abies cephalonica*, and *Abies borisii-regis* (http://www.eufgis.org/). Seed stands located in areas that were proved to harbor a great amount of genetic and adaptive variation, and are included in the National Catalogue, were designated as GCUs for the five above mentioned species. Given the excessive number of forest tree species and their populations growing in the country the designation of GCUs is rather limited. For 12 species though of high economic and ecological value, the conservation effort can be counterbalanced to a certain extend by the existing seed stands. Management plans of seed stands are aiming, besides seed production, also to the protection of the stand and its genetic material. Effort needs to be intensified though, as a huge forest genetic resources capital exists in the forests of the country that is of high current and potential future value, due to the extent of the genetic variation and the specific adaptations it bears, given climate change.

17.4 Conclusions

Hellenic ecosystems are considered to be among the most diverse across the globe due to their great variation, the number of flora and fauna species they harbor and the high levels of within species genetic variation. The value of conserving biodiversity in natural ecosystems was recognized quite early in the country and the first protected areas were recognized at a national level in 1937. Ever since, a great part of terrestrial and marine area of the country has been protected under national and regional (EU) legislation, as well as by international Conventions, most of which involve nature conservation. Following the participation of the country in EUFORGEN and EUFGIS, the conservation of genetic resources on a population basis has been initiated by the establishment of dynamic gene conservation units (GCUs). Fifteen GCUs has been designated for five coniferous species so far, that were exhaustively studied regarding their genetic and adaptive variation, in areas that were proved to harbor genetic resources with high genetic and adaptive variation. Given the extensive number of forest tree species, the levels of genetic variation harbored within species and their potential current and future value, intensification of the effort is needed at the national level.

References

Alizoti, P. G. (2007). Conservation of forest genetic resources and climate change. Measures-perspectives-suggestions. In *Proceedings of the 13th Conference of Hellenic Forestry Society*. Kastoria, 7–10 October 2005, pp. 406–412.

Alizoti, P. G. (2010, October 13–15). Forest ecosystems and genetic resources in a multiple changing environment: Risks-challenges-perspectives. In *Proceedings of the 13th Conference of Hellenic Scientific Society for Plant Genetics and Breeding on 'Genetic erosion'*, Kalamata, p. 218.

Alizoti, P. G. (2011). Human impact on forest genetic resources. In *Proceedings of the 14th Conference of Hellenic Forestry Society*. Patra, November 1–4, 2009, pp. 867–875.

Aravanopoulos, F. A., Drouzas, A. D., & Alizoti, P. G. (2001). Electrophoretic and quantitative variation in chestnut (*Castanea sativa* Mill.) in Hellenic populations in old-growth natural and coppice stands. *Forest Snow and Landscape Research, 76*(3), 429–434.

Blondel, J., & Aronson, J. (1999). *Biology and wildlife of the Mediterranean region*. New York: Oxford University Press.

Conservation International, Biodiversity hotspots website https://www.conservation.org/How/Pages/Hotspots.aspx

Dudley, N. (2008). *Guidelines for applying protected area management categories*. Gland: IUCN.

European Commision. (2008). *Natura 2000*. Protecting Europe's Biodiversity. European Commission. Directorate – General for Environment. Publ. Information Press, Oxford, U.K.

FAO. (1975). *Forest genetic resources information No. 4*. Rome: Food and Agriculture Organization of the United Nations.

FAO. (1989). *Plant genetic resources: Their conservation in situ for human use*. Rome: Food and Agriculture Organization of the United Nations.

Friedman, S. (1997). Forest genetics for ecosystem management. In K. Kohm & J. F. Franklin (Eds.), *Creating a forestry for the 21st century – the science of ecosystem management* (pp. 203–212). Washington, DC: Island Press.

Ganopoulos, I., Aravanopoulos, F. A., Argiriou, A., Kalivas, A., & Tsaftaris, A. (2011). Is the genetic diversity of small scattered forest tree populations at the southern limits of their range more prone to stochastic events? A wild cherry case study by microsatellite-based markers. *Tree Genetics & Genomes, 7*, 1299–1313.

Geburek, T., & Turok, J. (2005). Conservation and sustainable management of forest genetic resources in Europe-an introduction. In: *Conservation and sustainable management of forest genetic resources in Europe* (pp. 3–9). Zvolen: Arbora Publishers.

Hellenic Minstry of Environment. http://www.ypeka.gr/Default.aspx?tabid=433

Hobbs, R. J., Richardson, D. M., & Davis, G. W. (1995). Mediterranean-type ecosystems: opportunities and constrains for studying the function of biodiversity. In G. W. Davis & D. M. Richardson (Eds.), *Mediterranean-type ecosystems* (pp. 1–42). New York: Springer.

IUCN (1994). Guidelines for protected areas management categories. IUCN-WCPA and WCMC, Gland.

Koskela, J., Lefevre, F., Schueler, S., Kraigher, H., Olrik, D. C., Hubert, J., Longauer, R., Bozzano, M., Yrjana, L., Alizoti, P., Rotach, P., Vietto, L., Bordacs, S., Myking, T., Eysteinsson, T., Souvannavong, O., Fady, B., De Cuyper, B., Heinze, B., von Wuehlisch, G., Ducousso, A., & Ditlevsen, B. (2012). Translating conservation genetics into management: Pan-European minimum requirements for dynamic conservation units of forest tree genetic diversity. *Biological Conservation, 157*(1), 39–49.

Ledig, F. T. (1986). Heterozygosity, heterosis, and fitness in outbreeding plants. In M. Soule (Ed.), *Conservation biology – The science of scarcity and diversity* (pp. 77–104). Sunderland: Sincauer.

Lefevre, L., Koskela, J., Hubert, J., Kraiger, H., Longauer, R., Olrik, D. S., Schueler, S., Bozzano, M., Alizoti, P., Bakys, R., Baldwin, C., Bailian, D., Samuelsson, S. B., Bednarova, D., Bordacs, S., Collin, E., De Cuyper, B., De Vries, S., Eysteinsson, T., Frydl, J., Haverkamp, M., Ivankovic, M., Konrad, H., Koziol, C., Maaten, T., Notivol Paino, E., Ozturk, H., Pandeva, I. D., Parnuta, G., Pilipovic, A., Prostolache, D., Ryan, C., Steffenrem, A., Varela, M. C., Vessella, F., Volosyanchuk, R. T., Westergren, M., Wolter, F., Yrjana, L., & Zarina, I. (2012). Dynamic conservation of forest genetic resources in 33 European countries. *Conservation Biology, 27*(2), 373–384.

Morgante, M., Felice, N., & Vendramin, G. G. (1998). Analysis of hypervariable chloroplast microsatellites in *Pinus Halepensis* reveals a dramatic genetic bottleneck. In A. Karp, P. G. Isaac, & D. S. Ingram (Eds.), *Molecular tools for screening biodiversity*. Dordrecht: Springer.

Myers, N. (1988). Threatened biotas: "hot spots" in tropical forests. *Environmentalist, 8*, 187–208.

Myers, N., Mittermeier, R. A., Mittermeier, C. G., da Fonseca, G. A. B., & Kent, J. (2000). Biodiversity hotspots for conservation priorities. *Nature, 403*, 853–858.

Namkoong, G. (1997). A gene conservation plan for loblolly pine. *Canadian Journal of Forest Research, 27*, 433–437.

Official Government Gazette – Hellenic Parliament. (2014). Approval of the national strategy for biodiversity for the years 2014–2029 and five years plan of action. *FEK, 2383*(B'), 29615–29734 (http://www.et.gr).

Official Government Gazette – Hellenic Parliament. http://www.et.gr

Ramanatha, V., & Hodgkin, T. (2002). Genetic diversity and conservation and utilization of plant genetic resources. *Plant Cell, Tissue and Organ Culture, 68*, 1–19.

Ryder, O. A. (1986). Species conservation and systematics: The dilemma of subspecies. *Trends in Ecology & Evolution, 1*, 9–10.

Shaffer, M. L. (1981). Minimum population sizes for species conservation. *Bioscience, 31*, 131–134.

Thomson, L., & Theilade, I. (2001). Protected areas and their role in conservation of forest genetic resources. In FAO, DFSC, IPGRI (Ed.), *Forest genetic resources conservation and management. Vol. 2: In managed natural forests and protected areas (in situ)*. Rome: International Plant Genetic Resources Institute.

Chapter 18
In Situ Conservation of Forest Genetic Resources in Bulgaria

Alexander H. Alexandrov and Rumen D. Dobrev

Abstract The chapter presents information about in situ conservation of forest genetic resources in Bulgaria. The activities have resulted in establishment of many in situ units for gene conservation. A survey of all such object is done for 9 coniferous and 39 broad-leaved species. The units are classified into three categories: reserves, maintained reserves, and natural permanent seed production stands. The total number of the units is 6032 coniferous stands and 15,772 stands of broad-leaved species. All they contribute to gene resources conservation and their sustainable use in forest trees in Bulgaria.

Keywords Gene conservation units · Seed production stands

18.1 Introduction

The forest inventory in Bulgaria resulted in producing maps of different scales (most frequently 1:50,000 and 1:100,000). The maps contain information about the predominant tree species and many other stand characteristics, which makes them suitable for using in the gene conservation activities of forest trees.

Gene conservation of forest tree species is performed in many different territories, most important being 55 nature reserves and 35 maintained reserves covering total 59348.2 ha. These include also the reserves situated within the territory of the three national parks and eleven nature parks in Bulgaria.

Threatened tree species have different share in the conservation programmes – from 1% to 100%, depending on the particular conditions and other circumstances.

In situ conservation of forest genetic resources in Bulgaria started in 1952, when the forestry authorities laid down the foundations of scientifically based seed supply and founded two forest seed control station. These stations situated in the largest two

A. H. Alexandrov (✉) · R. D. Dobrev
Forest Research Institute, Bulgarian Academy of Sciences, Sofia, Bulgaria
e-mail: alexandrov_38@abv.bg

M. Šijačić-Nikolić et al. (eds.), *Forests of Southeast Europe Under a Changing Climate*, Advances in Global Change Research 65,
https://doi.org/10.1007/978-3-319-95267-3_18

cities in Bulgaria – Sofia and Plovdiv – had the task to organize the selection of permanent seed production stands (Dobrinov et al. 1965, 1982, 1992). This point marked the start of large-scale in situ conservation, performed in a cooperation between the experts in forestry practice on the one hand, and the academic community, represented by researchers from the Forest Research Institute and University of Forestry. It should be pointed out, however, that in situ conservation in fact started with declaring of the first protected territories in Bulgaria – Silkosia reserve in Strandzha Mountains in 1933 and the first national park – Vitosha Mountains in 1934.

Genetic resources in reserves and maintained reserves have the statute of populations conserved "in situ" according to the Protected Areas Act (1998). They contribute also to fulfilling the objectives of the Convention of Biological Diversity (CBD). Moreover, permanent seed producing stands were also included in this category in an appropriate Instruction of the Ministry of Agriculture (Anonymous 2012).

Table 18.1 is filled in based on data of the Ministry of Environment and Waters (2011), Markoff (2011), Central Balkan National Park (2004), Pirin National Park (2004), Rila National Park (2004), and the Executive Forest Agency (2011a, b).

Total 90 nature reserves of different status, situated at least partly within the three national and eleven nature parks, together with the permanent seed production stands (almost 4000) act as objects conserving the valuable forest genetic resources.

The total area aimed at in situ gene conservation is 90799.4 ha. This value is about 2.4% of the total forest cover in Bulgaria. The area of conifers is 38652.7 ha and that of broadleaved forests – 52146.7 ha. Mean stand size is 5.8 ha (6.4 ha for conifers and 5.4 ha for deciduous species), which reflects the highly diverse environmental conditions and the variable mountain landscape of Bulgaria (Alexandrov and Dobrev 2014).

Different tree species are protected to different extent in the in situ conservation units. Bosnian Pine (*Pinus heldreichii* Christ.) keeps first position with 88.4% of its stands conserved, followed by *Pinus mugo* Turra (36%), *Pinus peuce* Gris. (19.4%), *Abies alba* Mill. (10.4%) and *Picea abies* (L.) Karst. (6.9%). These figures indicate sufficient amount of in situ conservation territories for these species. The areas occupied by the broadleaved species represent lower, but still good percentage: *Fraxinus* sp. (15.6%), *Castanea sativa* Mill. (6.3%), *Acer* spp. (6.1%), *Fagus* spp. (4.6%), and *Betula pendula* Roth (3.6%). Oaks (*Quercus* spp.), Hornbeams (*Carpinus* spp.), poplars (including Aspen, *Populus* spp.) and some other genera are underrepresented.

Total area of the permanent seed production stands (NPSPS) is 31451.2 ha, or 0.8% of the total forest area, composed by 13109.4 ha (1.2%) of coniferous and 18341.8 ha (0.7%) of deciduous forests. Conifers are enough represented in the NPSPS area, as estimated by the percentage of the species' area of distribution in Bulgaria (*Abies alba* Mill. – 2.3%, *Picea abies* (L.) Karst – 2.5%, *Pinus sylvestris* L. – 1.3%). An exception is *Pinus nigra* Arn., with only 0.4%. Of broadleaved species, the most represented are *Castanea sativa* Mill. – 1.6%, *Fraxinus* spp. – 1.8%, *Betula pendula* Roth. – 1.6%, *Fagus* spp. – 1.6%. Isufficient area of NPSPS was recorded in some species like *Quercus* spp., *Carpinus betulus* L., *Juglans regia* L., *Sorbus aucuparia* L., *Sorbus aria* (L.) Crantz., and *Pyrus communis* L.

Table 18.1 Forest tree species included in the in situ conservation programs/units

Species (scientific name in alphabetical order)	Type of in situ conservation unit	Number of stands conserved	Total area ha
Gymnosperms			
Abies alba Mill. and *A. borisii-regis* Mattf.	Reserves	774	2617.1
	Maintained reserves	19	34.8
	NPSPS[a]	153	752.4
	Σ	**946**	**3404.3**
Juniperus communis L.	Reserves	4	331.0
Juniperus excelsa Bieb.	Reserves	40	244.4
Picea abies (L.) Karst.	Reserves	1282	7034.7
	Maintained reserves	15	37.6
	NPSPS	667	4023.7
	Σ	**1964**	**11096.0**
Pinus heldreichii Christ.	Reserves	178	1107.1
	NPSPS	2	9.8
	Σ	**180**	**1116.9**
Pinus mugo Turra	Reserves	332	8558.8
Pinus nigra Arn.	Reserves	348	1239.6
	Maintained reserves	35	188.0
	NPSPS	180	1026.5
	Σ	**563**	**2454.1**
Pinus peuce Grsb.	Reserves	523	2710.5
Pinus sylvestris L.	Reserves	559	1438.0
	Maintained reserves	2	1.7
	NPSPS	919	7297.0
	Σ	**1480**	**8736.7**
Total gymnosperms	Reserves	4040	25281.2
	Maintained reserves	71	262.1
	NPSPS	1921	13109.4
	Σ	**6032**	**38652.7**
Angiosperms			
Acer campestre L.	Reserves	120	84.3
	Maintained reserves	16	20.8
	Σ	**136**	**105.1**
Acer platanoides L.	Maintained reserves	1	2.1
	NPSPS	2	3.4
	Σ	**3**	**5.5**
Acer pseudoplatanus L.	Reserves	155	137.3
	NPSPS	7	15.2
	Σ	**162**	**152.5**
Acer tataricum L.	Maintained reserves	1	0.4
Aesculus hippocastanum L.	Maintained reserves	10	18.6

(continued)

Table 18.1 (continued)

Species (scientific name in alphabetical order)	Type of in situ conservation unit	Number of stands conserved	Total area ha
Alnus glutinosa (L.) Gaertn.	Maintained reserves	5	28.5
	NPSPS	7	1.1
	Σ	**12**	**29.6**
Betula pendula Roth.	Reserves	145	177.2
	NPSPS	22	146.9
	Σ	**167**	**324.1**
Carpinus betulus L.	Reserves	548	1519.7
	Maintained reserves	34	129.0
	NPSPS	19	108.8
	Σ	**601**	**1757.5**
Carpinus orientalis Mill.	Reserves	410	2188.9
	Maintained reserves	29	70.4
	Σ	**439**	**2259.3**
Castanea sativa Mill.	Reserves	15	120.5
	NPSPS	8	40.6
	Σ	**23**	**161.1**
Fagus sylvatica L.	Reserves	2107	16644.8
	Maintained reserves	58	348.3
	NPSPS	940	9751.1
	Σ	**3105**	**26744.2**
Fagus orientalis Lipsky	Reserves	403	1720.5
	Maintained reserves	3	9.8
	Σ	**406**	**1730.3**
Fraxinus excelsior L.	Reserves	81	224.9
	Maintained reserves	2	6.5
	NPSPS	21	43.9
	Σ	**104**	**275.3**
Fraxinus ornus L.	Reserves	323	585.5
	Maintained reserves	17	9.8
	Σ	**340**	**595.3**
Fraxinus oxycarpa Willd. (*F. angustifolia* Vahl.)	Reserves	282	850.0
	Maintained reserves	50	391.5
	NPSPS	16	226.3
	Σ	**348**	**1467.8**
Juglans regia L.	Reserves	7	7.7
	Maintained reserves	2	0.2
	Σ	**9**	**7.9**
Ostrya carpinifolia Scop	Reserves	144	523.0
	Maintained reserves	2	1.7
	Σ	**146**	**524.7**

(continued)

Table 18.1 (continued)

Species (scientific name in alphabetical order)	Type of in situ conservation unit	Number of stands conserved	Total area ha
Populus alba L.	Reserves	1	2.7
	Maintained reserves	13	22.1
	NPSPS	6	3.1
	Σ	**20**	**27.9**
Populus nigra L.	Maintained reserves	6	14.0
	NPSPS	2	1.1
	Σ	**8**	**15.1**
Populus tremula L.	Reserves	176	268.4
	NPSPS	8	49.1
	Σ	**184**	**317.5**
Cerasus avium (L.) Moench.	NPSPS	10	16.4
Pyrus communis L.	Maintained reserves	1	0.8
Quercus cerris L.	Reserves	329	474.2
	Maintained reserves	30	26.9
	NPSPS	255	1934.7
	Σ	**614**	**2435.8**
Quercus frainetto Ten.	Reserves	635	2081.6
	Maintained reserves	35	139.7
	Σ	**670**	**2221.3**
Quercus petraea (Matt.) Liebl.	Reserves	1087	3968.4
	Maintained reserves	47	110.5
	NPSPS	637	5463.3
	Σ	**1771**	**9542.2**
	Reserves	71	265.7
Quercus pubescens Willd.	Maintained reserves	10	23.5
	NPSPS	16	103.2
	Σ	**97**	**392.4**
Quercus robur L.	Reserves	68	194.4
	Maintained reserves	36	166.0
	NPSPS	37	280.0
	Σ	**141**	**640.4**
Salix alba L.	Reserves	1	2.7
	Maintained reserves	10	11.4
	Σ	**11**	**14.1**
Salix caprea L.	Reserves	25	12.6
Sorbus aria (L.) Crantz.	Maintained reserves	3	3.0
Sorbus aucuparia L.	Reserves	3	7.5
Sorbus torminalis (L.) Crantz.	Reserves	33	9.9
Tilia cordata Mill.	Reserve	1	0.5
Tilia platyphyllos Scop.	NPSPS	20	153.6

(continued)

Table 18.1 (continued)

Species (scientific name in alphabetical order)	Type of in situ conservation unit	Number of stands conserved	Total area ha
Tilia tomentosa Moench.	Reserves	64	87.6
	Maintained reserves	8	2.1
	Σ	**72**	**89.7**
Ulmus glabra Huds.	Reserve	1	0.5
Ulmus laevis Pall.	Maintained reserves	10	20.5
Ulmus minor Mill.	Maintained reserves	31	62.7
Viburnum opulus L.	Reserves	2	3.1
Total angiosperms	Reserves	7237	32164.1
	Maintained reserves	470	1640.8
	NPSPS	2033	18341.8
	Σ	**9740**	**52146.7**
Total gymnosperms + angiosperms	Reserves	11,277	57445.3
	Maintained reserves	541	1902.9
	NPSPS	3954	31451.2
	Σ	**15,772**	**90799.4**

[a]*NPSPS* natural permanent seed production stands

The survey on the forest seed production revealed that for a long time the policy of forest authorities aimed at producing seeds from seed stands, rather than from seed orchards. Indigenous tree species were treated with priority, which is due that most afforestations were performed using local species. It could be hypothesized that using natural seed production stands of indigenous species has contributed to the establishment of ecologically sound plantations with a potential to adapt to future environmental changes.

The main constraint to improving in situ genetic conservation programmes is related more to economic, rather than to forestry issues.

The priorities of Bulgaria for the conservation in situ of genetic resources can be summarized as follows:

- Poviding financial support for forest gene conservation actions, including in situ activities;
- Keeping and development of forest seed control stations;
- Strengthenning of capacity-building activities in the forest service and support to research activities.

18.1.1 Special Issues Related to In Situ Conservation

Bulgaria has developed enough in situ measures for conservation of forest genetic resources. And given the fact that to date the efforts were directed to ensuring enough area for each species, now the focus could be transferred to the better management of the established conservation units, even at the expence of their area reduction.

Due to the specific issues in gene conservation of some species, it should be outlined that still conservation of *Juglans regia* L. and *Castanea sativa* Mill. Is being performed following on-farm practice.

Acknowledgements Thanks are due to the administration of the Ministry of Agriculture, Food and Forests, Executive Forest Agency, and Ministry of Environment and Waters of Bulgaria for providing the necessary information.

References

Alexandrov, A., & Dobrev, R. (2014). *The state of forest genetic resources in Bulgaria* (pp. 1–85). Scholar's Press.

Anonymous. (2012). Instruction № 21 of 12.11.2012 for the conditions and rules for identification, approval, registration and cancellation of the sources of forest seed-production base, collection and harvesting of forest reproductive materials, their testing, trade and import. Executive Forest Agency, MAFF. (36 pp).

CBD. (1992). Convention on Biological Diversity (pp. 1–28).

Dobrinov, I., Bodzakov, P., & Michev, B. (1965). *Temporary instruction for selection, management and utilization of forest seed producing stands in natural stands, Committee for Forests and Forest Industry at the Council of Ministers* (pp. 1–40). Sofia: Zemizdat (in Bulgarian).

Dobrinov, I., Dojkov, G., & Gagov, V. (1982). *Forest gene fund in P. R. Bulgaria* (pp. 1–259). Sofia: Zemizdat.

Dobrinov, I., Gagov, V., Alexandrov, A., Tsankov, G., Garilov, C., & Bouzov, B. (1992). *Instruction for building up, management and utilization of forest seed production base* (pp. 1–54). Sofia: Council of Ministers – Committee of Forests (in Bulgarian).

Executive Forest Agency (EFA). (2011a). *List target forest species included in ex situ conservation programmes/units*. Internal report (in Bulgarian).

Executive Forest Agency (EFA). (2011b). *Generalized list of the approved sources of forest reproductive materials* (in Bulgarian).

FAO, FLD, IPGRI. (2004). *Forest genetic resources conservation and management. Vol. 3: In plantations and gene banks (ex situ)*. Rome: International Plant Genetic Resources Institute.

Markoff, I. (2011). Data base for stands included in reserves and maintained reserves. Personal communication.

Ministry of Agriculture. (2012). Instruction 4/2012 for the conditions and order of registration of forest nurseries and the production of seedlings in forest nurseries – State property (State Gazette, No. 17, 2012).

MOEW. (2011). *Data base for stands included in the reserves and maintained reserves*.

National Park Central Balkan – Data base for stands included in the park. (2004). MS-Excel format.

National Park Pirin – Data base for stands included in the park. (2004). MS-Excel format.

National Park Rila – Data base for stands included in the park. (2004). MS-Excel format.

Protected Areas Act. State Gazette, No 133/11.11.1998.

Part IV
Ex Situ Conservation Case Studies

Chapter 19
Ex Situ Conservation of Forest Genetic Resources in Serbia

Srđan Stojnić, Saša Orlović, and Andrej Pilipović

Abstract Ex situ conservation of forest genetic resources (FGR) in Serbia is predominantly based on its preservation through the establishment of field trials, such as seed orchards, provenance trials and clonal archives. In the present chapter we made an overview of ex situ objects founded in Serbia so far, as well as research results gained within them. All the objects have been established by individual selection of plant material from phenotypically superior genotypes which characterized by high stem quality, and above-average height and diameter increments. Nevertheless, present situation on FGR conservation in Serbia is unsatisfactory due to neither sufficient number of objects nor tree species covered. Thus, future activities on ex situ conservation should be directed towards the establishment of new field objects of the most important tree species for Serbian forestry, as well as species that are projected to be the most endangered by climatic change. Likewise, taking into account climate change projections for the region of South-East Europe, a modified approach in tree breeding is required to meet new demands in a rapidly changing environment.

Keywords Forest genetic resources · Ex situ conservation · Climate change · Republic of Serbia

19.1 Introduction

Forest genetic resources (FGR) in Serbia are broadly affected by different biotic (non-native tree species, invasive pest and diseases) and abiotic (forest fires, drought, etc.) threats, as well as negative anthropogenic influence (tourism development, fragmentation of natural forests, poor state of private forests, negative forest practice, illegal logging, etc.) (Aleksić and Jančić 2011; Lazarević et al. 2012; Popović et al. 2017; Stojanović et al. 2015). In addition, climate changes are supposed to be

S. Stojnić (✉) · S. Orlović · A. Pilipović
Institute of Lowland Forestry and Environment, University of Novi Sad, Novi Sad, Serbia
e-mail: srdjan.stojnic@uns.ac.rs; sasao@uns.ac.rs; andrejp@uns.ac.rs

M. Šijačić-Nikolić et al. (eds.), *Forests of Southeast Europe Under a Changing Climate*, Advances in Global Change Research 65,
https://doi.org/10.1007/978-3-319-95267-3_19

additional challenges for sustainable forest management and conservation of FGR, especially in the region of South-East Europe.

Genetic diversity has a central role in survival and adaptive evolution of forest tree species (Koskela et al. 2007). Due to most of the genetic variation at the species level lays within rather than amongst different populations, presence of a high level of genetic diversity within population is a key prerequisite that allows more rapid adaptation of forest trees to altered environmental conditions (Milad et al. 2011). For that reason, long-term conservation of FGR represents the basis for sustainable forest management (Koskela et al. 2007).

Genetic diversity might be conserved throughout various in situ and ex situ conservation forms. Predominant form of FGR conservation in Serbia is in situ, mainly through the establishment of seed stands. Ex situ conservation is particularly important in the areas of high forest lost, areas strongly affected by different threats, as well as forest areas that are expected to be highly endangered in the future. Therefore, in the present chapter an overview of existing ex situ objects in Serbia has been made. Also, studies conducted in these objects, as well as brief recommendations concerning improvement of ex situ conservation in Serbia have been listed shortly.

19.2 Seed Orchards

19.2.1 Pedunculate Oak Seed Orchards

Growth and vitality of pedunculated oak mature stands in Serbia is negatively affected due to long-term decline of Danube and the Sava River water levels. Besides mature stands, this problem is also evident at the newly regenerated areas where the reforestation attempt failures are widely present, especially during dry years (Stojanović et al. 2015). Finally, pedunculate oak is projected to be the most endangered tree species in Serbia due to global warming and expected habitat changes (Stojanović et al. 2014). In order to survive such intensive environmental changes, high within-species genetic diversity is required (De Rogatis et al. 2013). Therefore, pedunculate oak programs aimed to the conservation of FGR have been focusing dominantly on addressing the issue of maintenance of this diversity (Stojnić et al. 2014).

For this purpose, three seed orchards of pedunculated oak have been established at the territory of PE Vojvodinašume – two vegetative (clonal) and one generative seed orchard. Joint for all of them is that are composed of all four pedunculated oak varieties native in the valley of the Sava River: early – *praecox*, typical – *typica* and two late varieties – *tardiflora* and *tardissima* (Orlović et al. 2000).

The first vegetative seed orchard was founded between 1979 and 1983 (Erdeši 1996). The material is collected from the area of 42,000 ha and the main criteria for selection of mother trees were the stem quality and resistance to *Erysiphe alphitoides* (Griffon & Maubl.) U. Braun & S. Takam. (Orlović et al. 2014). The seed orchard is composed of 86 genotypes (multiplied by grafting into 2520 remets) and cover the area of 7 ha. Another vegetative seed orchard was founded in 1986 at the area of 2.5 ha, from 43 genotypes collected from the previously established clonal seed orchard. Numerous researches have been carried out in these orchards and were primarily focused on the examination of phenotypic variation of acorn (Nikolić and Orlović 2002; Nikolić et al. 2010), leaf morphological (Nikolić et al. 2006) and leaf anatomical (Nikolić et al. 2003, 2005) traits. Finally, studying the genetic structure related to different phenology of oak varieties existing in the orchards, Galović et al. (2014) documented presence of high allelic polymorphism between and within varieties.

Generative seed orchard was established at the beginning of twenty-first century (2000–2004), from the 129 half-sib lines originating from seed collected in clonal seed orchards and plus trees from natural populations. In general, the orchard is composed of 2585 genotypes, planted at the area of 10 ha (Orlović et al. 2014). In comparison to the clonal seed orchards, where the different varieties are arranged randomly across the plot, in this orchard the oak varieties were grouped together into four larger sub-plots to facilitate the process of pollination.

19.2.2 Balkan Maple Seed Orchard

Generative seed orchard of Balkan maple (*Acer heldreichii* Orph.) was established in 1994, from the seedlings age 2 + 0, at the area of 1.05 ha. The seedlings were produced from seed collected from 26 phenotypically superior mother trees, selected at Golija Mountain (Isajev et al. 1994). The total number of plants in the orchard is 2962. The altitude of the orchard is 950–1020 m a.s.l. and the exposition is north-eastern. The orchard is situated at the territory of PE Srbijašume, FE Golija from Ivanjica, on the site *Fagetum montanum* Rud.

The orchard is established by the principle of metapopulation structure, where the seedlings were planted in six more-less similar blocks/sub-orchards (Isajev et al. 2010). This object presents the basis for testing of the maternal populations gene pool (Orlović et al. 2014). The first insight into variability of macroscopic and microscopic traits of half-sib lines in the orchard is provided by Isajev et al. (1994).

19.2.3 Serbian Spruce Seed Orchard

Serbian spruce seed orchard was established in 1987, on the basis of previously conducted multi-annual researches on the species genetic variability in Serbia (Isajev et al. 2010). The seed was collected from 50 dominant mother trees from natural populations. The orchard is situated in Western Serbia, on the area of 2.70 ha. Totally, 5959 seedlings, age 2 + 3, were planted in the blockes, with the space 2 × 2 m. Due to seedlings of same half-sib line were planted into one block, it provides opportunity for simultaneous application of different reproduction types, such as: inbreeding, outbreeding and uniparental (Isajev 1987).

In order to perform successful hybridization, the researches on variability of certain vegetative and reproductive characters at the intra- and inter-half-sib levels were performed. In another phase, controlled hybridization was conducted and the studies on a part of parent genotypes and their hybrids were performed. Based on the following researches it was possible to identify hybrid combinations characterized by the highest mean values of the studied morphometric traits of the cones, as well as those parental individuals with desirable general and specific combining capacity (Isajev 1987).

19.2.4 Austrian Pine Seed Orchard

Given the extent of forest degradation that has occurred in Serbia in the past, artificial forest regeneration is required to protect these sites from further disturbances and enable natural successional processes (Alvarez-Aquino et al. 2004). Due to well known ability of Austrian pine to achieve solid results at poor sites, there was a necessity for organized seed production.

Austrian pine seed orchard was established at the mountain Jelova Gora, of 5422 2-year-old seedlings originating from seed collected from 40 mother trees. The orchard is founded in 1991, on the site *Fagetum montanum* Rud. and cover the area of 2.70 ha. As in the case of Balkan maple, it was established by the principle of metapopulation structure, consisting of five replicated sub-orchards, According to Isajev et al. (2010), such an approach guaranty better adaptation, stability and production of quality seed.

To date, several researches dealing with influence of climatic factors on height and diameter growth (Vukin and Isajev 2004, 2006), variability of the photosynthetic pigments concentration in the needles (Mataruga et al. 2000), and variability of morphological characteristic of the germinators (Mataruga et al. 2003) have been carried out in the orchard.

19.3 Provenance Trials

Apart of gene pool conservation, provenance trails may provide a good basis for evaluation of genetic diversity of various provenances, as well as studying of population phenotypic responses to climate change. Provenance trials in Serbia were established of Norway spruce, Douglas fir and European beech. The trials of each tree species have been established across two or three trial sites, thus enabling studying of provenance by site interaction. Testing of provenances at only one site provides only limited information on the provenances adaptive potential due to superior provenance in one environment may not be regarded as the superior in another. Thus, studying of provenance performances across several sites might provide valuable information on their stability and adaptability.

19.3.1 Norway Spruce Provenance Trials

Norway spruce provenance trials were established in 1993, from five provenances originating from Serbia, and three provenances originating from Slovenia. The trials were established from 4-year-old seedlings on three sites (Rača, Rabrovica and Pavićev Laz) covering an altitudinal range between 600 and 1600 m a.s.l. (Ivetić et al. 2005). The aim of the research was to examine the spruce variability in the part of its natural range in the South-East Europe, as well as to evidence the production capacity and the differences between individual provenances (Isajev et al. 2010). Given the heterogeneity of habitat conditions at the sites where the plots were founded, the results from these trials represent a significant contribution to understanding of spruce provenances adaptive potential.

The first results from the spruce provenance trials established in Serbia are given by Šijačić-Nikolić (1995) and Šijačić-Nikolić et al. (2000), who studied variability of certain morphometric traits (height, height increment, diameter, thickness, number of branches and crown width) in the juvenile phase of seedlings development. Parallel studies in these trials showed significant effect of provenance, locality, and interaction provenance by locality on the analyzed traits. The authors concluded that in the southern parts of the region, spruce provenances originating from Southern Europe (i.e. Serbia) showed better productivity, in comparison to those from Slovenia (Šijačić-Nikolić et al. 2000; Ivetić et al. 2005). Furthermore, Isajev et al. (2010) reported that spruce provenances from Serbia are characterised by high adaptive potential to very different ecological conditions, indicating that, in addition to the natural optimum within their ecological niche, these provenances could successfully adapt and grow at the sites outside of its present natural range.

19.3.2 Douglas Fir Provenance Trials

Douglas fir provenance trials were established in 1982, from the seed imported from USA. The trials were founded by the Institute of Forestry at the Juhor Mountain, in the zone of beech forests (*Fagetum moesiacae montanum* Jov. 1976), and Tanda locality, at the typical oak site (*Quercetum farnetto cerris* Rud. 1949) (Lavadinović et al. 2008). Totally 29 provenances originating from Oregon, Washington and New Mexico were planted with 2 × 2 m spacing, and arranged in a randomized complete block design.

A number of studies aimed to assessment of Douglas fir provenances adaptability and phenotypic variability have been conducted to date. Significant variability between provenances was observed in terms of height growth (Lavadinović et al.

2008), phenology (Lavadinović et al. 2004), morphological traits (Lavadinović and Marković 2012), and physical and mechanical wood properties (Popović and Lavadinović 2003). Height increment largely depended on longitude and altitude of provenances origin sites (Lavadinović et al. 2008), since the best performances in Serbian climate conditions demonstrated maritime provenances originating from the longitudes above 122°W, and from altitudes below 1000 m a.s.l. (Lavadinović et al. 2008). Several studies also evidenced significant differences in needle concentrations of nitrogen, magnesium and phosphorus, as the essential components of plants mineral nutrition (Lavadinović et al. 2011a, b). Finally, the differences in the fresh needles concentration of the essential oils, such as monoterpenes, a-pinene, camphene, b-pinene, citronenal and bornyl acetate, were observed between different provenances (Tešević et al. 2002).

19.3.3 European Beech Provenance Trials

Climate change is supposed to negatively affect beech ecosystems in Serbia. Recent study demonstrated that until the end of the century almost 90% of present beech forests might fall outside their bioclimatic niche, whereas approximately 50% of these forests may experience mass mortality (Stojanović et al. 2013). Therefore, studies on beech trees adaptation strategies, as well as identification of beech provenances with high genetic diversity and growth potential are necessary to establish forest regeneration capable to adjust to the changing climate.

Beech provenance trials in Serbia were established during spring 2007, in the scope of COST Action E52: Evaluation of Beech Genetic Resources for Sustainable Forestry, by planting 2- and 3-year-old seedlings. This was the sixth international network of beech provenance trials, due to five series have been established during last century (1986, 1987, 1988, 1995, and 1998). However, in comparison to the series founded during 1980s and 1990s, where the provenances from Balkan were omitted, this was the first time that the majority of the provenances originate from the Balkan region. The main research objectives in all trial series are: tree improvement, gene conservation, studying of evolution biology and stimulation of pan-European cooperation (von Wuehlisch 2004).

Beech provenance trials in Serbia were founded at two localities: Fruska Gora Mountain (an isolated, marginal site for beech with low elevation, acidic soil and the low annual sum of precipitation), and Debeli Lug (site at a higher altitude, calcareous soil and within the mean humidity range in the pure beech stands of Serbia) (Stojnić et al. 2012a). The trials consist of 25 provenances that cover latitudinal range from the North to the South of Europe, and altitudinal gradient from 35 m to 860 m a.s.l.

The studies in Serbian beech provenance trials have been focused mainly on the investigation of various functional traits variability within and among different provenances, and researches related to phenotypic plasticity of leaf and wood anatomical traits. The results evidenced the presence of significant variability both within and between different provenances (Stojnić et al. 2010, 2012b, 2016; Šijačić-Nikolić et al. 2013; Štajner et al. 2013). Also, the studies indicated ecotypic pattern of genetic variation (Stojnić et al. 2016b; Šijačić-Nikolić et al. 2012), as well as possibility of plastic provenance response to alterations in environmental conditions (Stojnić et al. 2015a, b). Certain studies demonstrated the ability of provenances originating from mesic sites to adapt to drier site conditions (Stojnić et al. 2013). Finally, the studies evidenced that beech provenances from Balkan Peninsula might be potential targets in the search for drought-resistant ecotypes (Stojnić et al. 2015c). This is especially important from the perspective of production, marketing and transfer of forest reproductive material that is better adjusted to changing environmental conditions.

19.4 Clonal Archives

Clonal archives of the forest tree species in Serbia include genotype collections of fast-growing species – poplars, willows and black locust, which are located at the experimental estate of the Institute of Lowland Forestry and Environment. Clonal archives of poplars and willows count more than 1000 different genotypes, whereas black locust is represented with 39 clones selected at the Institute during 1980s and 1990s. Among poplars, most clones belong to the American black poplar (*Populus deltoides* Bartr.) and hybrids of European and American black poplar (*Populus* x *euramericana*). Recently, a clonal archive consisting of 39 poplar clones, originating from 18 European countries was founded at the Institute within EUFORGEN programme aimed to conservation of black poplar gene fund. Besides conservation of gene pool, these objects are also utilized for research activities focused on breeding of clones for multipurpose use (e.g. short rotation coppice, phytoremediation, reforestation of degraded sites, etc.).

19.5 Other Forms of Forest Genetic Resources Conservation

Besides already described conservation objects, it is also worth to mention a few other trials established across Serbia. A large-scale pedunculate oak breeding program through the progeny testing was initiated at the end of the 1990s of the last century. Four parallel progeny trials were established in the spring of 1999 at the territory of PE Vojvodinašume, FE Sremska Mitrovica from acorn collected in vegetative seed orchards and natural populations. The trials were established on various sites within forest associations *Carpino – Quercetum roboris*, *Carpino – Quercetum robori-cerris typicum* and *Fraxino – Quercetum roboris typicum*, covering an area of 9.2 ha, in total. The project is designed to determine primarily the parameters of stem quality, such as diameter, height, trunk straightness, number and thickness of the branches (Orlović et al. 2001).

Scots pine (*Pinus sylvestris* L.) field trial was founded in 2011, as a result of studies on population and individual biological and ecological variability of this species in Serbia (Lučić et al. 2011a, b, c). The object is established at the territory of PE Srbijašume, FE Užice, at the area of 0.9 ha, within the forest association *Abieti-fagetum moeasiacae* B. Jovanović 1953 s.l. (Lučić 2011). The trial is designed in the form of metapopulation, with four sub-populations. The main objectives in this object are studying of inter-population genetic variability and conservation of scots pine genetic diversity. In the next period the trial is planned to be registered as seed orchard.

Finally, a generative collection of wild-fruit tree species was founded by the Institute of Forestry in 2007. The collection is established from 490 genotypes of 16 tree species at the area of 1.9 ha.

19.6 Conclusions

Ex situ conservation of FGR in Serbia mostly relies on the foundation of seed orchards, provenance trials and clonal archives. Likewise, the majority of tree species relevant for Serbian forestry were involved in this programme, while the main criteria in selection of mother trees were phenotypical superiority. Nevertheless, there is a necessity for intensified scientific and expert activities on the conservation of forest genetic, primarily by founding of more field trials, as well as by involvement of more tree species. Besides this, attention should be also paid to marginal tree populations, since these populations might be better adapted to future climate conditions. Similarly, recent studies that involved the most abundant tree species in Serbia evidenced that climate change is supposed to negatively influences their ecosystems and sustainable wood production in them. Having this in mind, it seems that conventional breeding strategies, based on selection of dominantly stem quality and vigorous growth rate, will need to be revised and adjusted to the new climate requirements (e.g. selection for improved water-use efficiency, drought resistance, etc.).

References

Aleksić, P., & Jančić, G. (2011). Zaštita šuma od šumskih požara u Javnom preduzeću "Srbijašume". *Šumarstvo, 1–2*, 95–110.

Alvarez-Aquino, C., Williams-Linera, G., & Newton, A. C. (2004). Experimental native tree seedling establishment for the restoration of a Mexican cloud forest. *Restoration Ecology, 12*, 412–418.

De Rogatis, A., Ferrazzini, D., Ducci, F., Guerri, S., Carnevale, S., & Belletti, P. (2013). Genetic variation in Italian wild cherry (*Prunus avium* L.) as characterized by nSSR markers. *Forestry, 86*, 391–400.

Erdeši, J. (1996). *Vegetativna semenska plantaža hrasta lužnjaka (Quercus robur L.) – izvođački projekat*. Sremska Mitrovica: JP "Srbijašume" Beograd, Šumsko gazdinstvo Sremska Mitrovica.

Galović, V., Orlović, S., Zorić, M., Kovačević, B., & Vasić, S. (2014). Different phenology induced genotype diversity of *Q. robur* in the seed orchard in Srem provenance, Republic of Serbia. In M. Cvjetko Bubalo, I. Radojčić Radovniković, T. Jakovljević, M. Vuković, & D. Erdec Hendrih (Eds.), *Natural resources, green technology and sustainable development* (p. 133). Zagreb: University of Zagreb, Faculty of Food Technology and Biotechnology.

Isajev, V. (1987). Oplemenjivanje omorike (*Picea omorika*/Panč./Purkyne) na genetičko selekcionim osnovama. Doctoral thesis, University of Belgrade, Faculty of Forestry, Serbia.

Isajev, V., Šijačić, M., & Vilotić, D. (1994). Varijabilnost makroskopskih i mikroskopskih karakteristika dvogodišnjih sadnica 26 familija polusrodnika planinskog javora (*Acer heldreichii* Orph.). *Šumarstvo, 3–4*, 21–28.

Isajev, V., Mataruga, M., Ivetić, V., & Lučić, A. (2010). Ex situ conserving, testing and utilization gene pool of endemic-relic tree species in Serbia. In R. Ristić, M. Medarević, & Z. Popović (Eds.), *First Serbian forestry congress – future with forests* (pp. 2–11). Belgrade: University of Belgrade, Faculty of Forestry.

Ivetić, V., Isajev, V., & Šijačić-Nikolić, M. (2005). Results of fourtheen year's old Norway spruce provenance test in Serbia. In I. V. Abrudan, G. Sparchez, G. Ignea, D. Simon, G. Ionascu, & G. Chitea (Eds.), *Proceedings of the symposium forest and sustainable development* (pp. 65–71). Brasov: Universitătii Transilvania Braşov.

Koskela, J., Buck, A., & Teissier du Cros, E. (Eds.). (2007). *Climate change and forest genetic diversity: Implications for sustainable forest management in Europe* (pp. 1–111). Rome: Bioversity International Available at http://www.euforgen.org/fileadmin/bioversity/publications/pdfs/1216.pdf.

Lavadinović, V., & Marković, N. (2012). The differences of needle length of douglas-fir provenances at two sites in test plantations. *Sustainable Forestry, 65–66*, 7–14.

Lavadinović, V., Koprivica, M., & Isajev, V. (2004). Phenological characters of Douglas-fir provenances in Serbia. *Silva Balcanica, 4*, 89–94.

Lavadinović, V., Isajev, V., Rakonjac, L., & Marković, N. (2008). Effect of altitude and continentality of Douglas fir provenances on height increment in test plantations in Serbia. *Lesnícky Časopis – Forestry Journal, 54*, 53–59.

Lavadinović, V., Miletić, Z., Lavadinović, V., & Isajev, V. (2011a). Variability in magnesium concentration in needles of different douglas-fir provenances. *Forestry ideas, 17*, 74–79.

Lavadinović, V., Isajev, V., Miletić, Z., & Krstić, M. (2011b). Variability of nitrogen content in the needles of douglas-fir (*Pseudotsuga menziesii* /Mirb./Franco) provenance. *Genetika, 43*, 407–417.

Lazarević, P., Stojanović, V., Jelić, I., Perić, R., Krsteski, B., Ajtić, R., Sekulić, N., Branković, S., Sekulić, G., & Bjedov, V. (2012). Preliminarni spisak invazivnih vrsta u Republici Srbiji sa opštim merama kontrole i suzbijanja kao potpora budućim zakonskim aktima. *Zaštita Prirode, 62*, 5–32.

Lučić, A. (2011). Podizanje šuma belog bora (*Pinus sylvestris* L.) u Srbiji na ekološko-genetičkim osnovama. Doctoral thesis, University of Belgrade, Faculty of Forestry, Serbia.

Lučić, A., Isajev, V., Rakonjac, L., & Živadinović, V. (2011a). Varijabilnost morfometrijskih svojstava šišarica belog bora (*Pinus sylvestris* L.) u Srbiji. *Šumarstvo, 1–2*, 83–94.

Lučić, A., Isajev, V., Cvetićanin, R., Rakonjac, L., Novaković, M., Nikolić, A., & Mladenović-Drinić, S. (2011b). Interpopulation genetic-ecological variation of Scots pine (*Pinus sylvestris* L.) in Serbia. *Genetika, 43*, 1–18.

Lučić, A., Isajev, V., Rakonjac, L., Ristić, D., Kostadinović, M., Babić, V., & Nikolić, A. (2011c). Genetic divergence of Scots pine (*Pinus sylvestris* L.) populations in Serbia revealed by RAPD. *Archives of Biological Sciences, 63*, 371–380.

Mataruga, M., Isajev, V., & Tucović, A. (2000). Međulinijski varijabilitet fotosintetičkih pigmenata crnog bora (*Pinus nigra* Arn.). *Glasnik Šumarskog Fakulteta, 82*, 107–118.

Mataruga, M., Isajev, V., & Tucović, A. (2003). Varijabilnost morfoloških krarakteristika klijavaca 40 linija polusrodnika crnog bora (*Pinus nigra* Arn.). In J. Berenji (Ed.), *Drugi simpozijum za oplemenjivanje organizama* (p. 122). Vrnjačka Banja: Društvo genetičara Srbije.
Milad, M., Schaich, H., Bürgi, M., & Konold, W. (2011). Climate change and nature conservation in Central European forests: A review of consequences, concepts and challenges. *Forest Ecology and Management, 261*, 829–843.
Nikolić, N. P., & Orlović, S. (2002). Genotipska varijabilnost morfoloških osobina žira hrasta lužnjaka (*Quercus robur* L.). *Zbornik Matice Srpske za Prirodne Nauke, 102*, 53–58.
Nikolić, N., Merkulov, L. J., Krstić, B., & Orlović, S. (2003). A comparative analysis of stomata and leaf trichome characteristics in *Quercus robur* L. genotypes. *Zbornik Matice Srpske za Prirodne Nauke, 105*, 51–59.
Nikolić, N., Merkulov, L., Pajević, S., & Krstić, B. (2005). Variability of leaf anatomical characteristics in Pedunculate oak genotypes (*Quercus robur* L.). In B. Gruev, M. Nikolova, & A. Donev (Eds.), *Balkan scientific conference of biology* (pp. 240–247). Plovdiv: University of Plovdiv "Paisii Hilendarski", Faculty of Biology, Union of scientists in Bulgaria.
Nikolić, N., Krstić, B., Pajević, S., & Orlović, S. (2006). Varijabilnost osobina lista kod različitih genotipova hrasta lužnjaka (*Quercus robur* L.). *Zbornik Matice Srpske za Prirodne Nauke, 111*, 95–105.
Nikolić, N. P., Merkulov, L. S., Krstić, B. Đ., Pajević, S. P., Borišev, M. K., & Orlović, S. (2010). Varijabilnost anatomskih osobina žira kod genotipova hrasta lužnjaka (*Quercus robur* L.). *Zbornik Matice Srpske za Prirodne Nauke, 118*, 47–58.
Orlović, S., Radivojević, S., Erdeši, J., Obućina, Z., & Janjatović, G. (2000). Seed orchards of intra-species forms as a way for maintain and increasing genetic variability of Pedunculate oak in Yugoslavia. In T. Oszako & C. Delatour (Eds.), *Recent advances on oak health in Europe* (p. 31). Warsaw: Forest Research Institute.
Orlović, S., Erdeši, J., Radivojević, S., Obućina, Z., & Janjatović, G. (2001). Strategija i rezultati oplemenjivanja hrasta lužnjaka (*Quercus robur* L.) u Sremu. *Šumarstvo, 1-2*, 11–16.
Orlović, S., Ivanković, M., Andonoski, V., Stojnić, S., & Isajev, V. (2014). Forest genetic resources to support global bioeconomy. *Annals of Silvicultural Research, 38*, 51–61.
Popović, Z., & Lavadinović, V. (2003). Main physical and mechanical wood properties of different Douglas-fir provenances. In Kúdela, J., & S. Kurjatko (Eds.), *Proceedings of the 4th international symposium "Wood structure and properties"* (pp. 217–219), Zvolen: Arbora Publishers.
Popović, V., Lučić, A., & Rakonjac, L. (2017). Stanje šumskih genetičkih resursa u Srbiji i pregled aktivnosti na njihovoj konzervaciji. *Selekcija i Semenarstvo, 23*, 1–13.
Šijačić-Nikolić, M. (1995). Procena genetskog potencijala osam provenijencija smrče (*Picea abies*/L./Karst) iz test kultura kod Ivanjice. Master thesis, University of Belgrade, Faculty of Forestry, Serbia.
Šijačić-Nikolić, M., Isajev, V., & Mataruga, M. (2000). Evaluation of morhometric properties of several spruce (*Picea abies*/L./Karsten) provenances in monocultures in Serbia. In E. Klimo, H. Hager, & J. Kulhavý (Eds.), *Spruce monocultures in Central Europe – problems and prospects* (pp. 145–153). Joensuu: European Forest Institute.
Šijačić-Nikolić, M., Milovanović, J., Nonić, M., Knežević, R., & Babić, V. (2012). Ekotipska karakterizacija genetičke varijabilnosti provenijencija bukve iz jugoistočne Evrope na osnovu morfometrijskih svojstava listova. *Glasnik Šumarskog Fakulteta, 106*, 197–214.
Šijačić-Nikolić, M., Milovanović, J., Nonić, M., Knežević, R., & Stanković, D. (2013). Leaf morphometric characteristics variability of different beech provenances in juvenile development stage. *Genetika, 45*, 369–380.
Štajner, D., Orlović, S., Popović, B., Kebert, M., Stojnić, S., & Klašnja, B. (2013). Chemical parameters of oxidative stress adaptability in beech. *Journal of Chemistry*. https://doi.org/10.1155/2013/592695.
Stojanović, B. D., Krzić, A., Matović, B., Orlović, S., Duputie, A., Djurdjević, V., Galić, Z., & Stojnić, S. (2013). Prediction of the European beech (*Fagus sylvatica* L.) xeric limit

usinga regional climate model: An example from southeast Europe. *Agricultural and Forest Meteorology, 176*, 94–103.

Stojanović, B. D., Matović, B., Orlović, S., Kržić, A., Trudić, B., Galić, Z., Stojnić, S., & Pekeč, S. (2014). Future of the main important forest tree species in Serbia from the climate change perspective. *South-East European Forestry, 5*, 117–124.

Stojanović, D., Levanič, T., Matović, B., & Orlović, S. (2015). Growth decrease and mortality of oak floodplain forests as a response to change of water regime and climate. *European Journal of Forest Research, 134*, 555–567.

Stojnić, S., Orlović, S., Pilipović, A., Kebert, M., Šijačić-Nikolić, M., & Vilotić, D. (2010). Variability of physiological parameters of different European beech provenances in international provenance trials in Serbia. *Acta Silvatica & Lignaria Hungarica, 6*, 135–142.

Stojnić, S., Orlović, S., Galić, Z., Vasić, V., Vilotić, D., Knežević, M., & Šijačić-Nikolić, M. (2012a). Stanišne i klimatske karakteristike u provenijeničnim testovima bukve na Fruškoj gori i u Debelom Lugu. *Topola, 189/190*, 145–162.

Stojnić, S., Orlović, S., Pilipović, A., Vilotić, D., Šijačić-Nikolić, M., & Miljković, D. (2012b). Variation in leaf physiology among three provenances of European beech (*Fagus sylvatica* L.) in provenance trial in Serbia. *Genetika, 44*, 341–353.

Stojnić, S., Sass-Klaassen, U., Orlović, S., Matović, B., & Eilmann, B. (2013). Plastic growth response of European beech provenances to dry site conditions. *IAWA Journal, 34*, 475–484.

Stojnić, S., Trudić, B., Galović, V., Šimunovački, Đ., Đorđević, B., Rađević, V., & Orlović, S. (2014). Očuvanje genetičkih resursa hrasta lužnjaka (*Quercus robur* L.) na području Javnog preduzeća "Vojvodinašume". *Topola, 193/194*, 47–71.

Stojnić, S., Orlović, S., Živković, U., von Wuehlisch, G., & Miljković, D. (2015a). Phenotypic plasticity of European beech (*Fagus sylvatica* L.) stomatal features under water deficit assessed in provenance trial. *Dendrobiology, 73*, 163–173.

Stojnić, S., Orlović, S., Miljković, D., Galić, Z., Kebert, M., & von Wuehlisch, G. (2015b). Provenance plasticity of European beech leaf traits under differing environmental conditions at two Serbian common garden sites. *European Journal of Forest Research, 134*, 1109–1125.

Stojnić, S., Orlović, S., Ballian, D., Ivanković, M., Šijačić-Nikolić, M., Pilipović, A., Bogdan, S., Kvesić, S., Mataruga, M., Daničić, V., Cvjetković, B., Miljković, D., & von Wuehlisch, G. (2015c). Provenance by site interaction and stability analysis of European beech (*Fagus sylvatica* L.) provenances grown in common garden experiments. *Silvae Genetica, 64*, 133–147.

Stojnić, S., Orlović, S., Trudić, T., Kesić, L., Stanković, M., & Šijačić-Nikolić, M. (2016). Height and root-collar diameter growth variability of european beech provenances from Southeast Europe. *Topola/Poplar, 197–198*, 5–14.

Stojnić, S., Orlović, S., Miljković, D., & von Wuehlisch, G. (2016b). Intra- and inter-provenance variation of leaf morphometric traits in European beech (*Fagus sylvatica* L.) provenances. *Archive of. Biological Sciences, 68*, 781–788.

Tešević, V., Doković, D., Milosavljević, S., Lavadinović, V., Matović, M., & Vujević, D. (2002). Etarsko ulje četina duglazije (*Pseudotsuga menziesii* Mirb. Franco) iz različitih provenijencija. *Lekovite Sirovine, 22*, 53–57.

Vukin, M., & Isajev, V. (2004). Istraživanje uticaja određenih klimatskih činilaca na visinski prirast crnog bora u semenskoj plantaži na Jelovoj Gori. *Glasnik Šumarskog Fakulteta, 90*, 91–107.

Vukin, M., & Isajev, V. (2006). Uticaj klimatskog indeksa na debljinski prirast 40 linija polusrodnika crnog bora. *Glasnik Šumarskog Fakulteta, 93*, 31–48.

von Wuehlisch, G. (2004). Series of international provenance trials of European beech. In K. Sagheb-Talebi, P. Madsen, & K. Terazawa (Eds.), *Proceedings from the 7th international beech symposium IUFRO Research Group 1.10.00 improvement and silviculture of beech* (pp. 135–144). Tehran: Research Insitute of Forests and Rangelands.

Chapter 20
Selection and Breeding of Fast Growing Trees for Multiple Purposes in Serbia

Andrej Pilipović, Saša Orlović, Branislav Kovačević, Vladislava Galović, and Srđan Stojnić

Abstract Fast growing tree species play significant role in timber production in Serbia with emphasis on poplars and willows as most dominant species. Use of poplar and willow for establishment of highly productive plantations requires reproductive material which originates from long term breeding and selection processes which are conducted both in Serbia and worldwide. Besides coping with traditional issues such are productivity and resistance to pest and diseases, current circumstances (climate change, degraded lands etc.) pose new challenges in breeding of poplars and willows. This article presents review of both traditional and contemporary breeding and selection processes conducted in Serbia with aim of production new cultivars/clones suitable to obtain various desired properties.

Keywords Poplar and willow clones · Productivity · Rooting · Phytoremediation · Molecular markers

20.1 Introduction

As one of the most often fast growing woody plants of temperate continental climate, genus *Populus* is widely spread naturally in the northern hemisphere in Europe, Asia and North America and also introduced and grown in plantations in South America. The wide altitudinal zone and ability to perform spontaneous and controlled intra and inter species hybridization, enabled the formation of many subspecies as well as a series of transitional forms, i.e. simple and complex hybrids (Orlovic et al. 2010). Above mentioned characteristics resulted in great natural variability enabling poplars to inhabit different sites ranging from riparian zones to mountain forests. In addition to all these advantages, their fast growth and easy vegetative propagations promoted poplars as ideal candidates for tree breeding.

A. Pilipović (✉) · S. Orlović · B. Kovačević · V. Galović · S. Stojnić
Institute of Lowland Forestry and Environment, University of Novi Sad, Novi Sad, Serbia
e-mail: andrejp@uns.ac.rs; sasao@uns.ac.rs; branek@uns.ac.rs; galovic@uns.ac.rs; srdjan.stojnic@uns.ac.rs

M. Šijačić-Nikolić et al. (eds.), *Forests of Southeast Europe Under a Changing Climate*, Advances in Global Change Research 65,
https://doi.org/10.1007/978-3-319-95267-3_20

The importance of poplars and its use for timber and biomass production worldwide reflect in establishment of International Poplar Commission under UN FAO Forestry group which was founded in 1947 by nine European countries in the aftermath of WWII destructions, when poplar and willow culture was considered a priority to supporting reconstruction of rural and industrial economies. Up to date, International Poplar Commission (IPC) comprises of 38 member states worldwide and also involves plants from genus *Salix* with perspective of broadening its scope to other fast growing tree species in the future. These activities resulted in area of 91.9 million ha under poplars and 9.02 million ha under willow forests and plantations worldwide (IPC 2016).

Most significant contribution to widespread growing and use of fast growing poplars and willows refers to more than 70 years of breeding, selection and seeking for new cultivars and clones with improved characteristics in many countries. Following international contemporary trends, government of Yugoslavia formed National Poplar Commission in 1957 and established Poplar Research Institute in Novi Sad in 1958.

20.2 Selection and Breeding Goals and Strategies

Experiences in wood production using fast growing tree species such are poplars and willows until 1950s worldwide, showed that use on one or few cultivars and clones in large areas poses great risk which can be mitigated by selection and breeding of new cultivars and clones with resistance or lower susceptibility to diseases. Feasibility of such strategy can be secured only by obtaining broad genepool for further hybridizing and re-combinations to obtain offspring with desired properties. Growing of poplars and willows plays significant role in overall forestry production in former Yugoslavia. According to unpublished data, in the period from forming of National Commission for poplars and establishment of Poplar research Institute in Novi Sad, Serbia until 2000s (approximately 50-year period), the poplar wood volume production recorded tenfold mass and 18-fold value increase.

Modern breeding programs in Serbia started at the end of 1990s including selection and breeding based on creating of cultivars and clones with multifunctional characteristics (Orlovic et al. 1998, 2001, 2002, 2003, 2006a, b, 2010).

The aim of these breeding programs was to select fast growing cultivars/clones also characterized by resistance to both biotic (pest and diseases) and abiotic factors. Besides resistance and tolerance to pests and diseases, desired properties included rooting ability of clones, wood production and use of new cultivars/clones in different environments (i.e. drought, marginal sites, contaminated soils and groundwater). Investigations in primary phase of selection focused on genetic variability of clones from the point of their anatomy, physiology and growth parameters, such are net photosynthesis, and dark respiration, content of photosynthetic pigments and number and size of stomata (Orlovic 1996, 1998; Orlovic et al. 2006a Topic et al. 2013), and also molecular analysis (Orlovic et al. 2009, 2010; Galovic et al. 2015).

With certainty we can conclude that achievements of improvement in poplar cultivation mentioned above were obtained through use of new cultivars, modern silvicutural practices but also by afforestation of barren land in some extent. Substantial contribution can also be addressed to selection of proper *g x e* (clone x site) interaction, where selection process focuses on investigation of candidate cultivars/clones grown on different soils with the aim of obtaining best cultivar/clone for each soil type regardless the type of the end product. Likewise, a number of research have been focused on timber production where growth vigor of seven poplar clones was investigated with five candidate clones for registration and two clonal cultivars registered earlier (Orlovic et al. 2006b, 2010) or in established biomass plantations (Klasnja et al. 2008, 2012). Regarding willows investigation of productivity follows same pattern with use of physiological, anatomical and growth parameters in selection of cultivars/clones for production of willow timber (Orlovic et al. 2006a, b).

Specificity of plantation forestry refers to application of "cultivar technology" with aim of obtaining maximum performance of selected cultivars/clones. Such demands directed selection process to investigate the effect of application of herbicides to different poplar clones (Vasic et al. 2010, 2015) and also the effect on soil microbiota (Vasic et al. 2010) on sites where herbicides were used.

The improvement of cuttings rooting ability is important task within process of any modern selection and breeding project in poplars (Teissier Du Cros 1984; Guzina 1987; Zalesny et al. 2005). In Serbia, main effort is invested in Eastern cottonwood (*Populus deltoides* Bartr. ex Marsh), due to well-known problems in rooting of this species. The importance of this trait is based on fact that successful rooting of hardwood cuttings of desired genotypes was and still is the ground for utilization of genetic potential of poplars in modern forestry (Zalesny et al. 2005; Kovacevic et al. 2008).

The base of successful cutting rooting in poplars are primordia, and their initiation and activation is the important task in rooting of cuttings in poplar nursery and plantation production as well as in selection and breeding of new poplar cultivars (Smith and Wareing 1972). Preformed primordia, formed during previous growing period, give rise to "preformed" or "bark" roots, while wound primordia, initiated on cut after its formation, give rise to "wound" roots. Due the fact that wound primordia are initiated after wound formation, preformed primordia represents considerable advantage since the period between appearance of preformed and wound roots can be as long as 4 weeks (Kovacevic et al. 2009). According to Heilman et al. (1994), at the end of growing season of *Populus deltoides* partition of wound roots in the total root mass was greater than the mass of preformed roots. That was confirmed by Kovacevic et al. (2008) at the end growing season in *P. deltoides* plants grown from 1-year old rooted cuttings. Though, roots from preformed primordia are less important for growth and development of rooted cuttings, they are of principal importance for their survival, since they are the first roots on that plant can rely at the beginning of growing season (Tschaplinski and Blake 1989; Zalesny et al. 2005; Kovacevic et al. 2008b, 2009, 2011). According to Kovacevic et al. (2009) there is high positive correlation between total number of roots on 20th day after the planting of a cutting and its survival. They described a gap in the increase of total number of roots and even total root length between 20th and 40th day after the planting, while there was no gap in the formation of new leafs in the same period. They assumed that in that period rooted cuttings were under stress due to imbalanced root and shoot growth, due to which the first 80 days from cuttings' planting are crucial for their survival.

Studies on variability of rooting characters in poplars mostly included total number and length of first order roots in poplars in field conditions, in soil substrate and hydroponics, and on different taxonomic and genetic levels (Wilcox and Farmer 1968; Ying and Bagley 1977; Guzina 1987; Zalesny et al. 2005; Kovacevic et al. 2008). There are difficulties of such survey, due to strong influence of random variation and many environmental and within-genotype characteristics, thoroughly reviewed by Kovacevic et al. (2009). Nevertheless, many authors found high heritability of examined rooting characters. According to Kovacevic et al. (2007), in four different terms of measurement during the spring and early summer in 3 years most of examined rooting characters except number of wound roots had relatively high broad sense heritabilites (H_{BS} >0.70), while heritability of number of leaves and shoot's height had high values only 40 days after planting. Kovacevic et al. (2007) also found that due to collinearity, 14 examined characters can be grouped in just two groups by principal component analysis. Shoot characters were in the same group with total number and length of roots, while in the second group were number of roots at the middle and upper portion of cuttings and their partition in total number of roots.

Studies on relationship between rooting characters and cutting survival are also rare. According to Kovacevic et al. (2009) the difference in number of roots that emerged in the first 20 days and mortality between 40th and 80th day after planting made difference between easy-to-root and difficult-to-root genotypes of *Populus deltoides*. Also, the number of leaves that appeared between 20th and 40th day after planting, as well as increment of shoot height in the same period had high positive correlations with cuttings' survival. High positive correlation of shoot characters with survival was presented by Kovacevic et al. (2011) too, while Kovacevic et al. (2008a, b) found by stepwise regression analysis and path coefficient analysis that shoot height at 40th day after planting is the most correlated examined character with cutting survival, with the strongest direct effect. These results are strongly support implementation of shoot characteristics in selection process, allowing the possibility of efficient testing of cuttings' rooting without damaging the plant.

Nowadays, cuttings' rooting is still considered as one of the most important events in the establishment of poplar nursery production and plantations, and it is necessary to continue work on the improvement of rooting abilities of new cultivars, as well as on improvement of the selection and breeding methodologies.

Climate change trends worldwide stresses out inevitable change in forestry practice in the future due to the effect on forests and forest tree species. These changes will influence both site conditions and precipitation and ultimately species occurrence and distribution. Considering previously mentioned facts, forestry practice should focus on adaptation of forest trees and mitigation of climate change impact through increased carbon assimilation. From the fast growing tree species perspective, the mitigation should be primary way to cope with the climate change where afforestation of marginal lands and contaminated sites present new challenge. Therefore, the use of fast growing poplars and willows in climate change mitigation includes additional desired properties of candidate cultivars/clones such are drought tolerance or phytoremediation potential.

Phytoremediation is the use of plants and their associated microorganisms in environmental cleanup (Salt et al. 1995; Raskin et al. 1997). This technology makes use of the naturally occurring processes by which plants and their microbial rhizosphere flora degrade and sequester organic and inorganic pollutants (Pillon-Smits 2005). Amongst various plants that can be used for this purpose poplars and willows present most widely used wooden species in northern hemisphere due to its characteristics such as fast growth, the conductivity of water and minerals throughout the entire section of the tree (diffuse porous species) and easy vegetative propagation by cuttings (Gordon et al. 1997). Due to their biology of pioneer species of emerging alluvial soils, characterized by very rapid growth and highly developed root capable to uptake large amounts of water, it makes them ideal candidates for phytoremediation (Licht and Isebrands 2005). Phytoremediation studies related to the use of poplars and willows in Serbia were directed towards three groups of contaminants: (I) heavy metals, (II) excessive nutrients (nitrates) and (III) petroleum hydrocarbons using clones of *P. nigra*, *P. deltoides* and their hybrids from genus *Populus* and *S.alba*, *S nigra*, *S.matsudana* and *S.viminalis* from genus *Salix*. Results showed different potential of poplar and willow clones for phytoextraction of heavy metals both poplar (Pilipovic et al. 2005) and willow clones (Borisev et al. 2009) with emphasis on clones selected from *P.deltoides* (Nikolic et al. 2017) and *S. nigra* and *S. matsudana* (Nikolic et al. 2015). Phytoremediation of nitrates and petroleum hydrocarbons potential of investigated clones showed also supremacy of *P. deltoides* amongst poplar genotypes and *S.alba* clones amongst willows (Pilipovic et al. 2012). Obtained results of breeding of fast growing species in Serbia showed significant potential of genepool used in selection for phytoremediation.

20.3 Use of New Tools and Technologies for Breeding Improvement

The conventional breeding and selection methods of morphological characterization in forestry represent the basis and irreplaceable part in the process of introducing the genetic diversity of forest species. However, the phenotypic level of characterization (Pilipovic et al. 2012; Stojnic et al. 2015), opens a new chapter and questions about the processes that take place at the genome level and the consideration of the expression of a specific genome for a particular phenotype.That is why marker assisted selection (MAS) and discovery and use of quantitative trait loci (QTL) were used in breeding to reduce the long breeding cycles. It is given the great importance in cooperation with classical breeding and so called "associating" the marker genes that are related to a particular phenotypic trait, inherited together existing in the next generations, which makes possible to select a candidate genes for a particular phenotypic trait. However, those techniques have not been successful in forest trees (Haggman et al. 2016) due to that QTL in experimental population explain a small percentage of phenotypic variation and also the genotype by environment interaction makes MAS application more complicated (Isik 2016). In the other hand, that can be changed using integrative approach of genotyping platforms using molecular markers (Orlovic et al. 2009; Galovic and Orlovic 2007; Pakull et al. 2015; Fladung et al. 2016; Konstantinov et al. 2005; Nonic et al. 2014) to be utilized in high-density genetic maps.

Breeding programs implies gathering knowledge about genome of the species of interest using conventional breeding and selection methods revealing its morphological, phenological, physiological and genetic characteristics. This way, the genetic potential data of the gene pools of forest species would be transparent and ready to use for further utilisation in biochemical (Popovic et al. 2016) and molecular monitoring (Galovic et al. 2010). This way significant information about the existence and level of inter and intraspecific genetic variability and relations was collected (Galovic and Orlovic 2007; Galovic et al. 2009). The method of choice in genotyping fast growing species like poplars, were dominant and codominant also nuclear and organelar molecular markers like RFLPs, AFLPs, SSRs, cpSSRs, mtSSRs, ISSRs, AFLPs, ITS, RAPDs, SNPs (Schroeder and Fladung 2014; Kersten et al. 2015). Besides measuring existance of genetic variability these methods are used to reveal molecular taxonomy and phylogeny as well, undoubtedly confirm the taxonomic affiliation of unknown genes to a particular species, both in plants and in the world of pathogens (Galovic et al. 2010a) and insects. The results showed that the applied molecular marker systems are reliable in determining the genetic structure of this species therefore they represent specific and reliable ways of determining the polymorphism of each individual in the existing poplar collection. Through SSRs, ITS analysis, SNPs approach (Galovic et al. 2015a) and expression assays, DNA profiles of different poplar clones and hybred genomeswererevealed. This was valuable achievement to the process of protection of the breeders rights e.g. the process of registering of the new genotype. Genomic approach was applied to several other forest species as well: beech (Galovic et al. 2013), oak (Galovic et al. 2014), sorbus (Galovic et al. 2012), spruce (Galovic et al. 2015). Transcriptomics and genetic engineering was imployed in abiotic stress related investigation of multiple candidate genes and their differential expression in different poplar clones (Galovic et al. 2017).

Even being slow, conventional breeding can cause notable changes in the tree genome while genetic engineering technologies affect moderately by changing the genome organization in only few basis (Haggman et al. 2016). Nowdays genetic engineering became the techniques of choice that can accelerate tree breeding. Involved new engineered nucleases techniques of genome editing, like zinc finger nucleases (ZNF), transcription activator-like effector nuclease (TALEN), clustered regularly interspaced short palindromic repeats (CRISPR Cas) systems, RNA interference (RNAi) that are promising to be more precise in genome reconstruction in the near future for shaping the genomes of forest tree species being the "molecular scissors" in comparison to conventionally used *Agrobacterium* mediated or biolistic methods. An aspect of a new generation sequencing technologies of parcial (Birol et al. 2013; Zimin et al. 2014) or "de novo" whole genome sequencing (Tuskan et al. 2006; Nystedt et al. 2013) is increasingly present where international teams work on decoding complete, often complicated and abundant genomes of forest species in order that manipulation with genes and its transfers would be in favor of preserving forest species in ecological escalations.

On the basis of above mentioned research in this field, reliable, reproducible molecular techniques were introduced in the program of breeding and conservation of forest genetic resources. They are necessary tools for detailed characterization and establishment of the specific structure of adaptive and reproductive potentials of forest species genomes. Genetic variability that is revealed in breeding programs is aimed to conserve the forest species gene pools. The future involving of genetic engineering in conventional forest tree breeding will fasten up gathering scientific knowledge of tree genomes and therefore help in genome reconstruction neccessary for new demanding ecological changes.

20.4 Conclusions

Modern forestry demands worldwide require holistic approach during process of breeding of forest trees. Besides traditional goals such are growth and vitality, new selected materials must consider both climate change and other environmental issues that limit growth of trees (i.e. degraded and contaminated soils, increase of carbon assimillation) what indicates necessity of upgrading conventional breeding methods. Above described review of breeding of fast growing species shows that state of selection processes in Serbia includes both traditional and contemporary methods what insures sufficient background for future research and improvement of forestry practice.

References

Birol, I., Raymond, A., Jackman, S. D., Pleasance, S., Coope, R., Taylor, G. A., Yuen, M. M. S., Keeling, C. I., Brand, D., Vandervalk, B. P., Kirk, H., Pandoh, P., Moore, R. A., Zhao, Y., Mungall, A. J., Jaquish, B., Yanchuk, A., Ritland, C., Boyle, B., Bousquet, J., Ritland, K., MacKay, J., Bohlmann, J., & Jones, S. J. M. (2013). Assembling the 20 Gb white spruce (*Picea glauca*) genome from whole-genome shotgun sequencing data. *Bioinformatics, 29*, 1492–1497.

Borisev, M., Pajevic, S., Nikolic, N., Pilipovic, A., Krstic, B., & Orlovic, S. (2009). Phytoextraction of Cd, Ni, and Pb using four willow clones (Salix spp.). *Polish Journal of Environmental Studies, 18*, 553–561.

Fladung, M., Schroeder, H., Wehenkel, C. A., & Kersten, B. (2016). Differentiation of Eucalyptusclones grown in Mexico by ITS and chloroplast barcoding markers. *Silvae Genetica, 64*, 121–130.

Galovic, V., & Orlovic, S. (2007). Overview of DNA based studies of genetic variability in poplars. *Topola, 179/180*, 71–79.

Galovic, V., Orlovic, S., Kovacevic, B., &Pilipovic, A. (2009). Microsatellite genetic relationships of Populus clones, cultivars and hybrids from two sections of genus Populus. In: *Proceedings of the International Scientific Conference "Forestry in achieving millennium goals"*, (Serbia). pp. 325–331.

Galovic, V., Orlovic, S., Pap, P., Zgonjanin-Bosic, D., Mataruga, M., Poljakovic-Pajnik, L., & Drekic, M. (2010a, September 20–25). Revealing the genetic background of poplar breeding material. *Fifth International poplar symposium "Poplars and willows: from research models to multipurpose trees for a bio-based society"*, Orvieto, Italy. p. 64.

Galovic, V., Orlovic, S., Pap, P., Kovacevic, B., & Markovic, M. (2010b). Specificity of SSR loci for Melampsora species on poplars. *Genetika, 42*(3), 513–520.

Galovic, V., Orlovic, S., Trudic, B., Pekec, S., & Vasic, S. (2012). Unutarlokusni polimorfizam Sorbus spp. na teritoriji Republike Srbije. *Topola, 189/190*, 87–98.

Galovic, V., Orlovic, S., Trudic, B., Pekec, S., Pilipovic, A., & Stojnic, S. (2013, October 1–2). Sequence polymorphisms of beech provenances in north and central Serbia. *Book of abstracts of the International Scientific Conference*, Sofia, Bulgaria, p. 47.

Galovic, V., Orlovic, S., Zoric, M., Kovacevic, B., & Vasic, S. (2014, November 26–28). Different phenology induced genotype diversity of Q. robur L. in the seed orchard in Srem provenance, Republic of Serbia. *Book of Abstracts of the International Scientific Conference "Natural resources green technologies & sustainable development"*, Zagreb, Croatia. p. 133.

Galovic, V., Sijacic-Nikolic, M., Safhauzer, R., Cortan, D., & Orlovic, S. (2015a). Genetic differentiation of Norway spruce (*Picea abies* (L.) Karst.) trees with diferent crown types from the mountain Golija. *Genetika, 47*, 849–861.

Galovic, V., Orlovic, S., & Fladung, M. (2015b). Characterization of two poplar homologs of the GRAS/SCL gene, which encodes a transcription factor putatively associated with salt tolerance. *iForest, 8*, 780–785.

Galovic, V., Orlovic, S., Szabados, L., Perez, I., Prathiba, M., & Fladung, M. (2017, September 18–22). Stress related overview of two poplar homologues of GRAS/SCL TF: From characterization, cloning to differential expression. *IUFRO 125th Anniversary Congress*, Freiburg, Germany, p. 570.

Gordon, M., Choe, N., Duffy, J., Ekuan, G., Heilman, P., Muiznieks, I., Newman, L., Raszaj, M., Shurtleff, B., Strand, S., & Wilmoth, J. (1997). Phytoremediation of trichloroethylene with hybrid poplars. In *Phytoremediation of soil and water contaminants* (pp. 177–185). Washington, DC: American Chemical Society.

Guzina, V. (1987). The variability of poplar clones in their capacity of rooting of their cuttings. *Topola/Poplar, 151/152*, 13–24 [In Serbian].

Haggman, H., Sutela, S., & Fladung, M. (2016). Genetic engineering contribution to forest tree breeding efforts. In: *Biosafety of forest transgenic trees* (pp. 11–29). Heidelberg: Springer.

Heilman, P. E., Ekuang, G., & Fogle, D. B. (1994). First order root development from cuttings of *Populus trichocarpa* x *P. deltoides* hybrids. *Tree Physiology, 14*, 911–920.

IPC. (2016, September 12–16). Trends & perspectives in poplar & willow cultivation: Global synthesis. *25th Session n of FAO International Poplar Commission*, Berlin. Available online at: http://ipc25berlin2016.com/fileadmin/allgemein/pdf/veranstaltungen/IPC25_2016/Plenary_4/Carle_JB_Global_Synthesis_Keynote_Plenary_16_Sept_2016.pdf

Isik, F. (2016). Genomic selection in forest tree breeding: The concept and an outlook to the future. *New Forests, 45*, 379–401.

Kersten, B., Voss, M. M., & Fladung, M. (2015). Development of mitochondrial SNP markers in different Populus species. *Trees, 29*, 575–582.

Klasnja, B., Orlovic, S., Galic, Z., Drekic, M., Vasic, V., & Pilipovic, A. (2008). Poplar biomass of high density short rotation plantations as raw material for energy production. *Wood Research (Drevarsky Vyskum), 53*, 27–38.

Klasnja, B., Orlovic, S., & Galic, Z. (2012). Energy potential of poplar plantations in two spacings and two rotations. *Sumarski list, 3–4*(CXXXVI), 161–167.

Konstantinov, K., Mladenovic-Drinic, S., Sijacic-Nikolic, M., Vasilije, I., & Matagura, M. (2005). Molecular markers application for genetic resourses characterization of different plant species. *International workshop: The role of Biotehnology, Book of proceedings*, Turin, Italy. pp. 181–183.

Kovacevic, B., Guzina, V., Kraljevic-Balalic, M., & Ivanovic, M. (2007). Variability and relationship among cuttings' rooting characters for eastern cottonwood. *Genetika, 39*(1), 29–38.

Kovacevic, B., Orlovic, S., Katanic, M., Pekec, S., & Stojnic, S. (2008a, November 13–15). Variability of rooting characters in one-year old black poplar (Aigeiros Duby) rooted cuttings. In: *Proceedings of International Scientific Conference "Forestry in Achieving Millennium Goals"*, Novi Sad. pp. 427–432.

Kovacevic, B., Guzina, V., Kraljevic-Balalic, M., Ivanovic, M., & Nikolic-DJoric, E. (2008b). Evaluation of early rooting traits of eastern cottonwood that are important for selection tests. *Silvae Genetica, 57*(1), 13–21.

Kovacevic, B., Roncevic, S., Miladinovic, D., Ivanisevic, P., & Katanic, M. (2009). Early shoot and root growth dynamics as indicators for the survival of black poplar cuttings. *New Forests, 38*, 177–185.

Kovacevic, B., Orlovic, S., Ivanovic, I., Cobanovic, K., Nikolic-Doric, E., Katanic, M., & Galovic, V. (2011). Relationship among eastern cottonwood genotypes according to early rooting traits. *Genetika, 43*(2), 307–320.

Licht, L. A., & Isebrands, J. G. (2005). Linking phytoremediated pollutant removal to biomass economic opportunities. *Biomass and Bioenergy, 28*, 203–218.

Nikolic, N., Borisev, M., Pajevic, S., Arsenov, D., Zupunski, M., Orlovic, S., & Pilipovic, A. (2015). Photosynthetic response and tolerance of three willow species to cadmium exposure in hydroponic culture. *Archives of Biological Sciiences, 67*(4), 1411–1420.

Nikolic, N., Zoric, L., Cvetkovic, I., Pajevic, S., Borisev, M., Orlovic, S., & Pilipovic, A. (2017). Assessment of cadmium tolerance and phytoextraction ability in young *Populus deltoides* L. and *Populus x euramericana* plants through morpho-anatomical and physiological responses to growth in cadmium enfiched soil. *iForest, 10*, 635–644.

Nonic, M., Heinze, B., Ivetic, V., Mengl, M., & Slunsky, R. (2014, September 28–October 2). Assessment of inter-population genetic variability of beech by microsatellite markers. *Book of abstracts of the V Congress of the Serbian Genetic Society*, Kladovo – Belgrade, Serbia, p. 366.

Nystedt, B., Street, N. R., Wetterbom, A., Zuccolo, A., Lin, Y.-C., Scofield, D. G., Vezzi, F., Delhomme, N., Giacomello, S., Alexeyenko, A., Vicedomini, R., Sahlin, K., Sherwood, E., Elfstrand, M., Gramzow, L., Holmberg, K., Hällman, J., Keech, O., Klasson, L., Koriabine, M., Kucukoglu, M., Käller, M., Luthman, J., Lysholm, F., Niittylä, T., Olson, Å., Rilakovic, N., Ritland, C., Rosselló, J. A., Sena, J., Svensson, T., Talavera-López, C., Theißen, G., Tuominen, H., Vanneste, K., Wu, Z.-Q., Zhang, B., Zerbe, P., Arvestad, L., Bhalerao, R., Bohlmann, J., Bousquet, J., Garcia Gil, R., Hvidsten, T. R., de Jong, P., MacKay, J., Morgante, M., Ritland, K., Sundberg, B., Lee Thompson, S., Van de Peer, Y., Andersson, B., Nilsson, O., Ingvarsson, P. K., Lundeberg, J., & Jansson, S. (2013). The Norway spruce genome sequence and conifer genome evolution. *Nature, 497*, 579–584.

Orlovic, S. (1996). *Proucavanje svojstava crnih topola znacajnih za unapredjenje selekcije na bujnost.* Doktorska disertacija, Univerzitet u Beogradu Sumarski fakultet, Beograd, Srbija.

Orlovic, S., Guzina, V., Krstic, B., & Merkulov, L. J. (1998). Genetic variability in anatomical, physiological and growth characteristics of hybrid poplar (Populus x euramericana DODE (GUINIER)) and eastern cottonwood (*Populus deltoides* BARTR.) clones. *Silvae Genetica, 47*, 183–190.

Orlovic, S., Pajevic, S., Krstic, B., & Kovacevic, B. (2001). Genetic and phenotypic correlation of physiological and growth parameters of poplar clones. *Genetika, 33*, 53–64.

Orlovic, S., Pajevic, S., & Krstic, B. (2002). Selection of black poplars for water use efficiency. *Zbornik Matice Srpske za Prirodne Nauke, 102*, 45–51.

Orlovic, S., Klasnja, B., Pilipovic, A., Radosavljevic, N., & Markovic, M. (2003). A possibility of early selection of black poplars (Section Aigeiros DUBY) for biomass production on the basis of anatomical and physiological properties. *Topola, 171/172*, 35–44.

Orlovic, S., Pilipovic, A., Galic, Z., Ivanisevic, P., & Radosavljevic, N. (2006a). Results of poplar clone testing in field experiments. *Genetika, 38*(3), 257–264.

Orlovic, S., Pajevic, S., Klasnja, B., Galic, Z., & Markovic, M. (2006b). Variability of physiological and growth characteristics of white willow (*Salix alba* L.) clones. *Genetika, 38*, 145–152.

Orlovic, S., Galovic, V., Zoric, M., Kovacevic, B., Pilipovic, A., & Galic, Z. (2009). Evaluation of interspecific DNA variability in poplars using AFLP and SSR markers. *African Journal of Biotechnology, 8*, 5241–5247.

Orlovic, S., Pilipovic, A., Galovic, V., Klasnja, B., Pap, P., & Radosavljevic, N. (2010). The role of new poplar strains in the alleviation of the effects of climate change. In L. J. Rakonjac (Ed.), *Proceedings from the international scientific conference: Forest ecosystems and climate changes* (pp. 57–70). Belgrade: Institute of Forestry.

Pakull, B., Kersten, B., Luneburg, J., & Fladung, M. (2015). A simple PCR-based marker to determine sex in aspen. *Plant Biology, 17*, 256–261.

Pilipovic, A., Nikolic, N., Orlovic, S., Petrovic, N., & Krstic, B. (2005). Cadmium phytoextraction potential of poplar clones (Populus spp.). *Zeitschrift für Naturforschung, 60c*, 247–251.

Pilipovic, A., Orlovic, S., Nikolic, N., Borisev, M., Krstic, B., & Roncevic, S. (2012). Growth and plant physiological parameters as markers for selection of poplar clones for crude oil phytoremediation. *Sumarski List, 3–4*(CXXXVI), 273–281.

Pillon-Smits, E. (2005). Phytoremediation. *Annual Review of Plant Biology, 56*, 15–39.

Popovic, B. M., Stajner, D., Zdero-Pavlovic, R. A., Tumbas-Saponjac, V., Canadanovic-Brunet, J., & Orlovic, S. (2016). Water stress induces changes in polyphenol profile and antioxidant capacity in poplar plants (Populus spp.). *Plant Physiology and Biochemistry, 105*, 242–250.

Raskin, I., Smith, R. D., & Salt, D. E. (1997). Phytoremediation of metals: Using plants to remove pollutants from the environment. *Current Opinion in Biotechnology, 8*, 221–226.

Salt, D. E., Blaylock, M., Nanda Kumar, P. B., Dushenkov, A. V., Ensley, B. D., Chet, I., & Raskin, I. (1995). Phytoremediation: A novel strategy for the removal of toxic metals from the environment using plants. *Biotechnology, 13*, 468–474.

Schroeder, H., & Fladung, M. (2014). Differentiation of Populus species by chloroplast SNP markers for barcoding and breeding approaches. *iForest – Biogeosciences and Forestry, 8*, 49–53

Smith, N. G., & Wareing, P. F. (1972). The distribution of latent root primordia in stems of Populus x robusta and factors affecting the emergence of preformed roots from cuttings. *Forestry, 45*, 197–209.

Stojnic, S., Orlovic, S., Trudic, B., Zivkovic, U., von Wuehlisch, G., & Miljkovic, D. (2015). Phenotypic plasticity of European beech (Fagus sylvatica L.) stomatal features underwater deficit assessed in provenance trial. *Dendrobiology, 73*, 163–173.

Teissier du Cros, E. (1984). Breeding strategies with poplars in Europe. *Forest Ecology and Management, 8*, 23–39.

Topic, M., Borisev, M., Orlovic, S., Tomicic, M., Zupunski, M., Nikolic, N., Pajevic, S., Krstic, B., & Pilipovic, A. (2013). Clonal differences of black poplar cuttings for morpho-physiological and biochemical responses to soil water deficits. *Journal of Animal and Plant Sciences, 23*, 1725–1732.

Tschaplinski, T. J., & Blake, T. J. (1989). Correlation between early root production, carbohydrate metabolism and subsequent biomass production in hybrid poplar. *Canadian Journal of Botany, 67*, 2168–2174.

Tuskan, G. A., DiFazio, S., Jansson, S., Bohlmann, J., Grigoriev, I., Hellsten, U., Putnam, N., Ralph, S., Rombauts, S., Salamov, A., Schein, J., Sterck, L., Aerts, A., Bhalerao, R. R.,

Bhaterao, R. P., Blaudez, D., Boerjan, W., Brun, A., Brunner, A., & Busov, V. (2006). The genome of black Cottonwood, Populustrichocarpa (Torr. & Gray). *Science, 313*, 1596–1604.
Vasic, V., Galic, Z., & Drekic, M. (2010). Ucinkovitost i selektivnost nekih herbicida u rasadnickoj proizvodnji sadnica topola. *Sumarski List, 134*, 395–401.
Vasic, V., Orlovic, S., Pap, P., Kovacevic, B., Drekic, M., Poljakovic-Pajnik, L., & Galic, Z. (2015). Application of pre-emergence herbicides in poplar nursery production. *Journal of Forestry Research, 26*, 143–151.
Wilcox, R. J., & Farmer, E. R., Jr. (1968). Heritability and effects in early root growth of Eastern cottoonwood cuttings. *Heredity, 23*, 239–245.
Ying, C., & Bagley, W. T. (1977). Variation in rooting capability of *Populus deltoides*. *Silvae Genetica, 26*, 204–207.
Zalesny, R. S., Hall, R. B., Bauer, E. O., & Reimenshneider, D. E. (2005). Soil temperature and precipitation affect the rooting ability of dormant hardwood cuttings of Populus. *Silvae Genetica, 54*, 47–58.
Zimin, A., Stevens, K. A., Crepeau, M. W., Holtz-Morris, A., Koriabine, M., Marcais, G., Puiu, D., Roberts, M., Wegrzyn, J. L., de Jong, P. J., Neale, D. B., Salzberg, S. L., Yorke, J. A., & Langley, C. H. (2014). Sequencing and assembly of the 22-Gb loblolly pine genome. *Genetics, 196*, 875–890.

Chapter 21
Ex Situ Conservation – Case Study in Bosnia and Herzegovina

Vanja Daničić, Dalibor Ballian, Milan Mataruga, Branislav Cvjetković, and Rifet Terzić

Abstract Ex situ conservation is the preservation of components of biological diversity outside their natural habitats. Such ex situ work is an increasingly important component of conservation and restoration efforts. Its primary objective is to support conservation by ensuring the survival of threatened species and the maintenance of associated genetic diversity. To do so, ex situ institutions preserve the genetic or reproductive material of target species, or take care of the living target species for the purpose of reintroduction. Ex situ conservation measures should support in situ conservation measures. Organized ex situ conservation activities in Bosnia and Herzegovina (BiH) started in the middle of the last century. Activities were stopped during the 1990s because of the civil war, and then after 2000, the ex situ conservation of species continued. Ex situ cultures in BiH are established within arboreta, live archives, provenance tests, progeny tests, seed orchards and clone tests.

Keywords Ex situ · Conservation · Bosnia and Herzegovina

V. Daničić (✉) · M. Mataruga · B. Cvjetković
Faculty of Forestry, University of Banja Luka, Banja Luka, Bosnia and Herzegovina
e-mail: vanja.danicic@sf.unibl.org; milan.mataruga@sf.unibl.org; branislav.cvjetkovic@sf.unibl.org

D. Ballian
Faculty of Forestry, University of Sarajevo, Sarajevo, Bosnia and Herzegovina
e-mail: balliandalibor9@gmail.com

R. Terzić
Faculty of Natural Sciences and Mathematics, University of Tuzla, Tuzla, Bosnia and Herzegovina
e-mail: rifet.terzic@untz.ba

M. Šijačić-Nikolić et al. (eds.), *Forests of Southeast Europe Under a Changing Climate*, Advances in Global Change Research 65,
https://doi.org/10.1007/978-3-319-95267-3_21

21.1 Introduction

Ex situ conservation in Bosnia and Herzegovina (BiH) is preformed within: arboreta, live archives, provenance tests, progeny tests, seed orchards, clone tests.

Organized ex situ conservation activities in BiH started in the middle of the last century. Conservation activities included several domestic and introduced species. Activities were stopped during the 1990s because of the civil war, and then after 2000, the ex situ conservation of species continued.

Growth and phenology parameters were observed and measured in newly established provenance tests, progeny tests and seed orchards.

21.2 Arboreta and High Mountain Protected Areas

The arboreta in BiH are mainly found in nurseries which produce forest and horticultural planting material and within the university campus.

21.3 Clone Collections

A clone archive of black poplars with 163 clones of autochthonous black poplar and an archive of cheque tree (*Sorbus domestica*) were established in Bosnia and Herzegovina.

21.4 Provenance Tests

In Bosnia and Herzegovina, the outdoor tests were performed with domestic and introduced species. In the earlier period during the twentieth century, the focus was placed on coniferous species such as Norway spruce, douglas fir, larch, the Sitka spruce (*Picea sitchensis*) etc. In recent years, mainly domestic decidous tree species such as beech and common oak have been established. In addition, progeny tests of the Norway spruce have been established in several locations in BiH. The aim of the later tests is to monitor the influence of GxE interaction, that is, reactions of different provenances to climate change and the possibilities of transfering the forest reproductive material in a safe way.

21.4.1 Douglas Fir Provenance Tests

Douglas fir, as an introduced species, was presented in provenance tests the most. This is the result of the international co-operation through the IUFRO organization, when during 1962, the tests were established with American and Canadian provenances of Coast douglas fir (*Pseudotsuga menziesii* var. *menziesii*) in several locations in BiH. Douglas fir is tested on several sites: Goleš near Travnik, Gradiška and Batalovo Brdo.

Pintarić (1991) writes about the survival percentage of 8 douglas fir provenances from four sites in BiH: Gradiška, Zavidovići, Kreševo and Visoko. The percentage of survived individuals varies from 47% (Zavidovići) to 82% (Gradiška), as well as morphometric parameters: height, diameter, base, volume and quality of trees. The recommended populations are those from Washington zone, since the reproductive material originating from this zone showed best results in terms of survival and growth.

21.4.2 Larch Provenance Tests

Larch tests were established in 1959 and monitored until 1990 when the last measurements were taken. The research activities were carried out within the Second international experiments of larch originating from different provenance and applying the methodology written by Schober. In Europe and in the United States 46 trials were established, and one was established in BiH. Unfortunately, the provenance test was devastated in the period 1992–1995 and therefore was not suitable for further observation activities. The provenance test included ten provenances from the Alps, High and Low Tatras, Sudetan, Scotland and Japan.

The larch research focusing on the increment, showed that provenances from the eastern part of the natural distribution area (Sudetes and Carpathians) achieved better growth, however the provenances of the eastern part of the Alps (Vienna woods) are not lagging behind the most productive provenances of the Carpathian area. The best productivity among all the provenances was shown by larch, although at the end of testing period the differences were reduced and provenances with the best increment were not eventually of the best quality (Pintarić and Koprivica 1980).

The provenance tests of the Sitka spruce (*Picea sitchensis*) were established in 1979 on 9 permanent field tests with 6–10 provenances 19 and 99 in "Kružića" site in the area of "Forestry" Enterprise in Travnik (Pintarić 1998).

Scots pine field tests were established in 1989 in the following locations: Romanija, Teslić, Bosanski Petrovac and the first findings revealed significant variation of morphometric parameters, the best results were performed by local provenances (Ballian et al. 2009; Cvjetković et al. 2014).

21.4.3 Scots Pine Provenance Tests

In Bosnia and Herzegovina, four Scots pine provenance tests were established, a species of disjuncted distribution area and ecotype differentiation (Stefanović 1980). Tests were established in 1989 at four sites: Romanija – Sokolac, Zavidovići, Teslić and Bosanski Petrovac. The tests in Romanija Sokolac monitored the survival and morphometric parameters (Ballian et al. 2009; Cvjetković et al. 2015a, b, c). Measurements have been made in other tests but the findings have not been published yet. The findings indicate significant variability among the tested provenances in terms of morphometric parameters. In addition to these four provenance tests, there are two more in Žepče and Kupres, which are still under research process.

21.4.4 Fir Provenance Tests

Fir provenance test was established in 1990 and included 9 fir populations from the Dinarides area, namely Bugojno, Bosanski Petrovac, Olovo-Palež, Pale, Konjic, Fojnica, Sokolac, Olovo-Klis and Prozor. Five year old (3 + 2) seedlings were incorporated in a test located within the unit "Krivaja" in the "Delimusa" site. The test was established at the altitude of 970 m with the north exposure. The variability of fir morphological properties within the provenance test was researched by Ballian and Halilović (2016).

21.4.5 Beech Provenance Test in Kakanj

The international beech provenance test in Kakanj is part of a larger European research activities with trials established in several European countries. An international beech provenance test was established in Kakanj. Ballian and Zukić (2011) studied morphometric parameters of different provenances while Ballian et all. (2015) observed phenology processes within the same provenance test and identified significant differences. The findings can be used in terms of references to select appropriate provenances in the light of expected climate changes.

21.4.6 Common Oak Provenance Test

Tests were performed on common oak, a species which is fragmented in BiH and covers smaller forest areas. Large populations can be found in nonspecific habitats that are karstified. The research findings of such material can refer to the application of new sources of reproductive material that would be adaptive to climate change.

The common oak seed material that was used to establish the test was collected in 2007 from 28 sites (populations) in Bosnia and Herzegovina. At plant age 1 + 0, a permanent provenance test was established. The field test was established in 2009.

The first findings revealed that some of the populations that are not in the river Sava belt, a place of common oak distribution area, showed very good results and vice versa (Memišević-Hodžić et al. 2016).

21.5 Spruce Progeny Tests

Due to the seed objects delineation in the Republic of Srpska (RS), spruce progeny tests were established in 2009 at four locations in BiH (Drinić, Derventa, Srebrenica and Nevesinje). Progeny tests include six populations from RS. The progeny tests monitored survival and the height increment for 1 year after they were transplanted in Srebrenica and Drinić, and significant differences between the populations and the half-sib lines were observed (Mataruga et al. 2010). Afterwards, research activities were carried in two progeny tests: Drinić and Srebrenica. The findings showed significant differences between tests, populations, and half-sib lines in terms of survival, growth dynamics, and physiological parameters which are important in the context of climate change. The earliest and the laest opening buds were observed, as well as the population that are characterized by the fastest growth (Cvjetković et al. 2015a, b, c, 2016).

21.6 Seed Orchards

In order to produce a high quality reproductive material, several Scots pine seed orchards have been established. It is a species of a disjuncted distribution area and at the same time of great importance for BiH forestry.

21.6.1 Scots Pine Seed Orchard in Doboj

The Scots pine seed orchard was established in a vegetative way. The clone plantation was established in autumn of 1968 in Ozimica, whereas in the spring of 1972, it was transplanted to Stanova near Doboj. It included 20 clones on the area of 1 ha at an altitude of 155 m. The mountain Romanija, locality Knežinski palež is the origin of these clones. Detailed phenology studies were carried out by Daničić et al. (2012a, b). Te differences among clones in phenophases were observed, the strobiles phenomenon was detected and therefore it is necessary to continue the research activities of phenological variations in order to select the best clones that can endure climate changes, as well as the demand to manage hybridization to direct the production of reproductive material with desirable adaptive properties.

21.6.2 Seed Orchard "Koziji grm"

The Scots pine seed orchard "Kozji Grm" is a clone seed orchard in the immediate vicinity of Sarajevo. The orchard included 40 clones originating from five different provenances within Bosnia and Herzegovina (Gornji Janj 10 clones, Klekovača 7 clones, Kalinjina Bioštica 16 clones, Romanija 8, Glasinac 6 Clones, Igman 1 clone). Each of the 40 clones was represented by 25 copies, and when the seed orchard was establised at the moment there were altogether 1000 ramets (Ballian and Božič 2004; Ballian et al. 2005).

21.6.3 Seed Orchard "Šamin gaj"

The Scots pine seed orchard "Šamin gaj" is located in the immediate vicinity of Rakovice, around 20 km northwest from Sarajevo. The seed orchard included 20 clones originating from one provenance, Igman site. Each of the 20 clones was represented by 20 copies, and at the moment of planting activities there were altogether 400 ramets. According to the climate data of the nearest weather station (Stefanović et al. 1983), it can be concluded that conditions for flowering, pollination and seed ripening were very good for common pine, and along with the application of appropriate agrotechnical measures the seed yield could have been increased (Ballian et al. 2006).

21.6.4 Seed Orchard in Kakanj

The seed plantation of *Picea omorika* covers an area of 8.2 ha. The analyses of plant origin was carried out by Ballian (2006) and he discovered that the omorika originating from Kakanj did not identify with any other known omorika population growing on natural habitats. According to the author, the cause of this situation is that beside the alleles that were identified in many omorika population, one of the identified alleles was not identified in natural populations, but originates from populations which have not been analyzed yet or from a smaller group of trees.

References

Ballian, D. (2006). Kontrola podrijetla Pančićeve omorike (*Picea omorika*/Panč./purk.) iz plantaže kod Kaknja pomoću izoenzimskih biljega. *Šumarski List, 7–8*, 295–304.

Ballian, D., & Božič, G. (2004). The control of morphologic identification of the clones from the seed orchard of Scots pine "Koziji grm" by using the isoezyme markers. *Works of Faculty of Forestry Sarajevo, 1*, 47–56.

Ballian, D., & Halilović, V. (2016). Varijabilnost obične jele (*Abies alba* Mill.) u Bosni i Hercegovini, Udruženje inženjera i tehničara šumarstva Federacije Bosne i Hercegovine, *Silva Slovenica-izdavački centar Šumarskog institutaSlovenije*, Sarajevo, Ljubljana 2016.

Ballian, D., & Zukić, N. (2011). Analysis of the growth of common beech provenances (*Fagus sylvatica* L.) in the international experiment near Kakanj. *Works of the Faculty of Forestry University of Sarajevo, 2*, 75–91.

Ballian, D., Bogunić, F., & Konnert, M. (2005). Usporedba molekularno genetičkih svojstava sjemenskih plantaža običnog bora (*Pinus sylvestris* L.) u Bosni i Hercegovini. *Radovi Šumarskog Instituta Jastrebarsko, 41*(2), 7–16.

Ballian, D., Konnert, M., & Bogunić, F. (2006). Comparison of molecular and genetic properties of pine (*Pinus sylvestris* L.) seed plantations in Bosnia and Hercegovina. *Radovi Šumarskog Instituta Jastrebarsko, 41*(1–2), 7–16.

Ballian, D., Mujanović, E., & Čabarabdić, A. (2009). Varijabilnosti običnog bora (*Pinus sylvestris* L.) u pokusu provenijencija Glasinac – Sokolac (Bosna i Hercegovina) Šumarski list br. *11–12*(CXXXIII), 577–588.

Ballian, D., Jukić, B., Balić, B., Kajba, D., & von Wüehlisch, G. (2015). Phenological variability of European beech (*Fagus sylvatica* L.) in the international provenance trial. *Šumarski List, 11–12*, 521–533.

Cvjetković, B., Mataruga, M., Dukić, V., Daničić, V., & Lučić, A. (2014). The variability of Scots pine (*Pinus sylvestris* L.) in the provenance test in Bosnia and Herzegovina. In: *Proceedings of the Biennial International Symposium "Forest and sustainable development"*. Transylvania University Press, pp 79–88.

Cvjetković, B., Mataruga, M., Šijačić-Nikolić, M., Ivetić, V., Daničić, V., Stojnić, S., & Stojanović, M. (2015a). Norway spruce (*Picea abies* (L.) Karst.) Seedlings survival in progeny test "Drinić". *Bulletin of Faculty of Forestry Banja Luka, 22*, 5–14.

Cvjetković, B., Mataruga, M., Šijačić-Nikolić, M., Daničić, V., & Stojnić, S. (2015b). Survival of Norway spruce seedlings (*Picea abies* (L.) Karst.) in progeny test "Srebrenica" in purpose of transfer of forest reproductive material of this species. The Third Republic of Srpska Biologist and Ecologist Symposium, *Book of abstracts*, pp. 136–137.

Cvjetković, B., Mataruga, M., Šijačić-Nikolić, M., Daničić V., & Lučić A. (2015c). Bud burst and height increment of Norway spruce (*Picea abies* Karst.) in progeny tests in Bosnia and Herzegovina. International Conference "Reforestation Challenges", *Proceedings*, pp. 251–259.

Cvjetković, B., Mataruga, M., Šijačić-Nikolić, Dukić, V., & Popović, V. (2016). Variability of Norway spruce morphometric characteristics in progeny tests in Bosnia and Herzegovina. *Bulletin of Faculty of Forestry Belgrade, 113*, 11–35.

Daničić, V., Isajev, V., Mataruga, M., Cvjetković, B., & Milijević, I. (2012a). Varijabilitet sadržaja fotosintetičkih pigmenata belog bora (*Pinus silvestris L.*) u semenskoj plantaži „Stanovi" kod Doboja, Forest science and practice of sustainable development of forestry – 20 years of the Faculty of Forestry in Banja Luka, pp. 687–694.

Daničić, V., Isajev, V., Mataruga, M., Cvjetković, B. (2012b, October 4–5). Variability of flowering and seed crops in a clone seed orchard of scots pine (*Pinus sylvestris* L.). International Scientific Conference "Forests in the Future – Sustainable Use, Risks and Challenges", Belgrade, *Proceedings*, pp. 441–450.

Mataruga, M., Isajev, V., Balotić, P., Burlica, Č., Cvjetković, B. (2010). *Progeny tests of Norway spruce (Picea abies Karst.) in Bosnia and Herzegovina – contribution to the European ex situ conservation*. First Serbian Congress Future with Forests, Proceedings, pp. 378–389.

Memišević-Hodžić, M., Murlin, I., Ballian, D. (2016) Variability of eight years old plants of pedunculate oak in provenance test Žepče. *Bosnia and Herzegovina Natural Resources, Green Technology and Sustainable Development-Green Proceesidings* pp. 106–111.

Pintarić, K. (1991). Proučavanje prirašcivanja IUFRO-duglazije različitih provenijencija na nekoliko oglednih ploha u Bosni. *Šumarski List, CXV*(1–2), 5–22.

Pintarić, K. (1998). Ritam priraš́ćivanja u visinu sitkanske smrče (Picea sitchenssi L.) u ovisnosti od provenijencije – Height increment Rate in Sitka Spruze (*Picea sitchensis* L.) of Various Provenances (301). *Šumarski List, CXII*(7–8), 301–316.

Pintarić, K., & Koprivica, M. (1980). A study of the crown form and construction of Larch (Larix sp.) of various provenances. *Šumarski List, CIV*(1–2), 29–43.

Stefanović, V. (1980). Ekotipska diferencijacija bijelog bora (*Pinus sylvestris* L.) u Bosni i Hercegovini. Posebna izdanja br. 13. Sarajevo.

Stefanović, V., Beus, V., Burlica, Č., Dizdarević, H., & Vukorep, I. (1983). Ekološko-vegetacijska rejonizacija Bosne i Herzegovine. *Posebna izdanja br. 18*. Šumarski fakultet u Sarajevu, 1–49.

Chapter 22
Ex Situ Conservation – Case Study Croatia

Davorin Kajba and Ivan Andrić

Abstract Genetic diversity of Croatian forest tree species is being conserved using the ex situ static method. Clonal seed orchards are the nucleus of forest genetic resource conservation, since the relationship between the size of the population and the percentage of preserved heterozygocity is thus reduced to minimal loss of total additive genetic variability. In accordance with the division of forests in Croatia into ecogeographic seed regions and zones, the following productive clonal seed orchards have been established: three orchards of pedunculate oak (*Quercus robur* L.), two of narrow-leaved ash (*Fraxinus angustifolia*Vahl), one of sessile oak (*Quercus petraea* (Matt.) Liebl,), one of wild cherry (*Prunus avium* L.), and one of black pine (*Pinus nigra* J. F. Arnold). Phenotypical selection and heterovegetative propagation of plus trees, as well as the establishment of clonal seed orchards were initiated with the goal of controlling more regular yield periodicity and obtaining forest seed of good genetic quality in the categories of Qualified and Tested. The orchards are regularly subjected to pomotechnical treatments, protection and other agrotechnical measures. The evaluation of genetic values of mother trees in progeny tests and genotypic selection was also initiated for the purpose of obtaining increased genetic gain. Climate changes and new site conditions will pose additional challenges to seed production and forest management; in turn, this will influence their economic and social benefits, as well as biological diversity of forest ecosystems.

Keywords Clonal seed orchards · Forest reproductive material

D. Kajba (✉) · I. Andrić
Faculty of Forestry, University of Zagreb, Zagreb, Croatia
e-mail: davorin.kajba@zg.t-com.hr; iandric@sumfak.hr

M. Šijačić-Nikolić et al. (eds.), *Forests of Southeast Europe Under a Changing Climate*, Advances in Global Change Research 65,
https://doi.org/10.1007/978-3-319-95267-3_22

22.1 Introduction

In the countries with developed forestry, seed orchards represent a source of seed reproductive material for the majority of forest species. They are used as a reliable and renewable source of seed, but with the development of breeding, increasing emphasis is placed on the conservation of forest genetic resources and on genetic gain. Seed orchards are by far the most important link between the operative forestry practice, breeding and the conservation of genetic resources. Their importance is multiple: generation of the sources of forest reproductive material which will be necessary for future generations, adapted to the altered climate, habitat and economic conditions, and which will serve as an important factor in the conservation of genetic variability. Over the recent decades, the management of most economically important forest tree species in Croatia has faced problems of natural regeneration, repair planting on partially naturally regenerated areas and afforestation of open sites. There is no regular periodicity of seed crops in stands and the time period between mast yield years is prolonged. Even though Croatian forests are supposed to be naturally regenerated, this cannot be fully accomplished. The most common reason is the lack of yield in those stands that should be naturally reforested. The seed amount of economically important forest tree species required in 1 year is higher than its yearly production. To reduce or annul this negative difference between the seed required and collected, and to encourage the production of genetically improved seed, clonal seed orchards were established (Kajba et al. 2011b, 2012; Kajba and Andrić 2015). It is particularly important to fulfil the needs for forest seed in the nurseries which have had over the past period the need and task to produce and deliver larger quantities of forest seedlings. The production of seed in seed orchards should provide better solutions, in terms of a regular yield of quality and genetically improved seed, and for the needs of producing forest seedlings in the nurseries the annual production of which has ranged between 18 and 26 million seedlings in the past 10 years. To reduce or annul this negative difference between the seed needed and collected, the production of seed should be better organized. In that regard, the production of seed in seed orchards should provide better solutions (Kajba et al. 2011b).

Genetic variability is the result of differences in genetic sequences, as opposed to the differences caused by the environment. Genetic variability of a species includes between and within population variability, and can be categorized into two types: neutral and adaptive. Neutral variability is the result of differences between genotypes which do not impact the ability to survive or reproduce, while adaptive variability represents those differences that impact the viability of an organism or population. In order for a natural population to be viable, it must have the ability to respond to changes in the environment, competition with other species and the capacity to reproduce. In order to conserve the genetic resources in forest tree species, one should protect the existing genetic variability, its adaptability to natural evolution processes and to breeding, and enhance the knowledge and identification of units tolerant to certain diseases and pests, while avoiding the reduction of the size of genetic resources of endangered species. The ex situ method represents the conservation of forest tree species outside their natural habitats. This method is applied in parallel with the in situ method, and is especially used in the species in which the conservation of parts or the entire population is not possible (Kajba et al. 2006b).

The assessment of the size of the sample for ex situ conservation of genes is particularly important and there are usually several approaches to attaining its appropriate size. So far, several methods which were used in the past have been tested, and certain authors have considered and explained the sampling in a way which will preserve at least 95% of genetic variability. The results have demonstrated that for the sake of total efficiency, it is generally better to take samples from several localities and from a smaller number of plants per locality than the other way around. It has been concluded that in the majority of cases, the sample size should not exceed 50 plants per locality or population, which was also what guided us when establishing seed orchards (Lande and Barowclough 1987). Naturally, the population structure and geographic distribution of the species should be considered before planning the sampling. Another approach is based on the view that as many allelomorphic genes as possible should be preserved. Rare allelomorphs which are neutral, with regard to the natural selection in certain environments, most probably have a very important adaptive value. Therefore, in many studies a great deal of attention was focused on rare allelomorphs, but it is difficult to associate them directly with adaptation. There is very little precise information on rare allelomorphs which control adaptation in forest trees (Kajba et al. 2006b; Kajba and Hrašovec 2009).

The conservation of genetic resources of forest tree species represents a continuous and long-term process that requires significant investment, time, personnel, as well as machines and tools. The reasons for conservation should be viewed through the prism of logical, scientific and economic criteria as necessary and valuable, and through which at the end of the day the conservation of genetic resources is feasible. The assumption is that seed orchards make an appropriate basis for ex situ conservation of genetic variability of a species, as well as for the production of genetically good quality seed for the restoration of stands or assistance in natural restoration. It is also assumed that the offspring of selected plus trees is superior to the offspring of average trees from natural stands, which should be corroborated by studies in established progeny trials (Bogdan et al. 2004; Kajba et al. 2006a, b, 2011a).

22.2 Establishment and Maintenance of Clonal Seed Orchards in Croatia

In Croatia, the establishment of clonal seed orchards commenced some 50 years ago, primarily for scientific purposes in order to gain experience (Vidaković 1996). They were established as experimental seed orchards on small areas mostly from

Table 22.1 Clonal seed orchards established from 1966 to 1997

No.	Species	Seed region	Forest range office	Year of establishment	No. of clones	Area (ha)
1.	*Tilia cordata*	–	Otok	1972	13	0.72
2.	*Alnus glutinosa*	1.1.2	Đurđevac	1976–1981	31	0.90
3.	*Alnus glutinosa*	1.2.1	Kutina	1977–1978	30	0.80
4.	*Larix decidua*	–	Bjelovar	1965	28	2.00
5.	*Larix kaempferi* and *Larix decidua*	–	Poreč	1997	47	2.30
6.	*Pinus strobus*	–	S. Brod	1966	27	1.50
7.	*Pinus strobus*	–	Ogulin	1972	32	1.00
8.	*Pinus strobus*	–	Karlovac	1968	32	3.70
9.	*Pinus strobus*	–	Bjelovar	1966	28	1.00
10.	*Pinus sylvestris*	3.2.1	Gospić	1966	30	1.50
11.	*Pinus sylvestris*	2.1.1	S. Brod	1966	34	1.50
12.	*Pinus nigra*	3.3.1	Gospić	1969	41	1.50
13.	*Pseudotsuga menziesii*	–	Karlovac	1966	26	2.00

Table 22.2 Clonal seed orchards established from 1996 to 2008

No.	Species	Seed region	Forest range office	Year of establishment	No. of clones	Area (ha)
1.	*Quercus robur*	1.1.2	Orahovica	1996	40	16.0
2.	*Quercus robur*	1.2.1	Otok	2000	57	25.0
3.	*Quercus robur*	1.2.3	Čazma	2001	53	26.0
4.	*Quercus petraea*	2.1.1	Požega	2008	52	7.0
5.	*Fraxinus angustifolia*	1.2.2	Nova Gradiška	2005	56	3.5
6.	*Fraxinus angustifolia*	1.2.3	Čazma	2007	45	6.0
7.	*Prunus avium*	–	Kutina	2002	26	3.0
8.	*Pinus nigra*	3.3.1	Krk	2006	32	2.0

conifers, and to a lesser extent from deciduous trees (Table 22.1). More recent production clonal seed orchards were established on larger areas (from 15 ha to more than 20 ha), in the period from 1996 to 2008 (Table 22.2). Plantations of the pedunculate and sessile oak, narrow-leaved ash, wild cherry and black pine were established. In addition to the above clonal seed orchards of the sessile oak, also established were progeny trials to determine the genetic quality of selected plus trees (Vidaković et al. 2000).

All clonal seed orchards in Croatia have been regularly maintained by pruning and other agrotechnical treatments. The purpose of successful fruit techniques is maintaining the balance between the vegetative and generative activity. These techniques are also applied on forest tree grafts in clonal seed orchards (Kajba et al.

2008). The forming pruning is used to bring the grafts into the required training shape, whereas pruning for higher yield maintains the adequate balance between the growth and the seed production (Figs. 22.1 and 22.2). Tree pruning and training started right after the planting with the goal of obtaining the desired canopy shape with well-deployed scaffold branches in the next 7 or 8 years. Grafts were reduced to the desired canopy height. Some of the branches were pruned to induce further branching from the side buds.

The main function of pruning is the removal of competing shoots to enforce the growth of the remaining desired ones. Pruning intensity, i.e. the relation between vegetative and generative buds in the canopy determines the tree's condition, density and yield. The balance between vegetative and generative buds can only be achieved by appropriate underground and aboveground tree parts pruning. Knowing the production morphology of each tree species is very important. For instance, oak – the fruits called acorns are on the long stems of terminal buds on 1-year branches; ash flowers in dense clusters – inflorescences; cherry fruits on 1-year branches of various length, long shoots or short bearing branches. A different training system is applied to each tree species. Training systems should provide a strong framework and good light penetration in the crown canopy (Kajba et al. 2007, 2008).

With the possibility of stimulating flowering and better protection of the flowers and fruits from biotic and abiotic factors, the production of seed in plantations should be greater and more frequent than in natural stands and cultures. For that reason, the planned needs for seed can be better fulfilled. The collection of seed is simpler because these are trees with a modified training shape and height for better

Fig. 22.1 Clonal seed orchard of narrow-leaved ash, Forest range office Nova Gradiška

Fig. 22.2 Clonal seed orchard of pedunculate oak, Forest range office Čazma

fruit production and easier harvest, and the collection is managed on a foreseen and prepared area.

Excessive exploitation of forests or the occurrence of invasive pathogens which under new habitat conditions cause massive dying out, can lead to fragmentation and reduction in the size of populations, to the point where it endangers the survival of an individual species. Molecular genetic markers, in combination with the principles of population genetics, can greatly facilitate programs of genetic resource conservation (Rajora and Mosseler 2001). The importance of certain types of molecular markers in the research of seed orchards can be understood if we mention the basic preconditions for the complete fulfillment of the function of a seed orchard: conservation of genetic variability, maximum phenological synchronicity of clones and their equal contribution through their free pollination, minimum inbreeding and introduction of foreign pollen (Kang et al. 2005). A turning point in population genetics, and at the same time in the study of plantations, was brought about by the application of molecular methods, using DNA sequences and markers, in the observation, monitoring and management of genetic sources. All of the clones in seed orchards of the pedunculate oak, narrow-leaved ash and wild cherry were genotyped using nuclear microsatellite markers, and an analysis of genetic variability was conducted (Katičić et al. 2010; Katičić-Bogdan 2012; Tančeva Crmarić et al. 2011). The accelerated development of genomics yields new methods which will simplify and enable their application in the conservation of genetic variability of forest trees (Grattapaglia et al. 2010).

22.3 Challenges and Perspectives for Clonal Seed Orchards

The methods of breeding forest trees in seed orchards must at the same time take into account their genetic variability, the breeding value of clones or families obtained by free or controlled pollination, time, costs and technical requirements. Optimum alignment of all those factors into an efficient and cost-effective system is no simple task and is a challenging subject of many recent studies.

The impact of climate changes on forest ecosystems and forest objects, such as seed orchards, is a topic which has in recent times caused great interest in the scientific circles. Thirty years ago, global climate change did not represent a serious problem, the long-term impact of which would be taken into consideration when introducing reproductive material into stands or planting new stands. However, today it is becoming clear that the global climate is changing, moving towards warmer and most frequently drier conditions in many forest ecosystems of the world. Moving the borders of the area of forest trees, expansion of the impact of certain pests, and increased possibility of forest fires are only some of the possible challenges that forest ecosystems management will face in the future. More optimistic forecasts foresee a significant capacity of forest trees to adapt to the newly created conditions. Account is taken of evolutionary changes that can take place on the species level through the migration of genes, population, where phonological, morphological and growth characteristics are formed by natural selection, but also on the level of units, in the sense of phenotypic variations of the same or similar genotypes, due to the different expression of genes in altered conditions. There is evidence, based on empirical data of long-term studies, that evolution conditioned by climate changes has already begun and that trees on the level of units and population are already adapting to e.g. increased concentrations of carbon dioxide in the atmosphere. The results of some studies of provenance tests or introduction of exotic species into new areas suggest an adaptation of forest trees already within a few or even one generation, thus contradicting theoretical models which anticipate that the change of adaptive properties in a population of forest trees in different climate scenarios would take about 20 generations (Kremer 2007).

Optimistic prognoses of a significant capacity of forest trees to adapt to altered bioclimatic conditions are opposed by the possibility of changes happening too fast, due to frequent extreme weather events, pollution, and other factors that exert additional selection pressure. In the future unpredictable conditions, one cannot claim with certainty that the possibility of adaptation will not reach certain limits, or that evolutionary processes will completely follow the principles of causality. Regardless of the fact that the results from provenance tests suggest a relatively quick adaptation of forest trees to the local conditions and the development of cline diversity of certain adaptive properties, climate changes could overtake adaptation processes, especially in the conditions of habitats broken up by geographic and anthropogenic factors, which greatly complicate the migration of forest trees (Savolainen et al. 2007). Adaptation is definitely a very complex process, encompassing many properties simultaneously, so based on the research to date, which mostly deals with a single property, it is difficult to reach conclusions on how affected forest ecosystems will be by climate change. It is definitely certain that the greatest danger threatens the species of a broken up or limited area, especially if problems in reproduction occur and if migration to new areas is restricted (Bosselman et al. 2008).

Climate change also brings about problems in the management of existing orchards. Until recently, the most important and somewhat stable factors affecting the economic and genetic significance of yield from seed orchards were thought to be the time and duration of blooming, diversity in fertility, and the total number of included clones. However, changes in climate conditions are beginning to exert a decisive influence on the time of flowering and synchronization of clones, and thus also on the quantity and quality of the yield (Alizoti et al. 2010).

The research conducted on 30-year-long phenological monitoring in Europe has shown that flushing begins 6 days earlier, and that the autumn coloration is prolonged by 4.8 days, which finally demonstrates that the vegetation season has extended by 10.8 days (Menzel and Fabian 1999). Parmesan and Yohe (2003) report that 62% out of 677 studied plant species demonstrated a trend of earlier phenological manifestations with regard to the beginning of flushing and blooming. Menzel et al. (2006) estimate over 125,000 phenological monitorings from 21 European countries and come to the conclusion that 78% of flushing, blooming and fruit production was subject to the trend of earlier onset, and only 3% exhibited the trend of later onset in the period from 1971 to 2000, with the average of 2.5 days per decade. The extension of the vegetation season is directly associated with an increase in the average air temperature, which has increased over the last three decades by 0.6 °C (Hansen et al. 2006). In addition to the increase in air temperature, we can also observe an increase in the amount of CO_2 in the atmosphere, which in synergy results in the increase of photosynthetic activity. The extension of the vegetation season has its advantages and disadvantages for forest conservation units. The advantage is manifested in the extended period of photosynthetic activity, where the amount of nutrients accumulated in the plants is also greater. This results in a more abundant fruit production the next year. On the other hand, a negative consequence of the extended vegetation season or of the trend of earlier flushing for forest conservation objects is possible damage caused by late spring frosts (Andrić et al. 2016; Andrić and Kajba 2017). The increase of CO_2 in the atmosphere will currently not have negative consequences for forest conversation units because the enzyme ribulose in the cells of forest trees is not saturated by the current quantities of CO_2 in the atmosphere (Ainsworth and Long 2005). There are evident changes in the rainfall regime. Increasingly frequent is the trend of a lack of rainfall during the summer months, and more abundant precipitation during the fall and winter months, which will in many species cause physiological stresses and the weakening of certain genotypes, making them more susceptible to attacks and damage caused by numerous pathogens (Huberty and Denno 2004; Diminić et al. 2017).

The management of forest conservation units in the coming decades will present a major challenge both for science and for the operative level. The altered trends of climate parameters that forest trees are facing will require intensive care for forest conservation units in order for genetic material to be conserved as long as possible, to gain time, and especially for the species that we conserve to adapt to the new conditions.

According to experts, the importance of maintaining genetic variability of forest trees in mitigating the effect of climate change on the forestry activity should be recognized and legally regulated on the state and pan-European level (Koskela et al. 2007). It is necessary to encourage such forest management that will enable evolutionary processes to take place and support the biological restoration of forests, where possible. In doing so, it should be established whether the population in question is well adapted to the local conditions and whether it possesses sufficient genetic variability. In artificial or assisted natural restoration, the reproductive material should also be well adapted and genetically diverse. It is recommended to define pan-European guidelines for the transfer of forest reproductive material, as the result of a strategy to accelerate the adaptation of trees to climate changes through breeding and transfer of suitable material to new areas. The strategy must be scientifically founded on interdisciplinary research, which would integrate the areas of genetics, physiology, phytopathology, entomology, forest management and economics, and modeling. The establishment of clonal seed orchards protects the genetic resource of a given species, because they make up an archive of different genotypes and represent well the genetic variability of a species in its natural environment.

References

Ainsworth, E. A., & Long, S. P. (2005). What have we learned from 15 years of free-air CO_2 enrichment (FACE)? A meta-analytic review of the responses of photosynthesis, canopy properties and plant production to rising CO_2. *New Phytologist, 165*(2), 351–372.

Alizoti, P. G., Kilimis, K., & Gallios, P. (2010). Temporal and spatial variation of flowering among *Pinus nigra* Arn. clones under changing climatic conditions. *Forest Ecology and Management, 259*(4), 786–797.

Andrić, I., & Kajba, D. (2017). The impact of environmental drivers on narrow-leaved ash (*Fraxinus angustifolia* Vahl) budburst dates. *Šumarski list, 141*(1–2), 7–13.

Andrić, I., Poljak, I., Milotić, M., Idžojtić, M., & Kajba, D. (2016). Fenološka svojstva listanja poljskog jasena (*Fraxinus angustifolia* Vahl) u klonskoj sjemenskoj plantaži. *Šumarski list, 140*(3–4), 117–126.

Bogdan, S., Katicic-Trupcevic, I., & Kajba, D. (2004). Genetic variation in growth traits in a *Quercus robur* L. open-pollinated progeny test of the Slavonian provenance. *Silvae Genetica, 53*(5–6), 198–201.

Bosselmann, A. S., Jacobsen, J. B., Kjær, E. D., & Thorsen, B. J. (2008). Climate change, uncertainty and the economic value of genetic diversity: A pilot study on methodologies. *Forest & Landscape, 31*, 58.

Diminić, D., Kajba, D., Milotić, M., Andrić, I., & Kranjec, J. (2017). Susceptibility of *Fraxinus angustifolia* clones to *Hymenoscyphus fraxineus* in Lowland Croatia. *Baltic Forestry, 23*(1), 233–243.

Grattapaglia, D., Sansaloni, C. P., Petroli, C. D., Resende Junior, M. F. R., Faria, D. A., Missiaggia, A. A., Takahashi, E. K., Zamprogno, K. C., Kilian, A., & de Resende, M. D. V. (2010).

Genomic selection in eucalyptus: Marker assisted selection coming to reality in forest trees. In *Proceedings of the Embrapa Florestas-Resumo em anais de congresso (ALICE)*, USA, San Diego CA.
Hansen, J., Sato, M., Ruedy, R., Lo, K., Lea, D. W., & Medina-Elizade, M. (2006). Global temperature change. *Proceedings of the National Academy of Sciences, USA, 103*, 14288–14293.
Huberty, A. F., & Denno, R. F. (2004). Plant water stress and its consequences for herbivorous insects: A new synthesis. *Ecology, 85*(5), 1383–1398.
Kajba, D., & Andrić, I. (2015). Forest reproductive material and conservation of forest genetic resources in Croatia. *Open Journal of Forestry, 5*(02), 117–128.
Kajba, D., & Hrašovec, B. (2009). Klonske sjemenske plantaže hrasta lužnjaka (Quercus robur L.) i njihova uloga u očuvanju genofonda u uvjetima klimatskih promjena i povećanih rizika od napada šumskih kukaca. In *Proceedings of the forests of pedunculate oak in changed site and management conditions*, pp, 143–152.
Kajba, D., Bogdan, S., & Katičić, I. (2006a). Estimation of genetic gain for vigorous growth by clonal seed orchards of pedunculate oak (*Quercus robur* L.). *Glasnik za šumske pokuse, pos izd, 5*, 251–260.
Kajba, D., Gračan, J., Ivanković, M., Bogdan, S., Gradečki-Poštenjak, M., Littvay, T., & Katičić, I. (2006b). Očuvanje genofonda šumskih vrsta drveća u Hrvatskoj. *Glasnik za šumske pokuse, pos izd, 5*, 235–249.
Kajba, D., Pavičić, N., Bogdan, S., & Katičić, I. (2007). Pomotehnički zahvati u klonskim sjemenskim plantažama listača. *Šumarski list, 131*(11–12), 523–528.
Kajba, D., Pavičić, N., Bogdan, S., & Katičić, I. (2008). Pomotechnical treatments in the broadleave clonal seed orchards. In *Proceedings of theseed orchard conference*, Umeå, Sweden, pp. 95–103.
Kajba, D., Katičić, I., & Bogdan, S. (2011a). Procjena genetskih parametara u testovima polusrodnika hrasta lužnjaka (*Quercus robur* L.) iz sjemenskih zona Posavine, Podravine i Podunavlja. *Croatian Journal of Forest Engineering, 32*(1), 177–190.
Kajba, D., Katičić, I., Šumanovac, I., & Žgela, M. (2011b). Važnost klonskih sjemenskih plantaža u sjemenarstvu i očuvanju genofonda šumskih vrsta drveća u Hrvatskoj. *Radovi (Hrvatski šumarski institut), 44*(1), 37–51.
Kajba, D., Katičić, I., Bogdan, S., & Tančeva Crmarić, O. (2012). Management, genetic gain and genetic diversity in clonal seed orchards in Croatia. In *Proceedings of the seed orchards and breeding theory conference*, Antalya, Turkey, pp. 42–44.
Kang, K. S., El-Kassaby, Y. A., Han, S. U., & Kim, C. S. (2005). Genetic gain and diversity under different thinning scenarios in a breeding seed orchard of *Quercus accutissima. Forest Ecology and Management, 212*(1–3), 405–410.
Katičić, I., Bogdan, S., Sever, K., Šatović, Z., & Kajba, D. (2010). Genetic structure and variability of phenological forms of pedunculate oak (Quercus robur L.) from clonal seed orchards in Croatia. In *Proceedings of the Forest ecosystem genomics and adaptation*. Book of Abstracts, Bioversity International (Rome, Italy) and INIA (Madrid, Spain), 181 p.
Katičić-Bogdan, I. (2012). Genetska raznolikost hrasta lužnjaka (*Quercus robur* L.) u klonskim sjemenskim plantažama u Hrvatskoj. Doctoral thesis. Šumarski fakultet Sveučilište u Zagrebu, 165 pp.
Koskela, J., Buck, A., & du Cros, E. T. (2007). *EUFORGEN climate change and forest genetic diversity*. Rome: Bioversity International 111 pp.
Kremer, A. (2007). How well can existing forests withstand climate change? In *Proceedings of the climate change and forest genetic diversity: Implications for sustainable forest management in Europe*, Rome, Italy, pp. 3–17.
Lande, R., & Barrowclough, G. F. (1987). Effective population size, genetic variation, and their use in population management. *Viable Populations for Conservation, 87*, 88–123.
Menzel, A., & Fabian, P. (1999). Growing season extended in Europe. *Nature, 397*(6721), 659–659.
Menzel, A., Sparks, T. H., Estrella, N., Koch, E., Aasa, A., Ahas, R., Alm-Kübler, K., Bissolli, P., Braslavská, O., Briede, A., Chmielewski, F. M., Crepinsek, Z., Curnel, Y., Dahl, Ĺ., Defila, C.,

Donnelly, A., Filella, Y., Jatczak, K., Mĺge, F., Mestre, A., Nordli, O., Peńuelas, J., Pirinen, P., Remišová, V., Scheifinger, H., Striz, M., Susnik, A., van Vliet, A. J. H., Wielgolaski, F., Zach, S., & Zust, A. (2006). European phenological response to climate change matches the warming pattern. *Global Change Biology, 12*(10), 1969–1976.

Parmesan, C., & Yohe, G. (2003). A globally coherent fingerprint of climate change impacts across natural systems. *Nature, 421*(6918), 37–42.

Rajora, O., & Mosseler, A. (2001). Molecular markers in conservation, restoration and sustainable management of forest genetic resources. *Forest Science, 70*, 187–201.

Savolainen, O., Bokma, F., Knürr, T., Kärkkäinen, K., Pyhäjärvi, T., & Wachowiak W. (2007). Adaptation of forest trees to climate change. In *Proceedings of the climate change and forest genetic diversity: Implications for sustainable forest management in Europe*, Rome, Italy, pp. 19–28.

Tančeva Crmarić, O., Štambuk, S., & Kajba, D. (2011). Genotipska raznolikost divlje trešnje (*Prunus avium* L.) u dijelu prirodne rasprostranjenosti u Hrvatskoj. *Šumarski list, 135*(11–12), 543–554.

Vidaković, M. (1996). Podizanje klonske sjemenske plantaže hrasta lužnjaka. In *Proceedings of the Hrast lužnjak (Quercus robur L.) u Hrvatskoj*, Vinkovci – Zagreb, pp. 127–138.

Vidaković, M., Kajba, D., Bogdan, S., Podnar, V., & Bećarević, J. (2000). Estimation of genetic gain in a progeny trial of pedunculate oak (*Quercus robur* L.). *Glasnik za šumske pokuse, 37*, 375–381.

Chapter 23
Register of Wild Cherry (*Prunus avium* L.) Plus Trees in Slovenia

Robert Brus, Gregor Božič, Hojka Kraigher, and Kristjan Jarni

Abstract Larger number of wild cherry candidate plus trees in the entire area of Slovenia were identified and recorded. Detailed seven-scale evaluation of their vitality, stem straightness, crown size, constriction, cylindrical trunk growth, trunk warping, branch diameter, branching angle and forking resulted in the selection of 107 trees possessing the most desirable characteristics. Selected trees are relatively large and of phenotypically superior quality. The established register of wild cherry plus trees in Slovenia represents a selection of superior individual phenotypes that will be used as a source of parental trees to establish a first-generation seed orchard and then, if appropriate, as a source of superior material for the next breeding steps and further breeding programs.

Keywords *Prunus avium* · Wild cherry · Plus trees · Seed orchard · Forest tree breeding · Slovenia

Scattered broadleaves are an important component of European forests. They increase forest diversity and thus may play an important role in insuring against risks such as pests and pathogens, fire, and habitat loss. They can also contribute to the economic importance of forests. The genetic resources of scattered broadleaves have often suffered from forest management that reduces genetic variation and results in habitat deterioration and introgression. The management of genetic resources is therefore focused on the conservation of high genetic variation. However, even if local genetic resources are well adapted to the local environment, they may not necessarily be compatible with all socio-economic needs and

R. Brus (✉) · K. Jarni
Department for Forestry and Renewable Forest Resources, Biotechnical Faculty, University of Ljubljana, Ljubljana, Slovenia
e-mail: robert.brus@bf.uni-lj.si; kristjan.jarni@bf.uni-lj.si

G. Božič · H. Kraigher
Slovenian Forestry Institute, Ljubljana, Slovenia
e-mail: gregor.bozic@gozdis.si; hojka.kraigher@gozdis.si

M. Šijačić-Nikolić et al. (eds.), *Forests of Southeast Europe Under a Changing Climate*, Advances in Global Change Research 65,
https://doi.org/10.1007/978-3-319-95267-3_23

industrial requirements or be able to adapt to adverse climatic constraints. Forest tree breeding can be an effective method for improving the genetic resources of forest tree species and thus addressing these potential shortcomings (Pâques 2013).

Due to its economic and ecological importance and its possible contribution to the preservation and restoration of the ecological and cultural heritage of European landscapes, wild cherry (*Prunus avium* L.) is one of the main scattered broadleaf species that has been the focus of breeding programs across Europe in the last decades. Several European countries have established provenance, progeny and clonal trials as well as clonal or seedling seed orchards of wild cherry in order to obtain high quality forest reproductive material.

In Slovenia, wild cherry is a native tree species that occurs individually or in smaller groups in forest stands. Given its considerable economic and ecological potential, it would be beneficial to increase its share in Slovenian forests. Although only a small proportion of Slovenian forest is regenerated by planting, artificial regeneration can have a larger impact on the genetic gain and genetic structure of future forest. In Slovenia there are eight registered seed objects of wild cherry that are assigned for the production of forest reproductive material (FRM) (Seznam gozdnih… 2017). Six of them are seed stands; three stands belong to the category "selected" and three to the category "known origin." There are no seed objects in the categories "qualified" or "tested," which include seed plantations.

Genetic analysis of four Slovenian wild cherry seed stands indicated significant genetic variability ($H_E = 0.726$), a low degree of inbreeding ($F_{IS} = -0.029$) and low intra-population differentiation ($F_{ST} = 0.040$; $G'_{ST} = 0.137$) (Jarni et al. 2012). However, it also revealed that there is a significant amount of clonal reproduction in natural wild cherry stands (Jarni et al. 2015), which could be problematic in terms of ensuring sufficient genetic variability of progeny or forest reproductive material (FRM) originating from these stands. In addition, obtaining sufficient quantities of FRM of wild cherry is often problematic and expensive in existing seed stands in Slovenia due to irregular fruiting, difficult accessibility and difficult harvesting, as well as very low quantities of used seeds. A possible solution to these problems is a clone seed orchard composed of phenotypically and genetically superior tree clones. FRM from this orchard would be genetically improved, and its production would be simpler, safer and cheaper. The selection of superior trees and elaboration of a register of plus trees is the first step towards the establishment of such a seed orchard.

In collaboration with the Slovenian Forest Service, we identified and recorded more than a hundred wild cherry candidate trees (a total of 127 trees) in the entire area of Slovenia. Based on the first preliminary survey and acquired data, we prepared a broader set of criteria that we took into consideration during subsequent evaluation of the same trees. Using a seven-level scale, the following features were evaluated: vitality, stem straightness, crown size, constriction, cylindrical trunk growth, trunk warping, branch diameter, branching angle and forking.

Detailed evaluation of candidate trees resulted in the selection of 107 trees possessing the most desirable characteristics. The analysis showed that these selected trees are relatively large (average: DBH = 50.79 cm, height = 26.91 m) and of phenotypically superior quality. Trunk warping proved to be the least problematic and,

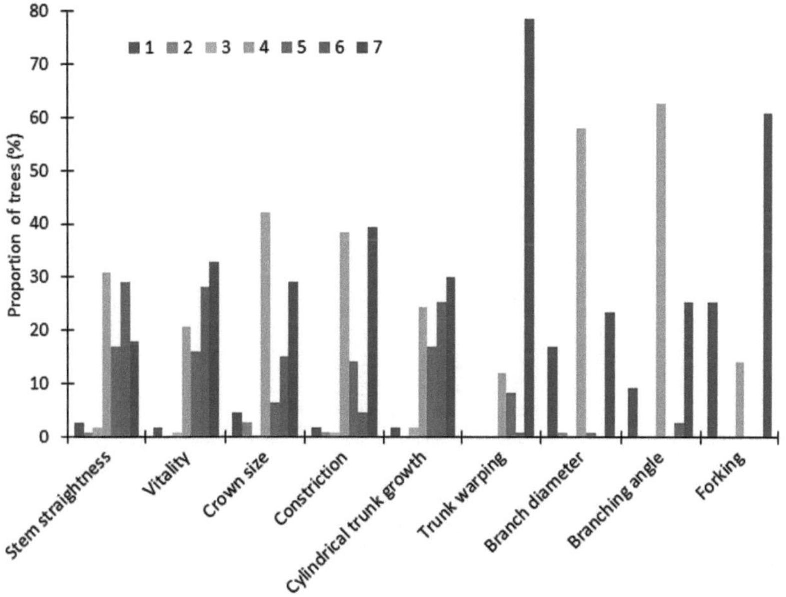

Fig. 23.1 Qualitative rating of characteristics of wild cherry plus trees. (1 = strongly undesirable, 7 = very desirable)

at the same time, the least variable characteristic, while the other evaluated characteristics were more variable (Fig. 23.1).

The analysis confirmed a positive correlation between crown size and forking; larger crowns also had larger diameter branches (Table 23.1). We also confirmed a positive correlation between straightness and cylindrical growth of the trunk.

All of the 107 trees evaluated in detail will be recorded as wild cherry plus trees in Slovenia. Ballian and Kajba (2011) argue that the maintenance of adaptation potential and population variability in most tree species can be achieved with 30–50 different genotypes (clones). Ongoing breeding programs in other European countries that fall within this number range include Croatia with 24 (Tančeva-Crmarić et al. 2011), Montenegro with 35 (Orlović et al. 2006), Belgium with 65 (De Cuyper et al. 2005), Greece with 36 (Avramidou et al. 2010) and Germany with 40 (Ducci et al. 2013).

The register of wild cherry plus trees will be the first register of its kind in Slovenia. It represents a selection of superior wild cherry individual phenotypes that will be used as a source of parental trees to establish a first-generation seed orchard and then, if appropriate, as a source of superior material for the next breeding steps and for further breeding programs. Wild cherry, as a scattered forest tree species with desirable ecological and economic characteristics, well-developed criteria for assessment of phenotypic properties, and highly supported genetic and propagation knowledge, provides a good model and foundation for breeding programs of other tree species, particularly for other scattered forest tree species.

Table 23.1 Spearman's correlation coefficient between evaluated characteristics

	Vitality	Crown size	Constriction	Cylindrical trunk growth	Trunk warping	Branch diameter	Branching angle	Forking
Stem straightness	−.487***	−.042	−.141	−.477***	−.391**	−.176	−.155	−.065
Vitality		−.112	−.119	−.322**	−.291	−.124	−.019	−.119
Crown size			−.185	−.054	−.130	−.323**	−.048	−.385**
Constriction				−.138	−.031	−.026	−.135	−.130
Cylindrical trunk growth					−.251	−.009	−.056	−.078
Trunk warping						−.121	−.114	−.004
Branch diameter							−.075	−.014
Branching angle								−.008

Significance levels are adjusted using sequential Bonferroni according to Rice (1989)
*$0.01 < P < 0.05$; **$0.001 < P < 0.01$; ***$P < 0.001$

References

Avramidou, E., Ganopoulos, I. V., & Aravanopoulos, F. A. (2010). DNA fingerprinting of elite Greek wild cherry (*Prunus avium* L.) genotypes using microsatellite markers. *Forestry, 83*(5), 527–533.

Ballian, D., & Kajba, D. (2011). *Oplemenjivanje šumskog drveća i očuvanje njegove genetske raznolikosti*. Sarajevo: Zagreb, Šumarski fakultet: 299 str.

De Cuyper, B., Sonneveld, T., & Tobutt, K. R. (2005). Determining self-incompatibility genotypes in Belgian wild cherries. *Molecular Ecology, 14*(4), 945–955.

Ducci, F., De Cuyper, B., De Rogatis, A., Dufour, J., & Santi, F. (2013). Wild cherry breeding (*Prunus avium* L.). In L. E. Pâques (Ed.), *Forest tree breeding in Europe. Managing forest ecosystems* (Vol. 25, pp. 463–511). New York: Springer.

Jarni, K., De Cuyper, B., & Brus, R. (2012). Genetic variability of wild cherry (*Prunus avium* L.) seed stands in Slovenia as revealed by nuclear microsatellite loci. *PLoS One, 7*, e41231. https://doi.org/10.1371/journal.pone.0041231.

Jarni, K., Jakše, J., & Brus, R. (2015). Vegetative propagation: Linear barriers and somatic mutation affect the genetic structure of a *Prunus avium* L. stand. *Forestry, 88*(5), 612–621.

Orlović, S., Matović, N., & Pilipović, A. (2006). Noble hardwoods in Serbia and Montenegro. In M. Bozzano, M. Rusanen, P. Rotach, & J. Koskela (Eds.), *EUFORGEN noble hardwoods network: Report of the sixth (9–11 June 2002, Alter do Chão, Portugal) and seventh meetings (22–24 April 2004, Arezzo, Italy)* (pp. 21–24). Rome: International Plant Genetic Resources Institute.

Pâques, L. E. (2013). *Forest tree breeding in Europe. Current state-of-the-art and perspectives* (527 p). Dordrecht: Springer.

Rice, W. R. (1989). Analyzing tables of statistical tests. *Evolution, 43*(1), 223–225.

Seznam gozdnih semenskih objektov – stanje na dan. 2017. 1. 1. 2017. Ur. l. RS št. 4/17.

Tančeva-Crmarić, O., Štambuk, S., Šatović, Z., & Kajba, D. (2011). Genotipska raznolikost divlje trešnje (*Prunus avium* L.) u dijelu prirodne rasprostranjenosti u Hrvatskoj. *Šumarski list, 135*(11–12), 543–555.

Chapter 24
International European Beech Provenance Trial Kamenski hrib/Straža in Slovenia

Gregor Božič and Hojka Kraigher

Abstract The present work summarized establishment and replanting of the beech provenance trial in Slovenia as a part of the international beech provenance trials launched by Institute of Forest Genetics and Forest Tree breeding from Grosshansdorf, Germany in 1996/1998. About the experiment area basic climatic and soil data are presented and for each provenance the name, seed lot number, geographical coordinates and altitude is provided. In the frame of the COST Action E52 the European beech genetic resources in wide scale were analyzed and databases harmonized. The main publications for the Slovenian trial are listed.

Keywords *Fagus sylvatica* (L). · International provenance trial · Slovenia · COST Action E52

24.1 Introduction

In last 40 years the largest network of provenance trials of beech (*Fagus sylvatica* L.) in Europe was established by the Institute of Forest Genetics and Forest Tree breeding at Grosshansdorf, Germany. The network consists of sets of field trials established in Germany in years 1983/1986, 1984/1987 and 1985/1988 and three series of international beech provenance trials established in different European countries in years 1993/1995, 1996/1998 and 2005/2007. In total more than 500 beech provenances are planted.

The aim of beech trials is to determine the site stability of economically important growth related phenological and morphological traits (survival, phenology), form (stem and branching habit), and yield (height and diameter), the variation in the time, adaptability and adaptation potential populations to important site characteristics on a species wide scale (von Wühlisch 2007) as well as (cit. Giannini and von Wühlisch 2009: 104) "to make predictions of the future distribution range of

G. Božič (✉) · H. Kraigher
Slovenian Forestry Institute, Ljubljana, Slovenia
e-mail: gregor.bozic@gozdis.si; hojka.kraigher@gozdis.si

M. Šijačić-Nikolić et al. (eds.), *Forests of Southeast Europe Under a Changing Climate*, Advances in Global Change Research 65,
https://doi.org/10.1007/978-3-319-95267-3_24

beech forest ecosystems under the assumption of certain scenarios of climate change, based on the analysis of the reaction pattern of European beech populations of defined origin (progenies of natural beech stands) under changed climate situations in sets of pan European field trials".

24.2 International Beech Provenance Trial in Slovenia

The Slovenian provenance trial Kamenski hrib/Straža is a part of the second series of the international beech provenance trials launched by Institute of Forest Genetics and Forest Tree breeding from Grosshansdorf, Germany in 28 sites in 15 European countries with 61 provenances. Seeds were harvested mainly in the year 1995. In spring of 1996 seeds were sown in the institute nurseries. Seedlings were labelled and lifted in autumn 1997 at an age of 1 + 1 and 2 + 0 and held during a winter in cold storage house in Hamburg. Plants were transported from Hamburg to collaborative institutions in European countries in the spring 1998 for trails establishment (von Wühlisch 2007).

The experiment in Slovenia was established by the Slovenian Forestry Institute in close cooperation with Slovenian Forestry Service, Regional unit Novo mesto and Forest Enterprise Novo mesto. The trial Kamenski hrib/Straža was established in SE part of Slovenia on the hilltop plateau of "Kamenski hrib" (latitude 45° 47′ 46″ N; longitude 15° 02′ 54″ E) in spring 1998 with 2-year old European beech seedling of 31 provenances (Table 24.1). The selected location is in the Pre-Dinaric region of provenance near the town "Straža". The altitude of the trial is 540–550 m above sea level. Before planting, the old mix forest stand with 60% share in growing stock of *Fagus sylvatica*, 30% of *Abies alba* and 10% of *Picea abies* was clear-cut, stems and residues removed. Ground rock material is limestone. Forest soils are eutric cambisols and haplic luvisols (FAO – UNESCO 1989) with good nutrient availability and average water capacity. The soils are usually deep, loamy to clayey, with a small skeleton, with shallow and sprstenous humus accumulation horizon A, biologically active, fresh, good fertility and favorable for beech. Main climatic variables were determined as 6.3 °C mean annual temperature and 1275 mm mean annual precipitation and based on 1971–2009 averages interpolated from the official weather stations in Novo mesto and Ambrus, both situated 20 km from the site. Late or early frosts can occur in the site from October till April. During winter, snow cover persists from 50 to 60 days.

A 1.3 ha trail is designed in three blocks divided into 31 plots of 10 × 10 m. Originally a total of 31 provenances of European beech from 15 European countries were planted in each block with 50 seedlings in each plot following a 1 × 2 m spacing (five rows by ten seedlings). The experimental design was proposed by dr. Mirko Liesebach, dr. Georg von Wühlisch and prof. dr. Hans Muhs from Institute of Forest Genetics and Forest Tree Breeding, Grosshansdorf, Germany. Each provenance originally represented 150 individuals, t.i. 4650 test seedlings in all were used. One border row with 2 years old beech seedlings with unknown provenances was planted

Table 24.1 Survey of 39 European beech provenances planted in the international trial Kamenski hrib/Straža

Name	ID seed collection number	Country	Provenance	Lat. N	Lon. E	Altitude (m a.s.l.)
P01	9235	F	Perche	48°25′	00°33′	205
P02	9236	F	Bordure Man	49°32′	00°46′	80
P05	9240	F	08-Bretagne	48°22′	−02°09′	180
P06	9241	F	11-Plateaux du	46°48′	05°50′	600
P08	9243	F	19-Pyreness Or	42°55′	02°19′	670
P11	9194	LUX	Heinerscheid	50°05′	06°07′	423
P13	9191	B	Soignes	50°50′	04°25′	110
P14	9170	NL	Aarnink	51°56′	06°44′	45
P15	9171	NL	Elspeet	52°17′	05°48′	33
P17	9172	GB	Westfield 2002	57°24′	−03°15′	10
P21	9184	DK	Grasten, F.413	54°55′	09°35′	45
P23	9183	S	Torup	55°34′	13°12′	40
P27	9209	D	Graf von Westf. (NW)	51°31′	08°47′	375
P31	9182	D	Urach 12A13 (BW)	48°28′	09°27′	760
P32	9245	D	Ebrach (BY)	49°51′	10°30′	406
P34	9112	CH	Oberwil	47°10′	07°27′	530
P35	9205	A	Hinterstoder	47°43′	14°06′	1250
P36	9206	A	Eisenerz	47°32′	14°51′	1100
P37	9114	I	Val di Sella	46°01′	13°30′	1150
P38	9214	PL	Bilowo 124a, 123 c	54°20′	18°10′	250
P39	9217	PL	Jaworze	49°50′	19°10′	450
P40	9221	PL	Tarnawa	49°28′	22°20′	540
P43	9226	PL	Jawornik, 92b	49°15′	22°49′	900
P46	9180	CZ	Domazlice-Vyhl.	49°24′	12°45′	760
P48	9200	CZ	Jablonec N.N.	50°48′	15°14′	760
P49	9202	CZ	Brumov-Sidonie	49°03′	18°03′	390
P51	9212	CZ	Horni Plana-Ce.	48°51′	14°00′	990
P53	9190	SI	Postojna Mašun	45°38′	14°23′	1000
P54	9188	SI	Idrija-II/2, 14	46°00′	13°54′	930
P55	9189	SI	Postojna Javorniki	45°44′	14°19′	1040
P57	9196	BG	Gramaticevo	42°04′	26°08′	200
P61	9223	PL	Tworylczuk, 180 F	49°15′	22°49′	320
P63	9238	F	06-Nord-Est Ma.	46°11′	03°01′	580
P64	9201	CZ	Nizbor	50°00′	14°00′	480
P65	9219	PL	Koino, 105F	49°55′	20°25′	400
P67	9215	PL	Bilowo 115, 116	54°20′	18°10′	250
P69	9218	PL	Sucha	49°40′	19°20′	400
P77	9120	I	Val Fondillo	41°46′	13°51′	1145
P84	9153	SK	Topolcianky	48°25′	18°24′	No data

Modified according to Liesebach (1998)

around the trial. The trail was fenced to prevent game damage. Costs for establishment, regular management, measurements end evaluations are covered by the the Slovenian Public Forest Service financed by the Ministry of Agriculture, Forestry and Food of Republic Slovenia. The experiment is planned for an observation period of 60 years.

Data of the Straža site 1 year after planting showed at age 3 the overall survival was 82%. The best survival rates had provenances P37 (98%), P8, P51 and P54 (97%), P36 (95%). The poorest were P57 (27%) and P11 (45%). Slovenian provenances Postojna Mašun (P53), Idrija-II/2, 14 (P54), Postojna Javorniki (P55) varied from 89% to 97%, Austrian 85–95%, Czech 83–97%, German 75–93%, Poland 65–95%, France 62–97%, The survival rate of single provenance presented from Great Britain, Belgium, Netherlands, Sweden and Switzerland varied from 73% to 93%. In first years after establishment all dead trees were removed and replaces with additional planting obtained from the rest seedlings stock from the Hamburg nursery. Many of dead plants were replaced in spring 1999 with the three years old seedlings except for the Bulgarian provenance Gramaticevo (P57) later determined as *Fagus orientalis* Lipsky which was completely removed and replaced with new 150 seedlings of a Czech provenance (P65). In total 447 seedlings were additionally planted in provenance plots. The second time the provenance trial was additionally replanted after the massive attack of rodent animals (vole, mice) which occurred during the winter 1999/2000. From 4233 trees survived only 2530. The mortality rate was 40%. In autumn 2000 the trial was prepared for the reconstruction in the way agreed with Dr. Georg von Wühlisch and Prof. Dr. Hans Muhs. The plots were replanted in spring 2001 with 5 years old seedlings. Two hundred two trees were obtained from trial plots in-situ and 290 from the rest stock of the testing provenances in Hamburg nursery. Empty plots were replanted with seven new provenances (P05, P15, P38, P61, P63, P77, P84) transported from Hamburg nursery with 100 seedlings per provenance. In total 1192 seedlings were planted in spring 2001. Therefore, the design of provenance trial changed from originally planted 31 provenances to 38 provenances (Table 24.1) presented in 1–3 repetitions.

The provenance trial is regularly managed and investigated by assessment and measurement of the phenological and morphological traits (survival, phenology), form (stem and branching habit), and yield (height and diameter) obtained from all trees. Height measurements and survival assessment were done at almost yearly basis until 2009 and later in a few years intervals. The genetic resources of beech in wild scale were jointly analyzed in the frame of the COST Action E52 "Evaluation of Beech Genetic Resources for Sustainable Forestry" (2006–2010) and all data harmonised (Robson et al. 2018). Main results from the Slovenian trial were published by Ivanković et al. 2008; Mátyás et al. 2009a, b; Sinjur and Bozic 2010; Robson et al. 2011; Alìa et al. 2011. Additionally, types of ectomycorrhizae on seedlings and root growth dynamics of selected beech provenances in a beech provenance trial Kamenski hrib/Straža were studied by Westergren et al. (2004) and Železnik et al. (2018).

Results of the study Robson et al. (2011) which jointly analysed the timing of leaf flush of 29 provenances from 13 European countries in 13 international beech provenance trials established in years 1995/1996 and 1996/1998 shown that in

Slovenia exist early and late flushed beech provenances. The results confirmed the provenance Postojna Masun (P53) as one of the earlier flushed provenances and Idrija-II/2, 14 (P54) as one of the latest flushed provenances in pan European scale. The provenance Postojna Masun flushed in average 3.9 days ±0.8 SE and the provenance from Idrija 12.0 days ±2.3 SE later than the first provenances in all analysed trials. The provenance Idrija differed from nearby tested provenances and was also the latest flushed provenance among 15 Slovenian beech provenances studied by Sittler (1981).

Acknowledgements The establishment and evaluation of the trial was supported by the Concerted Action of the Commission of the European Communities (AIR3 Programme, Grant No. CT94-2091), the Research Programme P4-0107 financed by the Slovenian Research Agency, Public Forest Service Programme financed by the Ministry for Agriculture, Forestry and Food of the Republic of Slovenia, and by the COST Action E52 (2006–2010).

References

Alía, R., Božič, G., Gömöry, D., Huber, G., Rasztovits, E., & von Wühlisch, G. (2011). The survival and performance of beech provenances over a Europe-wide gradient of climate. In *Genetic resources of European beech (Fagus sylvatica L.) for sustainable forestry: Proceedings of the COST E52 evaluation of beech genetic resources for sustainable forestry*. Final Meeting, 4–6 May 2010, Burgos, Spain. Madrid: Instituto Nacional de Investigación y TecnologíaAgraria y Alimentaria, 22, pp. 115–126

Giannini, R., & von Wühlisch, G. (2009). Workshop COST E52 "Evaluation of beech genetic resources for sustainable forestry". *iForest, 2*, 104. https://doi.org/10.3832/ifor0500-002.

Ivanković, M., Bogdan, S., & Božič, G. (2008). Varijabilnost visinskog rasta obične bukve (*Fagus sylvatica* L.) u testovima provenijencija u Hrvatskoj i Sloveniji. *Šumarski list, 132*(11/12), 529–541.

Liesebach, M. (1998). International beech provenance trial. Manual. 20 p.

Mátyás, C., Božič, G., Gömöry, D., Ivankovic, M., & Rasztovits, E. (2009a). Juvenile growth response of European beech (*Fagus sylvatica* L.) to sudden change of climatic environment in SE European trials. *IForest, 2*, 213–220. https://doi.org/10.3832/ifor0519-002.

Mátyás, C., Bozic, G., Gömöry, D., Ivankovic, M., & Rasztovits, E. (2009b). Transfer analysis of provenance trials reveals macroclimatic adaptedness of European beech (*Fagus sylvatica* L.). *Acta Silvatica Lignaria Hungarica, 5*, 47–62.

Robson, T. M., Alia, R., Bozic, G., Clark, J., Forstreuter, M., Gömöry, D., Liesebach, M., Mertens, P. Rasztovits, E., Zitová, M., & Von Wuehlisch, G. (2011). The timing of leaf flush in European beech (*Fagus sylvatica* L.) saplings. In *Genetic resources of European beech (Fagus sylvatica L.) for sustainable forestry: Proceedings of the COST E52 evaluation of beech genetic resources for sustainable forestry*. Final Meeting, 4–6 May 2010, Burgos, Spain. Madrid: Instituto Nacional de Investigación y Tecnología Agraria y Alimentaria, 22, pp. 61–79.

Robson, T. M., Benito-Garzón, M., Alia, R. M., Egido, D. B., Bogdan, S., Borovics, A., Božič, G., Brendel, O., Clark, J., De Vries, S. M. G., Delehan, I., Ducousso, A., Fady, B., Fennessy, J., Forstreuter, M., Frýdl, J., Geburek, T., Gömöry, D., Hauke-Kowalska, M., Huber, G., Ibañez, J. I., Ioniţă, L., Ivankovič, M., Kehlet-Hansen, J., Kóczán-Horváth, A., Kraigher, H., Lee, S., Liesebach, M., Mátyás, C., Mertens, P., Muhs, H. J., Novotný, P., Parnuţa, G., Paule, L., Picardo, A., Rasztovics, E., Rogge, M., Stener, L. G., Sułkowska, M., Urban, O., Von Wuehlisch, G., Vendramin, G. G., Vettori, C., & Wesoły, W. (2018). Phenotypic trait variation measured on European genetic trials of *Fagus sylvatica* L. *Scientific Data, 5*, 180149. https://www.nature.com/articles/sdata2018149.

Sinjur, I., & Božič, G. (2010). Air temperature in the international beech provenance trail: Measurements or interpolation? In K. A. Spanos (Ed.), *Beech genetic resources for sustainable forestry in Europe: Proceedings of the Workshop and MC Meeting of the COST Action E52 Evaluation of Beech Genetic Resources for Sustainable Forestry, Thessaloniki, May 5–7, 2009* (pp. 98–104). Thessaloniki: National Agricultural Research Foundation, Forest Research Institute.

Sittler, B. (1981). Experimentell ökologische Untersuchungen an 15 slowenischen Buchenprovenienzen zur Beurteilung ihrer Anbaufähigkeit in der Bundesrepublik Deutschland. Dissertation thesis. Albert-Ludwig University of Freiburg, 205 p.

Wühlisch, G. von. (2007). Series of international provenance trials of European beech. In *Proceedings from the 7th International Beech Symposium IUFRO Research Group 1.10.00 "Improvement and Silviculture of Beech"*. Teheran, Iran, May 2004, pp. 135–144.

Westergren, M., Grebenc, T., Božič, G., Brus, R., & Kraigher, H. (2004). Identification of types of ectomycorrhizae on seedlings in a beech provenance trial. *Zbornik gozdarstva in lesarstva, 75*, 87–104.

Železnik, P., Westergren, M., Božič, G., Eler, K., Bajc, M., Helmisaari, H.-S., Horvath, A., & Kraigher, H. (2018). Root growth dynamics of three beech (*Fagus sylvatica* L.) provenances. *Forest Ecology and Management*. https://doi.org/10.1016/j.foreco.2018.06.024.

Chapter 25
Ex Situ Conservation of Forest Genetic Resources in Romania

Ecaterina Nicoleta Apostol, Neculae Şofletea, and Alexandru Lucian Curtu

Abstract Forest tree breeding program started in Romania more than 55 years ago and the main objectives of the program consisted in improving wood production, wood quality, resistance to biotic and abiotic stress factors, development of biotechnological techniques for mass clonal propagation. During this period, more than 6400 plus-trees were selected within natural, and in some cases, artificial populations. By their multiplication, a total of around 643.4 ha of clonal, hybridization and seedling seed orchards (440.3 ha conifers and 203.1 ha of broadleaves) were established. Also, a number of 71 provenance trials (46 for conifers and 25 for broadleaves), 8 half-sib progenies trials (6 for conifers and 2 for broadleaves), 7 full-sib progeny trials for conifers, 18 interspecific hybrids trials (8 for conifers and 10 for broadleaves) and 3 genitor clone's collections for conifers were established. A total of 113 ex situ conservation core units, amounting to 964.2 hectares, were established in stands with non-native species and in plantations with native species located outside their natural range.

Keywords Forest genetic resources · Ex situ conservation · Romanian breeding program

E. N. Apostol (✉)
National Institute for Research and Development in Forestry "Marin Drăcea" Ilfov County, Voluntari, Romania
e-mail: ecaterina.apostol@icas.ro

N. Şofletea
Faculty of Silviculture and Forest Engineering, Transilvania University of Braşov, Braşov, Romania
e-mail: nic.sofletea@unitbv.ro; lucian.curtu@unitbv.ro

A. L. Curtu
Department of Forest Sciences, Transilvania University of Braşov, Braşov, Romania
e-mail: nic.sofletea@unitbv.ro; lucian.curtu@unitbv.ro

M. Šijačić-Nikolić et al. (eds.), *Forests of Southeast Europe Under a Changing Climate*, Advances in Global Change Research 65,
https://doi.org/10.1007/978-3-319-95267-3_25

25.1 Introduction

In the context of global warming, with increasing temperatures and decreasing precipitations (Dubrovský et al. 2015; Simota et al. 2014; Thomas et al. 2014), the forest tree species are likely to migrate northward or to a higher altitude. This is only possible where ecological corridors are linked together. The arising question is then what will happen in lowlands, where there are natural or anthropic barriers, particularly crops that use herbicides? Ex situ conservation of the valuable provenances, with higher plasticity and adaptability, and with good capacities for production is a prerequisite to ensure the stability of forest ecosystems in the context of climate change and anthropogenic influence.

As a future strategy for sustainable forest management, the breeding program aims to conserve the genetic diversity of forest trees and to provide quality seeds on a reliable basis for the afforestation. Even if it is a more expensive conservation method, for safeguarding of valuable, threatened and endangered species, ex situ conservation is more practical than in situ, because these species require more protective measures. Otherwise, the material conserved ex situ could be of great relevance to restoration of in situ sites as well.

25.2 Status of Forest Genetic Resources Conserved Ex Situ

In Romania, the ex situ conservation of the forest genetic resources is mainly performed in: (i) seed orchards; (ii) comparative trials and genitors clone collections; (iii) arboreta; (iv) core collections established in the stands with non-native species, or native species, established outside their natural range.

(i) Forest tree breeding program started in Romania more than 55 years ago and the main objectives of the program consisted in improving wood production, wood quality, resistance to biotic and abiotic stress factors, development of biotechnological techniques for mass clonal propagation.

During this period, more than 6400 plus-trees were selected within natural, and in some cases, artificial populations, of which, by their multiplication, in the present, a total of around 643.4 ha of clonal, hybridization and seedling seed orchards are available for seed harvest.

Concerning the forest structure by species, the main group tree species in the Romanian forest are: conifers (Norway spruce, Silver fir and others coniferous species) (26%), beech (32%), oak species (18%), hard wood broadleaves (17%) and soft wood broadleaves (7%). According to the forest structure by species and corroborated with the fact that the conifers are the most utilized in afforestation, the surface of the conifers seed orchards established is about 440.3 ha. The conifer species conserved ex situ are: *Picea abies*, *Abies alba*, *Larix decidua*, *Larix* x

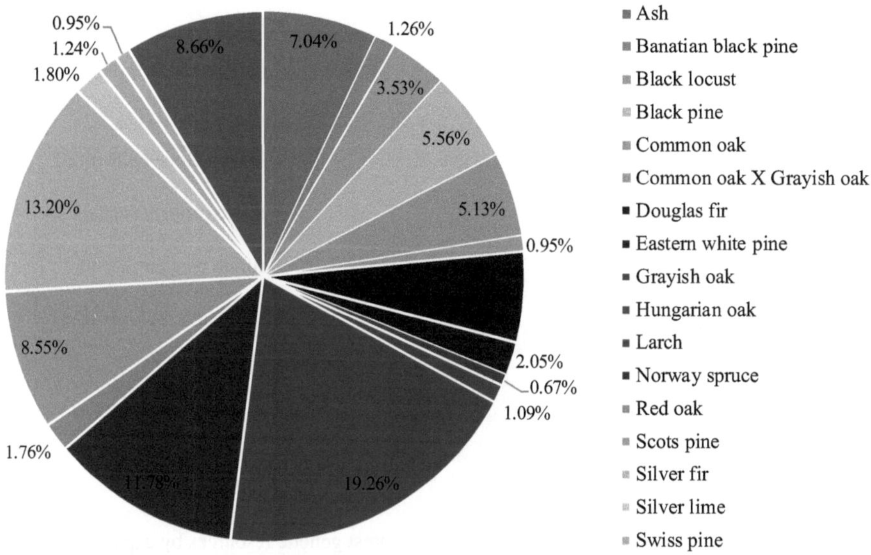

Fig. 25.1 Romanian seed orchards by species

eurolepis, Pinus nigra ssp. *nigra, Pinus nigra* ssp. *banatica, Pinus sylvestris, Pinus cembra, Pinus strobus, Pseudotsuga menziesii.*

The broadleaves seed orchards occupy about 203.1 ha and the conserved species are: *Quercus robur, Quercus rubra, Quercus pedunculiflora, Quercus robur* x *Quercus pedunculiflora, Quercus frainetto, Acer pseudoplatanus, Prunus avium, Tilia cordata, Tilia tomentosa, Fraxinus excelsior, Robinia pseudacacia*) (Pârnuţă et al. 2012) (Fig. 25.1).

The seed orchards, qualified and tested, are included in the National Catalogue of Basic Materials for Production of Forest Reproductive Materials (Parnuţă et al. 2012), excluding the newest ones established in the last 5 years, which will be included in the new version of the catalog.

(ii) A number of 71 provenance trials were established for: *Picea abies* (15), *Abies alba* (6), *Larix decidua* (7), *Pseudotsuga menziesii* (5), *Pinus sylvestris*(8), *Pinus nigra* ssp. *nigra* (3), *Pinus cembra* (2), *Quercus robur* (12), *Quercus petraea* (3), *Fraxinus excelsior* (4), *Fagus sylvatica* (6).

Within the same breeding program, there were established 8 half-sib progenies trials for *Pinus cembra* (3), *Abies alba (1), Picea abies* (2), *Fraxinus excelsior* (1) *and Quercus robur* (1), 7 full-sib progenies trials for *Pinus cembra* (2), *Picea abies (5),* 18 interspecific hybrids in trials for *Pinus* sp. (8) and *Populus* sp. (10) and 3 genitor clone's collections for *Picea abies* (narrow crown Norway spruce) (Fig. 25.2).

The clonal collections of genitors were established in field conditions, after the vegetative propagation works by grafting of narrow-crowned spruce ideotypes. This

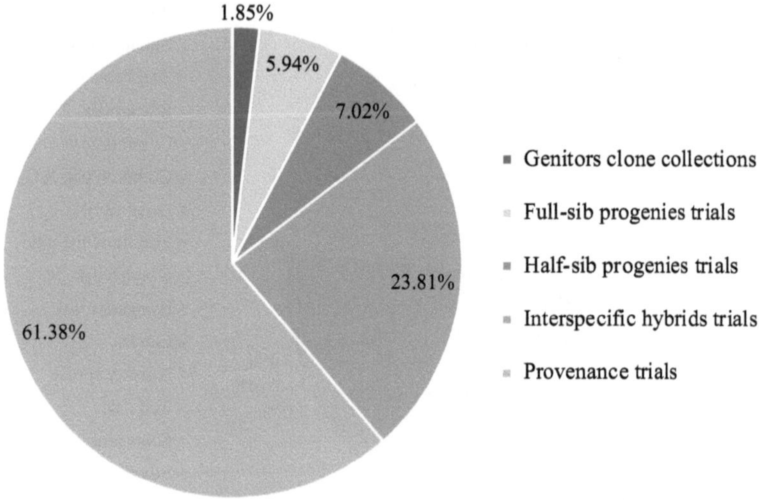

Fig. 25.2 The proportion of ex situ conservation of forest genetic resources by experiment type

acts the role of ex situ conservation of selected narrow-crowned spruce ideotypes from the principal spruce distribution areas, with different environmental conditions, creating a large gene pool. Also, that collection offers the possibility to continue the researches in the frame of breeding program to realize new control pollinations between narrow-crowned spruce trees and will also have the seed orchard role, based on generative propagation of ideotypes.

According to mating system particularities of spruce ideotypes and of genetic parameters established by full-sib progenies testing, it was underline the breeding program of narrow-crowned spruce which is based by generative and vegetative propagation of selected material having as objectives to increase the productivity and the resistance to the snow breaks of spruce stands.

All the forest genetic resources conserved ex situ in Romania are mapped (Fig. 25.3).

(iii) In Romania there are 5 arboreta: Simeria, Bazoş, Dofteana, Hemeiuşi and Mihăeşti. Arboretum of Simeria, the most valuable, oldest and picturesque collection of woody plants in Romania, has an area of 67 ha, and the collection of woody plants contains more than 2000 taxa, being conserved ex situ wood species from several geographical areas: 14% European species, 25% North American, 47% Asian, the rest of the species being common for 2 or even 3 continents.

In the Bazoş Arboretum, there are ex situ conserved about 720 taxa, 63 of conifers and 657 of broadleaves.

In the Dofteana Arboretum, there are ex situ conserved about 630 taxa, the conifers ones represent 22%, while the broadleaves species together with shrubs represent 78%.

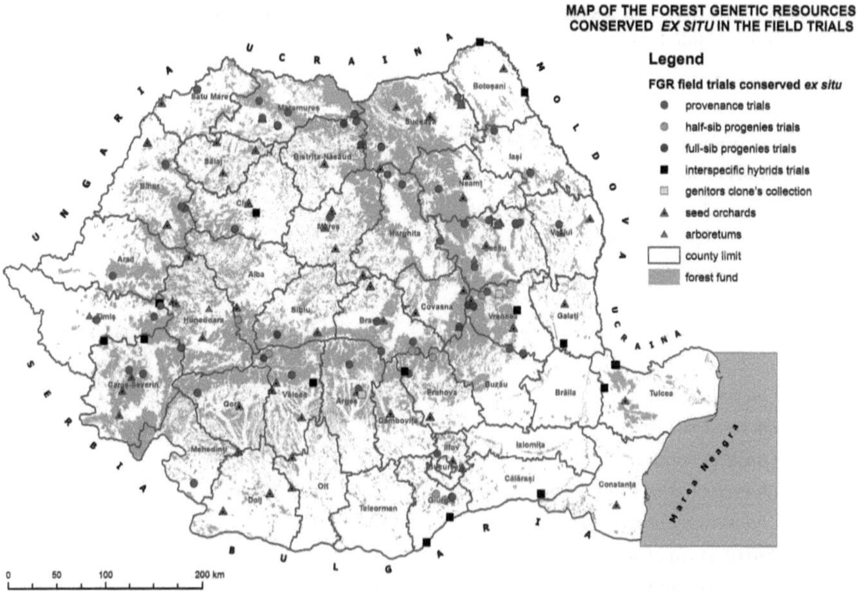

Fig. 25.3 Map of the forest genetic resources conserved ex situ in seed orchards, field trials and arboreta

The Hemeiuşi Arboretum has a rich information content, very valuable from the scientific, forestry, botanical, didactic, social, cultural, landscaping point of view, but mainly for ex situ conservation of the forest genetic resources. There are conserved about 1060 taxa, of which 82% are broadleaves and 18% are conifers.

In the Mihăeşti arboretum, even if there are ex situ conserved about 183 taxa, most valuable is the systematic units represented by those taxa.

(iv) The conservation within ex situ conserved core units was done for non-native species grown in Romania, respectively for native species grown outside their natural area (Table 25.1). There were established a total of 113 core conservation units, out of which 61 non-native and 52 native species grown outside their natural area, respectively.

(v) Other ways of ex situ conservation

- With regard to the *biotechnological methods* (in vitro culture of tissues and organs and somatic embryogenesis), there were established the protocols for the ex situ conservation of the forest genetic resources of *Alnus incana*, *Ulmus minor*. *Populus alba*, *Populus x canescens*, *Populus nigra*, *Prunus avium*, *Quercus frainetto*, *Quercus robur*, *Quercus pedunculiflora*, *Pseudotsuga menziesii*, *Fraxinus excelsior*, *Betula pendula*, *Robinia pseudacacia* (Ioniţă 2009; Ioniţă et al. 2017).

Table 25.1 Ex situ core conservation units for non-native tree species and for native species used in plantations outside their natural range

No.	Species	Number of core units	Surface area (ha)
1	*Abies grandis*	1	2.1
2	*Larix decidua*	21	162.4
3	*Pinus cembra*	1	8
4	*Pinus sylvestris*	17	186.3
5	*Pinus nigra* ssp. *nigra*	15	161.4
6	*Pinus strobus*	1	14.2
7	*Pseudotsuga menziesii* var. *menziesii*	9	54.4
8	*Pseudotsuga menziesii* var. *glauca*	1	5.2
9	*Taxodium distichum*	2	3.5
10	*Quercus rubra*	8	31
11	*Quercus palustris*	1	0.4
12	*Quercus petraea*	1	13.5
13	*Quercus robur*	2	47.4
14	*Quercus pedunculiflora*	2	50.2
15	*Fraxinus excelsior*	4	17.2
16	*Fraxinus angustifolia*	1	7
17	*Fraxinus pennsylvanica*	1	1.4
18	*Castanea sativa*	1	8.3
19	*Populus alba*	2	13.2
20	*Robinia pseodoacacia*	11	144.1
21	*Juglans nigra*	7	24.2
22	*Carya ovata*	1	2.6
23	*Ulmus pumila*	1	2.1
24	*Platanus hybrida*	1	1.2
25	*Ailanthus altissima*	1	2.9
Total		**113**	**964.2**

- With regard to the ex situ conservation in seed banks, in the seed storages of the National Forest Administration and of the National Institute for Research and Development in Forestry "*Marin Dracea*" there are preserved, for a short period, seeds from the main species of conifers (Norway spruce, silver fir, larch). In the last period, there are some research activities on oak seed preservation.
- Gene banks for the main forest tree species (silver fir, Norway spruce, beech, elm, ash) exist in the genetic laboratories of the National Institute for Research and Development in Forestry „*Marin Dracea*".

25.3 Conclusions

In Romania, there is a wide array of ex situ conservation units for both native and non-native forest trees. Most of the plantations were established in the seventies and eighties for breeding purposes but these are also valuable resources from the conservation point of view. The increasing demand for high-quality forest reproductive material can bring additional funding for the development of a new network of ex situ conservation units.

References

Dubrovský, M., Trnka, M., Holman, I. P., Svobodová, E., & Harrison, P. A. (2015). Developing a reduced-form ensemble of climate change scenarios for Europe and its application to selected impact indicators. *Climatic Change, 128*, 169–186. https://doi.org/10.1007/s10584-014-1297-7.

Ioniţă, L. (2009). Conservarea ex situ a resurselor genetice forestiere utilizând metode biotehnologice. *Revista pădurilor, 95*(124), 31–40.

Ioniţă, L., Mirancea, I., Apostol, E. N., & Budeanu, M. (2017). Conservarea resurselor genetice de stejar brumăriu prin metode biotehnologice. *Revista de silvicultură şi cinegetică, 40*(XXII), 16–20.

Pârnuţă, G., Budeanu, M., Stuparu, E. l.., Scărlătescu, V., Chesnoiu, E. N., Tudoroiu, M., Filat, M., Nica, M. S., Teodosiu, M., Lorent, A., Daia, M., & Dinu, G. C, (2012). Catalogul Naţional al Materialelor de Bază pentru producerea Materialelor Forestiere de Reproducere. Editura Silvică, Bucureşti www.icpa.ro/documente/ADER%20511_ghid.pdf. (in Romanian)

Simota, C., Dumitru, S., Vizitiu, O., Coteţ, V., Ignat, P., Mateescu, E., Alexandru, D., & Cofas, E. (2014). *Ghid de bune practici agricole pentru atenuarea efectului schimbărilor climatice asupra agriculturii*. Bucureşti: ICPA, ANM, USAMV.

Thomas, M., Kuster, T. M., Dobbertin, M., Günthardt-Goerg, M. S., Schaub, M., & Arend, M. (2014, February 24). A phenological timetable of oak growth under experimental drought and air warming. *PLoS One, 9*(2). https://doi.org/10.1371/journal.pone.0089724.

Chapter 26
Ex Situ Conservation of Forest Genetic Resources in Greece

Paraskevi G. Alizoti, Filippos A. Aravanopoulos, and Kostas Ioannidis

Abstract Hellenic ecosystems are biodiversity hotspots, as they harbor 22% of the species found across the Mediterranean basin, while 10% of the plants occurring in them are endemic. Forest tree species are keystone species of the forest ecosystems that harbor a great part of the above mentioned biodiversity and for this reason the conservation of their genetic resources should be an ultimate goal. *In situ* conservation of forest genetic resources is applied intensively at the global, regional and national level, *ex situ* conservation though can be proved of unique value in the case of rare, endemic, threatened species or populations of widely distributed species that could be eliminated due to stochastic events, biotic or abiotic stresses. *Ex situ* can be either static or dynamic. The *static ex situ* conservation plantations (i.e. seed orchards, provenance and provenance-progeny tests) established in Greece will be presented, as well as the potential for long-term seed storage. The reported plantations were established in the framework of breeding programs carried out by two organizations; namely the Aristotle University of Thessaloniki (Laboratory of Forest Genetics and Tree Improvement) and the Institute of Mediterranean and Forest Ecosystems (IMFE – Laboratory of Silviculture, Forest Genetics and Biotechnology).

Keywords Dynamic and static *ex-situ* conservation · Seed storage · Breeding genetic tests

P. G. Alizoti (✉) · F. A. Aravanopoulos
School of Forestry and Natural Environment, Aristotle University of Thessaloniki, Thessaloniki, Greece
e-mail: alizotp@for.auth.gr; aravanop@for.auth.gr

K. Ioannidis
Institute of Mediterranean Forest Ecosystems and Forest Products Technology, Hellenic Agricultural Organization "DEMETER", Athens, Greece
e-mail: ioko@fria.gr

M. Šijačić-Nikolić et al. (eds.), *Forests of Southeast Europe Under a Changing Climate*, Advances in Global Change Research 65,
https://doi.org/10.1007/978-3-319-95267-3_26

26.1 Introduction

Forest trees serve as keystone species in forest ecosystems and are of high significance and value especially in the Mediterranean Basin, a high ranking biodiversity hotspot across the globe (Blondel and Aronson 1999). The forests of the Mediterranean basin are growing in the transitional zone of three continents and harbor 10% of the globally registered plants in an area as large as 1.8% of the surface of the globe (Moussouris and Regato 1999). The biogeographical variation and the high biodiversity found in Greece can be explained by environmental, topographical and geological variation, as well as by the evolutionary history of the species. According the second report on the State of the World's Plant Genetic Resources for Food and Agriculture (Commission on Genetic Resources for Food and Agriculture 2010), out of all the species growing across the Mediterranean basin the 22% can be found in Greece, while almost 10% of the plant species are endemic. Great genetic and adaptive variation has also been reported at the population, as well as within population level, for all the species studied so far. Additionally, marginal/peripheral populations, that might harbor specific adaptations to stress conditions, can be recognized for all forest tree species growing in the country. Given the above, it can be easily recognized that the Hellenic forest genetic resources deserve special attention and their conservation should be an ultimate goal.

Conservation of genetic resources aims to the conservation of genetic diversity among and within the populations of a species. Species can be prioritized based on their environmental, ecological or economic importance, the risks they may face, their rarity, their marginality (geographical or ecological) and their viable population size (Rajora and Mosseler 2001). *In situ* conservation (aiming to the continuous maintenance of a population within the environment where it originally evolved and to which we assume it is adapted (Frankel 1976) can be evolutionarily dynamic (Finkeldey and Gregorius 1994). This case is true especially when one of the management goals is to safeguard the species genetic variation and the evolutionary processes and thus, secure that the species/populations can keep up with environmental change and remain adapted (Koskela et al. 2012). *Ex situ* conservation (conservation of components of biological diversity outside their natural habitats as stated in Article 2 of the Convention on Biological Diversity), is applied when conservation *in situ* is not possible (FAO 1989; Skrøppa 2005). *Ex situ* conservation will safeguard populations or individuals that are in danger of physical destruction or genetic deterioration, ensure an available supply of reproductive material (by either creating a production source or through storage), or allow commercial improvement and supply of the genetically improved reproductive material, given that the genetic base is not substantially restricted. *Ex situ* conservation can be *dynamic* or *static*.

According to Amaral et al. (2004), the *dynamic ex situ conservation* can occur in *ex situ* stands, established artificially by planting species, provenances or families, that regenerate naturally from seed, ensuring in this way that the genetic structure of the next generations is shaped by natural selection. In case the above stands fulfill the requirements of minimum viable population size (FAO 1992) and represent sufficiently the genetic structure of the original population, then they could also serve as sources of reproductive material for commercial forestry. To avoid losses of genetic material and succeed a most effective conservation, multiple stands/populations should be established in environmentally different areas, despite their high establishment cost (Eriksson et al. 1993; Geburek and Turok 2005).

Conservation seed orchards, not being part of a breeding program, may also serve the above conservation goal, as their aim is to conserve or even increase the genetic variability of collections of trees or small populations that have an endangered status or face reproduction issues in their natural environments (Skrøppa 2005). This seed orchard type concerns mainly species that are rare or have a scattered distribution and consists of a sufficient number of clones (to ensure ample genetic variation), originating from many stands within each ecological region, that will form the new breeding population which is expected to produce seed with high genetic diversity (Skrøppa 2005). The offspring can be planted in the forests and in environmentally different locations, so that they can be subjected to natural selection and evolutionary processes. Conservation seed orchards can also be established in the framework of the evacuation procedure for threatened species/populations (i.e. *Abies nebrodensis* seed orchard – CREA Arezzo, Italy).

Static ex situ conservation of genetic resources includes mainly populations established for tree breeding purposes, like clonal archives, seed orchards, provenance and progeny tests, as well as stored seed and pollen. The different breeding populations can be valuable genetic resources depending upon the extent they conserve genetic variability. This is related to the extent and procedure of initial sampling, the breeding population size, the selection procedure and intensity, the inbreeding levels and the long-term maintenance planning of the breeding populations (Skrøppa 2005).

Provenance tests can be planted within or outside the natural distribution of a species and may include genetic material sampled and tested at a national or international level. If the number of populations tested is large, covering adequately the distribution area of a species, and the number of the individuals per population is sufficient then the provenance trial could be of value as a genetic resource (Skrøppa 2005). Under certain conditions, provenance tests could be used to restore seed sources that are extinct in their native environment; i.e. a Norway spruce provenance of Polish origin, planted in 20 trials established in 13 countries (IUFRO 1964/68 experiment), was restored by establishing a 'restoration seed orchard' that included grafts of individual trees growing in 14 provenance tests and 11 countries (Giertych 1975; Skrøppa 2005).

Clonal seed orchards are established in a breeding program for the mass production of improved seed that will be used in commercial forestry. The first-generation clonal seed orchards could potentially play a role in conservation, if considered as a clonal archive of the progeny tested clones. Despite rare, there are cases where natural regeneration occurred at the seed orchard site and the orchard population developed into a new generation (Skrøppa 2005). Eriksson et al. (1993) also proposed that conservation and long-term breeding goals can be simultaneously fulfilled in the same populations, if the breeding designs applied are aiming also to retain high levels of genetic variation in the breeding populations, that may include even 50–100 individuals.

First generation progeny tests may include many families from different populations that grow within the same ecological zone, and could be potentially managed as genetic resources, if a systematic thinning is applied after the termination of selections for breeding, so that they could develop into stands containing a large part of the original genetic diversity (Skrøppa 2005).

Collections of trees in arboreta and botanic gardens represent *ex situ* collections, they include though only a limited number of individuals per species/population and are generally of narrow genetic base (White et al. 2007; Hurka 1994). Seed storage (in freezers, or cryopreservation) is used extensively for conserving threatened species/populations (Aravanopoulos and Műller-Starck 2003), as well as species in breeding programs. In the first case the seed stored may represent the sole genetic resource conserved for the threatened species/population (White et al. 2007). Stored seeds are *static* collections, unable to be subjected to the ongoing evolutionary processes and they require periodic outplanting and recollection of seed for the replenishment of genetic variation (White et al. 2007). This whole procedure is quite problematic in the case of forest trees, due to the long time, intensive labor and high cost required, as trees are long living organisms that reach reproductive maturity many years after planting (White et al. 2007; El-Kassaby 1992).

In the current chapter, the *static ex situ* conservation plantations (i.e. seed orchards, provenance and provenance-progeny tests) established in Greece will be presented, as well as the potential for long-term seed storage. The reported plantations were established in the framework of breeding programs carried out by two organizations; namely the Aristotle University of Thessaloniki (Laboratory of Forest Genetics and Tree Improvement) and the Institute of Mediterranean and Forest Ecosystems (IMFE – Laboratory of Silviculture, Forest Genetics and Biotechnology), will be presented.

26.2 Seed Orchards

In the framework of breeding programs, six first generation clonal seed orchards have been established in the decade 1978–1987 for *Pinus nigra (Arn.), Pinus halepensis (Mill.)* and *Abies borisii regis (Mattf.)*, as their economic value and ecological significance are of extreme importance for the country. These are ecologically keystone species extensively distributed in the country, while they produce wood and non-wood products of high economic value that support the rural economies.

The main effort was put on *Pinus nigra* resulting in the establishment of four clonal seed orchards located in different geographic zones. The aim was to cover the needs of each specific region for improved forest reproductive material that could be used either for commercial plantations or for reforestation purposes. The clonal seed orchard located in Northern Greece (Chalkidiki region) was established in 1980 and is comprised of 60 clones, derived from intensively selected plus trees in five natural populations of the species growing in the Central Macedonia region; namely Mt. Olympus, Mt. Pieria, Mt. Vermion, Chalkidiki peninsula and Thassos island (Alizoti et al. 2010). In Central Greece two clonal seed orchards have been established in 1981; specifically, in the Fthiotida and Trikala regions. The first seed orchard includes 91 *P. nigra* clones, derived from selected intensively plus trees growing in the natural populations of Pindos mountain range. The second seed orchard comprises of 49 phenotypically selected clones of the introduced black pine variety *P. nigra var. austriaca*. In Southern Greece a clonal seed orchard was established in 1978 in Northwest Peloponnese (Elea region). The seed orchard consists of 52 clones, selected as plus trees from four natural black pine populations of the Peloponnese; namely Mt. Parnon, Mt. Taygetos, Feneos and Zarouchla areas (Matziris 1993; Ioannidis et al. 2017).

A clonal seed orchard has also been established in 1987 for *Pinus halepensis* Mill. The seed orchard is located in the Central-Western part of the country (Aitoloakarnania region) and comprises of 76 clones, derived from selected plus trees growing in Euboea region (Papadima 2014). The provenances of the specific region are of great interest as they have been proved to be among the best performing and well adapted ones when tested in a range of environmental conditions (Alizoti 2000; Alizoti et al. 2000).

The *Abies borisii regis* clonal seed orchard, located in Northern Greece (Chalkidiki region), has been established during the 1980s with clones selected from the central Pindos mountain region.

All the above clonal seed orchards may serve as *static ex situ* conservation plantations, harboring elite phenotypes/genotypes originating from different natural populations, some of which are marginal/peripheral ones and potentially bearing specific adaptation alleles. As soon as the breeding programs advance to the next generation seed orchards, natural regeneration can be allowed in the first generation ones, so that new generations subjected to natural selection can develop. In that case the seed orchards could gradually form a *dynamic* ex situ *site* harboring high genetic variability, due to the diverse origin of the parental clones.

Four of the above mentioned seed orchards have been progeny tested; namely the *Pinus nigra* seed orchards located in Elea (Matziris 2004) and Chalkidiki, as well as the *Pinus halepensis* and *Abies borisii regis* seed orchards. These progeny trials, if thinned following the termination of selection for breeding, could develop into *ex situ* stands containing large part of the original genetic diversity.

26.3 Provenance, Provenance-Progeny Tests, Clonal Archives

The species being intensively studied in provenance and provenance-progeny trials are *Pinus halepensis (Mill.), Pinus brutia (Ten.), Pinus nigra (Arn.), Pinus sylvestris (L.), Abies sp.* and *Castanea sativa (Mill).* The majority of these trials were established in the 1970s and 1980s, except for the *Castanea sativa* one that was established in 2001.

A multi-environmental trial was established for *Pinus halepensis* in the early 1970s that included two provenance-progeny trials, located in Central Greece (Fthiotida and Viotia regions), and a provenance trial located in Northwest Peloponnese (Panetsos 1981). Fifteen Hellenic populations were tested, representing the whole distribution of the species in the country. In the provenance-progeny trials the fifteen Hellenic populations were represented by ten open pollinated families each, while in the provenance test, additionally to the Hellenic provenances, eighteen exotic ones originating from Italy (3 populations), Spain (1 provenance), Morocco (3 populations) and Tunisia (11 provenances) were also tested (Panetsos 1981; Alizoti 2000; Alizoti et al. 2000). Unfortunately, only the Fthiotida trial is still in place, as the the other two were destroyed quite recently, and to a large extent, by stochastic events. Data were recorded though continuously for all three trials, allowing the evaluation of survival and performance of the genetic material over time and across environments (Alizoti 2000; Alizoti et al. 2000).

A multi-environmental trial, that included five provenance tests and one provenance-progeny test, was established in the 1970s and 1980s for *Pinus brutia*, testing seven populations that represented almost all the natural range of the species in the country (Panetsos 1981). In the provenance-progeny test each population was represented by 10 open pollinated families. The trials were located in the North, Central East, Central West, Southeast and South parts of the country and were all located out of the natural distribution of the species, expect for the one located in the North part and specifically in the Evros region (Panetsos 1981; Alizoti 2000; Alizoti et al. 2000). An additional trial was established in the Central West part of the country and included the two northernmost and marginal Hellenic populations (growing at the edge of the species natural range to the West) of the species; namely the Alexandroupolis and Thassos ones, together with four populations of Turkish origin (Matziris and Tziovaras 1988).

Significant effort has been devoted in establishing provenance trials of *Pinus nigra* during the 1970s and 1980s. Seven provenance trials were established, four of which have been located in Northern Greece, two in Thrace and one in Central West Greece. Ten Hellenic black pine populations are tested in those trials, while three of them include also genetic material from France (three populations), Turkey (five populations), Italy (one population) and Spain (four populations) (Varelides 1997; Varelidis and Brofas 1997; Varelides et al. 2001).

Pinus sylvestris populations are tested in three provenance trials across the country; two of them are located in Northern Greece and the third one in Central Greece. In those trials, 14 Hellenic populations, representing the southernmost populations of the species distribution in the Europe and in the country, are tested together with two populations originating from the Former Yugoslav Republic of Macedonia and two from Bulgaria (Moulalis 1984).

An *Abies* species-provenance trial has been established in 1974 in Central Greece, testing four species and a hybrid (*A. alba, A. cephalonica, A. equi-trojani, A. bornmuelleriana, A. concolor* and the hybridogenous *A. x borisii regis*) and 21

Hellenic provenances of fir; namely 21 populations of *A. x borisii regis,* covering the whole range of the hybridogenous fir in the country, and one population of *A. cephalonica. A. alba* was represented by four populations of Italian origin, while *A. bornmuelleriana* and *A. equi-trojani* by one population of Turkish origin each (Panetsos 1990; Donta et al. 2010).

A *Castanea sativa* provenance-progeny trial was established in 2002 in the framework of the CASCADE EU Project in Central Greece (Chalkidiki region). In the trial, 6 provenances originating from contrasting environments (in terms of xerothermic index) of Greece, Spain and Italy are tested, while each population is represented by 26 open pollinated families (Fernandez et al. 2004; Tchatchoua and Aravanopoulos 2010a, b; Nikolaou et al. 2011).

Besides the above, provenance tests were established in the country in the framework of FAO/IUFRO network of trials for several exotic species, as for example *Pinus pinaster, Pinus radiata, Pinus muricata, Pseudotsuga menziesii, Pinus strobus* and *Pinus controrta. Pinus pinaster* was the species tested more intensively in nine provenance trials, established at seven sites across the country; two trials in North West Greece, two trials in Central Macedonia, one trial in Thrace, one trial in Euboea, two trials in Central West Greece and one trial in South Greece (Peloponnese). In those trials, provenances of French and Portuguese origin were tested for their growth and adaptive potential (Matziris 1981). All the above mentioned experimental trials, established in the framework of breeding activities, may serve as *static ex situ* conservation plantations, harboring an ample part of the existing within population variation at the places of origin during the time of sampling.

Also, clonal archives of *Populus sp., Platanus orientalis, Platanus occidentalis, Platanus acerifolia, Prunus avium* and *Cupressus sempervirens* clones have been established in North and Central Greece. In particular, clonal archives for *Populus sp.* and *Platanus sp.,* are present in the area of Central Macedonia (Malliarou et al. 2014). The clones have been DNA fingerprinted and possess a significant amount of genetic diversity given the respective sample sizes (Malliarou et al. 2014; Mitsiou et al. 2013; Vaxevanidou and Aravanopoulos 2002). A gene bank for wild cherry (*Prunus avium*) is present in the area of eastern Macedonia. Plus-trees from the most important wild cherry populations of northern Greece are located there and possess extensive genetic diversity, while their DNA fingerprints have been recorded (Avramidou et al. 2010). Moreover, *Populus* and *Platanus* clonal trials have been established for the evaluation of clonal biomass production under different mixtures and spacings (Aravanopoulos 2010; Panetsos et al. 1994, 1996; Panetsos and Alizoti 1996).

26.4 Seed Storage

The Central Forest Seed Bank of the Hellenic Ministry of Environment and Energy was established in 1992 and its goal is the collection, dressing, preservation, handling and quality control of forest seed for the production of high quality seedlings in the forest nurseries. Among the 46 forest tree species collected, handled and stored by the Central Forest Seed Bank, the most important ones are the *Pinus, Populus, Fraxinus and Quercus* species growing in the country, as well as other species like *Abies sp.* and *Cercis siliquastrum* (Paitaridou 2009). Annually 2 tons of seed of various forest tree species is collected, and following its proper handling is stored in fridges and freezers that can keep them viable for several decades (Paitaridou 2009). Stored seeds despite the fact that are *static* collections, can be proved of great value in the case of elimination of rare or endemic species in their native environments due to stochastic events or biotic/abiotic factors, while the same holds for threatened species/populations (Aravanopoulos and Mű ller_Starck 2003). Seed collections might be the only genetic material preserved in the above cases and this is why collection procedures should aim towards capturing the greatest possible part of the existing genetic variation.

26.5 Conclusions

Hellenic forest ecosystems are characterized by great diversity among and within species and for this reason they are considered to be among the most important ones at a European and global level. The value of conserving biodiversity in natural ecosystems was recognized at the national level as early as 1937 and ever since a notable part of the Hellenic ecosystems is conserved *in situ* under national and European legislation, as well under International Conventions. Since the early 1970s breeding activities were initiated in the country, aiming towards the genetic improvement of target species that were of importance for the economy of the country. In this framework a wide network of experimental plantations was established for several species (i.e. seed orchards, provenance and provenance-progeny testes, progeny tests, gene banks) at the national level or as part of the international test series coordinated by FAO and IUFRO. These trials can also be considered as *static ex situ* conservation plantations, conserving valuable genetic resources of populations of native and exotic species. They could be of great value for the potential reconstruction of populations with specific adaptations or interest, in case they are lost by stochastic events. Several trials, like the seed orchards and the progeny tests, can evolve into *dynamic ex situ* conservation stands, if managed appropriately, after the completion of selections for the breeding programs. Establishment of *ex situ conservation seed orchards* and *dynamic ex situ conservation stands* for marginal/peripheral populations or populations of great economic/ecological importance should become a top priority.

References

Alizoti, P. G. (2000). Time trends of genetic parameters in the group *halepensis*. PhD dissertation. Faculty of Forestry and Natural Environment, Aristotle University of Thessaloniki, p. 370.

Alizoti, P. G., Panetsos, K. P., Moulalis, D. (2000, October 23–25). Evaluation of *Pinus halepensis* (lMill.) genetic material and genetic parameters for growth traits. In *Proceedings of the 8th Conference of Hellenic Scientific Society for Plant Genetics and Breeding*, Arta, pp. 37–49.

Alizoti, P. G., Kilimis, K., & Gallios, P. (2010). Temporal and spatial variation of flowering among *Pinus nigra* Arn. Clones under changing climatic conditions. *Forest Ecology and Management, 259*, 768–797.

Amaral, W., Yanchuk, A., & Kjaer, E. (2004). Methodologies for ex situ conservation. In: FAO, FLD, IPGRI, *Forest genetic resources conservation and management*: In plantations and Genebanks (ex situ) (Vol. 3, pp. 3–7). Rome: International Plant Genetic Resources Institute.

Aravanopoulos, F. A. (2010). Breeding of fast growing forest tree species for biomass production in Greece. *Biomass & Bioenergy, 34*, 1531–1537.

Aravanopoulos, F. A., & Muller-Starck, G. (2003). Artificial seed transfer. In: Kremer, A., & Reviron, M. P. 2003. Dynamics and conservation of genetic diversity in forest ecosystems. *Forest Genetics, 9*, 333–340.

Avramidou, E., Ganopoulos, I. V., & Aravanopoulos, F. A. (2010). DNA fingerprinting of elite Greek wild cherry (*Prunus avium* L.) genotypes using microsatellite markers. *Forestry, 83*, 527–533.

Blondel, J., & Aronson, J. (1999). *Biology and wildlife of the Mediterranean region*. New York: Oxford University Press.

Commission on Genetic Resources for Food and Agriculture. (2010). *The second report on the state of the world's plant genetic resources for food and agriculture* (pp 399). Food and Agriculture Organization.

Donta, E., Alizoti, P. G., Scaltsoyiannes, A. (2010). Growth and adaptation of *Abies* species and provenances. In *Proceedings of the 12th Conference of Hellenic Scientific Society for Plant Genetics and Breeding*, Naousa, (pp 354–359.

El-Kassaby, Y. A. (1992). Domestication and genetic diversity – should we be concerned? *Forest Chronicle, 68*, 687–700.

Eriksson, G., Namkoong, G., & Roberts, J. (1993). Dynamic gene conservation for uncertain futures. *Forest Ecology Management, 62*, 15–37.

FAO (Food and Agriculture Organization). (1989). *Plant genetic resources: Their conservation in situ for human use*. Rome: FAO.

FAO (Food and Agriculture Organization). (1992). Establishment and management of ex-situ conservation stands. In: *Forest genetic resources information no.20* (pp. 7–10). Rome: FAO.

Fernández-López, J., Zas, R., Aravanopoulos, F. A., Alizoti, P. G., Botta, R., Mellano, M. G., Villani, F., Cheburini, M., & Eriksson, G. (2004). Geographic variability among extreme European wild chestnut populations. *Acta Horticulturae, 693*, 181–186.

Finkeldey, R., & Gregorius, F. H. (1994). Genetic resources: Selection criteria and design. In Z.-S. Kim & H. H. Hattemer (Eds.), *Conservation and manipulation of genetic resources in forestry* (pp. 322–347). Seoul: Kwang Moon Kag.

Frankel, O. H. (1976). Natural variation and its conservation. In A. Muhammed, R. Aksel, & R. C. von Borstel (Eds.), *Genetic diversity in plants* (pp. 21–44). New York: Plenum Press.

Geburek, T., & Turok, J. (2005). *Conservation and management of forest genetic resources in Europe* (pp. 567–583). Zvolen: Arbora Publishers.

Giertych, M. (1975). Seed orchard designs. In R. Faulkner (Ed.), *Seed orchards* (Forestry commission bulletin, Vol. 54, pp. 25–37). London.

Hurka, H. (1994). Conservation genetics and the role of botanical gardens. In O. T. Sandlund, K. Hindar, & A. H. D. Brown (Eds.), *Conservation genetics* (pp. 371–380). Basel: Birkhauser Verlag.

Ioannidis, K., Melliou, E., Alizoti, P., & Magiatis, P. (2017). Identification of black pine (*Pinus nigra* Arn.) heartwood as a rich source of bioactive stilbenes by qNMR. *Journal of the Science of Food and Agriculture, 97*(6), 1708–1716.

Koskela, J., Lefevre, F., Schueler, S., Kraigher, H., Olrik, D. C., Hubert, J., Longauer, R., Bozzano, M., Yrjana, L., Alizoti, P., Rotach, P., Vietto, L., Bordacs, S., Myking, T., Eysteinsson, T., Souvannavong, O., Fady, B., De Cuyper, B., Heinze, B., von Wuehlisch, G., Ducousso, A., & Ditlevsen, B. (2012). Translating conservation genetics into management: Pan-European minimum requirements for dynamic conservation units of forest tree genetic diversity. *Biological Conservation, 157*(1), 39–49.

Malliarou, E. S., Avramidou, E. V., & Aravanopoulos, F. A. (2014). DNA fingerprinting *Platanus* sp. and *Populus* sp. clones with nuclear microsatellites (nSSR). In *Proceedings of the 15th conference of Hellenic Scientific Society for Plant Genetics and Breeding*, Larisa, pp. 289–294.

Matziris, D. (1981). Performance of *Pinus pinaster* provenances at seven sites of Central and Southern Greece. *Dassiki Erevna, 2*(3), 261–312 (in Greek, summary in English).

Matziris, D. (1993). Variation in Cone Production in a clonal seed orchard of black pine. *Silver Genetics, 42*(2–3), 136–141.

Matziris, D. (1997). Variation in growth, flowering and cone production in a clonal seed orchard of Aleppo Pine grown in Greece. *Silver Genetics, 46*(4), 224–228.

Matziris, D. (2004). Genetic variation and realized genetic gain from black pine tree improvement. *Silver Genetics, 54*(3), 96–104.

Matziris, D., & Tziovaras, E. (1988). Variation in morphological and anatomical needle characteristics in brutia pine provenances. *Dasiki Erevna Journal, 3*(IX), 193–213 in Greek with English summary.

Mitsiou, M., Kiparissis, D., Papadima, A., & Aravanopoulos, F. A. (2013). DNA fingerprinting and genetic diversity of selected genetic material of *Platanus* sp. In *Proceedings of the 16th conference of Hellenic Forestry Society*, Thessaloniki, Greece, pp. 284–289.

Moulalis, D. (1984). Comparison of *Pinus sylvestris* provenances tested in the University forest of Pertouli. Publ. Lab. For. Gen. and Tree Improv., Aristotle University of Thessaloniki.

Moussouris, Y., & Regato, P. (1999). *Forest harvest: An overview of non-timber forest Products in the Mediterranean region*. WWF Mediterranean Programme. Arborvitae, IUCN. WWF, Supplement.

Nikolaou, G., Alizoti, P. G., Aravanopoulos, F. A. (2011, November 1–4). Adaptive potential and quantitative variation of European sweet chestnut (*Castanea sativa* Mill.) populations. In *Proceedings of the 14th Conference of Hellenic Forestry Society*, Patra, pp. 877–887.

Paitaridou, D. (2009, October 15–16). The role of the Central Forest Seed Bank in forest restoration. In *Proceedings of the International Conference on 'New approaches to the restoration of black pine forests'*, Sparta. LIFE 07 NAT/GR/000286 Project: 'Restoration of *Pinus nigra* forests on Mount Parnonas (GR 2520006) through a structured approach'.

Panetsos, K. P. (1981). Monograph of *Pinus halepensis* (Mill.) and *P. brutia* (Ten.). Annales Forestales Vol. IX/2. Yugoslavian Academy of Sciences and Arts. Zagreb, 1981.

Panetsos, K. P. (1990, June 11–15). Species – provenances test of Mediterranean firs. Seminar international. 'Sapins Mediterraneens – Adaptation, seleetion et sylviculture' INRA, Avignon, EU Report 13491:29–42.

Panetsos, K. P, & Alizoti, P.G. (1996, October 11–13). Results obtained from *Platanus* sp. plantations.. Possibilities and perspectives. In *Proceedings of the 7th Conference of Hellenic Forestry Society*, Karditsa, pp. 150–163.

Panetsos, K. P., Scaltsoyiannes, A. V., & Alizoti, P. G. (1994). Vegetative propagation of *Platanus orientalis x P. occidentalis* F1 hybrids by stem cuttings. *Forest Genetics, 1*(3), 125–130.

Panetsos, K. P., Alizoti, P. G., & Scaltsoyiannes, A. V. (1996). Comparison of poplar clones – Potential of their use in poplar culture. *Annals of Factulty of Forestry and Natural Environment, ΛΗ/1*, 160–168.

Papadima, A. (2014). Genetic diversity, mating systems and evaluation of the genetic material in a Pinus *halepensis* (Mill.) seed orchard. Doctoral thesis, Department of Forestry & Natural Environment, Aristotle University of Thessaloniki, Thessaloniki, Greece.

Rajora, O. P., & Mosseler, A. (2001). Challenges and opportunities for conservation of forest genetic resources. *Euphytica, 118*, 197–212.

Skrøppa, T. (2005). Ex situ conservation methods. In T. Geburek, & J. Turok (Eds.), *Conservation and Management of Forest Genetic Resources in Europe*. (pp 571–583). Arbora Publishers.

Tchatchoua, D. T., & Aravanopoulos, F. A. (2010a). Evaluation of selected European Chestnut (*Castanea sativa*) provenances I: inter-provenance genetic variation. *Acta Horticulturae, 866*, 203–213.

Tchatchoua, D. T., & Aravanopoulos, F. A. (2010b). Evaluation of selected European Chestnut (*Castanea sativa*) provenances II: intra-provenance family variation. *Acta Horticulturae, 866*, 215–224.

Varelides, Y. (1997). A first evaluation of a *Pinus nigra* provenance trial over three sites in northern Greece. B.Sc. (Hons) thesis, Aberdeen University, UK, 85pp.

Varelides, C., & Brofas, G. (1997). Results of a *Pinus nigra* Arn. provenance trial at Anthrakia site, northern Greece. *Geotechnical Scientific Issues*, *8*(1), 34–40 (in Greek, sum. in English).

Varelides, C., Brofas, G., & Varelides, Y. (2001). Provenance variation in *Pinus nigra* at three sites in Northern Greece. *Annals of Forest Science, 58*, 893–900.

Vaxevanidou, Z., & Aravanopoulos, F. A. (2002). DNA fingerprinting of poplar clones using random amplified polymorphic DNA markers. In *Proceedings of the 9th Conference of Hellenic Scientific Society for Plant Genetics and Breeding*, Thermi, Greece, pp. 420–426.

White, T. L., Adams, W. T., & Neale, D. B. (2007). *Forest genetics* (p. 702). Cambridge, MA: CABI Publishing.

Papageorgiou A. (2014) Genetic diversity, mating systems and evaluation of the genetic material in a *Pinus halepensis* Mill. seed orchard. Doctoral thesis, Department of Forestry & Natural Environment, Aristotle University of Thessaloniki, Thessaloniki, Greece
[illegible]
[illegible]
[illegible]
[illegible]
[illegible]
[illegible]
[illegible]
[illegible]
[illegible]

Chapter 27
Ex Situ Conservation of *Abies alba*: Bulgarian Case Study

Velichko Gagov, Petar Zhelev, and Ivan Evtimov

Abstract Silver Fir (*Abies alba*) is important forest tree species across Europe and in Bulgaria, in particular. Ex situ gene conservation of the species has long-term history in Bulgaria dating back to 1970s, when many provenance and progeny trials have been established. Besides these experiments and *ex situ* collections, clonal seed orchard was established in 1980s, and 20 years later it served as a source of tested clones for establishment of a second-generation seed orchard – one of the first second-generation orchards for the species in Europe. Of more than 400 clones included in the first generation orchard, 25 were selected for the second-generation orchard, based on a number of criteria. The approach for conservation of genepool of the species could be used as a pilot guide for *ex situ* conservation of other conifers in Bulgaria.

Keywords Silver Fir · Provenance trials · Progeny tests · Seed orchard

27.1 Introduction

Silver Fir (*Abies alba* Mill.) is considered one of the most important forest tree species of mountainous ecosystems in Bulgaria. It possesses high ecological plasticity and occurs on various soil substrates and at altitude from 400 m (Stara planina) to 2000 m (Rila and Pirin). The pure and mixed stands with domination or participation of Silver Fir in Bulgaria are about 32,000 ha, and 60% of them are older than 80 years. Ecology and silviculture of Silver Fir in Europe and in Bulgaria, in particular, are relatively well-studied (Penev et al. 1969; Kostov 1989; Dimitrov and Roussakova 2015; Diaci et al. 2011; Barbu and Barbu 2005; see the last review by Dobrowolska et al. 2017). There is also substantial information about the genetic diversity of the species (Gagov 1973a, b; Bergmann and Gagov 1998, 2000, 2003;

V. Gagov (✉) · P. Zhelev · I. Evtimov
University of Forestry, Sofia, Bulgaria
e-mail: abies@abv.bg; zhelev@ltu.bg; evtimov.i@ltu.bg

M. Šijačić-Nikolić et al. (eds.), *Forests of Southeast Europe Under a Changing Climate*, Advances in Global Change Research 65,
https://doi.org/10.1007/978-3-319-95267-3_27

Dobrinov and Gagov 1983; Korpel' et al. 1982; Wolf 2003; Longauer 1996; Ballian and Halilović 2016; Longauer et al. 2003).

A serious problem that emerged in 1980s was the Silver Fir decline in the whole area of the species natural distribution. There were numerous studies and decline was registered across almost the whole range of the species in Europe (Bernadzki 1983; Cailleret et al. 2014; Diaci et al. 2011; Gagov 1988, 1990; Bergmann et al. 1990; Kandler 1993; Larsen 1986; Thomas et al. 2002; Linares and Camarero 2012). The reasons for the species decline were traced in environmental pollution, climate change, the specific eco-physiological behavior of the species, insufficient genetic diversity in particular populations, improper management. Most probably, different combination of these (and some other unmentioned here) factors caused the Silver Fir decline, and they acted differently in different parts of the species distribution. However, during the last two decades the health conditions of Silver Fir stands in Europe improved markedly (Wolf 2003) and the species currently is not considered in danger. In a study in Spanish Pyrenees Oliva and Colinas (2007) found that "at stand level Silver Fir does not seem to be a species in decline".

However, complex decline syndrome is not the only threat to Silver Fir populations. Deforestation, over-exploitation and some other "classic" threats could pose problems to the normal development of populations and to integrity of the genepool. Therefore, conservation efforts and measures are always necessary, both *in situ* and *ex situ*, and better – in combination. Eriksson (1994) pointed out that classification of gene conservation approaches to *in situ* and *ex situ* is too static and does not reflect the whole complexity and dynamics of the issue. It is better to provide opportunities for future evolution of the species through combined and more sophisticated approaches. However, *ex situ* conservation approach is still an integral and important part of each conservation program. In this paper we describe in brief the specific moments and achievements in *ex situ* gene conservation of Silver Fir in Bulgaria.

27.2 *Ex Situ* Conservation of Silver Fir in Bulgaria

27.2.1 Survey on the Ex Situ *Objects*

The first focused studies on the ecotype differentiation and phenotypic variation of *Abies alba* in Bulgaria started in 1960s (Gagov 1973a, b, 1985; Dobrinov and Gagov 1975; Dobrinov et al. 1982). Together with the other studies, a program for gene resources conservation was launched (Dobrinov and Gagov 1985; Plugtschieva et al. 2003). It included establishment of provenance test, progeny tests, clone collections and seed orchards. Also, independently of our research team, the state forestry authorities are responsible to store seed lots under controlled temperature and humidity conditions in the buildings of the Seed Control Station situated in Plovdiv, and these seed lots are periodically renewed.

Table 27.1 Ex situ conservation units for *Abies alba* in Bulgaria

No	Locality – Forestry Service, compartment (where available), geographic coordinates and altitude	Year of establishment	Area [ha]
1	2	3	4
Provenance trials			
1	Yundola Experimental Forest Service 42°03′50″N, 23°52′22″E, 1530 m	1974	1.2
2	Yundola Experimental Forest Service (Ezerata) 42°04′39″N, 23°51′36″E, 1560 m	1974	3.9
3	Petrokhan Experimental Forest Service 43°10′54″N, 23° 09′08″E, 600 m	1974	1.0
4	State Forestry Service (SFS) Sherba (formerly Staro Oryahovo, near the village Solnik) 42°55′37″N, 27°39′49″E, 180 m	1974	1.0
5	SFS Gramatikovo – 1 42° 04′10″ N, 27°38′15″ E, 250 m	1976	0.6
6	SFS Gramatikovo – 2 (Rudnika locality) 42° 04′ 07″ N, 27°39′ 27″ E, 250 m	1976	1.7
7	SFS Borovetz 42°17′06″N, 23°38′10″ E, 1200 m	1976	1.0
8	SFS Sherba – 381-v 42°55′54″N, 27°40′52″E, 170 m	2005	3.5
9	SFS Sherba – 386-I 42°55′29″N, 27°39′48″E, 200 m	2010	1.0
Progeny tests			
10	SFS Sherba – 386-b 42°55′48″N, 27°39′39″E, 160 m	1989	4.1
11	SFS Novo Panicharevo, 42°16′38″N, 27°34′33″E, 120 m	1989	2.0
12	SFS Goce Delchev, 41°33′ N, 23°41′ E, 700 m	1989	2.0
13	SFS Sherba – 386-d 42°55′32″N, 27°39′54″E, 190 m	1992	2.0
14	SFS Sherba 386-d 42°55′32″N, 27°39′54″E, 190 m	1993	8.0
15	SFS Sherba 385–2 42°55′9″N, 27°40′1″E, 220 m	2000	3.0
16	SFS Sherba – 386-j 42°55′35″N, 27°39′43″E, 190 m	2002	4.0
17	SFS Sherba – 384-g 42°55′38″N, 27°40′29″E, 230 m	2003	1.1
18	SFS Sherba – 384-j 42°55′12″N, 27°40′42″E, 230 m	2003	2.2
19	SFS Sherba – 386-d 42°55′32″N, 27°39′54″E, 190 m	2003	3.0
20	SFS Sherba – 381-b 42°55′58"N, 27°40′39″E, 190 m	2006	2.7
21	SFS Sherba – 382-g 42°56′1″N, 27°40′5″E, 180 m	2006	2.9
22	SFS Sherba – 386-i and z 42°55′32″N, 27°39′42″E, 200 m	2010	1.8
Clonal seed orchards			
23	First-generation clonal seed orchard – 450 clones SFS Sherba – 386-e 42°55′31″N, 27°39′51″E, 190 m	1985	3.3
24	Second-generation clonal seed orchard – 25 clones SFS Sherba – 382 g, 42°56′2″N, 27°40′53″E, 160 m	2006	2

The *ex situ* objects designed and established for gene conservation of *Abies alba* in Bulgaria are shown in Table 27.1. There are 9 provenance tests, 13 progeny tests, one first-generation, and one second-generation seed orchard.

Seven provenance trials were established in 1970s. They include provenances from Bulgarian mountains: Smolyan and Rakitovo from Rhodopes, Borovetz from Rila, Kipilovo and Ribaritza from Stara planina, and Slavyanka. The last provenance

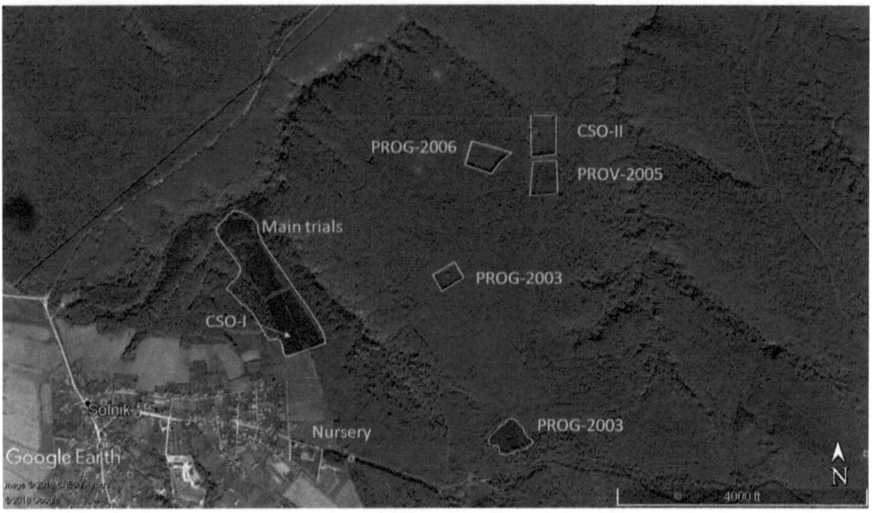

Fig. 27.1 The objects for *ex situ* conservation of Silver Fir genepool near the village Solnik, Eastern Stara planina (State Forestry Service "Sherba")
Legend: CSO-I means Clonal seed orchard – first-generation; CSO-II means Clonal seed orchard – second-generation; PROG-2006 indicates progeny test established 2006 (No 20 in Table 27.1); PROV-2005 indicates provenance test established 2005 (No 8 in Table 27.1); PROG-2003 are two progeny tests established in 2003 (NoNo 17 and 18 in Table 27.1) and Main trials indicates the locality of all other provenance and progeny tests established in SFS "Sherba", near the first-generation seed orchard

is considered as belonging to the taxon *Abies borisii-regis* Mattf. (Gagov 1973a, b; Dobrinov and Gagov 1981; Krajmerová et al. 2016).

The provenance trials were established in Yundola, Petrokhan, Borovetz, Gramatikovo and near the village Solnik, on the territory of State Forestry Service "Sherba" (formerly "Staro Oryahovo"). Most *ex situ* conservation objects are situated in the last region, close to the village Solnik. The region proved to be very suitable and is currently a center for *ex situ* gene conservation of *Abies alba* (Fig. 27.1). The provenances were subjected to extensive studies and yielded important results related to the use of respective provenances in the afforestation practice (Gagov 1979, 1985). Two new provenance trials were established only in 2005 and 2010. The former was established within the framework of the second IUFRO provenance series. It includes 23 provenances – 8 from Poland, 6 from Germany, three from Romania, and one from Austria, Bulgaria, Czech Republic, Macedonia, Serbia and Slovakia (Gagov et al. 2009).

Several progeny tests (using individual samples) were established in the period 1985–1993. These experiments were set in relation to the seed orchard establishment. The first-generation seed orchard included more than 100 clones. In fact, 152 plus trees were selected in the whole natural range of the species in Bulgaria (Dobrinov and Gagov 1985). The selected plus trees were marked and numbered in the field and scions and cones were collected for establishment of a seed orchard and progeny tests. The measurements of cone size and classification of seed colour forms contributed additionally to the study of variation of Silver Fir in Bulgaria. Also, additional parameters of the selected plus trees were scored – self-prunning, bark type, and crown form. These were used for establishment of a database (Gagov 1990).

The progeny tests established consisted of two types of half-sib families. The first series of five trials (1987, 1989, 1992, 1993) were established using half-sib families of plus trees growing in natural populations, which were selected for including in the first-generation seed orchard. Three of them were established in SF "Sherba" and the other two – in SF "Novo Panicharevo" (Strandzha Mts) and Goce Delchev (Pirin Mts).

The second series of trials consisted of open-pollinated families of the first-generation seed orchard crops of the clones included and was a part of testing the general combining ability of the clones.

27.2.2 Clonal Seed Orchards

The seed orchards are one of the most important sources of forest reproductive material (Faulkner 1975). They combine both functions of supplying high-quality seeds and at the same time serve as a tool for ex situ gene conservation (Gagov 2005; Gagov and Dobrinov 1987; Gagov et al. 2003a). It is expected that mating in the seed orchard will take place among the trees include, which represent phenotypically superior individuals. They are considered to be important both from economic and from conservation points of view. This consideration is especially important for the main tree species, like the Silver Fir (*Abies alba* Mill.).

There are several reports about seed orchards of *Abies* in many European countries (Gagov 1990; Bergmann and Gagov 1998; Maurer et al. 2001, 2002; Hansen 2008; Hansen and Kjaer 2006; Hosius et al. 2000; Kobliha and Janeček 2005). However, most authors report masculinization of the grafted plants during the 1st years of development, i.e., development of almost exclusively male strobili and lack of female cones. This phenomenon delays substantially the first crops of the seed orchards. However, Gagov et al. (2003b, 2013) have shown that this problem is avoidable by using appropriate techniques and some specific approaches.

The first-generation seed orchard of *Abies alba* in Bulgaria was established close to the village of Solnik (Eastern Stara planina) and served not only as a seed orchard but also as an extensive and large *ex situ* conservation genebank. Even though the locality of establishment of the seed orchard was outside the natural range of the species, the results in the next 30 years showed that it was appropriately selected. As mentioned above, initially more than 450 clones were selected and included in the seed orchard. After some preliminary selection thinning, there were 113 clones remaining in the seed orchard about the year 2000.

27.2.3 *Establishment of Second-Generation Seed Orchard*

The genetic efficiency of a seed orchard is defined as a degree at which seed crops correspond to the genetic quality of the trees included in the orchard. This efficiency could be decreased by (1) pollen contamination (i.e. background pollen migration), (2) self-fertilization and (3) deviations from the random mating among the trees selected for seed production (Friedman and Adams 1982).

There are usually two approaches in the establishment of the second-generation seed orchards. The first one consists of removing the clones that have proven to be of lower quality and thus only the selected clones will remain in the orchard, while the second approach, the more laborious one, considers establishment of a new orchard, at new place, where the best genotypes/clones tested in the first-generation orchard will be propagated and grown. In both cases, it is expected that the second-generation orchard will provide seeds with considerably better genetic quality (Lstibůrek and El-Kassaby 2008).

The second-generation seed orchards have the advantage to use tested genetic material. While the first-generation orchard relies on the phenotypical characters only, the second generation orchard considers several very important criteria related to fitness and quality: inheritance of the economically important traits, fecundity, genetic diversity (allelic richness and heterozygosity), mating system and outcrossing rates, background pollen contamination, resistance to pests and diseases, and some other important characteristics. Also, establishment of a second-generation seed orchard allows modeling of the number of clones, in order to reach the highest genetic gain, and to control the location of the different ramets and clones in the orchard (i.e., the orchard design). There were many theoretical studies and computer programs developed for the seed orchard designs (Vanclay 1991; Lstibůrek and El-Kassaby 2008).

In many studies there was an attempt to optimize between the gain and diversity, or seeking a compromise between the selection intensity and the background of the genetic variation. This is not an easy task, but it is implied by the modern concept of close-to-nature forest management and maintaining the natural genetic diversity. Moreover, it affects directly the objective of *ex situ* genepool conservation.

After careful consideration of all circumstances, finally 25 clones were selected for including in the second-generation seed orchard. They were selected based on the criteria, listed below.

27.2.3.1 Phenotypic Superiority

It is usually a character used in the establishment of the first-generation orchards, but here we have in mind not selection of superior trees in the natural stands, rather than studying the mode of inheritance of the economically important characters. Since the most such characters are quantitative traits, the progeny tests are very important tool for gaining important information of the genetic background of these characters. There were several reports on the progeny performance of clones, included in the orchard (Gagov and Evtimov 2007; Evtimov and Gagov 2008). The heritability of height-, diameter- and volume growth and increment, stem straightness and wood quality are considered as the most important characters. Different studies have proven that these characters differ in their heritability, for example wood properties having much higher individual heritability than height, diameter, and volume growth (Cornelius 1994). Evtimov et al. (2003) reported moderate level of heritability (0.20–0.24) of height growth, and Gagov et al. (2005) reported substantial superiority of the progeny produced by the same parents, but from seeds collected from seed orchard. Mihai and Mirancea (2016) reported that additive variance and heritability, respectively, increased with age, which was also found by Evtimov (2015).

27.2.3.2 Resistance to Pests and Diseases

This is very important character, since the lack of resistance could easily compromise all other advantages, related to growth and quality. Roques (2017) pointed out that individuals in seed orchards are more sensitive than the ones growing in the adjacent forests. However, even though more tolerant and more sensitive clones could be identified, in coniferous species, and in *Abies* in general, there were no drastic differences at individual level (Bergmann et al. 1990).

27.2.3.3 Fertility and Seed Production

This is one of the most important traits for clones included in a seed orchard, since one of the main purposes of the orchard is to produce large amount of high-quality seeds. Here the annual variation is very important, because the studies have identified clones that produce cones and seeds almost each year, while other clones produce cones rarely. Again, a compromise must be achieved between the other qualities and the fecundity.

There are some specific problems that should be considered when establishing a second-generation seed orchard.

27.2.3.4 Mating System and Outcrossing Rates

Most of the coniferous species, incl. Silver Fir exhibit mixed mating model, with some proportion of inbreeding. The outcrossing rate could differ substantially from orchard to orchard and from year to year, however, maintaining relatively high values. Self-fertilization also occurs but as a result of the inbreeding depression the selfed seeds are not vital (Adams and Birkes 1991, Sorensen 1999).

Many studies have shown that the mating among the clones in the seed orchards is not always random and this must be considered in the seed orchard design and establishment. The study of Bergmann and Gagov (1998) outlined the high diversity in the progeny of the seed orchard.

27.2.3.5 Phenological Synchrony

The phenological synchronization among the clones included in the seed orchard, and particularly in the second-generation is one of the most important factor promoting the outcrossing and random mating among the trees (Erickson and Adams 1989; Adams 1992; Burczyk and Prat 1997). Many studies showed that there is synchrony among the clones in the orchards, but sometimes even small difference could cause significant deviations from the random mating and will require supplemental mass pollination (El-Kassaby et al. 1993). However, our studies revealed that there is a very good synchrony among the clones included in the second-generation orchard.

27.2.3.6 Background Pollen Contamination

Pollen contamination is the pollination of seed orchard parents by pollen from outside of the seed orchard. It reduces the genetic gain of seed orchard crops and may adversely affect the adaptability of the resulting seedlings (Squillace and Long 1981; Friedman and Adams 1985; Adams and Burczyk 2000; Kang et al. 2001; Slavov et al. 2005).

We did not perform a special study on the background pollination, but in our case the pollen contamination does not seem to be an important problem, because the second-generation orchard is surrounded by natural oak forests and is relatively well isolated from other stands of the same species.

27.2.3.7 Genetic Diversity

Genetic diversity was measured in the first-generation seed orchard in Solnik by means of allozyme markers (Bergmann and Gagov 1998; Evtimov 2015). The seed orchard contained 94.7% of the alleles detected in all natural populations studied (18 of 19). Mean allele number in the orchard was 2.57 vs 2.53 in the natural populations. Effective allele number in the seed orchard was 1.38 (1.41 in the natural populations), expected heterozygosity was 0.275 and observed heterozygosity was 0.277 (inbreeding coefficient −0.007).

Genetic diversity in the second-generation orchard was measured also by SSR markers (Cremer et al. 2006) and compared to a Bavarian seed lot (Table 27.2). Results showed that despite of almost four times higher sample size (25 vs 96), the genetic diversity of Bulgarian clones was generally higher. The only exception was

Table 27.2 Comparison of genetic diversity between Bulgarian and Bavarian clones (Cremer et al. 2012)

	N	Na	Ne	Ho	He	F
BG clones	25	13.50	9.67	0.83	0.85	0.016
Bavarian seed sample	96	13.67	9.59	0.66	0.79	0.178

N sample size, *Na* mean allele number, *Ne* effective allele number, *Ho* observed heterozygosity, *He* expected heterozygosity, *F* inbreeding coefficient

the mean allele number, which is much more sensitive to the sample size (Cremer et al. 2012).

High genetic diversity is a good indicator of the quality of seed orchard crop, because there were many studies showing the effect of overdominance, i.e., superiority of heterozygotes over homozygotes in forest trees. Therefore, the studies of genetic diversity helped substantially the whole process of selecting the best clones and establishment of advanced-generation seed orchard.

The second-generation seed orchard was established in the period 2010–2013 and is expected to provide seeds of high genetic quality for afforestation. At the same time, the two seed orchards, together with the established numerous provenance and progeny tests will serve as a tool for *ex situ* conservation of Silver Fir genetic resources in Bulgaria.

Acknowledgements The financial support of the Executive Forest Agency – Ministry of Agriculture, Food and Forests of Bulgaria (Project No 334) for the establishment of second-generation seed orchard is very much appreciated. Authors express special thanks to Dr. Monika Konnert, Dr. Eva Cremer, Dr. Fitz Bergmann, and Dr. Bernhard Hosius for their fruitful cooperation in the course of years, and to numerous colleagues of the Forest Service that helped in the field studies and seed orchard establishment.

References

Adams, W. T. (1992). Gene dispersal within forest tree populations. *New Forests, 6*, 217–240.

Adams, W. T., & Birkes, D. S. (1991). Estimating mating patterns in forest tree populations. In S. Fineschi, M. E. Malvoltti, F. Cannata, & H. H. Hattemer (Eds.), *Biochemical markers in the population genetics of forest trees* (pp. 157–172). SPB Academic Publishing Bv, The Hague.

Adams, W. T., & Burczyk, J. (2000). Magnitude and implications of gene flow in gene conservation reserves. In A. Young, D. Boshier, & T. Boyle (Eds.), *Forest conservation genetics: Principles and practice* (pp. 215–244). Collingwood: Commonwealth Scientific and Industrial Research Organization (CSIRO) Publishing.

Ballian, D., & Halilović, V. (2016). Variabilnost Obične Jele (*Abies alba* Mill.) u Bosni i Hercegovini. Sarajevo-Ljubljana, 347 pp.

Barbu, I., & Barbu, C. (2005). Silver fir (*Abies alba* Mill.) in Romania, Editura Technica Silvica, 220 pp.

Bergman, F., & Gagov, V. (2000). Genetische Diversität und Differenzierung von Tannenpopulationen der Balkanhalbinesel. In: *Proceedings of the 9th International European Silver Fir Symposium in Skopje*, Macedonia, pp. 1–4.

Bergmann, F., & Gagov, V. (1998). Comparison of genetic structure between clones and their progenies (embryos) in Staro Oryahovo Silver Fir (*Abies alba* Mill.) seed orchard. *Forestry Ideas, 3*, 136–143.
Bergmann, F. & Gagov, V. (2003). The detection of hybrid populations between fir species in the southern Balkan Peninsula. In: I. Raev (Ed.), *Proceedings of the International Scientific Conference "Study, Conservation and Utilization of Forest Resources"* (01–05.10.2001), Sofia, pp. 97–102.
Bergmann, F., Gregorius, H.-R., & Larsen, J. B. (1990). Levels of genetic variation in European Silver Fir (*Abies alba* Mill.). Are they related to species' decline? *Genetica, 82*, 1–10.
Bernadzki, E. (1983). Fir decline in natural range of the species. In S. Białobok (Ed.), *Silver fir – Abies alba Mill. Nasze Drzewa Leśne* (pp. 483–501). Warszawa: PAN – Instytut Dendrologii.
Burczyk, J. L., & Prat, D. (1997). Male reproductive success in *Pseudotsuga menziesii* (Mirb.) Franco: The effects of spatial structure and flowering characteristics. *Heredity, 79*, 638–647.
Cailleret, M., Nourtier, M., Amm, A., Durand-Gillmann, M., & Davi, H. (2014). Drought-induced decline and mortality of silver fir differ among three sites in Southern France. *Annals of Forest Science, 71*(6), 643–657.
Cornelius, J. (1994). Heritabilities and additive genetic coefficients of variation in forest trees. *Canadian Journal of Forest Research, 24*, 372–379.
Cremer, E., Liepelt, S., Sebastiani, F., Buonamici, A., Michalczyk, I. M., Ziegenhagen, B., & Vendramin, G. G. (2006). Identification and characterization of nuclear microsatellite loci in *Abies alba* Mill. *Molecular Ecology Notes, 6*, 374–376.
Cremer, E., Konnert, M., Gagov, V. & Zhelev, P. (2012). Second generation seed orchard of Silver Fir in Bulgaria – clone selection. In: *14th International Fir Symposium*, Kastamonu, Turkey.
Diaci, J., Rozenbergar, D., Anic, I., Mikac, S., Saniga, M., Kucbel, S., Visnjic, C., & Ballian, D. (2011). Structural dynamics and synchronous silver fir decline in mixed old-growth mountain forests in Eastern and Southeastern Europe. *Forestry, 84*(5), 479–491.
Dimitrov, M. & Roussakova, V. (2015). Forests of Silver Fir. In: V. Bisserkov (Ed.) *Red Data Book of Bulgaria, vol 3*, Natural Habitats. S., BAS and MoEW.
Dobrinov, I. & Gagov, V. (1975). Study on the pollen of the Silver Fir (*Abies alba* Mill.) in Bulgaria. *Higher Institute of Forestry, Research Works, Series Forestry*, *XX*, 9–16.
Dobrinov, I., & Gagov, V. (1981). Distribution and some anatomic-morphological and biologic features of the fir *Abies alba* var. *acutifolia* Turill in Bulgaria. *Higher Institute of Forestry, Scientific Papers, Series Forestry, 26*, 103–109 (In Bulgarian, English summary).
Dobrinov, I. & Gagov, V. (1983). The role of individual selection in the Silver Fir (*Abies alba* Mill.). *Higher Institute of Forestry, Research Works, Series Forestry*, *XXVIII*, 59–64.
Dobrinov, I. & Gagov, V. (1985). Selection of the valuable genepool in the Silver Fir (*Abies alba* Mill.) and including it in a clonal seed orchard. Unpublished report, S., 63.
Dobrinov, I., Doykov, G. Gagov, V. (1982). The forest gene pool in Bulgaria. S., Zemizdat, 259.
Dobrowolska, D., Bončina, A., & Klumpp, R. (2017). Ecology and silviculture of silver fir (*Abies alba* Mill.): a review. *Journal of Forest Research, 22*(6), 326–335.
El-Kassaby, Y. A., Barnes, S., Cook, C., & MacLeod, D. A. (1993). Supplemental mass pollination success rate in a mature Douglas-fir seed orchard. *Canadian Journal of Forest Research, 23*(6), 1096–1099.
Erickson, V. J., & Adams, W. T. (1989). Mating success in a coastal Douglas-fir seed orchard as affected by distance and floral phenology. *Canadian Journal of Forest Research, 19*, 1248–1255.
Eriksson, G. (1994). Gene conservation and tree breeding, harmony or cacophony?: Tree breeding, population genetics and conservation of genes. Lectures given at the 1994 Marcus Wallenberg Prize Symposium held in Stockholm on September 22, 19–32.
Evtimov, I. (2015). Studies on *ex situ* conserved genepool of the Silver Fir (*Abies alba* Mill.) in Bulgaria. PhD thesis, University of Forestry, Sofia, 162 pp.

Evtimov, I., & Gagov, V. (2008). Studies on the growth of different provenances of the Silver Fir (*Abies alba* Mill.) in trial plantations. *Forestry Ideas, 36*, 3–11 (in Bulgarian with English summary).

Evtimov, I., Gagov, V., & Zhelev, P. (2003). Results of progeny experiments with silver fir (Abies alba Mill.) in Bulgaria. *Mitteilungen aus der Forschungsanstalt für Waldökologie und Forstwirtschaft Rheinland-Pfalz, 50*(03), 125–130.

Faulkner, R. (1975). Seed orchards. *Forestry Commission Bulletin, 54*, 149.

Friedman, S.T. Adams, W.T. (1982). Genetic efficiency in loblolly pine seed orchards. In: *Proceedings of the 16th Southern Forest Tree Improvement Conference*. Virginia Poly Tech and State University, Blacksburg, VA, pp. 213–220.

Friedman, S. T., & Adams, W. T. (1985). Estimation of gene flow into two seed orchards of loblolly pine (Pinus taeda L.). Theoretical and Applied Genetics, 69, 609–615.

Gagov, V. (1973a). Study on the variation of the natural populations of *Abies alba* Mill. In Bulgaria. PhD, thesis, University of Forestry, Sofia.

Gagov, V. (1973b). Variation of the seeds of Silver Fir (*Abies alba* Mill.) originating from different populations in Bulgaria. Higher Institute of Forestry, Research Works, Series Forestry, XIX, 51–60.

Gagov, V. (1979). Comparative studies on the growth of the Silver Fir (Abies alba Mill.) in experimental trials. *Higher Institute of Forestry, Research Works, Series Forestry, XXIV*, 99–104.

Gagov, V. (1985). Different origins silver fir (Abies alba Mill.) growth study in provenance trials. In: W. Kramer (Ed.): Ergebnisse des 4. Tannen-Symposium. Schriftenreihe aus der Forstlichen Fakultat der Universitat Gottingen und der Niedersachsischen Forstlichen Versuchsanstalt, 80, pp. 278–286.

Gagov, V. (1988). The Silver Fir (*Abies alba* Mill.) problem in Bulgaria. In L. Paule, & S. Korpel (Eds.), *Proceedings of 5th IUFRO Tannensymposium*, Zvolen, Slovakia, 159–162.

Gagov, V. (1990). *Genetic studies on the conserved genepool of Abies alba in Bulgaria and establishment of a seed orchard. Scientific Report*. Sofia: University of Forestry.

Gagov, V. (2005). Seed orchards conserve forest genetic resources. *Gora, 9*, 16–18 (in Bulgarian).

Gagov, V., & Dobrinov, I. (1987). Status and prospects for the forest seed orchards. *Gorsko stopanstvo & Gorska promishlenost, 43*(2), 1–4 (in Bulgarian).

Gagov, V., & Evtimov, I. (2007). Study on the growth of half-sibs of the Silver Fir (Abies alba Mill.) and King Boris Fir (Abies borisii-regis Mattf.). *Forestry Ideas, 35*(1–2), 3–13.

Gagov, V., Bergmann, F., & Hosius, B. (2003a). Seed orchards as a tool in tree improvement and conservation of forest gene resources. *Mitteilungen der Forschungsanstalt für Waldökologie und Forstwirtschaft Rheinland-Pfalz Nr, 50*, 59–70.

Gagov, V., Bergmann, F., Hosius, B., Zhelev, P., Mirchev, S., & Evtimov, I. (2003b). *Genetic assessment of the Silver Fir populations and clone collections in Bulgaria. Scientific Report* (p. 92). Sofia: University of Foresty.

Gagov, V., Evtimov, I. Zhelev, P. (2005). Comparative study on the growth of half-sib families of selected trees of Abies alba Mill. and Abies borisii-regis Mattf. from Bulgarian and Macedonian provenances. In: *Proceedings of the 11th IUFRO Tannensymposium*, Pojana Brasov, Romania, 05-09.09.2005.

Gagov, V., Evtimov, I. Zhelev, P. (2009). Survival and growth of European Silver Fir in the Second IUFRO Provenance experiments in Bulgaria. In: *12th international Silver fir symposium*. Hittisau, 31.08-04.09.2009.

Gagov, V., Zhelev, P. Evtimov, I. (2013). Genetic characteristics of the clones and progenies of the first-generation seed orchard of Silver Fir and establishment of a second-generation seed orchard. Report of the project 334. Executive Forest Agency, Ministry of Agriculture and Food, Sofia, 120 pp.

Hansen, O. K. (2008). Mating patterns, genetic composition and diversity levels in two seed orchards with few clones—impact on planting crop. *Forest Ecology and Management, 256*, 1167–1177.

Hansen, O. K., & Kjaer, E. D. (2006). Paternity analysis with microsatellites in a Danish *Abies nordmanniana* clonal seed orchard reveals dysfunctions. *Canadian Journal of Forest Research, 36*(4), 1054–1058.
Hosius, B., Bergmann, F., Konnert, M., & Henkel, W. (2000). A concept for seed orchards based on isoenzyme gene markers. *Forest Ecology and Management, 131*, 143–152.
Kandler, O. (1993). Development of the recent episode of Tannensterben (Fir decline) in Eastern Bavaria and Bavarian Alps. In R. F. Huettl & D. Mueller-Dombois (Eds.), *Forest decline in the Atlantic and Pacific Region* (pp. 216–228). Berlin: Springer.
Kang, K. S., Lindgren, D., & Mullin, T. J. (2001). Prediction of genetic gain and gene diversity in seed orchard crops under alternative management strategies. *Theoretical and Applied Genetics, 103*, 1099–1107.
Kobliha, J., & Janeček, V. (2005). Development of hybrid fir clonal material. *Journal of Forest Science, 51*(Special Issue), 3–12.
Korpel', Š., Paule, L., & Laffers, A. (1982). Genetics and breeding of the Silver Fir (*Abies alba* Mill.). *Annales Forestales, 9*(5), 151–184.
Kostov, G. (1989). Studies on the ecology and management of the mixed forests of Norway Spruce, Siver Fir and Beech in the Rila-Rhodopes massif. PhD thesis, University of Forestry, Sofia (in Bulgarian).
Krajmerová, D., Paule, L., Zhelev, P., Voleková, M., Evtimov, I., Gagov, V., & Gömöry, D. (2016). Natural hybridization in Eastern-Mediterranean firs: The case of *Abies borisii-regis. Plant Biosystems, 150*(6), 1189–1199.
Larsen, J. B. (1986). Das Tannensterben: Eine neue Hypothese zur Klärung des Hintergrundes dieser rätselhaften Komplexkrankheit der Weißtanne (*Abies alba* Mill.). *Forstwissenschaftliche Zentralblatt, 105*, 381–396.
Linares, J. C., & Camarero, J. J. (2012). Growth patterns and sensitivity to climate predict silver fir decline in the Spanish Pyrenees. *European Journal of Forest Research, 131*(4), 1001–1012.
Longauer, R. (1996). Genetic diversity of European silver fir (Abies albaMill.). PhD thesis, Technical University, Zvolen, Slovakia, 154 pp.
Longauer, R., Paule, L., & Andonoski, A. (2003). Genetic diversity of southern populations of *Abies alba* Mill. *Forest Genetics, 10*, 1–9.
Lstibůrek, M., & El-Kassaby, Y.A. (2008). Advanced Generation Seed Orchard Designs. In: D. Lindgren (Ed.), *Seed orchards: Proceedings from a conference at Umeå, Sweden*, September 26–28, 2007, pp. 155–160.
Maurer, W., Tabel, U., Hosius, B., Leinemann, L., Bergmann, F., Eder, W. Gagov, V. (2001). Zur Konzeption und Anlage von Generhaltungsplantagen der Weißtanne (*Abies alba* Mill.) in Rheinland-Pfalz. In: *IUFRO-Proceedings of the 9th International European Silver Fir Symposium*, Skopje, Macedonia, pp. 37–48.
Maurer, W.D., Tabel, U., Hosius, B., Leinemann, L., Bergmann, F., Gagov, V. Eder, W. (2002). Die Anlage von Generhaltungssamenplantagen für Weißtanne (Abies alba Mill.) in Rheinland-Pfalz: Isoenzymuntersuchungen als effiziente Entscheidungshilfen. In: W. D. Maurer (Ed.), *Vom genetischen Fingerabdruck zum gesicherten Vermehrungsgut: Untersuchungen zur Erhaltung und nachhaltigen Nutzung forstlicher Genressourcen in Rheinland-Pfalz.* Mitteilungen aus der Forschungsanstalt für Waldökologie und Forstwirtschaft Rheinland-Pfalz, Nr. 49/02, 59–74.
Mihai, G., & Mirancea, I. (2016). Age trends in genetic parameters for growth and quality traits in *Abies alba. iForest, 9*, 954–959.
Oliva, J., & Colinas, C. (2007). Decline of silver fir (*Abies alba* Mill.) stands in the Spanish Pyrenees: Role of management, historic dynamics and pathogens. *Forest Ecology and Management, 252*(1–3), 84–97.
Penev, N., Garelkov, D., Marinov, M. Naumov, Z. (1969). Forest Types in Bulgaria. S., BAS, 353.
Plugtschieva, M., Gagov, V., Simeonov, I., Bjalkov, S. Bardarov, D. (2003). Strategie zur Erhaltung der genetischen Ressourcen der Tannen Abies alba Mill. und *Abies borisii-regis Mattf. in Bulgarien. Mitteilungen der Forschungsanstalt für Waldökologie und Forstwirtschaft Rheinland-Pfalz* Nr 50: pp. 85–96.

Roques, A. (2017, September 4–6). Why are insect pests more impacting in seed orchards than in surrounding forests? The experience of long-term surveys of pest damage in France.: IUFRO Seed Orchard Conference 2017, Bålsta, Sweden, pp. 26–27.

Slavov, G. T., Howe, G. T., & Adams, W. T. (2005). Pollen contamination and mating patterns in a Douglas-fir seed orchard as measured by simple sequence repeat markers. *Canadian Journal of Forest Research, 35*(7), 1592–1603.

Sorensen, F. C. (1999). Relationship between self-fertility, allocation of growth, and inbreeding depression in three coniferous species. *Evolution, 53*, 417–425.

Squillace, A. E. Long, E. M. (1981). Proportion of pollen from non-orchard sources. In: E.C. Franklin (Ed.), *Pollen management handbook* (Agriculture handbook (United States, Department of Agriculture), no. 587).Washington, DC, pp. 15–19.

Thomas, A. L., Gégout, J. C., Landmann, G., Dambrine, É., & King, D. (2002). Relation between ecological conditions and fir decline in a sandstone region of the Vosges mountains (northeastern France). *Annals of Forest Science, 59*, 265–273.

Vanclay, J. K. (1991). Seed orchard designs by computer. *Silvae Genetica, 40*, 89–91.

Wolf, H. (2003). *EUFORGEN technical guidelines for genetic conservation and use for silver fir (Abies alba)* (p. 6). Rome: International Plant Genetic Resources Institute.

Part V
Climate Change Aspects in Forest Genetic Resources Conservation

Chapter 28
Climate Change Aspects in Forest Genetic Resources Conservation in Serbia

Jelena Milovanović, Mirjana Šijačić-Nikolić, and Marina Nonić

Abstract Serbia's current adaptive capacity to climate change is fairly modest. However, with the beginning of the process of Serbia's accession to the European Union, the capacities for strategic consideration of climate change issues began to strengthen. Activities are focused primarily on the processes necessary to reduce the causes of climate change (mitigation), and the consideration of adaptation is still neglected. Forest ecosystems play an important role in the global carbon cycle since they have a large potential for carbon sequestration within forest biomass and forest soils. However, conservation of forest genetic resources has not been recognized as a separate significant mechanism for adaptation to climate change within the national policy framework. Conservationists will face numerous challenges in their attempts to adapt conservation activities to climate change trends. According to the estimated speed of climate change, spontaneous natural selection will play a less important role than phenotypic plasticity. Plasticity will be the most desirable feature in the conservation of natural populations and artificial regeneration. An interesting example of the "carryover effect" from the territory of Serbia is the hypothesis of the phenotype groups formation of Serbian spruce and the high degree of intraspecific variability of this relict and endemic species. National conservation priorities, based on elaborated challenges and adaptive mechanisms of trees, are defined and elaborated through concrete measures and activities.

Keywords Adaptation · Mitigation · Conservation challenges · National priorities

J. Milovanović (✉)
Faculty of Applied Ecology FUTURA, Singidunum University, Belgrade, Serbia
e-mail: jelena.milovanovic@futura.edu.rs

M. Šijačić-Nikolić · M. Nonić
Faculty of Forestry, University of Belgrade, Belgrade, Serbia
e-mail: mirjana.sijacic-nikolic@sfb.bg.ac.rs; marina.nonic@sfb.bg.ac.rs

M. Šijačić-Nikolić et al. (eds.), *Forests of Southeast Europe Under a Changing Climate*, Advances in Global Change Research 65,
https://doi.org/10.1007/978-3-319-95267-3_28

28.1 Introduction

Despite its strong technological development, human civilization is still essentially tied to natural systems and is directly dependent on the processes that take place in them. The direct consequences of climate change such as temperature rise, melting ice, sea and ocean level rise and changes in precipitation regime can cause significant problems in the functioning of human society. The production and availability of food and water, human health, transport, energy supply are just some of the elements on which the functioning of human communities is based, which are highly dependent on climatic conditions and which can be significantly destabilized by climate change.

Although climate change is a global problem in itself, the impacts will not be felt equally across the Planet, but will vary by intensity and rate of change in the different regions. The ability to adapt to climate change can influence the effect of climate change on individual, community, national, and global level.

Serbia's current adaptive capacity to climate change is fairly modest. The development of the system of adaptation to new climate trends has not yet been recognized as a priority in official policies. Political and economic elite seem to still fail to recognize the seriousness of the situation and the degree of vulnerability of the entire society in an unstable climate situation. Several factors influenced this situation. Longstanding political instability in the Balkans has led to the degradation of the state systems necessary for acting in such situations. Reviewing the future trends and dangers they carry for society as a whole is simply overlooked and has lost priority over the current political situation. The poor economic situation caused the impoverishment of the necessary scientific and professional capacities and resources that are of crucial importance in such situations. Consequently, strategic planning in terms of adaptation to climate change is hardly feasible.

However, with the beginning of the process of Serbia's accession to the European Union, the capacities for strategic consideration of climate change issues began to strengthen. Activities are focused primarily on the processes necessary to reduce the causes of climate change (mitigation), and the consideration of adaptation is still neglected. Serbia has significantly improved its capacities in the field of climate change monitoring. First of all, there is the work of the Republic Hydrometeorological Service of Serbia, which, along with state aid and foreign donations, has improved its capacities and has become a reputable institution in the region and in Europe. As such, it represents an important source of information. The Republic Hydrometeorological Service regularly provides a weather forecast (Meteoalarm), a hydrological forecast (Hidroalarm), a UV index forecast, announcements of hot or cold waves and other information.

Unfortunately, this information does not seem to be finding the way to the appropriate decision-makers to incorporate them into strategic documents. It is surprising that many sectors that are very sensitive to climate change in their strategic documents still do not recognize the need for better and more prompt adaptation planning.

Having in mind high natural capacity of this area for adaptation, it can be said that Serbia can do a lot to adjust its sectoral policies in order to reduce the negative effect of climate change. Much can be changed by applying a sustainable way of natural resources management.

28.2 Forest Ecosystems in Climate Change Policy in Serbia

The Republic of Serbia is a signatory of the United Nations Framework Convention on Climate Change and the Kyoto Protocol ("The Law on the Confirmation of the Kyoto Protocol along with the United Nations Framework Convention on Climate Change", Official Gazette of the Republic of Serbia, 88/07) in the status of "Non-Annex I Party". Serbia has shown its interest in combating climate change by developing the First National Communication (the First Report of the Republic of Serbia under the UN Framework Convention on Climate Change), adopted in 2010, while the Second National Communication is being prepared and is expected to be adopted in recent times.

Serbia successfully presented the First Biennial Update Report (FBUR) to the UN Framework Convention on Climate Change during the 23rd session of the Conference of Parties (COP23) in November 2017 in Bonn, during the Facilitative Sharing of Views workshop. According to this report, sector of Agriculture, Forestry and Land Use (hereinafter: AFOLU) in 2013 participated in total Greenhouse Gases (GHG) emissions with 9.116,10 Gg CO2eq. Sources of emissions in 2013 within the AFOLU sector contributed with 6257.79 Gg CO2eq, which makes 47.5% of emissions of the entire sector, i.e. 3142.50 Gg CO2eq, which is mainly due to direct and indirect emissions of CH4 and N2O due to enteric fermentation and manure management (biochemical processes from animal husbandry). On the other hand, the removal of emissions through sinks, namely forest land (forestry) and the use of wood products, in 2013 amounted to −15.735,64 Gg CO2eq, or – 1,42 Gg CO2eq, respectively in the subsectors (MAEP 2016).

Forest ecosystems have a large potential for carbon sequestration within forest biomass and forest soils. However they can also represent a source of carbon due to natural processes such as ageing, seasonal variations in carbon dynamics, and anthropogenic impacts related to forest management or forest degradation (Milovanović et al. 2017).

In order to reduce GHG emissions, the aforementioned FBUR defines the following national priorities in the field of forestry: afforestation, restoration of even-aged high forests, reconstruction of high degraded forests, indirect conversion of coppice forests, direct conversion of coppice forests, rehabilitation of heavily damaged stands, rehabilitation of fire affected stands, forest certification, preparation of planning documents in forestry, national inventory of forests and development of capacities for research activities in forestry.

The draft of the Second Report of the Republic of Serbia under the UN Framework Convention on Climate Change recognizes the following impacts of climate change on forest resources (Orlović et al. 2014): negative impact due to drought intensification; increasing the number of fires and increasing the area affected by the fire; increasing areas affected by diseases and pests; increased mortality of forests; change of distribution and structure of forests; enlargement of areas at risk of wind; enlargement of areas at risk from ice. In order to mitigate the consequences of the mentioned climate change impacts on forests, the following adaptation measures have been proposed:

- Reducing the risk of biotic and abiotic factors by strengthening adaptive forest capacity;
- Preparation of fire belts/Timely application of measures to combat pests;
- Promotion of mixed-species forests/Promotion of all-aged forests;
- Selection of appropriate woody species, provenances, populations and genotypes which are able to adapt to changed climate or specialized for potential habitat traits expected in the future;
- Develop a framework for better implementation of the afforestation process between different groups of participants;
- Changing the practice of forest management and promotion of the "close-to-nature" management concept;
- Adaptation measures for lowland forests – forests of English oak and Turkey oak are the most threatened: improvement of water management; increase underground water level during drought seasons, if it is possible; adjust thinning and cutting to altered environmental conditions; promote "close-to-nature" regeneration concept; promote mixed-species oak forests (optimization of mixture with ash and hornbeam);
- Adaptation measures for mountain forests – beech, fir and spruce as potentially highly endangered in the future: promotion of selective cutting and the "close-to-nature" management approach; promotion of mixed-species forests; proceed with private forests inventure;
- Capacity building of public enterprises in the forestry sector through providing of appropriate legal, organizational and financial framework;
- Awareness raising on importance of forests for societal adaptation to climate change;
- Awareness raising on ecosystem services provided by forests and its multi- functionality.

For the needs of the first assessment of the necessary long-term investments in the priority actions for adaptation to the changed climatic conditions, including risk reduction, the Ministry of Environmental Protection of the Republic of Serbia initiated the drafting of the National Adaptation Plan (NAP), which will establish the basis for the forthcoming development of the Adaptation Strategy. The main objectives of the NAP draft are: (1) Reducing vulnerability to climate change impacts by building adaptability and (2) Integration of adaptation to changed climatic conditions into relevant existing and new policies, activities and regulations, in particular strategies and development plans, at all levels and in all sectors.

The draft NAP for the Republic of Serbia describes the situation and needs in the field of forestry in the following way (Stojanović and Orlović 2014):

1. Raising new forests (especialy protection forests – wind barriers and forests which prevent erosion and torrents, with other multiple benefits;
2. The Republic of Serbia has increased the area under the forests for 6000 ha, or only 0.001% in the period 2000–2013 (Stojanović et al. 2015a);
3. In 2/3 of the districts in our country there is a decrease in forest cover, including all the districts in Vojvodina. One of the main causes is the increased number of sanitary cuts due to intensive forest dying (Stojanović et al. 2015b). Forests dying is associated with the occurrence of extreme events, primarily extreme droughts (2000, 2003, 2007, 2011 and 2012), as well as wind, ice, fire and pest and disease attacks;
4. Construction of new forest roads in order to improve fire protection and management of forests in general;
5. Coppice forests conversion.

The conservation of forest genetic resources has not been recognized as a separate significant mechanism for adaptation to climate change.

28.3 Natural Adaptation Mechanisms of Forest Ecosystems

It can be assumed that, as in the past, species and communities will adapt to new changes in the environment by aligning the frequency of genes within populations and by colonizing new habitats. However, effective migrations of species must follow the flow of expected changes. According to paleobotanical data, the natural velocity of the seed dispersion process ranges between 100 and 250 m per year for species with large seeds (*Quercus* – oak, *Fagus* – beech) and 200–400 m per year for small seed species (*Betula* – birch, *Pinus* – pine). In European conditions, according to the scenario of temperature rise of 2 °C by the end of the twenty-first century, which is a likely increase in the case of the RCP6.0 and RCP8.5 scenarios for the GHG emission, the isotherms can move in the north direction at an average speed from 50 km per °C, i.e. 3 km per year and vertical 200 m per °C, i.e. 12 m per year. By comparing the data on the rate of seed dispersion and the isothermal displacement velocity, possible problems of species due to the upcoming climate change can be defined. From these data it can be noticed that, even in the case of migration corridors and favorable conditions in ecosystems, the rates of horizontal migration of species through seeds are not in accordance with the predicted climate change rates.

On the other hand, there is no reliable information on the effectiveness of gene transfer across pollen in contact areas. According to indirect indicators, the rate of adaptation of the genome through pollen fluctuations can be estimated as significantly higher than through the dispersion of seeds. Within continuous distribution areas, adjusting the gene frequency through pollen transfer can be sufficient for species surviving in the future changes. Prerequisites for this are: continuous distribution of species, natural population structure and application of natural regeneration techniques. Certainly, none of these preconditions are guaranteed (Milovanović et al. 2015).

Phenotypic plasticity represents the phenotype variability of a genotype depending on the environment conditions. In the strict sense of the word, it refers to a single genotype, which means it can only be estimated for genotypes that can be propagated using vegetative techniques. However, the concept can be extended to families, populations or any other level of genetic hierarchy (Eriksson 2005). "*Carryover effect*" involves existence of the same phenotypes in several generations despite the changing environment conditions that caused their formation. An interesting example of the "carryover effect" from the territory of Serbia is the hypothesis of the phenotype groups formation of Serbian spruce (*Picea omorika* Panč./Purkynie) and the high degree of intraspecific variability of this relict and endemic species (Milovanović 2007), since strict genetic determination has not been established (Window 28.1).

Adaptability studies are primarily based on selection processes and changes in the genetic structure. Due to the difficulty in conducting experimental analyzes, phenotypic plasticity, as an adaptation mechanism, was significantly less explored.

However, plasticity will be the most desirable feature in the conservation of natural populations and artificial regeneration. In recent years in Serbia, great attention has been devoted to the research of the phenotypic plasticity of leaf and wood anatomical structure of beech provenances, in order to monitor the level of the plastic response of provenances to different habitat conditions (Stojnić et al. 2013, 2015a, b). Research on the efficiency of water use in Serbia was carried out only on populus species.

A survey conducted on nine most prevalent tree species in Serbia has shown that not all species will be equally threatened. Considering that there has been a continuous decline in the level of groundwater in pedunculate oak forests, as well as the fact that these forests are in the lowland area, which is most exposed to the influence of high temperatures, it is considered that pedunculate oak will be most endangered by the environmental changes. The same study showed that among the endangered species of trees will also be Sessile oak, Turkey oak, Norway spruce, fir and beech, while species such as Austrian pine and Hungarian oak will be less endangered (Stojanović et al. 2014). Stojanović et al. (2013) also show that by the end of the twenty-first century, 90% of beech forests in Serbia will be outside their bioclimatic niche, while even a half of beech forests will be exposed to massive extinction.

Indigenous populations of woody species have an unusually high level of genetic variability, compared to 1-year-old plants, and therefore almost ideal preconditions for genetic adaptation, i.e. for the selection of phenotypes adapted to the new environment conditions.

Window 28.1. Serbian Spruce Phenotype Groups Hypothesis

The phenomenon of the "carryover effect" can serve as an explanation for the intraspecific differentiation of Serbian spruce and the formation of the phenotype groups, since strict genetic determination, by analysis of the nad1 gene, has not been established. Serbian spruce phenotype groups (A, B, C, D, E, F and G) separated according to the observed characteristics (Isajev 1987), served as the basis for the selection of test trees whose seed was harvested in order to produce seedlings for the establishment of a generative seed orchard in Godovik village in Serbia (Tucović and Isajev 1988).

The intraspecific differentiation of Serbian spruce can be compared with the process of allopathic speciation (Chiras 1991). Due to geographical isolation and sudden changes in the environment, the former large populations of Serbian spruce began to disintegrate. Small populations were completely fragmented and settled in the most suitable habitats. After a certain isolation time, fragmentation into even less populations (*demes*) occurred (Gilmour and Gregor 1939). Genetic structure of demes was similar to ancestor population, but the environmental conditions between demes were different. Demes began to differ from one another. This scenario can serve, precisely, to explain the appearance of Serbian spruce phenotype groups.

The phenotypic characteristics, which are listed as crucial for the selection of Serbian spruce varieties, indicate the environmental conditions in which separated populations were exposed. Individuals from "C" phenotype group was retained on typical habitats of Serbian spruce, steep cliffs and deep gorges with a high snow cover, and they retained traits of typical tree habit, high stem, narrow, pyramidal crown and short branches. Phenotype group "A" is represented by individuals with a wide crown and branching type similar to Norway spruce to be able to compete with subsidiary species. The double-topped tree from the phenotype group "B" can be associated with the former strong influence of biotic or abiotic factor, which caused dying of terminal bud, and the trees developed a defensive mechanism and retained it after the disappearance of the causative agent. Phenotype group "D" is characterized with semi-dwarf individuals which has been caused by very unfavorable habitat. The silver appearance of the crown of the individuals from phenotype group "F", which is a consequence of upward oriented needles on 1-year and 2-year branches, is most likely created in response to high air humidity, high precipitations and high snow cover. The largest number of stoma is found on the lower epidermis, and there has been a modification of the direction of the growth of needles.

28.4 Conservation Challenges Related to Climate Change

The direction and intensity of the expected changes can disrupt the naturally balanced effects of the selection. Some important features necessary for adaptation to different environmental conditions are not sufficiently covered by forest species breeding programs (Loo et al. 2011): *resistance to drought* (changes in soil and air humidity regime will be much more important for many forest species than temperature changes); *resistance to disease and pests* (increasing the rate of disease and pest infestations due to climate change becomes an essential problem of forestry); *fire resistance/tolerance* (due to reduction of precipitation and increase of air temperature, accompanied by negative effects of man on forest ecosystems, there is an increased frequency of occurrence of forest fires); *phenotypic plasticity* (phenotype variability of a single genotype depending on environmental conditions); „*carry-over effect* "(the presence of phenotypes in several generations despite the changing environment conditions that caused their formation).

The climate change mitigation program should focus on the selection of targeted populations in highly vulnerable areas and the establishment of new populations in less vulnerable areas (Table 28.1).

It is precisely the approach that the MPBS (*Multiple Population Breeding System*) method is based (Milovanović and Šijacić-Nikolić 2006, 2007, 2008, 2009; Milovanović et al. 2012), which represents a conservation method in which the population with genetic conservation function is divided to subpopulations, each of which has an effective size of approximately 50 individuals (Namkoong 1984). A good example of this approach is the model of genetic conservation of Serbian spruce (*Picea omorika* Panč./Purkyine), using the MPBS method (Šijačić-Nikolić et al. 2009).

According to authors, the MPBS method of conserving the total variability and adaptability of Serbian spruce would involve the establishment of three subpopulations at three different altitudes: about 500, 1000 and 1500 m. Also, the structure, density and management measures for proposed subpopulations are defined. The proposed model of conservation of the Serbian spruce total genetic diversity using the MPBS method presents an adequate solution for the long-term planning of conservation of genetic resources of rare and endangered species.

Table 28.1 Level of urgency due to the species traits (Mátyás 2005)

Low urgency	High urgency
Continuous distribution	Fragmented distribution
High density	Small populations
Natural regeneration	Artificial regeneration
Efficient gene flow	Limited gene flow
Spontaneous spreading	Low dispersion capacity
Spacious, zonal habitats	Extreme habitats
Reproduction is not limited	Blooming and seed formation disorders

28.5 National Forest Genetic Resources Conservation Priorities Related to Climate Change

Priority 1	
Expanding knowledge and information on the impact of climate change on forest genetic resources in Serbia	Measure 1.1
	Develop a long-term program to study the impact of climate change on forest genetic resources
	Measure 1.2
	Develop guidelines for the inclusion of priority species in intensive programs to study the impact of climate change on forest genetic resources
	Measure 1.3
	Make an assessment of the vulnerability of rare and endangered forest tree species to climate change
	Measure 1.4
	To conduct an impact assessment of different management systems on the conservation of forest genetic resources in accordance with climate change
	Measure 1.5
	Explore the possibilities of applying climate scenarios and models for the needs of assessing the vulnerability of forest genetic resources
Priority 2	
Extension of knowledge and information on the importance and role of forest genetic resources to mitigate the effects of climate change in Serbia	Measure 2.1
	Develop a long-term program of research on the adaptability of forest species to environmental changes
	Measure 2.2
	Explore the potentials of forest species and forest types for carbon storage
	Measure 2.3
	Explore the neutral and adaptive variability of relict forest species in order to define adaptation mechanisms in a wide range of different environmental conditions
	Measure 2.4
	Comparatively analyze the degree of genetic variability and the vulnerability of forest species populations to disease and damage
	Measure 2.5
	Comparatively analyze the degree of genetic variability of forest species populations and the quality of other habitat characteristics (soil, water, air and biodiversity)

(continued)

Priority 3	
Improvement of existing strategic and legal frameworks in conservation of forest genetic resources in accordance with the effects of climate change	Measure 3.1
	Improve and harmonize the strategic goals of the development of forestry in Serbia with the new strategic document of the European Union, emphasizing the holistic approach to forest management
	Measure 3.2
	Assess the existing system of forest species provenance regions in relation to the needs of adaptation and mitigation of the consequences of climate change
	Measure 3.3
	To harmonize the legal framework related to forest reproductive material management with the expected needs of assisted migration of genetic material in accordance with the intensity of climate change
Priority 4	
Improvement of in situ conservation of forest genetic resources in accordance with climate change	Measure 4.1
	Develop programs of genetic conservation of forest species in protected natural areas and on specific habitats
	Measure 4.2
	Establish networks of conservation areas of forest species within protected natural areas and include them in management systems
	Measure 4.3
	Develop a monitoring system for the impact of climate change on forest genetic resources within conservation area networks
Priority 5	
Improvement of ex situ conservation system for forest genetic resources in accordance with climate change	Measure 5.1
	Develop a long-term program of ex situ conservation of forest genetic resources in accordance with climate change
	Measure 5.2
	Identify the list of priorities for forest tree species and their characteristics for inclusion in intensive ex situ conservation and breeding programs in line with climate change
	Measure 5.3
	Assess the adaptive ability of priority forest tree species at the level of test trees in selected populations
	Measure 5.4
	Develop guidelines for the establishment of experimental plantations and gene banks in order to preserve the adaptive variability of forest tree species

(continued)

Priority 6	
Improving an integral approach to conservation of forest genetic resources in order to mitigate the effects of climate change	Measure 6.1
	Develop a long-term program of an integral approach to conservation of forest genetic resources based on an adequate combination of in situ and ex situ methods and intersectoral cooperation
	Measure 6.2
	Develop guidelines for introduction of Multiple Population Breeding System (MPBS) conservation method as an efficient approach to climate change mitigation
	Measure 6.3
	Introduction of modern software packages for modeling ecosystem processes in relation to environmental changes in the system of forest management planning
Priority 7	
Inclusion of the approach of conservation of forest genetic resources in line with climate change in the system of regular forest management in Serbia	Measure 7.1
	Develop guidelines for inclusion of vulnerability and adaptability assessment of priority forest tree species into the regular process of forest management planning
	Measure 7.2
	Develop guidelines for the allocation and management of genetic zones in forest ecosystem in order to improve the adaptability of species
	Measure 7.3
	Develop guidelines for assessing and conserving genetic variability in private forests and to bring closer the importance of forest genetic resources to forest owners and users
Priority 8	
Strengthening the capacity of institutions and organizations to understand the importance of conserving forest genetic resources in line with climate change	Measure 8.1
	Improve the knowledge of employees in state administration and public enterprises about the necessity of harmonizing forest management with climate change through conservation of forest genetic resources
	Measure 8.2
	Improve the skills of employees in public enterprises for the application of modern software packages for modeling ecosystem processes in relation to changes in the environment
	Measure 8.3
	Encourage intersectoral cooperation in order to better understand the importance and planning of activities in conservation of forest genetic resources in accordance with climate change

(continued)

Priority 9	
Awareness raising of forest genetic resources conservation importance in line with climate change	Measure 9.1
	Clarify the concept of forest genetic resources to private forest owners and develop awareness of the importance of their conservation in order to mitigate the effects of climate change
	Measure 9.2
	Develop informative and educational material on networks of conservation areas and the importance of in situ conservation of forest genetic resources
	Measure 9.3
	Integrate forest conservation areas and conservation programs into educational and tourist activities of protected natural areas
	Measure 9.4
	Map ex situ conservation points and develop educational and information program for acquaintance with this form of conservation of genetic variability of forest species
	Measure 9.5
	Develop an internet portal on the importance of forest genetic resources in combating climate change with the presentation of all conservation capacities
Priority 10	
Ecological and economic aspects of conservation of forest genetic resources in accordance with climate change	Measure 10.1
	Develop guidelines for ecological and economic evaluation of forest genetic resources and their conservation in accordance with climate change
	Measure 10.2
	Develop guidelines for estimating costs and revenues of conservation programs and selecting optimal forest diversity conservation models
	Measure 10.3
	Develop a long-term program for financing the conservation of forest genetic resources and identify possible sources of financing

Acknowledgements Authors acknowledge the Ministry of Education, Science and Technological Development of the Republic of Serbia for support through following projects: TR31078 and TR31041.

References

Ahuja, M. R. (2011). Strategies for conservation of germplasm in endemic redwoods in the face of climate change: A review. *Plant Genetic Resources: Characterization and Utilization, 9*(3), 411–422.

Chiras, D. (1991). *Environmental sciences: Action for a sustainable future*. Textbook.

Eriksson, G. (2005). Evolution and evolutionary factors, adaptation and adaptability. In T. H. Geburek & J. Turok (Eds.), *Conservation and management of forest genetic resources in Europe* (pp. 199–211). Zvolen: Arbora Publishers.

Gilmour, J. S. L., & Gregor, J. W. (1939). Demes: A suggested new terminology. *Nature*, 144–333.

Isajev, V. (1987). *Serbian spruce breeding on genetic and selection basis* (321 p.). PhD thesis. Faculty of Forestry University of Belgrade (in Serbian).

Jablonka, E., Oborny, B., Molnar, I., Kisdi, E., Hofbauer, J., & Czaran, T. (1995). The adaptive advantage of phenotypic memory in changing environments. *Philosophical Transactions: Biological Sciences, 350*(1332), 133–141.

Koskela, J. (2007). European forest genetic resources programme: A European approach to gene conservation of forest trees. In J. D. Simpson (compiler) Proceedings of the forum on conservation of forest genetic resources: Challenges, issues and solutions (28–29 July 2006, Charlottetown, PEI). Natural Resources Canada, Canadian Forest Service – Atlantic Forestry Centre, Fredericton. Information Report M-X-220, 23–30.

Ledig, F. T. (1988). The conservation of diversity in forest trees. *Bioscience, 38*, 471–479.

Loo, J., Fady, B., Dawson, I., Vinceti, B., & Baldinelli, G. (2011). Climate change and forest genetic resources – State of knowledge, risks and opportunities. In *Commission on genetic resources for food and agriculture* (29p.).

MAEP (Ministry of Agriculture and Environmental Protection). (2016). *First biennial update report (FBUR) to the UN framework convention on climate change* (p. 93). Belgrade: Djura print (in Serbian).

Mátyás, C. S. (2005). Expected climate instability and its consequences for conservation of forest genetic resources. In T. H. Geburek & J. Turok (Eds.), *Conservation and management of forest genetic resources in Europe* (pp. 465–476). Slovakia: Arbora Publishers.

Mátyás, C. (2006). The missing link: Synthesis of forest genetics and ecological research in view of challenges of environmental change. In G. von Wühlisch (Ed.), *Forest genetics and its contribution to sustainability in forestry and to consumer protection* (Mitt. BFH, Nr 221, pp. 1–14). Hamburg: Kommissionsverlag.

Milovanović, J. (2007). *Investigation of Serbian spruce intraspecific variability applying genetic markers* (86 p.). MSc thesis. Faculty of Forestry University of Belgrade (in Serbian).

Milovanović, J., Arandjelović, M., & Radojević, U. (2015). Climate change adaptations – biolgical resources conservation challenge. In: Milovanović, J., & Djordjević, S. (Eds.) *Conservation and enhancement of biological resources in the service of ecoremediation, Ministry of Education, Science and technological Development of the Republic of Serbia and Faculty of Applied Ecology Futura Singidunum University Belgrade* (pp. 322–354). (in Serbian).

Milovanović, J., & Šijačić-Nikolić, M. (2006). Global climate changes and forest genetic resources conservation, international scientific conference in occasion of 60 year of operation of Institute of Forestry, Belgrade, Serbia «Sustainable use of forest ecosystems – The challenge of the 21st century. 8–10th November. Donji Milanovac, Serbia, The Book of Abstracts, 142.

Milovanović, J., & Šijačić-Nikolić, M. (2007). Serbian spruce genetic conservation model applying MPBS method. *Bulletin of the Faculty of Forestry, 95*, 119–126 (in Serbian).

Milovanović, J., & Šijačić-Nikolić, M. (2008). Forest tree species natural genetic diversity assesment as a tool for conservation and sustainable use. In *III international symposium of ecologists of the Republic of Montenegro, Hotel Delfin, Bijela, Herceg Novi, 08–12.10.* Book of abstacts and programe, 81.

Milovanović, J., & Šijačić-Nikolić, M. (2009). Eco-genetic phenomenon as a possible explanation of Serbian spruce intraspecific diversity pattern. In *V Balkan botanical congress. 7–11. September. Belgrade. Serbia*. Book of abstracts, 62.

Milovanović, J., Radojević, U., & Šijačić-Nikolić, M. (2012). Conservation of Serbian spruce genetic resources applying environmental modeling. In *International conference role of research in sustainable development of agriculture and rural areas. May 23–26, Montenegro*. Book of abstracts, 116.

Milovanović, J., Radojević, U., & Šijačić-Nikolić, M. (2017). Linking ecosystem variability and carbon sequestration: Estimation of sequestered carbon in natural forests and perennial crops. In M. Lukac, P. Grenni, & M. Gamboni (Eds.), *Soil biological communities and ecosystem resilience,* (pp. 265–276). Springer International Publishing AG.

Namkoong, G. (1984). Strategies for gene conservation in forest tree breeding. In C. W. Yeatman, D. Kafton, & G. Wilkes (Eds.), *Plant genetic resources: A conservation imperative* (AAAS selected symp. 87, pp. 79–89). Boulder: Westview Press.

Orlović, S., Galić, Z., Matović, B., & Stojanović, D. (2014). Report on vulnerability and climate change adaptation plan in forestry. In *The draft of the second report of the Republic of Serbia under the UN framework convention on climate change* (in Serbian).

Šijačić-Nikolić, M., & Milovanović, J. (2010). *Conservation and sustainable use of forest genetic resources* (200 p.). Faculty of Forestry University of Belgrade, Planeta Print (in Serbian).

Šijačić-Nikolić, M., Milovanović, J., & Katičić-Trupčević, I. (2009). Model of Serbian spruce genetic diversity conservation applying MPBS method for adaptability improvement. *International Journal of Biodiversity and Conservation, 1*(1), 001–003.

Šijačić-Nikolić, M., Milovanović, J., & Nonić, M. (2014). Conservation of forest genetic resources. In M. R. Ahuja & Ramawat K. G. (Eds.), *Biotechnology and biodiversity,* (pp. 103–128). Springer.

St. Clair, J. B., & Howe, G. T. (2011). Strategies for conserving forest genetic resources in the face of climate change. *Turkish Journal of Botany, 35*, 403–409.

Stojanović, D., & Orlović, S. (2014). National adaptation plan – Forestry. In *The draft of the second report of the Republic of Serbia under the UN framework convention on climate change* (in Serbian).

Stojanović, D. B., Krzić, A., Matović, B., Orlović, S., Duputie, A., Djurdjević, V., Galić, Z., & Stojnić, S. (2013). Prediction of the European beech (*Fagus sylvatica* L.) xeric limit usinga regional climate model: An example from southeast Europe. *Agricultural and Forest Meteorology, 176*, 94–103.

Stojanović, D. B., Matović, B., Orlović, S., Kržič, A., Trudić, B., Galić, Z., Stojnić, S., & Pekeč, S. (2014). Future of the main important forest tree species in Serbia from the climate change perspective. *South-East European Forestry, 5*, 117–124.

Stojanović, D., Matović, B., & Orlović, S. (2015a). Trends of forest cover changes in the Republic of Serbia. *Šumarstvo, 3*, 89–98 (in Serbian).

Stojanović, D., Levanič, T., Matović, B., & Orlović, S. (2015b). Growth decrease and mortality of oak floodplain forests as a response to change of water regime and climate. *European Journal of Forest Research, 134*(3), 555–567.

Stojnić, S., Sass-Klaassen, U., Orlović, S., Matović, B., & Eilmann, B. (2013). Plastic growth response of European beech provenances to dry site conditions. *IAWA Journal, 34*, 475–484.

Stojnić, S., Orlović, S., Živković, U., von Wuehlisch, G., & Miljković, D. (2015a). Phenotypic plasticity of European beech (*Fagus sylvatica* L.) stomatal features under water deficit assessed in provenance trial. *Dendrobiology, 73*, 163–173.

Stojnić, S., Orlović, S., Miljković, D., Galić, Z., Kebert, M., & von Wuehlisch, G. (2015b). Provenance plasticity of European beech leaf traits under differing environmental conditions at two Serbian common garden sites. *European Journal of Forest Research, 134*, 1109–1125.

Tucović, A., & Isajev, V. (1988). *Generative seed orchard of Serbian spruce in Godovik* (Establishment project documentation, 10 p.). Faculty of Forestry University of Belgrade (in Serbian).

Chapter 29
Phenotypic Plasticity of European Beech from International Provenance Trial in Serbia

Dijana Čortan, Marina Nonić, and Mirjana Šijačić-Nikolić

Abstract European beech (*Fagus sylvatica* L.) is a common forest tree species in Europe, distributed across environmentally and climatically diverse regions, resulting in wide natural diversity. Considering its wide distribution and the fact that it is drought-susceptible species, beech received a lot of attention recently, in the light of global warming. Consequently, the provenance trial will provide valuable information about genetic variation patterns of the species, the level of local adaptation to changes in environmental conditions as well as information about the amount of adaptive traits plasticity of the studied species within two research periods. The plastic response of provenance morphological traits depends on differences in climatic conditions of the test site and seed origin, showing changes only if the shift happens towards warmer/drier climate, where higher plasticity were showed in older plants within drier conditions. The presented result also show the existence of considerable genetic variation among European beech provenances, even between provenances that originated from the same geographical region, where genetic differentiation among studied provenances showed an ecotypic pattern. The results show that phenotypic differentiation among populations is not always obvious. The existence of significant genetic variation within and among studied provenances could be crucial to beech adaptations processes in terms of changing climate conditions.

Keywords *Fagus sylvatica* L. · Provenance trial · Phenotypic plasticity · Morphological traits · Ecodistance transfer

D. Čortan (✉)
Faculty of Education, University of Novi Sad, Novi Sad, Sombor, Serbia
e-mail: dijanacortan@yahoo.com

M. Nonić · M. Šijačić-Nikolić
Faculty of Forestry, University of Belgrade, Belgrade, Serbia
e-mail: marina.nonic@sfb.bg.ac.rs; mirjana.sijacic-nikolic@sfb.bg.ac.rs

M. Šijačić-Nikolić et al. (eds.), *Forests of Southeast Europe Under a Changing Climate*, Advances in Global Change Research 65,
https://doi.org/10.1007/978-3-319-95267-3_29

29.1 Introduction

Given that magnitude of the expected increase in the frequency of extreme climate events, the rate of adaptation given by natural selection and migration may no longer keep pace with the climate change (Lamy et al. 2014). It is known that plants can adjust to the different environmental conditions through phenotypic plasticity, natural selection or they can migrate following conditions to which they are already adapted; however according to Nicotra et al. (2010) these options are not mutually exclusive. Adaptation to constant environmental changes is an important aspect of biotic response to rapid climate change. Forest tree species are specifically sensitive and vulnerable to climate change, considering their long life-span they does not allow a rapid adaptation to environmental changes (Lindner et al. 2010). Generally, it is known that higher level of genetic diversity within natural populations improves their adaptation ability underdisrupted climate conditions exhibiting their phenotypic plasticity.

Plasticity is usually applied in studies focusing on global climate changes impact on key plant species and their communities, where populations with low plasticity in selectively important traits, considered as narrowly adapted, might be at a higher risk of extinction (Valladares et al. 2007). Sultan (2000) noticed that many studies have shown plants plasticity for various ecologically important traits.

European beech (*Fagus sylvatica* L.) is one of the most important deciduous forest tree species in Europe. It is present under maritime and temperate climate conditions; while the pronounced cold, dry and continental climate seems to limit its wide distribution (Peuke et al. 2002; Bolte et al. 2007, 2016). Since beech has wide distribution across Europe, covering various climatic regions and due to its drought-susceptibility, beech was the focus of significant attention in the light of global warming (Fotelli et al. 2009). It is familiar that beech is highly correlated with annual mean temperature and precipitation, thus their growth and competitive ability may be strongly impacted, especially during intensive drought that occurs during the growing season. It is predicted (Kramer et al. 2010) that the beech distribution might potentially expand its northern edge and lose its southern edge habitats due to constant climate changes.

The wide distribution of climate-sensitive traits within species is studied in provenance trial, in which plants from different geographical origins are grown together at different locations indifferent climate conditions. Provenance trials provide valuable information about the level of local adaptation to changes in environmental conditions as well as information about the amount of adaptive traits plasticity within studied species.

The aim of this study was to (a) examine the amount and the patterns of inter-provenance genetic differentiation based on leaf morphological traits of European beech cultivated in provenance trial and (b) to evaluate leaf morphological plasticity of different provenances with reference to their adaptation to local climate conditions and to evaluate their phenotypic stability within two research periods.

29.2 Materials and Methods

29.2.1 Provenance Trial

Beech provenance trial at Debeli Lug (N 44°19′34.01″, E 21°52′20.39″, 742 m a.s.l.) is one of two provenance trials in Serbia, in the framework of the COST Action E52. The tests were established by planting 2 and 3-year-old seedlings of 22 beech provenances (Table 29.1, Fig. 29.1). The seedlings were planted in a spacing of 1 m between plants and 2 m between rows (details in Šijačić-Nikolić et al. 2013).

29.2.2 Weather Conditions in the Provenance Trial

Climate data were recorded at closest meteorological station Crni Vrh (N 44° 08′, E 21° 58′, 1027 m a.s.l.), for the year of bud set (2009, 2015) and for the growing seasons of interest for leaf measurements (2010 and 2017).

Mean air temperature fluctuation between studied years was about 1 °C in average (Graphic 29.1). While observing the crucial period of leaf development, spring and early summer (March–July), it was noted that the mean air temperature increased by about 2 °C from 2009 up to 2017 (10.78 °C, 10.12 °C, 11.54 °C and 12.7 °C, respectively), while precipitation patterns were not noted. Looking at whole vegetation period (April to September), the mean temperature was around 15.5 °C, without high fluctuations, while precipitation showed high fluctuations within this period, ranging from 283 to 471 mm. The 2017 was the hottest year and it was also a year with lowest sum of precipitation, 544.8 mm, while year before, 2016, recorded a really high sum of precipitation, 904.5 mm. However 2010 was the

Table 29.1 Basic data on the studied beech provenances

No	Proven. lable	Provenance name	Country	Year of seed collecting	Longitude	Latitude	Altitude (m)
1	CRO24	Sjeverni dilj Čaglinski	Croatia	2003	18°01′	45°17′	350
2	CRO25	Vrani Kamen	Croatia	2003	17°19′	45°37′	600
3	BIH32	Crni vrh, Tešanj	Bosnia and Hercegovina	2003	17°59′	44°33′	500
4	SRB36	Fruška Gora	Serbia	2003	19°55′	45°10′	370
5	SRB38	Kopaonik	Serbia	2003	20°50′	43°10′	820
6	HU42	Valkonya	Hungary	2003	16°45′	46°30′	300
7	DE47	Schelklingen	Germany	2003	10°00′	48°00′	650
8	DE48	Höllerbach	Germany	2003	13°14′	49°01′	755
9	DE49	Hasbruch	Germany	2003	08°26′	53°08′	35
10	AT56	Scharnstein, Mitterndorf	Austria	2003	13°58′	47°54′	480

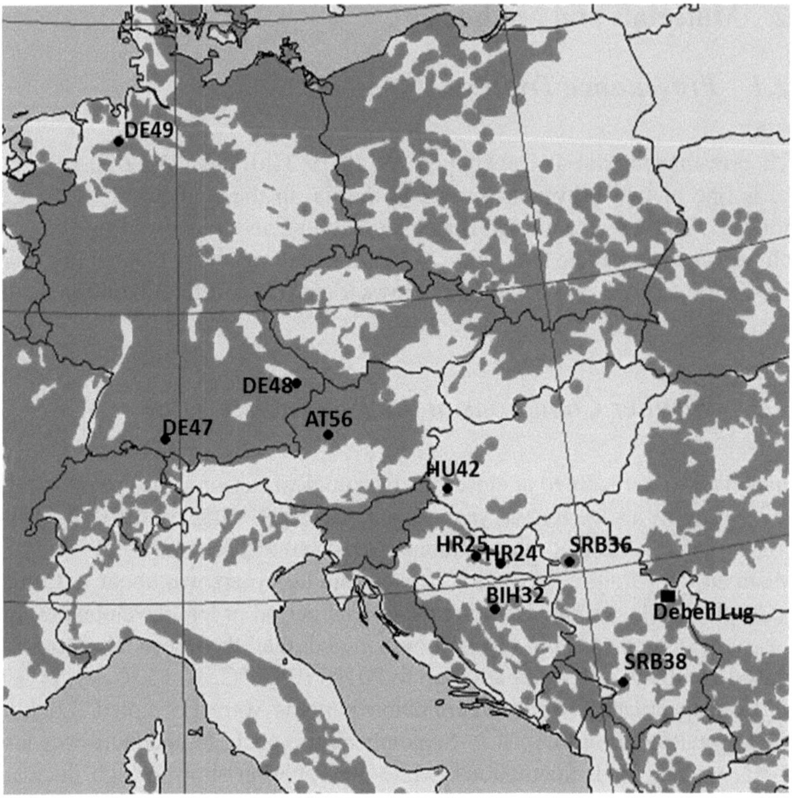

Fig. 29.1 Geographical distribution of the studied beech provenances
The symbol (•) represents provenances and symbol (■) the trial site at Debeli Lug, Serbia. The shaded area represents the natural distribution range of beech according to EUFORGEN. (von Wuelisch 2008)

year with highest precipitation, 1034.4 mm. The sum of precipitation in the dormant period (October to February) was higher in first research period (2009–2010) and amounted to 529.6 mm, while in the second period (2016–2017) it amounted only to 269.7 mm. The mean air temperature in dormant period reached 0.38 and −1.05 °C, respectively.

Furthermore Ellenberg's climate quotient (EQ), as a simple index denoting the joint effect of temperature and precipitation, is calculated for provenance origin localities (Table 29.2) and for provenance trial locality (Table 29.3). EQ has been generally used to express humidity conditions (Mátyás et al. 2010), and is defined as the mean temperature in the warmest month (July, T_{07}) divided by annual precipitation (P_a):

$$EQ = 1000\left(T_{07} / P_a\right)$$

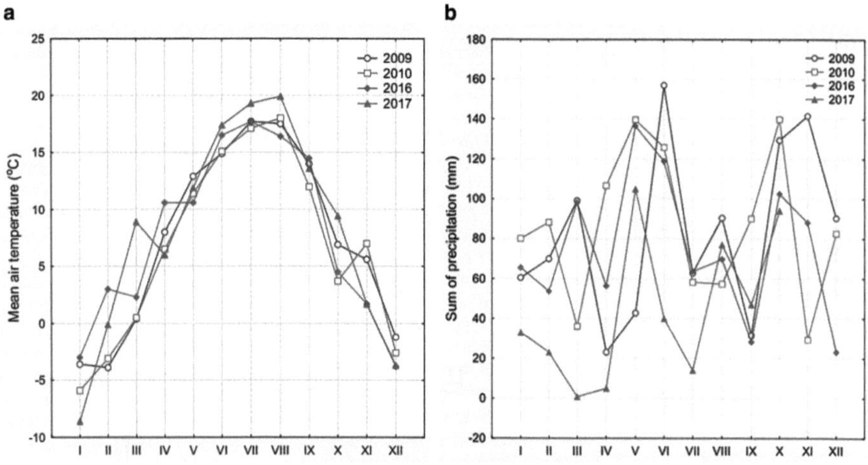

Graphic 29.1 (**a**) Mean air temperature (°C), (**b**) sum of precipitation (mm) during two periods 2009–2010 and 2016–2017

Table 29.2 Climate conditions and Ellenberg's climate index at the site of provenance origin

Provenance	Mean annual temp. T(°C)	Vegetation period mean temp. (°C)	July mean temp. (°C)	Sum of precipitation (mm)	Vegetation period precipitation (mm)	Ellenberg's index (EQ)	Mean ΔEQ
CRO 24	10.8	17.2	18.9	779	360	24.26	−1.95
CRO25	8.5	15.9	18.9	972	470	19.44	2.87
BIH 32	9.6	15.9	22.5	1069	579	21.05	1.26
SRB 36	11.2	18	21.5	782	351	27.49	−5.18
SRB 38	3.6	9	15.5	985	564	15.74	6.57
HU 42	9.5	17.8	20.1	800	500	25.13	−2.82
DE 47	6	14	16	840	260	19.05	3.26
DE 48	5	11.7	18.3	1200	580	15.25	7.06
DE 49	8.6	15	18	760	360	23.68	−1.37
AT 56	7.4	14.6	18.5	1263	706	14.65	7.66

29.2.3 Plant Material and Measurements

In the present study, ten provenances originating from six countries, Germany, Austria, Hungary, Croatia, Bosnia and Herzegovina, Serbia have been studied. Leaf samples were collected in mid summer in 2010 and 2017 from ten randomly selected trees per provenance. Analyses were carried out on 30 leaves per tree. Sampling was performed on shade-leafs from south-orientated branches at approximately first-thirds height of the tree. The analysis included the following six directly measured morphological traits (Fig. 29.2): LL – lamina length (cm), LW – lamina width (cm),

Table 29.3 Climate characteristics and Ellenberg's climate quotient of Debeli lug provenance trial

Year	Mean annual temp. T(°C)	Vegetation period mean temp. (°C)	July mean temp. (°C)	Sum of precipitation (mm)	Vegetation period precipitation (mm)	Ellenberg's index (EQ)
2009	7.5	15.4	17.7	998.3	384.4	17.73
2010	6.7	14.72	17.1	1034.4	471.4	16.53
2016	7.6	15.14	17.7	904.5	417.1	19.57
2017	8.0	16.42	19.3	544.8	283	35.43

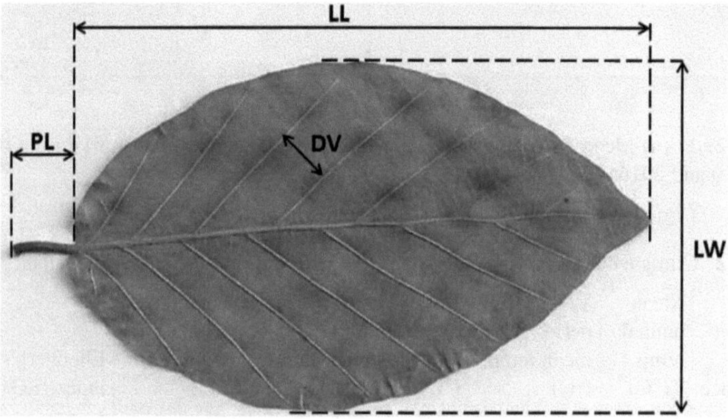

Fig. 29.2 Analyzed morphological leaf traits

PL = petiole length (cm), DV – distance between third and fourth vein – left (cm), NVL – number of veins on the left side of the leaf, NVR – number of veins on the right side of the leaf, as previously described in Šijačić-Nikolić et al. (2012).

Finally, the phenotypic plasticity index has been calculated for all measured leaf morphological traits in order to standardize the plasticity of various traits in present provenances and thereby allow for their objective comparison. Maximum and minimum values of leaf traits were extracted from each trait measurement. Plastic response of leaf morphological traits to different environmental condition of each research year was calculated as the difference between the maximum and the minimum values, divided by the maximum value (Valladares et al. 2000):

$$PI = \frac{\text{maximum value} - \text{minimum value}}{\text{maximum value}}$$

The index of phenotypic plasticity ranges from zero to one, in which zero represents no plasticity and one in maximum plasticity.

29.2.4 Data Analysis

All statistical data were processed with Statistica 10 (StatSoft, Inc.) data analysis software. Two-factorial ANOVA were used to estimate the statistical importance of different level sources of phenotypic variance of leaf morphological traits. A model with two phenotypic variance sources was produced for the analysis of factor in the total variation: year, provenance and "year by provenance" interaction.

Response to transfer has been analyzed through regression, calculating the individual response of provenances to respective ecodistance across origin sites and with mean ecodistance. Analysis of the polynomial regression included only traits that showed steep and relatively parallel reaction norms. These traits are considered as the ones with low level of genetic variability in their plasticity.

Relationships between provenances were examined through the principal component analysis (PCA). Principal components (PCs) with eigenvalues higher then 1.0 were used as they account for the majority of the variance in the whole data set (Dempster et al. 2013). Furthermore UPGMA (Unweighted Pair Group Method) cluster analysis (CA) was done in order to identify homogenous groups of present provenances using the square of the Euclidean distance.

29.3 Results and Discussion

29.3.1 Provenance Diversity

The evident rise of temperature, and especially summer rainfall deficits anticipated in the twenty-first century, may strongly affect species. Plants and their populations may significantly vary in the amount of their plastic responses to various environmental changes, which have been proved in many studies so far (Valladares et al. 2000, 2002, 2007; Sultan 2001; Balaguer et al. 2001). The results of ANOVA analysis showed significant difference in all levels of research. There are significant differences between different years of measurements, among studied provenances and among provenances within each of studied years (Table 29.4). The significant effect of a given year and "provenance by year interaction" on studied morphological traits suggests that beech provenances are capable to modify the leaf morphology according to present environmental conditions of provenance trial locality. The high genetic variability has been crucial factor for species long-term survival (Čortan et al. 2015).

The highest values of measured leaf size traits (LL, LW, PL) were noted in 2017, while the lowest values were noted in 2010, showing gradual growth of these traits during this research period (Graphic 29.2). In both years the provenances from Croatia (CRO24 and CRO25) were characterized by largest leaves, while the Austrian provenance AT56 had the smallest leaves in both studied years.

Table 29.4 F values of two-factorial ANOVA for morphological traits

	df	LL	PL	LW	DV	NVL	NVR	Total
Year	2	701.6***	1406.03***	526.88***	0.05***	5.2***	4.9***	287.91***
Provenance	9	58.3***	24.10***	35.37***	23.25***	6.4***	6.4***	19.33***
Year*provenance	18	6.4***	16.58***	10.14***	2.08***	7.2***	8.6***	7.35***

ns nonsignificant
$*p < 0.05$; $**p < 0.01$; $***p < 0.001$

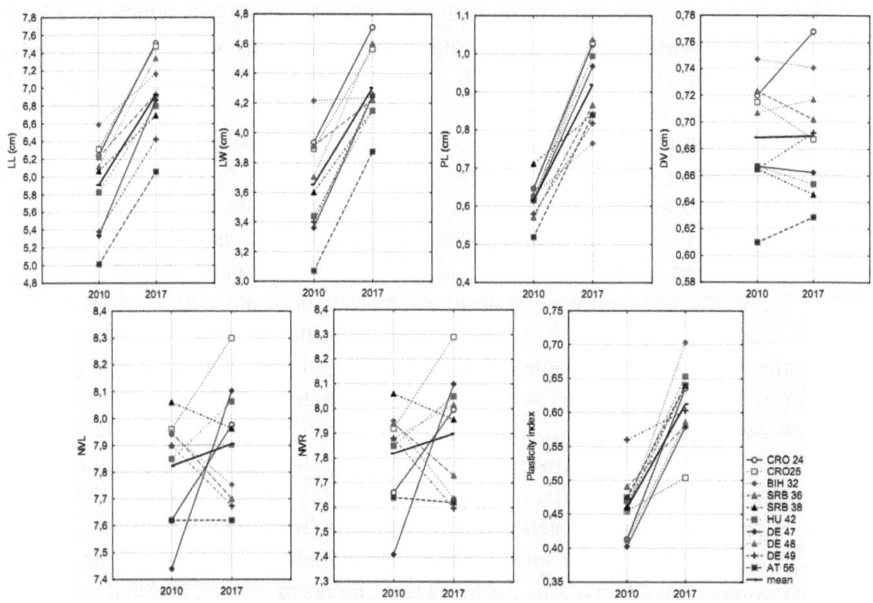

Graphic 29.2 Reaction norm plots for six measured morphological traits and calculated plasticity index of ten European beech provenances

29.3.2 Provenance Phenotypic Plasticity

Reaction norm is a specific way of representing a genotype expression variety across a range of environments, visualized as a line on two dimensional plots of environmental values versus the phenotypic values (Gratani 2014). The reaction norms (Graphic 29.2) are plotted together to present the pattern of genotypic and phenotypic diversity of studied provenances within and across different environments in two studied years (2010 and 2017). The average reaction norms were steep for each of the measured leaf size trait (LL, PL and LW) and for plasticity index (PI) within studied provenances, indicating a possible plastic response of each provenance to different conditions of two studied years. Relatively parallel reaction norms of these morphological traits indicate a low level of genetic variability in their plasticity. However, crossing reaction norms exhibit a particularly strong genotype-by-environment interaction (Fusco and Minelli 2010) and have an important effect on the amount of genetic variation perceived to be present in a trial. Looking at the reaction norms of these traits by each provenance, crossing reaction norms of different provenances could be noted, revealing the existence of low genetic variability in plasticity of these traits (DV, NVL and NVR). Also this crossing shows that different provenances dominate in different environments. Even though the environmental conditions of these two studied years are not significantly different, environmental variaty still may contribute to the maintenance of genetic variation (Ackerly et al. 2000).

The leaf size traits are highly sensitive to local site conditions which have been reported as the reason behind high morphological diversity of many tree species (Hatziskakis et al. 2011; Jarni et al. 2011; Paridari et al. 2013; Šijačić-Nikolić et al. 2013; Devetaković and Šijačić-Nikolić 2013; Čortan et al. 2013, 2014, 2015; Stojnić et al. 2016). Plant morphology diverges as a plant grows and develops from germination to maturity (Valladares et al. 2007; Valluru et al. 2012). Plasticity index has been widely used and is considered an important instrumentin understanding the occurrence and survival of plant species in heterogeneous and variable environments (Balaguer et al. 2001; Gratani et al. 2003; Valladares et al. 2006; Vieira et al. 2014). Plasticity index showed mean plasticity values of about 0.5 for all studied provenances. However, there were two provenances that showed higher mean plasticity index than other, provenance BIH32 (0.580) and provenance GER49 (0.582). In general, the plasticity index was higher in 2017 than in 2010, about 0.15, which could be explianed by the individuals maturity and different climate conditions in these 2 years. The studied traits in the present research which showed highest plasticity index were leaf size traits, while number of veins on both sides had the lowest values of plasticity index. In general, leaf size traits (width and length) are characteristics considered to be the most plastic ones, which are partially under genetic control (Čortan et al. 2015). This confirms reaction norm results, that leaf size traits are more under environmental influence, while the other traits are more genetically determined.

Physiological, morphological and anatomical plasticity may have a different role in plant adaptation to environmental changes. The climate, rainfall, altitude, relief, soil and light can impact plant morphological traits. Leaf morphology is a highly adaptive in highly stressed environment. For example, during the summer drought, leaf morphological plasticity plays an important role in resource supply and can limit photo-damages to the photosynthetic apparatus (Catoni et al. 2012). It could be considered as adaptive plasticity that favors the plant's fitness over time (Nicotra et al. 2010). Physiological leaf plasticity appears to be linked to an increased capacity to survive and grow in areas of potentially intense solar radiation (Valladares et al. 2002). Stojnić et al. (2015) pointed that low plasticity index in beech leaf anatomical traits shows limited plasticity of these traits in adapting to local environmental conditions of the studied provenance trial. Valladares et al. (2002) noted in the research on beech that the leaf morphological plasticity index showed higher plasticity (about 0.5) over anatomical (0.17) and physiological (0.34) traits that were quite low in shade tolerance research. The same authors (Valladares et al. 2007) also noted that low or even no plasticity might be adaptive when environmental change is not predictable or when no phenotypic response can overcome the environmental challenge.

29.3.3 Plasticity Response to Microclimatic Change

Mátyás et al. (2009) believes that "the phenotypic response to climate changes depends not only on the climate conditions where the population is actually grown or tested, but also on the ecodistance of transfer, i.e. on the magnitude and direction of environmental change experienced due to the transplanting to the test site, in relation to the macroclimate they had been adapted to originally". Ellenberg's climate quotient (EQ) has been selected to calculate ecodistance.

The response to transfer has been analyzed through regression, where only traits that showed steep and relatively parallel reaction norm were used (LL, PL, LW). These traits are the ones which indicate a low level of genetic variability in their plasticity, meaning that they are highly impacted by the environment. Regression was used to calculate the provenance response to respective ecodistance across origin stands and to mean ecodistance (ΔEQ), which presents the difference between the average of EQ value of trial site and the EQ of provenance origin (Mátyás et al. 2009). Polynomial regression of the leaf morphological traits from different provenances and their origin populations' EQ showed the positive slope for LL and LW, while for PL it was negative. The results did not show significant correlation for any of measured leaf size traits (results not shown).

Further more mean plasticity index of studied provenances were correlated with provenances EQ and ΔEQ (Graphic 29.3). Correlation between phenotypic plasticity index and EQ of provenance origin showed positive correlation index, $r = 0.5303$, showing that plasticity grows with higher EQ of provenances origin sites. The data and regression from Graphic 29.3 showed negative slope ($r = -0.5303$), indicating a decline of the plasticity index towards increasing warming. Provenances with lower altitude and higher EQ values had higher adaptation potential, while provenances with lower EQ from higher elevations, over 500 m a.s.l., showed the opposite trend. However these results did not show significant correlation ($p > 0.05$).

The provenances EQ ranged from 14 (AT56) to 27 (SRB36). The origin sites below 500 m a.s.l. had the EQ above 20, while sites above 500 m a.s.l. had the EQ below 20. The Austrian provenance AT56 was an exception, situated at 480 m a.s.l., which showed extremely low EQ (14.65), considering the high precipitation sum (1263 mm). The German provenance from the north edge of distribution, DE49 (760 mm) was a complete surprise with an extremely high EQ (23.68), which is most probably adapted to lower water deficit. The provenance trial site at Debeli lug, situated at 300 m a.s.l., had mean EQ for the last 10 years at about 24 (ranging from 14 up to 35), showing relatively humid conditions. However, 2017 had EQ at trial site over 35 – extremely xeric for beech, while 2010 had just 16 – humid climate. An EQ values less then 30 are characterized by a humid climate where beech dominates, while EQ greater then 30 is characterizes by dryer and warmer climate regions and are dominated by oak species (Jahn 1991; Fang and Lechowicz 2006; Stojanović et al. 2013). According to Jahn (1991), values below 20 indicates a pure beech climate in the "prevailing range"; the competitive vigor of beech gradually decrease with an increase from 20 to 30 in the "prevailing range"; and in "exclusion

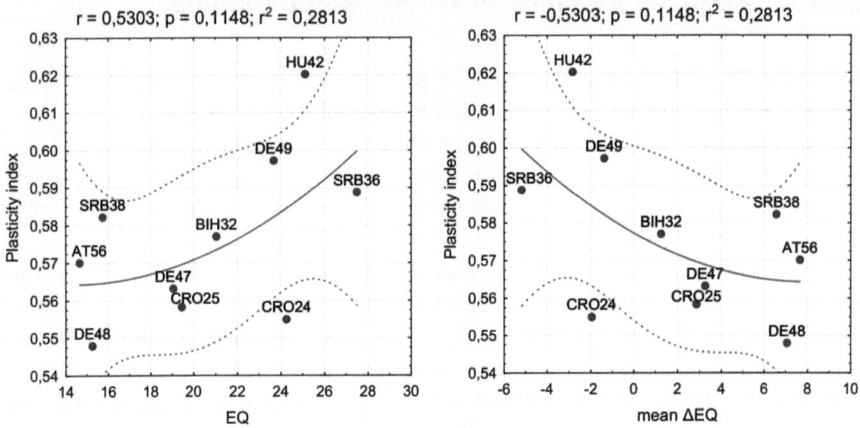

Graphic 29.3 General response regression for plasticity index and EQ of provenance origin sites; plasticity index and ΔEQ for both measured years ($p < 0.05$)

range" over 30 beech presence is strongly reduced. It is noted that beech xeric limit in Serbia for the period of 1961–1990 varied between EQ values of 20.1 and 29.5 (Stojanović et al. 2013). According to presented results six provenances have been transferred into an environment characterized with higher average temperatures and higher drought conditon then their original climate (EQ < 20), while other four provenances came into a less stressful environment than their original one (EQ > 20). In general forest stands, survive single extreme summers and recover merely with minimum yield loss, due to deep rooting of trees with the utilization of deeper soil water resources. However, Mátyás et al. (2010) found that summer droughts of three to four consecutive years, recurrent in short period of time, in which the EQ value was above 40–42, resulted in mass beech mortality, in pest and disease attacks. Even though the EQ for trial locality was on average 24 for last 10 years, showing "prevailing range", there is no more than two out of four consecutive dry years in this period (EQ over 30). Despite extreme weather conditions over last decade no mass mortality of beech has been recorded in Serbia (Stojanović et al. 2013).

The regression of plasticity vs. the mean ecodistance of transfer (ΔEQ) showed the decline of plasticity potential of leaf morphological traits with increasing warming (Graphic 29.3). The results correspond to a relationship between the origin climate and the character of response to the changing environment. The provenances that had positive ΔEQ values indicate decline due to a warming effect following the transfer. European beech provenances from dryer regions were better adapted to drought than those from more humid regions, which was also confirmed by Bolte et al. (2007). Matyas et al. (2010) also confirmed that negative response to changing environmental conditions is triggered only if the transfer happens towards warmer/ drier climate. Present provenances with positive ΔEQ value, the one from higher altitudes, showed lower plasticity index considering that they transfer in more xeric climate conditions than those with negative ΔEQ values. However this regression did not show significant results.

For many tree species growth rates are highest at test sites with temperatures similar to the source locations, while growth tends seems to be less and mortality greater in considerable warmer or colder locations (Davis et al. 2005). Height growth depends on the climatic distance which the adapted populations crossed (ΔEQ) and on microclimatic environment of locality where they are tested (trial EQ) (Mátyás et al. 2009). So the increasing, positive EQ values, that show warming and more arid conditions lead to height growth and vitality decrease. Horváth et al. (2016) noted that studied provenances within their trial which had high or very high positive ecodistance values (above 9 ΔEQ), indicated a serious decline of beech seedlings vitality towards increasing warming. Transferring provenances to warmer and drier conditions Mátyás et al. (2009) noted clear decline of height growth starting from ΔEQ values 4. It was already stated in several researches (Rose et al. 2009; Fotelli et al. 2009) that marginal beech provenances may express better drought adaptation capabilities. Fotelli et al. (2009) noted that southern beech populations showed greater eco-physiological adaptive capacity as answer to drought conditions than populations in Central Europe. Furthermore, the southern provenances of European beech have tendency to flush earlier than northern ones (Jazbec et al. 2007; Ivanković et al. 2011), showing their adaptation to warmer climatic conditions. These results partially corresponded to the results obtained for height and for survival. Bolte et al. (2016) believe that beech range marginality may be due to climatic conditions and that genetic variation has a major impact on the variable adaptive potential of the studied populations. Moreover, in research done by Peuke et al. (2002) all provenances from wet habitats responded like drought-sensitive ecotypes, but not all provenances from dry habitats responded like drought-tolerant ecotypes. They suggested that only provenances from dry habitats with qualities of drought-tolerant ecotypes could be used for reforestation considering that their provenances from wet habitat responded like drought-sensitive.

29.3.4 Provenance Distribution Along Climatic Gradient

Relationships between provenances were investigated within PCA for both studied years, where total variance has been explained by six PCs. However, we have used only PCs with eigenvalues greater than one as measurement for the significance. Eigenvalues indicate the amount of variance explained by each component. First tree PCs showed eigenvalues greater then one (3.08, 1.61 and 1.15, respectively). The three PCs explained approximately 97% of the total variation among studied European beech provenances, where the first two PCs accounted 87.17% of the multi-trait variation.

Table 29.5 Eigenvalues, proportion of total variability and correlation between the original variables and the first three principal components

Trait	Factor 1	Factor 2	Factor 3
LL	−0.352	**−0.618**	**0.687**
PL	**−0.800**	0.514	−0.140
LW	**−0.830**	−0.432	−0.340
DV	**−0.628**	−0.592	−0.501
NVL	**−0.820**	0.061	0.543
NVR	**−0.746**	0.653	0.030
Eigenvalue	**3.08**	**1.61**	**1.15**
% Total	**51.29**	**26.88**	**19.25**
Cum %	3.08	4.69	5.85
Cum. Eigenval	51.29	78.17	97.42

Considering everything stated we have presented only the first three PCs in Table 29.5 and in the rest of analysis. PC1 accounted for 52.29% of the total variance. The highest contribution on PC1 corresponded to variables related to leaf size: PL (−0.800) and LW (−0.830), while LL (−0.618 and 0.687) contributed with relatively high scores to PC2 and PC3 which accounted for 26.88% and 19.25% of the total variancerespectively. High scores on PC1 were also determined for the number of veins on both leaf sides (NVL = −0.820 and NVR = −0.746), and relatively high score was determined for DV (−0.628).

The scatterplots of first tree and first two PCs (Graphic 29.4) show geometrical distances among European beech provenances, which indicates a genetic similarity and relations among provenances at the level of measured leaf morphological traits. Having in mind that first tree PCs explained the majority of total variation among studied provenances, approximately 97%, it is considered that multivariate diversity of the provenances on the three PC axes is acceptable for the most important distinguishing traits. However considering that tree-dimensional scatterplot does not show clear distribution of the provenances, provenance projection was presented on two-dimensional plane with PC1 and PC2, accounting for 78.18% of total variance.

The provenance distribution on the PC1 and PC2 scatter plot showed that starting from the negative values toward positive values of PC1, the beech provenances showed a general decrease in leaf size traits (LW, PL) and number of veins (NVL and NVR). Starting from negative values toward positive values of PC2, the provenances decreased in leaf length values (LL).

Cluster analysis was conducted on the basis of mean values of studied traits for both measured years. UPGMA cluster analysis was applied to standardized data of six leaf morphological traits in order to group provenances into homogeneous groups according to their similarities. However there is no grouping pattern consistent with geographical origin of studied provenances within the studied years. Projection of provenances on a two-dimensional plane, based on the first two PCs, substantially confirmed the results of the cluster analysis (Graphic 29.5).

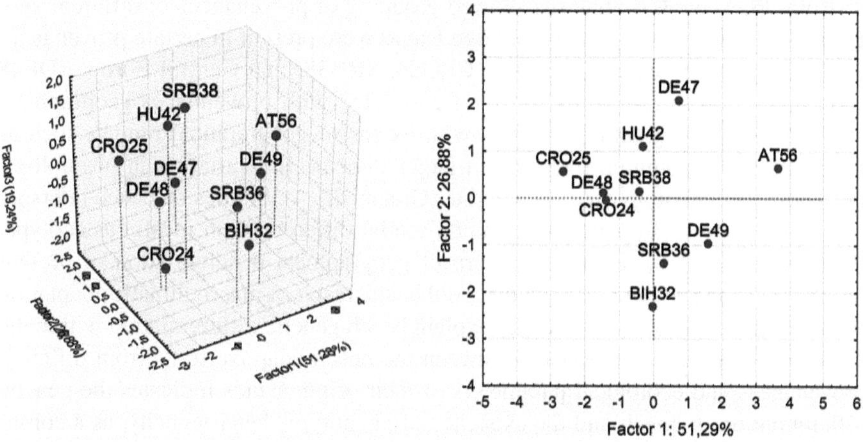

Graphic 29.4 Distribution of ten studied beech provenances (mean 2010–2017) in three dimensional (left) and two-dimensional (right) scatter plot showing relationship between provenances and PCs

Graphic 29.5 Dendrogram for the studied European beech provenances based on squared Euclidian distance using UPGMA method

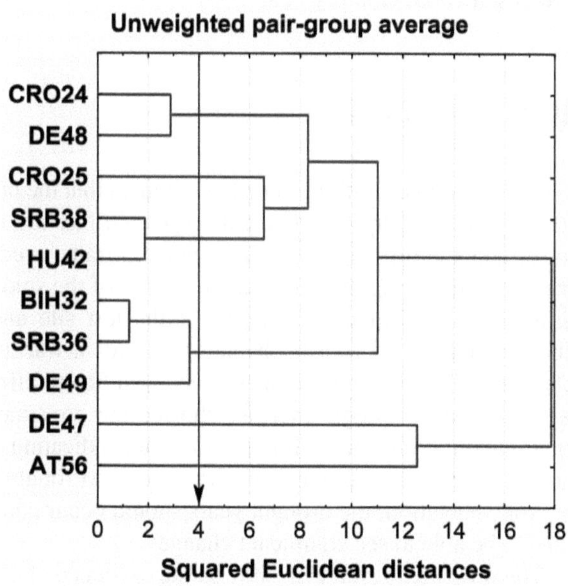

Both, PCA and CA analyses showed grouping of provenances of different geographical origin. It was found that large leaves were present in certain provenances from the Western Balkans (CRO25, CRO24, SRB38) and Central Europe (DE48 and HU42). These size traits were not correlated with provenances geographical origin and their EQ values. Species with extensive geographical ranges, such as beech, have the potential to exhibit a larger intraspecific variation in physiology, morphology, phenology, and growth rate (Gratani 2014). Studies on beech interspecific genetic variation patterns explained spatial differentiation mostly as ecotypic (Mátyás et al. 2010). Ecotypic pattern of geographical differentiation of present beech provenances within this study could be the result of macroclimatic adaptation to local site conditions which could explain beech genetic differentiation within the studied regions. The relationship between the beech adaptive traits from different provenances and ecological parameters of their origin stands indicates the genetic differentiation of beech and the existence of phenotypic heterogeneity, as a consequence of the population adaptation to local environmental conditions (Šijačić-Nikolić et al. 2012). Ecotypic variation has been present in many researches on height growth variability, survival, flushing phenology, leaf morphology (Chmura and Rozkowski 2002; Ivanković et al. 2008; Jazbec et al. 2007; Šijačić-Nikolić et al. 2012; Stojnić et al. 2016), while clinal genetic differentiation of beech populations was noted in researches mostly based on phenology or genetic markers (Nielsen and Jorgensen 2003; Gömöry et al. 2007).

29.4 Conclusion

Having in mind everything stated it is evident that the presented provenance that had transfer to more dry environment responded like drought-sensitive ones, showing lower adaptability, then those provenances transferred into less stressful environment then their original. Thus, the response of the studied provenances depends on differences of climatic conditions of the test site and the seed origin, showing changes only if the transfer happens towards a warmer/drier climate. Phenotypic plasticity in two different years showed significant differences where second period with significant drought increase showed the provenance ability to adapt to local environmental changes within a trial site, indicating drought resistance of these provenances. However, the response of trees to drought is not immediate, as in grass or crop vegetation; the drought years should occur consecutively in short period of time to be able to see significant changes.

The presented result also showed the existence of considerable genetic variation within and among European beech provenances, even between provenances which originated from the same geographical region. PCA revealed great genetic distance between the provenances from the same geographic regions revealing an ecotypic pattern of provenance geographical differentiation, explaining provenance specificity. The existence of significant genetic variation within and among studied provenances could be crucial for beech adaptations processes in relation to changing climate conditions.

Acknowledgements The performed research was conducted within the project "*Establishment of forest plantations to increase the afforested areas in Serbia*" (TR31041) and project "*Studying climate change and its influence on the environment: impacts, adaptation and mitigation*" (43007) financed by the Ministry of Education, Science and Technological Development of the Republic of Serbia. Many thanks to English Professor Jelena Mladjenović for language revision.

References

Ackerly, D. D., Dudley, S. A., Sultan, S. E., Schmitt, J., Coleman, J. S., Linder, C. R., et al. (2000). The evolution of plant ecophysiological traits: Recent advances and future directions: New research addresses natural selection, genetic constraints, and the adaptive evolution of plant ecophysiological traits. *BioScience, 50*(11), 979–995.

Balaguer, L., Martínez-Ferri, E., Valladares, F., Pérez-Corona, M. E., Baquedano, F. J., Castillo, F. J., & Manrique, E. (2001). Population divergence in the plasticity of the response of Quercus coccifera to the light environment. *Functional Ecology, 15*(1), 124–135.

Bolte, A., Czajkowski, T., & Kompa, T. (2007). The north-eastern distribution range of European beech—A review. *Forestry, 80*(4), 413–429.

Bolte, A., Czajkowski, T., Cocozza, C., Tognetti, R., de Miguel, M., Pšidová, E., et al. (2016). Desiccation and mortality dynamics in seedlings of different European beech (*Fagus sylvatica* L.) populations under extreme drought conditions. *Frontiers in Plant Science, 7*, 751.

Catoni, R., Gratani, L., & Varone, L. (2012). Physiological, morphological and anatomical trait variations between winter and summer leaves of Cistus species. *Flora-Morphology, Distribution, Functional Ecology of Plants, 207*(6), 442–449.

Chmura, D. J., & Rozkowski, R. (2002). Variability of beech provenances in spring and autumn phenology. *Silvae Genetica, 51*(2–3), 123–127.

Čortan, D., Šijačić-Nikolić, M., & Knežević, R. (2013). Variability of leaves morphological traits in black poplar (*Populus nigra* L.) from two populations in Vojvodina. *Šumarstvo, 65*, 193–202.

Čortan, D., Šijačić-Nikolić, M., & Knežević, R. (2014). Variability of morphometric leaf characteristics of Black poplar from the area of Vojvodina. *Glasnik Sumarskog Fakulteta, 109*, 63–72.

Čortan, D., Tubić, B., Šijačić-Nikolić, M., & Borota, D. (2015). Variability of black poplar (*Populus nigra* L.) leaf morphology in Vojvodina, Serbia. *Šumarski List, 139*(5–6), 245–251.

Davis, M. B., Shaw, R. G., & Etterson, J. R. (2005). Evolutionary responses to changing climate. *Ecology, 86*(7), 1704–1714.

Dempster, M., Dunlop, P., Scheib, A., & Cooper, M. (2013). Principal component analysis of the geochemistry of soil developed on till in Northern Ireland. *Journal of Maps, 9*(3), 373–389.

Devetaković, J., & Šijačić-Nikolić, M. (2013). Variability of morphometric characteristics of the leaves of European white elm from the area of great war island. *Glasnik Šumarskog Fakulteta, 107*, 57–69.

Fang, J., & Lechowicz, M. J. (2006). Climatic limits for the present distribution of beech (*Fagus* L.) species in the world. *Journal of Biogeography, 33*(10), 1804–1819.

Fotelli, M. N., Nahm, M., Radoglou, K., Rennenberg, H., Halyvopoulos, G., & Matzarakis, A. (2009). Seasonal and interannual ecophysiological responses of beech (*Fagus sylvatica*) at its south-eastern distribution limit in Europe. *Forest Ecology and Management, 257*(3), 1157–1164.

Fusco, G., & Minelli, A. (2010). Phenotypic plasticity in development and evolution: Facts and concepts. *Philosophical Transactions of the Royal Society B, 365*, 547–556.

Gessler, A., Keitel, C., Kreuzwieser, J., Matyssek, R., Seiler, W., & Rennenberg, H. (2007). Potential risks for European beech (*Fagus sylvatica* L.) in a changing climate. *Trees, 21*(1), 1–11.

Gömöry, D. (2009). *Geographic patterns in the reactions of beech provenances to transfer*. Report for COST E52 meeting, Thessaloniki, Greece, p. 8.

Gömöry, D., Paule, L., & Vysny, J. (2007). Patterns of allozyme variation in western Eurasian Fagus. *Botanical Journal of the Linnean Society, 154*, 165–174.

Gratani, L., Meneghini, M., Pesoli, P., & Crescente, M. F. (2003). Structural and functional plasticity of *Quercus ilex* seedlings of different provenances in Italy. *Trees, 17*(6), 515–521.

Gratani, L. (2014). Plant phenotypic plasticity in response to environmental factors. *Advances in Botany, 2014*, 208747.

Hansen, J., Ruedy, R., Sato, M., & Lo, K. (2006). *GISS surface temperature analysis. Global temperature trends: 2005 summation*. New York: NASA Goddard Institute for Space Studies and Columbia University Earth Institute.

Hatziskakis, S., Tsiripidis, I., & Papageorgiou, A. C. (2011). Leaf morphological variation in beech (*Fagus sylvatica* L.) populations in Greece and its relation to their post-glacial origin. *Botanical Journal of the Linnean Society, 165*(4), 422–436.

Horváth, A., & Mátyás, C. (2016). The decline of vitality caused by increasing drought in a beech provenance trial predicted by juvenile growth. *South-east European forestry, 7*(1), 21–28.

Ivanković, M., Bogdan, S., & Božič, G. (2008). European Beech (*Fagus Sylvatica* L.) height growth variability in Croatian and Slovenian provenance trials. *Šumarski List, 132*(11–12), 529–541.

Ivanković, M., Popović, M., Katičić, I., Wuehlisch, G. V., & Bogdan, S. (2011). Kvantitativna genetska varijabilnost provenijencija obične bukve (*Fagus sylvatica* L.) iz jugoistočne Europe. *Šumarski List, 135*(13), 25–36.

Jahn, G. (1991). Temperate deciduous forests of Europe. In E. Rohrig & B. Ulrich (Eds.), *Ecosystems of the world 7. Temperate deciduous forests* (pp. 377–502). London: Elsevier.

Jarni, K., Westergren, M., Kraigher, H., & Brus, R. (2011). Morphological variability of Fraxinus angustifolia Vahl in the north-western Balkans. *Acta Societatis Botanicorum Poloniae, 80*(3), 245–252.

Jazbec, A., Šegotić, K., Ivanković, M., Marjanović, H., & Perić, S. (2007). Ranking of European beech provenances in Croatia using statistical analysis and analytical hierarchy process. *Forestry, 80*(2), 151–162.

Kramer, K., Degen, B., Buschbom, J., Hickler, T., Thuiller, W., Sykes, M. T., & de Winter, W. (2010). Modelling exploration of the future of European beech (*Fagussylvatica* L.) under climate change—Range, abundance, genetic diversity and adaptive response. *Forest Ecology and Management, 259*(11), 2213–2222.

Lamy, J. B., Delzon, S., Bouche, P. S., Alia, R., Vendramin, G. G., Cochard, H., & Plomion, C. (2014). Limited genetic variability and phenotypic plasticity detected for cavitation resistance in a Mediterranean pine. *New Phytologist, 201*(3), 874–886.

Lindner, M., Maroschek, M., Netherer, S., Kremer, A., Barbati, A., Garcia-Gonzalo, J., et al. (2010). Climate change impacts, adaptive capacity, and vulnerability of European forest ecosystems. *Forest Ecology and Management, 259*(4), 698–709.

Mátyás, C., Bozic, G., Gömöry, D., Ivankovic, M., & Rasztovits, E. (2009). Transfer analysis reveals macroclimatic adaptation of European beech (*Fagus sylvatica* L.). *Acta Silvatica et Lignaria Hungarica, 5*, 47–62.

Mátyás, C., Berki, I., Czúcz, B., Gálos, B., Móricz, N., & Rasztovits, E. (2010). Future of beech in Southeast Europe from the perspective of evolutionary ecology. *Acta Silvatica et Lignaria Hungarica, 6*, 91–110.

Nicotra, A. B., Atkin, O. K., Bonser, S. P., Davidson, A. M., Finnegan, E. J., Mathesius, U., et al. (2010). Plant phenotypic plasticity in a changing climate. *Trends in Plant Science, 15*(12), 684–692.

Nielsen, C. N., & Jørgensen, F. V. (2003). Phenology and diameter increment in seedlings of European beech (*Fagus sylvatica* L.) as affected by different soil water contents: Variation between and within provenances. *Forest Ecology and Management, 174*(1), 233–249.

Paridari, I. C., Jalali, S. G., Sonboli, A., Zarafshar, M., & Bruschi, P. (2013). Leaf macro-and micro-morphological altitudinal variability of *Carpinus betulus* in the Hyrcanian forest (Iran). *Journal of Forestry Research, 24*(2), 301–307.
Peuke, A. D., Schraml, C., Hartung, W., & Rennenberg, H. (2002). Identification of drought-sensitive beech ecotypes by physiological parameters. *New Phytologist, 154*(2), 373–387.
Rose, L., Leuschner, C., Köckemann, B., & Buschmann, H. (2009). Are marginal beech (*Fagus sylvatica* L.) provenances a source for drought tolerant ecotypes? *European Journal of Forest Research, 128*(4), 335–343.
Šijačić-Nikolić, M., Milovanović, J., Nonić, M., Knežević, R., & Babić, V. (2012). Ecotypical characterization of genetic variation of beech provenances from South-Eastern Europe based on the morphometric characteristics of leaves. *Glasnik Sumarskog Fakulteta, 106*, 197–214.
Šijačić-Nikolić, M., Milovanović, J., Nonić, M., Knežević, R., & Stanković, D. (2013). Leaf morphometric characteristics variability of different beech provenances in juvenile development stage. *Genetika, 45*(2), 369–380.
Stojanović, D. B., Kržič, A., Matović, B., Orlović, S., Duputie, A., Djurdjević, V., Galic, Z., & Stojnić, S. (2013). Prediction of the European beech (*Fagus sylvatica* L.) xeric limit using a regional climate model: An example from southeast Europe. *Agricultural and Forest Meteorology, 176*, 94–103.
Stojnić, S., Orlović, S., Miljković, D., Galić, Z., Kebert, M., & von Wuehlisch, G. (2015). Provenance plasticity of European beech leaf traits under differing environmental conditions at two Serbian common garden sites. *European Journal of Forest Research, 134*(6), 1109–1125.
Stojnić, S., Orlović, S., & Miljković, D. (2016). Intra-and interprovenance variations in leaf morphometric traits in European beech (*Fagus sylvatica* L.). *Archives of Biological Sciences, 68*, 64.
Sultan, S. E. (2000). Phenotypic plasticity for plant development, function and life history. *Trends in Plant Science, 5*(12), 537–542.
Sultan, S. E. (2001). Phenotypic plasticity for fitness components in Polygonum species of contrasting ecological breadth. *Ecology, 82*(2), 328–343.
Valladares, F., Wright, S. J., Lasso, E., Kitajima, K., & Pearcy, R. W. (2000). Plastic phenotypic response to light of 16 congeneric shrubs from a Panamanian rainforest. *Ecology, 81*(7), 1925–1936.
Valladares, F., Chico, J., Aranda, I., Balaguer, L., Dizengremel, P., Manrique, E., & Dreyer, E. (2002). The greater seedling high-light tolerance of *Quercus robur* over *Fagus sylvatica* is linked to a greater physiological plasticity. *Trees, 16*(6), 395–403.
Valladares, F., Sanchez-Gomez, D., & Zavala, M. A. (2006). Quantitative estimation of phenotypic plasticity: Bridging the gap between the evolutionary concept and its ecological applications. *Journal of Ecology, 94*(6), 1103–1116.
Valladares, F., Gianoli, E., & Gómez, J. M. (2007). Ecological limits to plant phenotypic plasticity. *New Phytologist, 176*(4), 749–763.
Valluru, R., Link, J., & Claupein, W. (2012). Consequences of early chilling stress in two Triticum species: Plastic responses and adaptive significance. *Plant Biology, 14*(4), 641–651.
Vieira, W. D. L., Boeger, M. R. T., Cosmo, N. L., & Coan, A. I. (2014). Leaf morphological plasticity of tree species from two developmental stages in araucaria forest. *Brazilian Archives of Biology and Technology, 57*(4), 476–485.
von Wuehlisch, G. (2008). *EUFORGEN technical guidelines for genetic conservation and use for European beech (Fagus sylvatica)* (p. 6). Rome: Bioversity International.
Żółkoś, K., & Meissner, W. (2009). A simple technique of random leaf collecting for biometric studies in a tree stand. *Biodiversity: Research and Conservation, 15*, 29–34.

Chapter 30
Serbian Spruce and Climate Change: Possible Outcomes and Conservation Strategy

Vladan Ivetić and Jelena M. Aleksić

Abstract Serbian spruce is an old, relict and currently rare and endangered spruce species particularly susceptible to climate change that attracts a lot of attention in the academic community since its discovery in 1875. Today, Serbian spruce is limited to the area of ~100 km^2 localized around the mid-course of the Drina River in the mountainous central Balkans, at the border of Republic of Serbia and Republic of Srpska, Bosnia and Herzegovina. Within this small area, ~30 remnant populations of various sizes are scattered at north-to-northwest orientated slopes of hills. Given the current state of Serbian spruce populations as well as extreme climate events and poor natural regeneration in this species, applied "do not touch" conservation approach is severely questioned, and some alternative protection and in situ and ex situ conservation actions have been put forward. In situ actions should involve: (1) facilitation of natural regeneration by selective removal of individual competitor trees, (2) assisted natural regeneration by planting high quality seedlings throughout the species current natural range or at nearby suitable sites, and (3) assisted natural regeneration by direct seeding. Although in situ actions may provide short-term persistence of Serbian spruce, ex situ actions, i.e., assisted migration, achieved by both assisted range expansion and assisted species migration, are inevitable, since it is very likely that Serbian spruce will disappear from its natural habitats in the near future due to the rapid climate change.

Keywords Serbian spruce · *Picea omorika* · Climate change · Assisted migration

V. Ivetić (✉)
Faculty of Forestry, University of Belgrade, Belgrade, Serbia
e-mail: vladan.ivetic@sfb.bg.ac.rs

J. M. Aleksić
Institute of Molecular Genetics and Genetic Engineering, University of Belgrade, Belgrade, Serbia
e-mail: aleksic_jelena@yahoo.com.au

M. Šijačić-Nikolić et al. (eds.), *Forests of Southeast Europe Under a Changing Climate*, Advances in Global Change Research 65,
https://doi.org/10.1007/978-3-319-95267-3_30

30.1 Introduction

Serbian spruce (*Picea omorika* (Panč.) Purk.) is a extremely narrow ranged tertiary relict tree species growing naturally in the central Balkans (Ivetić and Aleksić 2016; Aleksić et al. 2017a). Since its discovery in the late nineteenth century (Pančić 1887), it attracts a lot of attention in the academic community, and that has led to more than 600 scientific reports on this species published until 1996 which elaborate various aspects of its taxonomy, systematics, ecology, genetics, conservation, etc. (Fukarek 1967; Šijak and Dinić 1996). Nowadays, the main scientific focus has shifted towards response of this rare and endangered species to the global climate change.

Serbian spruce natural range and the number of mature individuals declines constantly, and nowadays, ~30 remnant fragmented populations of various sizes is present around the mid-course of the river Drina on border between Serbia and Republic of Srpska, Bosnia and Herzegovina. This decline is a result of various causes like illegal harvest, fires, poor natural regeneration, and competition (Kolarović 1951). The illegal harvest, which posed a severe threat in the past, stopped during the mid-twentieth century when all populations and individuals of Serbian spruce have been protected by the law. Forest fires still represents a threat, and there are two documented events that have caused drastic reduction in the number of Serbian spruce populations. The first one was more severe because 13 stands in Republic of Srpska, Bosnia and Herzegovina and one in Serbia have been totally devastated in 1947 (Fukarek 1956), and the second one occurred in 1992–1993 when population at location Strugovi has been deliberately burnt during the war conflict (Aleksić et al. 2017a). Although populations of Serbian spruce usually re-establish rather quickly at sites devastated by fire, the recovery has not been recorded at all sites (Čolić 1987). The remaining threats which make Serbian spruce particularly susceptible to the climate change are poor natural regeneration and low competing ability. In addition, in recent years, extreme weather events have led to the drought related dieback, and that caused such reduction in number of individuals which cannot be compensated by natural regeneration (Ivetić and Aleksić 2016).

The response of forest tree populations to rapidly changing environment is extirpation or persistence achieved by migration or adaptation (Aitken et al. 2008). Programs for reforestation and conservation of Serbian spruce in terms of the rapidly changing climate have to integrate carefully both alternatives, migration and adaptation (Kremer et al. 2012), especially a number of adaptation strategies are currently available (Millar et al. 2007).

In this chapter, we present a current status of natural populations, we discuss actions aimed at species conservation, and finally, we suggest strategies for the long-term survival of Serbian spruce in a uncertany of achanging environment.

30.2 Serbian Spruce

30.2.1 Discovery and Evolutionary History of Serbian Spruce

The discovery of the Serbian spruce (*Picea omorika* (Panč.) Purk.) near the village of Zaovine at Mt. Tara in 1875 (Pančić 1887) attracted a lot of attention in the academic community in the late nineteenth century, and the interest in this species has not diminished yet. Among more than 600 publications on Serbian spruce taxonomy and systematics, morphology, ecology, genetic features, reproduction, etc. available until 1996 (Fukarek 1967; Šijak and Dinić 1996) and those published over the past 20 years, there are also surveys on Serbian spruce evolutionary relations which, however, appear to be rather contradictory.

According to Lockwood et al. (2013), Serbian spruce emerged in Asia ~16 million years ago. According to these authors, the closest living relatives of Serbian spruce currently found in the central Balkans are Caucasian *Picea orientalis* (L.) Peterm. and two spruce species endemic to Japan, *P. alcoquiana* (Veitch ex Lindl.) Carrière and *P. maximowiczii* Regel ex. Mast. These findings are concordant with the outcomes of the first study on spruce relations which was based on species crossability in relation to their taxonomy and distribution (Wright 1955). However, the North American origin of the Serbian spruce and its close relations with the two North American spruce species, *P. mariana* (Mill.) Britton, Sterns & Poggenb. and *P. rubens* Sarg., have been reported in a number of studies including those based on molecular data (e.g., Sigurgeirsson and Szmidt 1993; Nkongolo 1999; Ledig et al. 2004; Ran et al. 2006; Bouillé et al. 2011). Furthermore, it has been suggested that Serbian spruce represents an important link between two North American spruce complexes, *P. mariana-rubens* complex and *P. glauca-engelmannii-sitchensis-mexicana* complex (Ledig et al. 2004). In addition, a much younger age of this species has been reported (2.32 million years) as well as its migration from the North America to Eurasia throughout the Beringian bridge during the Pliocene (Ran et al. 2015). Altogether, these data suggest that further studies on evolutionary history of spruces are required, and they may help in resolving the above mentioned contradictions regarding the origin, age and relations of the Serbian spruce as well.

30.2.2 Natural Range of Serbian Spruce

Although scarce, the fossil findings of *Picea omorikoides*, a fossil relative of the Serbian spruce, indicate that this species was widely distributed in boreal regions of Asia and Europe during the Tertiary (Ravazzi 2002). Macrofossils of *Picea omorikoides/omorika* found in Pliocene sediments in former Czechoslovakia as well as Pleistocene fossils found in the northern Alpine foreland and the Aegean coast (Ravazzi 2002) indirectly indicate that Serbian spruce was present in the Balkans as early as the Pliocene/Pleistocene (it is worth mentioning that some fossils of this species are questionable). Southward migration of this species from its European habitats to the Balkans, i.e., to the region in which it persisted to the present days, would indeed be facilitated by the global climate cooling at the end of the Tertiary (Dizdarević et al. 1984). Therefore, Aleksić and Geburek (2014) stressed that a small area where Serbian spruce is currently found represents the long-term, cryptic and last refugium of this cold-adapted species.

Recently, Finsinger et al. (2017) studied the Holocene history of vegetation in the Balkans by analyzing sediments at one site with Serbian spruce, and they found that high forest cover was present in this region at the time when frequent anthropogenic openings were rather common in the rest of Europe. Thus, it is possible that Serbian spruce was indeed present within the same area over a long time. In this area of ~100 km^2, localized around the mid-course of the Drina river in western Serbia and eastern Bosnia and Herzegovina, ~30 remnant populations of Serbian spruce of various sizes are currently present (Ivetić and Aleksić 2016: Fig. 30.1; Aleksić et al. 2017a). These populations are found predominantly on limestone but also on serpentine (populations at sites Zmajevački Potok and Petlova stena), always on steep north-to-northwest facing slopes at altitudes ranging from 400 to 1800 m a.s.l. Serbian spruce was also present at one non-specific site, i.e. on peat (locality Crveni Potok), however, population at this site has been lost recently. Less then 10 years ago, there were six remaining trees at this site recorded (Ostojić and Dinić 2009, a couple of years later only three (Aleksić and Geburek 2010), while in 2016, we did not find Serbian spruce trees at this site (author's personal observation).

Fig. 30.1 Natural range of Serbian spruce, mainly around the mid-course of the river Drina at the border between Serbia and Republic of Srpska, Bosnia and Herzegovina. (Source: Ivetić and Aleksić 2016)

Climate in a region where the Serbian spruce is currently present is characterized by a low average annual temperature (5–7 °C) and high yearly sum of precipitation (~1000 mm). These conditions represents the lower limit for fir and spruce (Gajić et al. 1994), high snow cover and abundant fogs that, among other factors, may account for the low dispersal ability of this species (Aleksić and Geburek 2014).

30.2.3 Reproductive Strategy, Genetic and Other Important Features of Serbian Spruce

Serbian spruce is characterized by outcrossing (Kuittinen and Savolainen 1992; Aleksić and Geburek 2014) and also by high self-fertility (Langner 1959; Tucović and Isajev 1982; Schemske and Lande 1985; Kuittinen and Savolainen 1992), which is rather rare feature among conifers. Kuittinen and Savolainen (1992) found that in contrast to other conifers, Serbian spruce employs protogyny and spatial isolation of male from female strobili rather than early acting inbreeding depression to avoid selfing. Selfed individuals, however, may be obtained by artificial crossings (Tucović and Isajev 1982), and they most likely may be found in nature among abundant germinants produced by this pioneer species which is capable to generate large quantity of seeds and to occupy quickly newly available areas (Čolić 1957). Geburek (1986) reported decreased fitness in inbreeds during later ontogenetic stages, and this would suggest that selfed germinants are most likely eliminated early on in the establishment phase by natural selection. Thus, if Serbian spruce individuals are characterized by high levels of heterozygosity, adult selfed individuals are not to be expected in natural populations. The analysis of 499 adult Serbian spruce individuals from ten populations with nuclear microsatellites (Aleksić and Geburek 2014) as well as genotyping of one whole population (418 individuals, Aleksić et al. 2017b) revealed that trees produced by selfing or by mating among relatives were not present despite almost complete lack of the gene flow of this species (pollen and seed flow).

On the other hand, Serbian spruce may be crossed artificially with several spruce species (Wright 1955). Crossability studies in spruces were rather common during the second half of the nineteenth century, and several interspecific hybrids of Serbian spruce and other spruce have been produced, for instance, hybrids with *Picea sitchensis* (Bong.) Carr. (Langner 1959; Vidaković 1963; Roulund 1971), *Picea glauca* (Moench) Voss (Roulund 1971; Mikkola 1972; Nienstaedt 1977; David and Keathley 1996), and *Picea mariana* (Mill.) Britton, Sterns & Poggenb. (Nienstaedt 1977; Fowler 1980). Nielsen and Roulund (1992) carried out backcrosses of *Picea sitchensis* x *omorika* with Sitka spruce, and demonstrated that these backcrossed individuals are more frost resistant and vigourous at poor sites. However, findings on the performance of hybrids of Serbian spruce and other spruce species are inconclusive. It is important to mention that hybridization among currently sympatric species, Serbian spruce and Norway spruce, has not been observed in nature, and that studies on nuclear DNA content, base composition, heterochromatin and rDNA in these two species demonstrate that such events would be unlikely (Siljak-Yakovlev et al. 2002).

Genetic features of Serbian spruce natural populations, i.e., genetic diversity and genetic structure, have been studied to date by various genetic/molecular markers, and outcomes of these studies are seemingly contradictory. Low levels of genetic diversity have been reported at allozyme loci in 13 natural populations of Serbian spruce from the westward part of the species current natural range (i.e., from the Republic of Srpska, Bosnia and Herzegovina, Ballian et al. 2006) as well as in six natural populations from the eastward part of the species current natural range (i.e., from the Republic of Serbia) analyzed with plastid microsatellites (cpDNA SSRs) (Nasri et al. 2008). Although these findings are in accordance with expectations on levels of genetic diversity in species with small-sized populations (Leimu et al. 2006; Honnay and Jacquemyn 2007 and other references in those articles), Kuittinen et al. (1991) found previously exceptionally high levels of genetic diversity in one population of Serbian spruce based on allozyme variation, and the same pattern was observed at nuclear EST-SSR loci in ten populations of this species (Aleksić and Geburek 2014). Mitochondrial (mtDNA) diversity in Serbian spruce, on the other hand, was moderate (Aleksić and Geburek 2010, 2014). Underlying reasons for the mentioned contradictions regarding the genetic features of Serbian spruce natural populations and are discussed by Aleksić et al. (2009).

Available data on genetic structuring of Serbian spruce populations are contradictory as well. The lack of genetic trends among 13 westward populations of this species was inferred from allozyme data (Ballian et al. 2006) while clear geographic structure with two geographical groups based on cpDNA and SSRs loci was observed among six eastward natural populations from Serbia (Nasri et al. 2008). On the other hand, Aleksić and Geburek (2010, 2014) used nuclear EST-SSRs and one mtDNA marker and tested them in ten populations of Serbian spruce mainly from the eastward but also from the westward part of the species current natural range. The authors found two Evolutionary Significant Units among studied populations (ESU, sensu Moritz 1994) and furthermore, reported that each Serbian spruce population should be considered as an independent gene pool at the nuclear level. The authors also found that these highly differentiated populations are organized into higher levels of genetic structure, i.e., into five genetically distinct entities. In addition, Aleksić and Geburek (2014) reported substructure in several studied populations, while a recent study revealed substructure even in a population occupying only 0.34 ha (Aleksić et al. 2017b). The observed genetic structuring of Serbian spruce populations is mainly result of exceptionally limited gene flow and somewhat specific population dynamics over time. Aleksić and Geburek (2014) stated that "*more than three generations were required for the exchange of one seed between Serbian spruce populations distant 8 km in average*" and furthermore, demonstrated that the number of migrants per generation assessed using nuclear data is exceptionally low (~4%).

30.2.4 Ecology and Regeneration

Serbian spruce usually forms pure and dense stands, although it can also be found in less dense stands in association with a number of species such as *Abies alba* Mill., *Acer platanoides* L., *Carpinus betulus* L., *Fagus sylvatica* L., *Ostrya carpinifolia* Scop., *Picea abies* (L.) Karst., *Pinus sylvestris* L. and *Pinus nigra* Arnold and *Sorbus aucuparia* L. (Jovanović 2000). As a pioneer species, Serbian spruce may quickly occupy newly available sites (Čolić 1957). However, shade tolerant species usually suppress Serbian spruce from such sites over time.

Natural regeneration of Serbian spruce is rather poor (Čolić 1957). Predominant regeneration method is by seed, but vegetative regeneration is described as well (Vidaković 1982). According to Čolić (1959), rooting of lower branches covered with soil is frequent, especially on slopes facing north-east and on medium altitudes of its range (600–700 m a.s.l.). Although the germination rate of seeds in laboratory can exceed 80% (Ivetić and Milovanović 2005; Cvetkovic et al. 2013), germination rate in natural conditions is low and furthermore, a mortality of germinants during the first year in usually rather dense natural populations is exceptionally high (Ostojić and Dinić 2009). This is because additive genetic variance of Serbian spruce for shade tolerance is low (Tucić and Stojković 2001), and because successful competition of seedlings for radiant energy strongly depends on their capacity to become rapidly photoautotrophic (Tucić et al. 2005). Natural regeneration is most successful at sites with a sufficient amount of light at the ground level, such as those where Serbian spruce is associated with pines, while it is lacking on sites with dense herbaceous cover (Ostojić and Dinić 2012). However, the high mortality rate is present in nursery conditions as well, with limited and controlled shading and irrigating regimes, which indicates a need for further research on this aspect.

30.3 Serbian Spruce and Climate Change

30.3.1 The Threats

Numerous studies over the past decade focusing on effects of climate changes on living organisms reveal that environmental envelopes characterized by conditions suitable for forest populations will move to the north and to higher altitudes (Ledig and Kitzmiller 1992; Loarie et al. 2009; Gray and Hamann 2013). Furthermore, the pace of this movement will be rather fast (Kirschbaum and Fischlln 1996). Thus, it is possible that forest populations and their genes would not migrate fast enough to keep up with the pace of these rapid climate changes. Given the projected warming of 0.1–0.35 °C per decade, tree species will have to migrate 1.5–5.5 km per year northward and/or 1.5–5.5 m per year upwards to keep up with the movement of suitable environmental envelopes (Kirschbaum and Fischlln 1996). However, as reviewed by Ivetić and Devetaković (2016), migration rates of tree species during the past climate change (i.e., during the post-glacial recolonization) range from 1.26 to 1000 m per year, indicating the migration lag.

In order to provide predictions on temperature and precipitation changes for the area in which Serbian spruce is currently present, we used an extreme scenario of human-induced global warming, A2, according to which expectations on greenhouse gas emissions by 2100 are 250% larger compared to emission in the reference period, 1961–1990 (IPCC 2014). According to our projections, in the region with Serbian spruce, the temperature will increase between 2.4 and 7 °C, and precipitation will decrease in entire region with some local variations (Ivetić and Devetaković 2016). Such large changes in temperature and precipitation would suggest that the migration rates of Serbian spruce would have to be rather high in case that this species is supposed to track the movement of suitable environmental envelopes. However, the migration ability of Serbian spruce is exceptionally low (Aleksić and Geburek 2014), and thus, migration may not be a strategy which may enable survival of this species in the changing climate.

In addition, one of the largest hydrometeorological hazards in South-East Europe (SEE) are the frequent summer heat waves (Sippel and Otto 2014) like those that we have witnessed over the past 10 years (2007–2016). Summer heat waves may cause physiological weakening of trees which may become more susceptible to diverse pests and diseases. The hottest summer in SEE was in 2007 when the temperature reached 44.9 °C. Such high temperature has not been recorded in Serbia previously according to The Hydrometeorological Service of Serbia (2015). Another hot summer in Serbia was in 2012, when the highest measurements were recorded in 19 out of 29 main meteorological stations in this country. This was the hottest summer since 1848 when Serbia started with regular measurements of climate parameters (Hydrometeorological Service of Serbia 2013). In addition, rather low precipitation was recorded in Serbia in 2011 (only 63% of the average precipitation of the 30-years period, 1971–2000), (Ivetić 2015). On the other hand, in the spring 2014 we experienced heavy rains, which caused floods and landslides in both Serbia and Bosnia and Herzegovina.

30.3.2 The Response

The most obvious consequence of increased frequency, prolonged duration, and severity of extreme weather events is the dieback of trees and forests. After a 3-year dry period in Serbia (Ivetić 2015), dieback of individual trees or groups of trees is recorded on a total area of 13,885 ha (PE Srbijašume database). This is recorded in natural populations of Serbian spruce as well, particularly in Serbia. Similarly as in other tree species, Serbian spruce is firstly showing signs of the drought which are followed by physiological weakening, such weakened trees are then more prone to disease and pests, and they ultimately die within a short period of time. However, in comparison to the other tree species suffering from the dieback caused by the global warming, Serbian spruce is much more threatened (see bellow).

The dieback of Serbian spruce is also recorded in plantations outside of its natural range. The one and only Serbian spruce seed orchard, located in Godovik, Serbia (see Sect. 30.4.2), may serve as an excellent example for demonstrating a possible destiny of natural populations of this species in the near future. Serbian spruce is exceptionally susceptible to *Armillaria* infections (Keča 2010), and a mass infection by *Armillaria ostoyae* (Romang.) Herink. in seed orchard Godovik is recorded over the past 4 years (Jezdimirović 2016). Infected Serbian spruce trees die within a period of 1–3 years. The environment conditions in the Godovik area are very similar to those projected for the near future for area where Serbian spruce is currently distributed (Ivetić and Aleksić 2016), i.e., the climate in Godovik is warmer (annual average temperature of 9.7 °C in the orchard vs. projected annual average of 5–7 °C in natural range) and witl less precipitation (725 mm of annual average precipitation in the orchard during the period 1981–2010 vs. projected annual average precipitation of ~1.000 mm in natural range). In addition, temperature at seed orchard has been constantly higher than normal values, with precipitation under the normal values during a 5-year period (2011–2015, with the exception of the year 2014, as described above, Ivetić and Aleksić 2016). These unfavorable conditions have led to the physiological weakening of trees which became more susceptible to diseases and pests, leading ultimately to the lethal outcome – unfortunately, this is the most likely response of Serbian spruce to the ongoing rapid climate change.

30.4 Proposed Measures

30.4.1 On Site Measures: Dealing with Natural Populations

Serbian spruce was IUCN red-list under the category vulnerable (VU) in 1998 (Conifer Specialist Group 1998), while it is characterized as endangered (EN) at present (Aleksić et al. 2017a). In Serbia, this species has been subjected to different regimes of protection since 1950 (Čolić 1951). Current actions on Serbian spruce conservation comprise strict protection of all known populations and individual trees on an area of 3090 ha (Table 30.1). Thus, Serbian spruce populations, as untouched forests in forest reserves, are left for free development (Aleksić 2008). This level of protection corresponds to the category I of the Protected Area Management Categories of IUCN, and to the category 1.1 of the Ministerial

Table 30.1 Reserves with natural populations of Serbian spruce

Country	Total reserves area (ha)	Altitude range (m a.s.l.)	Slope range (°)	Exposure
Serbia	2846.2	220–1440	3–65	N, NE, SW, NW, W
Bosnia and Herzegovina	244	800–1500	NA	N, NE, NW

Adapted from Ivetić and Aleksić (2016)*

Conference on the Protection of Forests in Europe (FOREST EUROPE – MCPFE) classes of protected and protective forests.

Dominant trees in natural populations have an average age around 200 years and these forests are in the terminal phase of development (Ostojić and Dinić 2012). Current state of Serbian spruce populations combined with extreme climate events and poor natural regeneration, raised a question on suitability of the applied "do not touch" conservation approach, and the need for alternative protection and conservation actions has been put forward (Ivetić and Aleksić 2016).

The possible actions which may be applied immediately for the in situ protection and conservation of Serbian spruce are:

- Facilitation of natural regeneration by selective removal of individual trees which are competitors to Serbian spruce in dense forests (Ostojić and Dinić 2012; Aleksić and Geburek 2014). Despite the possible fatal outcome which may be expected due to changes of micro-ecological conditions at particular sites caused by felling of neighboring trees (Fukarek 1956), this practice, if carefully conducted, could be beneficial for Serbian spruce because it can facilitate natural regeneration at created gaps that have favorable light conditions required for seedling development.
- Enrichment planting of high quality seedlings inside, or nearby sites which were occupied by the Serbian spruce in the past (Ivetić and Aleksić 2016). Since such sites are nowadays characterized by somewhat different climate conditions in comparison to those which were present at the time when they were occupied by Serbian spruce, it is very likely that these newly established populations will have poor natural regeneration especially at the beginning (Spittlehouse and Stewart 2003), and thus, the survival of such stands should be enabled by artificial regeneration. Given the uncertainty of changing environments, the quality of seedlings becomes an imperative. A project specific seedling quality of Serbian spruce seedlings (Ivetić and Devetaković 2016; Dumroese et al. 2016), should be determined by the predicted future site conditions. Combined with appropriate site preparation, planting methods, and post planting protection and silviculture, the use of high quality Serbian spruce reproductive material should promote the success in term of initial survival rates. Serbian spruce seedlings originating from different population should be planted over several years and in an unregular – close to nature-like patterns, choosing the best microsites for seedling planting. In this way, even-aged structure and plantations will be avoided.
- Assisted natural regeneration by direct seeding (Grossnickle and Ivetić 2017). Direct seeding can be a usefull tool for reintroduction of endangered species like Serbian spruce (Laborde and Corrales-Ferrayola 2012; Atondo-Bueno et al. 2016). Such a procedure, carried out by variable seed distribution and a wide size range of established seedlings, has the potential to initiate a mixed structure and non-uniform forest stands (as reviewed by Grossnickle and Ivetić 2017). Any action which will result with increased size and the number of populations will promote the preserving of Serbian spruce.

30.4.2 Off-Site Measures: Creating New Populations by Translocation and Migration

One of the key factors for a long-term success of any artificial establishment of a new forest is making a distinction between site-specific and species-specific projects (Ivetić and Devetaković 2016). For site-specific projects it is important to select species which are best suited for the focal site, while species-specific projects are usually applied in cases of industrial tree plantations, but can also be used for species translocation for conservation purposes. Although Serbian spruce can be used for site-specific projects, only via species-specific projects it would be possible to achieve the main goal – species rescue in a rapidly changing environment and uncertainty of the climate change. For that purpose, it is important to select suitable sites for species translocation, and such sites are apparently in areas where Serbian spruce is not currently present.

The selection of sites suitable for Serbian spruce translocation represents the biggest challenge especially because of the ongoing rapid climate change. The sites for future introduction of Serbian spruce need to fulfill both spatial (soil characteristics) and temporal (projected future climate) criteria. Selected sites may be characterized by environmental conditions which may be unfavorable for the survival of seedlings if a short-term, but they should enable Serbian spruce persistence on the long run. On the other hand, seedlings survival and establishment at new sites can be promoted by a range of cultural practices, including the use of a high quality planting material (Ivetić and Devetaković 2016). Numerous successfully established stands of this species in Europe (e.g. Zehetmayr 1954; Meyer 1960; Nymoen 1978; Kuittinen et al. 1991; Widrlechner et al. 1992; Kasesalu 2002; Král 2002; Sander and Meikar 2009; Cejpek et al. 2013), support the view that for Serbian spruce fundamental niche overcomes its realized niche (as determined by Gray and Hamann 2011; see Ivetić and Aleksić 2016 for a review of Serbian spruce populations outside of its natural range).

It is worth mentioning that planting Serbian spruce outside the natural range has begun almost immediately after its discovery in the late nineteenth century (Meyer 1960; Mitchell 1975; Kasesalu 2002; Møller 2013; Kuittinen and Savolainen 1992; Král 2002, Table 30.2). Similarly as in case of majority of exotic species planted outside their natural range, Serbian spruce was initially used as an ornamental tree in parks, while trials aimed for the use in forestry have been established later on as well. These past movements of Serbian spruce can be considered as transfers (Ivetić and Aleksić 2016) that have been characterized by Koskela et al. (2014) as "*human-mediated movement of tree germplasm, regardless to geographical scales*". On the other hand, there are several plantations of Serbian spruce in Serbia and Bosnia and Herzegovina, and they occupy a total area of only 35 ha (Fig. 30.2). A few of them are older than 40 years, and, unfortunately, majority of them are in poor condition mainly because of the inappropriate selection of sites for their establishment (as reviewed by Ivetić and Aleksić 2016).

Table 30.2 Past transfers of Serbian spruce outside its natural range

1934	Czech Republic
1932	Finland, 2nd generation
1891	Denmark
1890	Estonia
1889	Britain
1881	Switzerland
1875	Discovery
Year	

Adapted from Ivetić and Aleksić (2016)

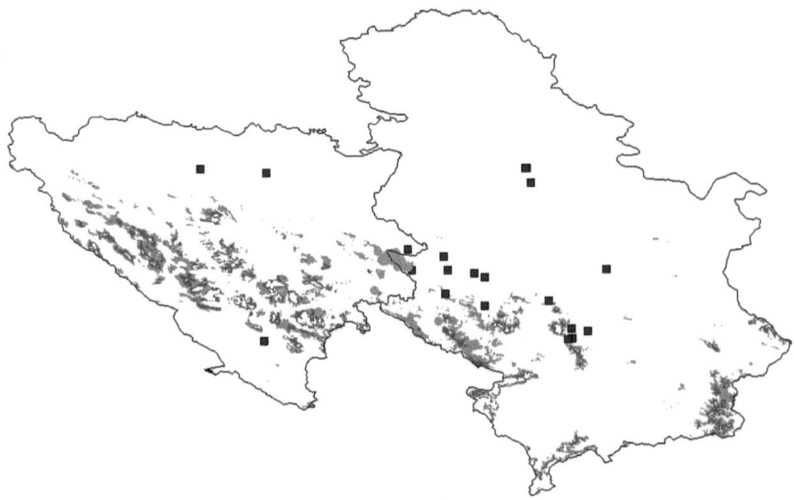

Fig. 30.2 Serbian spruce natural populations (green circles), plantations (blue squares), and sites in Serbia and Bosnia and Herzegovina suitable assisted migration of Serbian spruce (orange and red). Suitable future sites for Serbian spruce can be found beyond its current natural range. Previously established plantations are not on these sites

Although transferring of forest reproductive material has been practiced for centuries, assisted migration, which has a number of various definitions in recent literature (for detailed review see Dumroese et al. 2015), remains controversial (Pedlar et al. 2012). In this chapter, any intentional transfer of forest reproductive material as a response to projected climate change (Ivetić and Devetaković 2016) is considered as assisted migration. This strategy can be used in species-specific projects for species rescue (Pedlar et al. 2012). Since it is very likely that Serbian spruce will disappear from its natural habitats in the near future due to the rapid climate change, its rescue via assisted migration is inevitable. Both assisted range expansion, as a transfer just beyond current natural range, assisted species migration as a transfer on a long distance from its current range (as defined by Williams and Dumroese 2013), should be considered. The latter strategy is supported by the fact that Serbian spruce performs well far away from its natural range and thus, newly established stands are expected to survive as well. However, any assisted migration of Serbian spruce should be implemented in a small scale.

It is worth mentioning that seed orchards continue to be the major supplier for improved seed material (Lindgren 2016). Given the very strict regime of Serbian spruce protection according to which collection of seed from natural populations is prohibited, only seed from plantations (seed orchards and plantations transformed into seed stands) can be traded and used for seedlings production. At present, there are only four facilities for Serbian spruce seed collection: one seed orchard (in Serbia) and three seed stands (one located in Serbia and two in Republic of Srpska, Bosnia and Herzegovina).

Established in 1987 from 5956 genotypes from 50 half-sib lines (Tucović and Isajev 1988), the only seed orchard of Serbian spruce is nowadays in rather poor condition, due to unplanned felling of trees, difficulties in implementation of required management measures and high infection rate by *Armillaria ostoyae* (Romang.) Herink. However, maybe the most important obstacle for using seeds from Godovik for assisted migration is the lack of knowledge on the origin of individuals used for the production of half-sib lines, and questionable initial assumption that implemented seed orchard design captures genetic variability and adaptive potential of Serbian spruce on a high level (Ivetić and Aleksić 2016). Nevertheless, given the lack of other seed sources as well as findings on similar levels of expected heterozygosity (assessed using allozymes) in seed orchard and in natural and planted populations of Serbian spruce (Milovanović and Šijačić-Nikolić 2010), the role of this seed orchard is still essential for production of planting material. Serbian spruce seed is collected occasionally from seed stands as well (Mataruga et al. 2005; Ivetić and Aleksić 2016). The seed stands are between 30 and 50 years old, and, it appears that at least one of them (Bela Zemlja, located in Serbia) was established from seeds originating from Godovik.

Given the above mentioned obstacles regarding the quality of seed which may be obtained from existing Serbian spruce seed orchards and seed stands, it is evident that the establishment of new properly designed seed orchards is necessary. The selection of trees for their establishment should be based on outcomes on available studies on genetic features of Serbian spruce natural populations (Aleksić and Geburek 2010, 2014; Aleksić et al. 2017b), and any planting material for future Serbian spruce plantations must be produced from such seed orchards. The future sites with suitable environmental conditions for Serbian spruce in Serbia and Bosnia and Herzegovina can be found between 1200 and 1600 m a.s.l. (Fig. 30.2). For future planting of Serbian spruce in this region, it is necessary to select sites within this altitudinal range that are at north facing slopes, with appropriate levels of precipitations and soil types, and which can support the relevant plant species associations (Ivetić and Aleksić 2016).

30.5 Conclusions

Existing populations of Serbian spruce are subjected to both long-lasting and well-known threats (e.g., competition, poor natural regeneration and fires) as well as to a new threat (e.g., climate warming and more frequent extreme weather events), already causing a non-negligible dieback of individuals especially in Serbia. Small size of Serbian spruce natural populations and lack of natural regeneration make this species a much more threatened by the climate warming, and thus, actions for its preservation and rescue are required and urgent. Enrichment planting of high quality seedlings inside, or nearby sites which were occupied by the Serbian spruce in the past as well as assisted natural regeneration by direct seeding, combined with assisted migration to sites with suitable environmental conditions which may enable species persistence in the future, are suitable strategies facing climate change. These actions are supported by results of previous transfer of Serbian spruce outside its natural range. These populations established on various distances from natural range, are valuable resource for future research on Serbian spruce adaptive potential. Current conservation measures have to be revised, and improved by compulsory collection of seeds from natural populations. This seed collection should be storage in seed banks, used for seedling production and establishment of new seed orchards, and used for direct seeding. The new seed orchards should be designed in a way to capture a high level of genetic diversity founded in Serbian spruce natural populations, with a special target for conservation of rare alleles. The same aims should be considered for successful assisted migration of Serbian spruce.

References

Aitken, S. N., Yeaman, S., Holliday, J. A., Wang, T., & Curtis-McLane, S. (2008). Adaptation, migration or extirpation: Climatic changes outcomes for tree populations. *Evolutionary Applications, 1*, 95–111.

Aleksić, M. J. (2008). Genetic structure of natural populations of Serbian spruce [*Picea omorika* (Panč.) Purk.]. Dissertation, University of Natural Resources and Applied Life Sciences, Vienna, Austria.

Aleksić, M. J., & Geburek, T. (2010). Mitochondrial DNA reveals complex genetic structuring in a stenoendemic conifer *Picea omorika* [(Panč.) Purk.] caused by its long persistence within the refugial Balkan region. *Plant Systematics and Evolution, 285*, 1–11. https://doi.org/10.1007/s00606-009-0250-0.

Aleksić, M. J., & Geburek, T. (2014). Quaternary population dynamics of an endemic conifer, *Picea omorika*, and their conservation implications. *Conservation Genetics, 15*, 87–107. https://doi.org/10.1007/s10592-013-0523-6.

Aleksić, M. J., Schueler, S., Mengl, M., & Geburek, T. (2009). EST-SSRs developed for other *Picea* species amplify in *Picea omorika* and reveal high genetic variation in two natural populations. *Belgian Journal of Botany, 142*, 89–95.

Aleksić, J. M., Ballian, D., Isajev, D., Mataruga, M., Christian, T., Gardner, M. (2017a). *Picea omorika*. The IUCN Red List of Threatened Species 2017: e.T30313A84039544. https://doi.org/10.2305/IUCN.UK.2017-2.RLTS.T30313A84039544.en. Downloaded on 12 December 2017.

Aleksić, J. M., Piotti, A., Geburek, T., & Vendramin, G. G. (2017b). Exploring and conserving a "microcosm": Whole-population genetic characterisation within a refugial area of the endemic, relict conifer *Picea omorika*. *Conservation Genetics, 18*(4), 777–788.

Atondo-Bueno, E. J., López-Barrera, F., Bonilla-Moheno, M., Williams-Linera, G., & Ramírez, N. (2016). Direct seeding of *Oreomunnea mexicana*, a threatened tree species from Southeastern Mexico. *New Forests, 47*, 845–860. https://doi.org/10.1007/s11056-016-9548-2.

Ballian, D., Longauer, R., Mikić, T., Paule, L., Kajba, D., & Gömöry, D. (2006). Genetic structure of a rare European conifer, Serbian spruce (*Picea omorika* (Pančić) Purk.). *Plant Systematics and Evolution, 260*, 53–63. https://doi.org/10.1007/s00606-006-0431-z.

Bouillé, M., Senneville, S., & Bousquet, J. (2011). Discordant mtDNA and cpDNA phylogenies indicate geographic speciation and reticulation as driving factors for the diversification of the genus *Picea*. *Tree Genetics & Genomes, 7*(3), 469–484.

Cejpek, J., Kuráž, V., & Frouz, J. (2013). Hydrological properties of soils in reclaimed and unreclaimed sites after brown-coal mining. *Polish Journal of Environmental Studies, 22*(3), 645–652.

Čolić, D. (1951). Šumski rezervati *Picea omorica* Pančić = Forest reserves of *Picea omorica* Pančić. – Šumarstvo, IV (1); 20–34. Summ.

Čolić, D. (1957). Neki pionirski karakteri Pančićeve omorike i njena uloga u sukcesiji biljnih zajednica. *Archives des Sciences Bioloqiques, 9*(1–4), 51–60 (in Serbian with English and German summary).

Čolić, D. (1959). Prilog poznavanju ekologije vegetativnog razmnožavanja Pančićeve omorike (*Picea omorika* Panč). *Archive Bioloških Nauka, Beograd, 11*(1–4), 41–66.

Čolić, D. (1987). Spontana obnova Pančićeve omorike (*Picea omorika* Panč.) posle požara. *Zaštita Prirode, 40*, 37–56.

Conifer Specialist Group. (1998). *Picea omorika*. In IUCN 2007. 2007 IUCN Red list of threatened species.

Cvetkovic, B., Mataruga, M., Isajev, V., Levic, J., Lucic, A., Trkulja, V., & Kremenovic, Z. (2013). Variability in germination and germination dynamics of differently treated seeds of Serbian spruce (*Picea omorika* Pančić/Purkyně). *Genetika, 45*(1), 109–119.

David, A. J., & Keathley, D. E. (1996). Inheritance of mitochondrial DNA in interspecific crosses of *Picea glauca* and *Picea omorika*. *Canadian Journal of Forest Research, 26*(3), 428–432. https://doi.org/10.1139/x26-048.

Dizdarević, M., Lakušić, R., Grgić, P., Kutleša, L., Pavlović, B., & Jonlija, R. (1984). Ekološke osnove poimanja reliktnosti vrste *Picea omorika* Pančić. *Bilten Društva ekologa Bosne i Herzegovine, Ser A, 2*, 5–56 (in Bosnian with English abstract).

Dumroese, R. K., Williams, M. I., Stanturf, J. A., & St Clair, J. B. (2015). Considerations for restoring temperate forests of tomorrow: Forest restoration, assisted migration, and bioengineering. *New Forest, 46*, 947–964. https://doi.org/10.1007/s11056-015-9504-6.

Dumroese, R. K., Landis, T. D., Pinto, J. R., Haase, D. L., Wilkinson, K. W., & Davis, A. S. (2016). Meeting forest restoration challenges: Using the target plant concept. *Reforesta, 1*, 37–52.

Finsinger, W., Morales-Molino, C., Gałka, M., Valsecchi, V., Bojovic, S., & Tinner, W. (2017). Holocene vegetation and fire dynamics at Crveni Potok, a small mire in the Dinaric Alps (Tara National Park, Serbia). *Quaternary Science Reviews, 167*, 63–77.

Fowler, D. P. (1980). Hybridization of black spruce and Serbian spruce. Can For Serv Marit For Cent Inf Rep M-X-112.

Fukarek, P. (1956). Zaštita endemne Pančićeve omorike u NR Bosni i Hercegovini. Godišnjak Zemaljskog zavoda za zaštitu spomenika kulture i prirodnih rijetkosti NR Bosne i Hercegovine. Naše starine, III, 289–298.

Fukarek, P. (1967). Pančićevo otkriće omorike i njeno dalje proučavanje. In Josifović M (Ed.), *Pančićev zbornik u spomen 150-godišnjice njegovog rođenja. Srpska Akademija Nauka i Umetnosti, Odeljenje prirodno-matematičkih nauka, Beograd* (pp. 27–67).

Gajić, M., Vilotić, D., Karadžić, D., Mihajlović, L., & Isajev, V. (1994). *Serbian spruce—Picea omorika (Pančić) Purkyně on the territory of the National Park Tara*. Belgrade: The National Park Tara, Bajina Bašta and the Faculty of Forestry (in Serbian).

Geburek, T. (1986). Some results of inbreeding depression in Serbian spruce (Picea omorica (Panč.) Purk.). *Silvae Genetica, 35*, 169–172.

Gray, L. K., & Hamann, A. (2011). Strategies for reforestation under uncertain future climates: Guidelines for Alberta, Canada. *PLoS One, 6*(8), e22977. https://doi.org/10.1371/journal.pone.0022977.

Gray, L. K., & Hamann, A. (2013). Tracking suitable habitat for tree populations under climate change in western North America. *Climatic Change, 117*, 289–303. https://doi.org/10.1007/s10584-012-0548-8.

Grossnickle, S., & Ivetić, V. (2017). Direct seeding in reforestation – A field performance review. *Reforesta, 4*, 94–142. https://doi.org/10.21750/REFOR.4.07.46.

Honnay, O., & Jacquemyn, H. (2007). Susceptibility of common and rare plant species to the genetic consequences of habitat fragmentation. *Conservation Biology, 21*, 823–831.

Hydrometeorological Service of Serbia. (2013). http://www.hidmet.gov.rs/podaci/meteorologija/ciril/l2012.pdf. Accessed online: 09.11.2016.

Hydrometeorological Service of Serbia. (2015). http://www.hidmet.gov.rs/ciril/meteorologija/klimatologija_temp_rezim.php. Accessed online: 09.11.2016.

Ivetić, V. (2015). Reforestation in Serbia: Success or failure? In V. Ivetić, & D. Stanković (Eds.), *Proceedings: International conference reforestation challenges* (pp. 1–12). 03–06 June 2015, Belgrade, Serbia.

Ivetić, V., & Aleksić, J. (2016). Response of rare and endangered species *Picea omorika* to climate change – The need for speed. *Reforesta, 2*, 81–99. https://doi.org/10.21750/REFOR.2.09.24.

Ivetić, V., & Devetaković, J. (2016). Reforestation challenges in Southeast Europe facing climate change. *Reforesta, 1*, 178–220. https://doi.org/10.21750/10.21750/REFOR.1.10.10.

Ivetić, V., & Milovanović, J. (2005). Electrical conductivity test for Serbian spruce seed quality estimation. *Glasnik Šumarskog fakulteta, 2005*(91), 127–133.

Jezdimirović, J. (2016). *Reconstruction of Serbian spruce seed orchard in Godovik.* [In Serbian: Реконструкција семенске плантаже Панчићеве оморике у Годовику]. MSc thesis, University of Belgrade. 67 p.

Jovanović, B. (2000). *Dendrologija. Udžbenik, šesto dopunjeno izdanje*. Beograd: Univerzitet u Beogradu, Šumarski fakultet.

Kasesalu, H. (2002). Serbian spruce (*Picea omorika* (Panchic) Purkyne) in Estonia. In *Dendrological researches in Estonia III (Estonia). Estonian Agricultural University, Tartu (Estonia)* (pp. 171–177). Forest Research Institute. (no.3).

Keča, N. (2010). The test of eight tree species resistance to the attack of *Armillaria mellea* and *A. ostoyae* by artificial infection. *Bulletin of the Faculty of Forestry, 102*, 41–56. [In Serbian, Summary in English]. https://doi.org/10.2298/GSF1002041K.

Kirschbaum, M., & Fischlln, A. (1996). Climate change impacts on forests. In R. Watson, M. C. Zinyowera, & R. H. Moss (Eds.), *Climate change 1995 – impacts, adaptations and mitigation of climate change: Scientific-technical analysis*, *Contribution of Working Group to the Second Assessment Report of the Intergovernmental Panel on Climate Change* (pp. 95–129). Cambridge a.o.: Cambridge University Press.

Kolarović, S. (1951). Nalazišta i stanje Pančićeve omorike u NR Srbiji. *Šumarstvo Beograd, 4*(1), 27–34.

Koskela, J., Vinceti, B., Dvorak, W., Bush, D., Dawson, I. K., Loo, J., Kjaer, E. D., Navarro, C., Padolina, C., Bordács, S., Jamnadass, R., Graudal, L., & Ramamonjisoa, L. (2014). Utilization and transfer of forest genetic resources: A global review. *Forest Ecology and Management, 333*, 22–34. https://doi.org/10.1016/j.foreco.2014.07.017.

Král, D. (2002). Assessing the growth of *Picea omorika* [Panč.] Purkyně in the Masaryk forest training forest enterprise at Křtiny. *Journal of Forest Science, 48*(9), 388–398.

Kremer, A., Ronce, O., Robledo-Arnuncio, J. J., Guillaume, F., Bohrer, G., Nathan, R., Bridle, J. R., Gomulkiewicz, R., Klein, E. K., Ritland, K., Kuparinen, A., Gerber, S., & Schueler, S. (2012). Long-distance gene flow and adaptation of forest trees to rapid climate change. *Ecology Letters, 15*, 378–392.

Kuittinen, H., & Savolainen, O. (1992). *Picea omorika* is a self fertile but outcrossing conifer. *Heredity, 68*, 183–187. https://doi.org/10.1038/hdy.1992.27.

Kuittinen, H., Muona, O., Karkkainen, K., & Borzan, Ž. (1991). Serbian spruce, a narrow endemic, contains much genetic variation. *Canadian Journal of Forestry Resources, 21*, 363–367. https://doi.org/10.1139/x91-044.

Laborde, J., & Corrales-Ferrayola, I. (2012). Direct seeding of *Brosimum alicastrum* Sw. (*Moraceae*) and *Enterolobium cyclocarpum* (Jacq.) Griseb. (*Mimosaceae*) in different habitats in the dry tropics of central Veracruz. *Acta Botánica Mexicana, 100*(100), 107–134.

Langner, W. (1959). Ergebnisse Einiger Hybridisierrungsversuche Zwischen *Picea sitchensis* (Bong.) Carr. und *Picea omorika* (Pancic) Purkyne. *Silvae Genetica, 8*, 138–143.

Ledig, F. T., & Kitzmiller, L. H. (1992). Genetic strategies for reforestation in the face of global climate change. *Forest Ecology and Management, 50*, 153–169.

Ledig, F. T., Hodgskiss, P. D., Krutovskii, K. V., Neale, D. B., & Eguiluz-Piedra, T. (2004). Relationships among the spruces (Picea, Pinaceae) of southwestern North America. *Systematic Botany, 29*(2), 275–295.

Leimu, R., Mutikainen, P., Koricheva, J., & Fisher, M. (2006). How general are positive relationships between plant population size, fitness and genetic variation? *Journal of Ecology, 94*, 942–952.

Lindgren, D. (2016). The role of tree breeding in reforestation. *Reforesta, 1*, 221–237. https://doi.org/10.21750/REFOR.1.11.11.

Loarie, S. R., Duffy, P. B., Hamilton, H., Asner, G. P., Field, C. B., & Ackerly, D. D. (2009). The velocity of climate change. *Nature, 462*, 1052–1056.

Lockwood, J. D., Aleksić, J. M., Zou, J., Wang, J., Liu, J., & Renner, S. S. (2013). A new phylogeny for the genus Picea from plastid, mitochondrial, and nuclear sequences. *Molecular Phylogenetics and Evolution, 69*(3), 717–727.

Mataruga, M., Isajev, V., Lazarev, V., Balotić, P., & Daničić, V. (2005). *Registar šumskih sjemenskih objekata RS-osnova unapređenja sjemenske proizvodnje* (pp. 1–222). Banja Luka: Šumarski fakultet ISBN 99938-56-03-7.

Meyer, H. (1960). The Serbian Spruce, *P. omorika*, a species to supplement the scanty tree flora of Germany. *Archive Forstwesen, 9*(7), 595–614.

Mikkola, L. (1972). Crossability between *Picea omorika* (Pančič) Purkyne and *P. glauca* (Moench) Voss. *Annales Botanici Fennici, 9*(1), 33–36.

Millar, C. I., Stephenson, N. L., & Stephens, S. L. (2007). Climate change and forests of the future: Managing in the face of uncertainty. *Ecological Applications, 17*, 2145–2151.

Milovanović, J., & Šijačić-Nikolić, M. (2010). Characterization of Serbian spruce variability applying isoenzyme markers. *Biotechnology & Biotechnological Equipment, 24*(1), 1600–1605. https://doi.org/10.2478/V10133-010-0012-8.

Mitchell, A. F. (1975). *Conifers in the British Isles: A descriptive handbook*. London: HMSO 322 p.

Møller, P. F. (2013). *Opportunities and problems with introduced tree species in the Danish forests – in a historical perspective*. http://www.skogur.is/media/nordisk-skogshistorisk-konferanse/NSH-Reykjavik-Denmark.pdf.

Moritz, C. (1994). Defining 'evolutionary significant units' for conservation. *Trends in Ecology & Evolution, 9*, 373–375.

Nasri, N., Bojović, S., Vendramin, G. G., & Fady, B. (2008). Population genetic structure of the relict Serbian spruce, *Picea omorika*, inferred from plastid DNA. *Plant Systematics and Evolution, 271*, 1–7. https://doi.org/10.1007/s00606-007-0594-2.

Nielsen, U. B., & Roulund, H. (1992). Sitkahybrider [Sitkahybrids]. *Skoven, 24*, 72–75.

Nienstaedt, H. (1977). Mass production alternatives for fast-growing spruce hybrids. In *Proceedings of the Thirteenth Lake States Forest Tree Improvement Conference; Gen. Tech. Rep. NC-50* (pp. 56–71). St. Paul: U.S. Department of Agriculture, Forest Service, North Central Forest Experiment Station.

Nkongolo, K. K. (1999). RAPD and cytological analyses of Picea spp. from different provenances: Genomic relationships among taxa. *Hereditas, 130*(2), 137–144.
Nymoen, H. (1978). *Picea omorika* is not an alternative to *P. sitchensis* in afforestation. *Norsk Skogbruk, 24*(4), 15–16.
Ostojić, D., & Dinić, A. (2009). Experimental phytocoenological investigations of Serbian spruce (*Picea omorika*/Pančić/Purkyně) natural regeneration in the national park Tara. *Šumarstvo, 2009*(1–2), 23–35.
Ostojić, D., & Dinić, A. (2012). Rezervati prirode sa omorikom u Srbiji – osnovne karakteristike i zaštita. *Zaštita prirode, 62*(2), 5–17.
Pančić, J. (1887). Omorika nova fela četinara u Srbiji. *Težak, XVIII*(1), 1–8.
Pedlar, J. H., McKenney, D. W., Aubin, I., Beardmore, T., Beaulieu, J., Iverson, L., O'Neill, G. A., Winder, R. S., & Ste-Marie, C. (2012). Placing forestry in the assisted migration debate. *Bioscience, 62*, 835–842. https://doi.org/10.1525/bio.2012.62.9.10.
Pintarić, K. (1999). Forestry and forest reserves in Bosnia and Herzegovina. In J. Dijaci (Ed.), *Virgin forests and forest reserves in Central and East European Countries, COST Action E4: Forest Reserves Research Network* (pp. 1–15). Ljubljana: Department of Forestry and Renewable Forest Resources, Biotechnical Faculty.
Ran, J. H., Wei, X. X., & Wang, X. Q. (2006). Molecular phylogeny and biogeography of Picea (Pinaceae): Implications for phylogeographical studies using cytoplasmic haplotypes. *Molecular Phylogenetics and Evolution, 41*(2), 405–419.
Ran, J. H., Shen, T. T., Liu, W. J., Wang, P. P., & Wang, X. Q. (2015). Mitochondrial introgression and complex biogeographic history of the genus Picea. *Molecular Phylogenetics and Evolution, 93*, 63–76.
Ravazzi, C. (2002). Late Quaternary history of spruce in southern Europe. *Review of Paleobotany and Palinology, 120*, 131–177.
Roulund, H. (1971). Observations on spontaneous hybridization in *Picea omorika* (Pancic) Purkyne. *Forest Tree Improvement, 2*, 2–17.
Sander, H., & Meikar, T. Received January 2009. Exotic coniferous trees in Estonian forestry after 1918. Allg. Forst- u. J.-Ztg., 180. Jg., 7/8: 158–169.
Schemske, D. W., & Lande, R. (1985). The evolution of self-fertilization and inbreeding depression in plants. II. Empirical observations. *Evolution, 39*, 41–52. https://doi.org/10.2307/2408515.
Sigurgeirsson, A., & Szmidt, A. E. (1993). Phylogenetic and biogeographic implications of chloroplast DNA variation in Picea. *Nordic Journal of Botany, 13*(3), 233–246.
Šijak, M., & Dinić, A. (1996). Dodatak bibliografiji radova o Pančićevoj omorici (*Picea omorika* Pančić). *Ekologija, 30*(1), 165–178.
Siljak-Yakovlev, S., Cerbah, M., Coulaud, J., Stoian, V., Brown, S. C., Zoldos, V., Jelenic, S., & Papes, D. (2002). Nuclear DNA content, base composition, heterochromatin and rDNA in *Picea omorika* and *Picea abies*. *Theoretical and Applied Genetics, 104*(2), 505–512.
Sippel, S., & Otto, F. (2014). Beyond climatological extremes – assessing how the odds of hydrometeorological extreme events in South-East Europe change in a warming climate. *Climatic Change, 125*, 381–398. https://doi.org/10.1007/s10584-014-1153-9.
Spittlehouse, D. L., & Stewart, R. B. (2003). Adapting to climate change in forest management. *Journal of Ecosystems and Management, 4*, 7–17.
Tucić, B., & Stojković, B. (2001). Shade avoidance syndrome in *Picea omorika* seedlings: A growth-room experiment. *Journal of Evolutionary Biology, 14*, 444–455. https://doi.org/10.1046/j.1420-9101.2001.00291.x.
Tucić, B., Pemac, D., & Ducić, J. (2005). Life history responses to irradiance at the early seedling stage of *Picea omorika* (Pancic) Purkynhe: Adaptiveness and evolutionary limits. *Acta Oecologica-International Journal of Ecology, 27*, 185–195. https://doi.org/10.1016/j.actao.2004.12.004.
Tucović, A., & Isajev, V. (1982). The influence of different pollination types on some properties of Serbian spruce cones and seeds. *Bulletin Faculty of Forestry (Belgrade), 59*, 59–65.

Tucović, A., & Isajev, V. (1988). *Generativna semenska plantaža omorike u Godoviku* (pp. 1–40). Beograd: Izvođački projekat.

Vidaković, M. (1963). *Međuvrsno ukrštanje Pančićeve omorike (Picea omorika/Panč./Purkyne) sa sitkanskom smrčom (Picea sithensis (Bong.) Cariére)* (pp. 337–342). Beograd: Šumarstvo.

Vidaković, M. (1982). Četinjače – Morfologija i varijabilnost. JAZU & Liber, Zagreb. 711 pp.

Widrlechner, M. P., Hasselkus, E. R., Herman, D. E., lies, J. K., Pair, J. C., Paparozzi, E. T., Schutzki, R. E., & Wildung, D. K. (1992). Performance of landscape plants from Yugoslavia in the North Central United States. *Journal of Environmental Horticulture, 10*(4), 192–198.

Williams, M. I., & Dumroese, R. K. (2013). Preparing for climate change: Forestry and assisted migration. *Journal of Forestry, 111*, 287–297. https://doi.org/10.5849/jof.13-016.

Wright, J. W. (1955). Species crossability in spruce in relation to distribution and taxonomy. *Forest Science, 1*(4), 319–349.

Zehetmayr, J. W. L. (1954). Experiments in tree planting on peat. *Forestry Commission Bulletin, 22*, 110 p. London.

Chapter 31
Climate in Bosnia and Herzegovina, Its Changes and Impact on Forest Genetic Resources

Branislav Cvjetković, Milan Mataruga, Vanja Daničić, and Dalibor Ballian

Abstract Bosnia and Herzegovina is divided into three areas with different climatic conditions: the northern flat is characterized by continental climate, the central part characterized by moderate climate and alpine climate and the southern part characterized by Mediterranean. According to climate change models, it is expected that the average annual temperature will increase to an average between +0.8 and +1.0 °C during the 2010–2030 period and from 1 to 3 °C during the 2031–2060 period. Precipitation will decrease by 10% in the western parts of the country, but it will increase by 5% in the east. Autumn and winter will have the largest decrease in precipitation. The impact on forest ecosystems is inevitable. Negative consequences of extreme climate in forests and forest ecosystems are not easy to identify. Their disclosure requires short term and long-term research and monitoring. This is the only way to determine and identify the cumulative effects of temperature and precipitation. Some of very resistant populations of the most important tree species have been identified and there is need for further investigations and target utilization of such material. Bosnia and Herzegovina is taking steps to address climate change issues, both at state level and internationally. Within Bosnia and Herzegovina climate change is increasingly seen as an issue of key strategic importance, especially by local authorities and the academic community.

Keywords Climate changes · Forests · Bosnia and Herzegovina

B. Cvjetković (✉) · M. Mataruga · V. Daničić
Faculty of Forestry, University of Banja Luka, Banja Luka, Bosnia and Herzegovina
e-mail: branislav.cvjetkovic@sf.unibl.org; milan.mataruga@sf.unibl.org; vanja.danicic@sf.unibl.org

D. Ballian
Faculty of Forestry, University of Sarajevo, Sarajevo, Bosnia and Herzegovina
e-mail: balliandalibor9@gmail.com

M. Šijačić-Nikolić et al. (eds.), *Forests of Southeast Europe Under a Changing Climate*, Advances in Global Change Research 65,
https://doi.org/10.1007/978-3-319-95267-3_31

31.1 Climate Characteristics of Bosnia and Herzegovina

Climate in Bosnia and Herzegovina is a moderate continental in the northern part of the Pannonian Plain, the Alpine climate in the mountainous regions and the Mediterranean climate in the coastal and lowland regions of Bosnia and Herzegovina region in the south part and southeast part of the country. The climate of Bosnia and Herzegovina is determined by several geographical factors: the Atlantic Ocean in the west, the Mediterranean and the Adriatic Sea in the south and the vast continental masses in the northern part of Europe and northeastern and eastern Asia and Africa in the south.

The climate of Bosnia and Herzegovina shows significant diversity. Among the landscapes of moderate continental and altered Mediterranean (Adriatic) climate there are areas with continental, submediteranean, mountain and alpine climate.

In the north part of the country, the average air temperature varies between −1 and −2 °C in the coldest month (January) and between 18 and 20 °C in the warmest month (July). At altitudes above 1000 m the average air temperature varies from −4 to −7 °C in January, and from 9 to 14 °C in July. In the south part of the country, at the Adriatic coast (Mediateranean and submediteranean area) the air temperature varies from 3 to 9 °C in January, and from 22 to 25 °C in July (period 1961–1990). Recorded extreme temperatures were −41.8 °C (the lowest) and 42.2 °C (the highest). The lowlands of northern Bosnia and Herzegovina have an average annual temperature in range of 10–12 °C, and in the areas above 500 m the average temperature is below 10 °C. The mean annual air temperature in the coastal area is between 12 and 17 °C. The lowest temperatures are recorded in the highest mountains.

The annual amount of precipitation in Bosnia and Herzegovina (B&H) is unequally distributed. It rises from the south to the Dinaric Mountains, and then again it falls towards the Pannonia lowland area. The annual rainfall varies from 800 mm in the north part of the country along the River Sava, up to 2000 mm in the central and southeastern mountain regions (period 1961–1990). In the continental part of B&H, the major part of yearly precipitation occurs in the vegetation period, reaching the maximum in June. Concerning central and southern part of B&H, with many mountains and narrow coastal areas, is characterized by an altered Mediterranean participation regime which under the influence of the Adriatic Sea. Monthly maximum amounts are recorded in late autumn and early winter, mostly in two last months of the year – November and December.

Sunny periods duration periods decreases moving from the coast towards the interior and towards higher altitudes. The yearly sum of the sunshine hours in the central mountainous area is between 1700 and 1900 h, which is a consequence of the above-average cloudiness 60–70%. Frequent fogs during the cold season decrease solar irradiation in the interior (mountainous) part of the country in comparison to the same altitude on the coast. In the southern regions we have 1900–2300

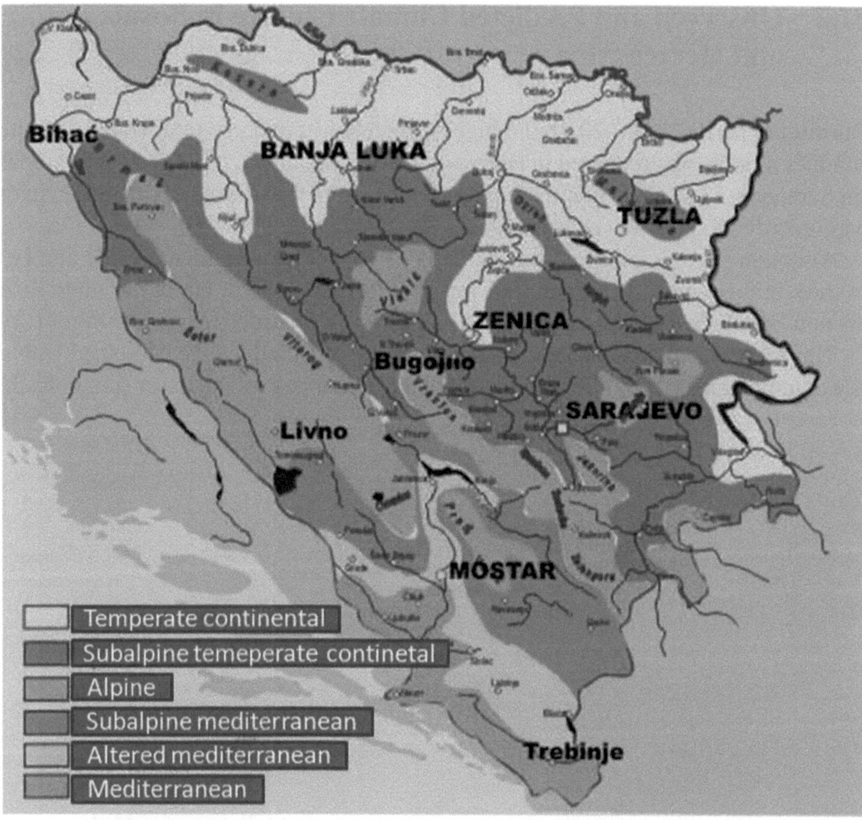

Fig. 31.1 Climatic map of B&H. (Source: hydrometeorological Institute of the Federation of Bosnia and Herzegovina)

sunny hours (Mostar = 2285 h). In northern Bosnia and Herzegovina the number of sunny hours is 1800–2000. Cloudiness decreases in the direction west to east (Second National Report of B&H 2013).

Snow is a regular winter phenomenon, and on mountain peaks it is maintained for 6 months a year. Bosnia and Herzegovina has complex climate characteristics. In the south, the climate is influenced by the Adriatic Sea. In the north, the effects of a moderately warm and humid climate with a warm summer span predominate. The central areas of Bosnia and Herzegovina are under the influence of a modified moderate warm and humid climate, which changes with altitude to the climate with fresh summers. Hyposometric levels over 700 m are characterized by mountain climate and the orographic units by alpine climate (Fig. 31.1).

The climate of Bosnia and Herzegovina is favorable for human life and health, and can be considered to be one of the important natural resources the country possesses.

31.2 Observed and Predicted Climate Change in Bosnia and Herzegovina

During the period 1981–2010, the highest increase in average temperatures in the summer months was recorded in Herzegovina (in Mostar – 1.2 °C) and central areas (in Sarajevo – 0.8 °C), while the highest temperature rise during spring and winter was recorded in northern central regions (in Banja Luka – 0.7 °C).

Although the amount of annual rainfall did not change significantly, the number of days in the year in which rainfall was recorded decreased, but at the same time the number of days during which intense rainfall was recorded increased (Fig. 31.2).

This represents a significant change in rainfall, especially when combined with a rise of temperature (Fig. 31.3). The result of these changes will be lower soil moisture

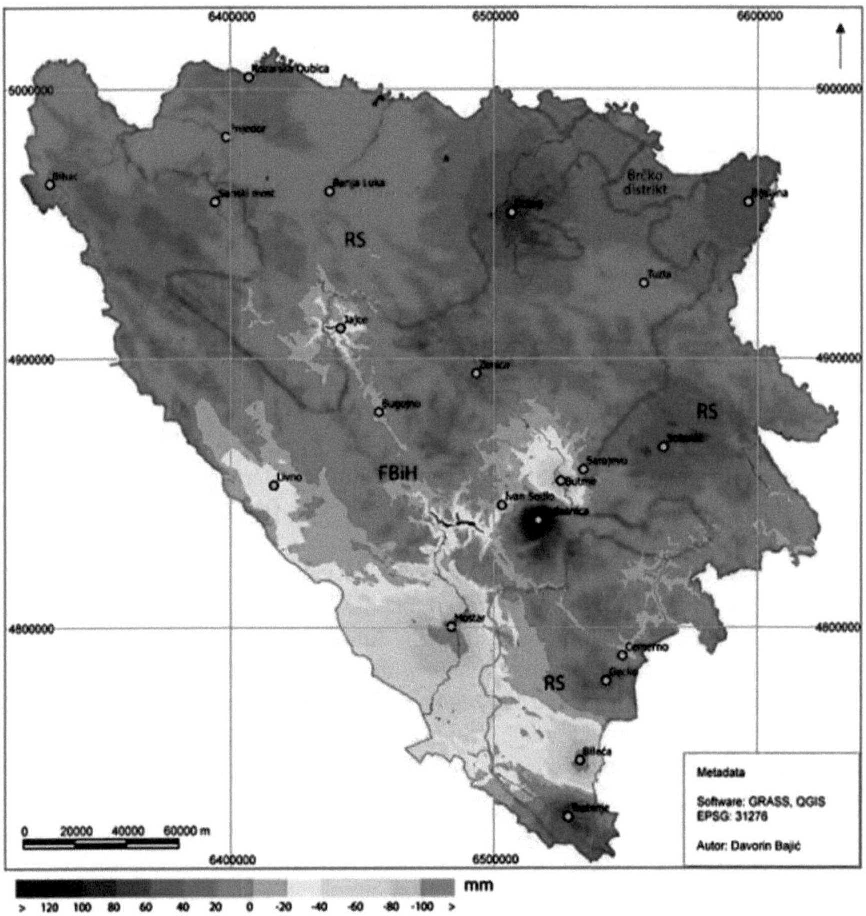

Fig. 31.2 Change of annual rainfall in B&H (1981–2010 in comparison to 1961–1990). (Source: The 2nd national report for UNFCCC)

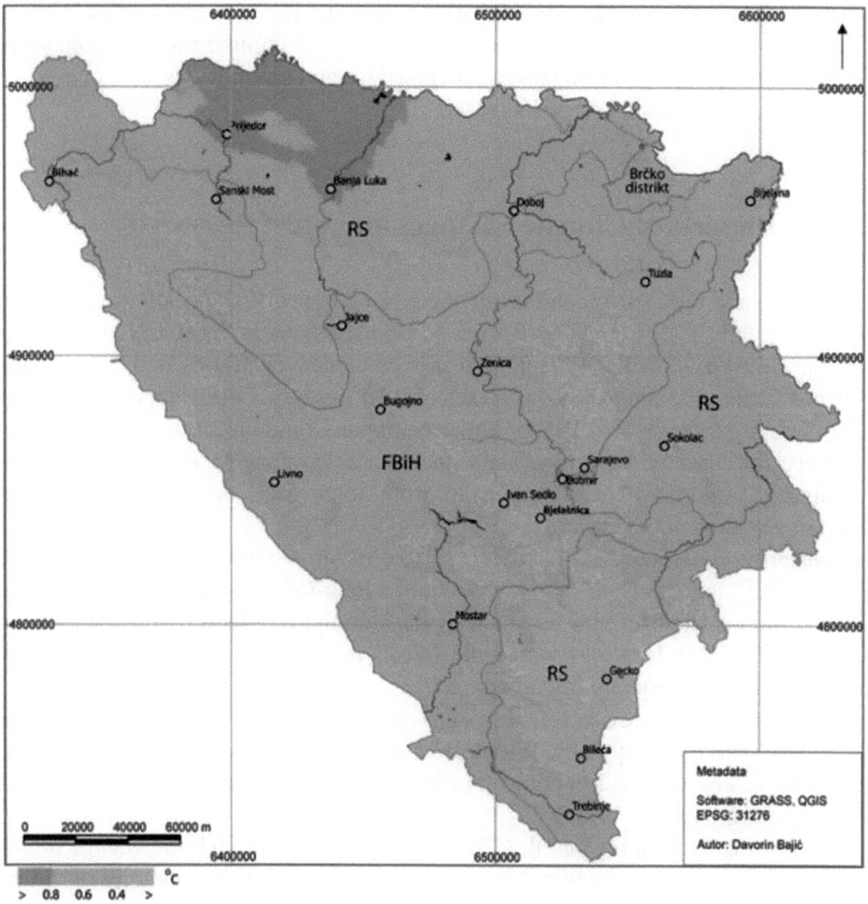

Fig. 31.3 Change of annual average temperatures in B&H (1981–2010 in comparison to 1961–1990). (Source: The 2nd national report for UNFCCC)

(potentially increasing frequency and magnitude of drought), as well as the increased likelihood of floods, given the increase in the frequency of intense rainfall.

According to forecasts for B&H based on climate models, it is expected that the average annual temperature will increase to between 0.8 °C up to 1.0 °C during the 2010–2030 period and from 1 to 3 °C during the 2031–2060 period. The biggest changes are expected in summer months: June, July and August. Precipitation will decrease by 10% in the western parts of the country, but it will increase by 5% in the east. Autumn and winter will have the largest decrease in precipitation. Climatological prognostic models applied to Bosnia and Herzegovina forecasted additional significant increases in temperature during the period 2031–2060, with predicted average temperature increase for 1–2 °C in coastal areas, and in the range from 2 to 3 °C in the interior of the country (Results of the EH50M prognostic model, First National Report (INC) BiH). In addition to temperature increase, climatological forecasting models predict that during the 2030–2060 period significant changes in the amount of precipitation will occur. It is expected that the Western Balkans will experience a reduction in the amount of summer precipitation, when precipitation levels could be halved compared to the current levels. The number of dry days in the interior will be increased.

31.3 Impacts of Climate Changes on Forest Resources

In accordance with the guidelines of the UN Framework Convention on Climate Change (UNFCCC), signed by Bosnia and Herzegovina in 2000, UNDPBH organized, while GEF financially supported an expert team which consisted of 45 experts from 14 scientific areas who were selected to develop First National Report of Bosnia and Herzegovina (INC – Initial National Communication) professionally and in accordance with international standards. It was done so, but without emphasizing the importance of forests, apart from some basic data. Second National Report was done in 2013.

Negative consequences of extreme climate in forests and forest ecosystems are not easy to identify. Their disclosure requires long-term research and monitoring. This is the way to determine and identify the cumulative effects of temperature and precipitation. The locations where the forests are located, i.e., geographical and climatic zones are equally important.

Bosnia and Herzegovina is taking steps to address climate change issues, both at state level and internationally. Within B&H climate change is increasingly seen as an issue of key strategic importance, especially by local authorities and the academic community.

According to Mataruga et al. (2013) direct changes through the change of climate parameters can be expected in the following: shifting the boundaries of certain types of forests in relation to latitude and altitude; different natural distribution of different types of forests in their mutual spatial arrangement; the withdrawal of certain plant communities under the environmental pressure of other species and their disappearance in the long run; different composition of plant communities with the disappearance of one species and the appearance of other species in relation to the floors and social position; the change of the relation of individual trees to light; incidence of drought periods and high temperatures, alongside with the effect of pollutants leads to the decrease in the vitality of trees, and this creates optimal conditions for the development of many pathogens.

Simulation of an average temperature increase by 2 °C predicts significant negative consequences for the distribution of coniferous forests (Second National Report of Bosnia and Herzegovina to the UNFCCC 2013). It is predicted that many habitats in southern Europe, as well as in B&H, will be affected by climate change and species will move towards the north (Falk and Hempelmann 2013). Climate change will be seen through: droughts, pest and disease occurrence, increased risk of fire and soil changes. This is in accordance with the research in the surrounding countries where, according to models of climate change, it can be expected that there will be a decrease in the areas under some forest species until 2100 (Matić 2011).

The destructive effects of natural forces on the Norway spruce have been manifested during the past years on the Balkan Peninsula. In the territory of Slovenia and Croatia, Norway spruce was affected by the appearance of extreme weather conditions that led to snowbreaks and windbreaks on tens of thousands of hectares (Matić 2011; Nagel et al. 2016). Nothing concrete has been done in practice on the impact of climate change on forests in Bosnia and Herzegovina, although there is some fundamental research that could serve as a solution to this problem. On the international level, in 2000 Bosnia and Herzegovina signed the Declaration of Climate Change (UN Convention on Climate Change – UNFCCC) and in 2009 the first national report on climate change was prepared.

During the past few years it has become obvious and clear that climate change is one of the biggest threats to the ecological stability of the planet Earth. As a member of the global community and a signatory to the UN Convention on Climate Change, B&H is obliged to meet the requirements of this Convention and become a part of the concentrated global efforts to stop and reduce the adverse effects of climate change.

As a consequence of global warming, more frequent occurrence of extremes through climate change is expected, threatening the functioning of forest ecosystems. The introduction of species from natural ranges such as marginal populations which are in the drier and warmer climates should be discussed as the options in order to adapt forest ecosystems to negative effects of climate change. High genetic diversity of certain species and hence the potentials in the diversity of tolerance to climate change distinguish certain species that have priority in terms of adaptive capacity. It is necessary to evaluate the response of different species and their provenances to climate changes in order to identify suitable populations or ecotypes which are better suited to the projected climate change.

Extreme temperatures and climatic conditions, such as frost and heat waves, as well as the changes in shape, weather and precipitation (for example, snow compared to rain, drought in comparison with floods) may have an impact on individual species, the location and the levels of forests because these changes can increase the susceptibility to biotic threats (Schlyter et al. 2006).

The positive and negative effects of climate change on forest growth have been observed in different parts of the world. However, much remains unknown in terms of the structure of forests, productivity, as well as the impact of climate change on the growth and the development of woody species in the future.

Climate change, reflected in the increase of average temperatures, change in rainfall patterns, and increase of carbon dioxide can be the "fertilizer" and can promote the growth and development of plants. Thus, forests can grow faster due to increased CO_2 levels in the atmosphere.

European silver fir forests in B&H can show a significant reaction to climate change, considering the fact that they have a very narrow ecological niche and can face a fall or loss. Due to its growth mainly in mixed forests with beech, which has a wider ecological niche and greater adaptability, beech trees have a greater chance to "push out" fir due to the changes in humidity and temperature. Species with narrow niches are likely to face declines and losses (Kirschbaum 2000) and may, in the case of B&H, begin to shift the boundaries of their habitats which show a change in vegetation due to climate change, thereby making other species more dominant (and potentially causing a decline in the economic value of these forests).

In the area of sub-Mediterranean forests in B&H there is a threat of the changes in the soil structure. This could cause a decrease in pH and lead to increased acidity, which will not be acceptable for the existing species. Moreover, mountainous forests and highland ecosystems are extremely vulnerable to temperature changes. The biggest threat will be for specific oak species, which mostly grow at lower altitudes. Threats can undoubtedly cause the migration of species.

In Bosnia and Herzegovina, there are no permanent measuring stations in which the changes and the reactions of the most important forest ecosystems to climate change are monitored and recorded. As for the institutional framework, there is a lack of the integration of problems and climate change issues in forestry policies and strategies, as well as the lack of coordination among the managers and the users of forest resources.

Based on the assumed models of climate change, the comparison with familiar general data on climate parameters for individual forest communities was carried out (Bertović 1975). The average temperatures in different forest ecosystems in B&H are currently ranging from beech forests in the Dinarides (with the average annual temperature of 7.2–7.7 °C) to and hornbeam forests (with the average annual temperature of 12.7–13. 5 °C). Also, Table 31.1 gives an overview of the average annual temperatures in individual locations from the lowest to the highest within the four ecological vegetation areas in B&H.

Predicted changes in forest ecosystems suggest that drastic changes will occur even in the moderate climate change scenario of 2 °C till 2100. Most of the forests characteristic of mountain (Dinaric) regions would evolve into alpine beech forests. Based on this assumption, at the end of this century the dominance of the thermophilic forests of oak, sessile oak with hornbeam, Downy oak and Holm oak is expected.

Table 31.1 Mean annual temperatures and precipitation rates at measuring stations with the lowest and highest values within the four ecological and vegetational areas (Stefanović et al. 1983)

	Annual air temperature[°C]		Annual sum of precitipation [mm]		
Area	Min	Max	Min	Max	IV–IX
Prirpanoinan area	9.6 (Tešanj)	10.9 (Modriča)	831 (Gradiška)	1139 (Sanski Most)	457–613
Transitional ilicri-moesiac area	4.8 (VreloPrače)	11.2 (Brčko)	719 (Višegrad)	1147 (Čajniče)	373–579
Inner dinarides	1.3 (Bjelašnica)	10.7 (Bihać)	804 (Sokola)	1787 (Vaganj)	416–711
Mediterranean	6.0 (Čemerno)	15.0 (Čapljina)	1112 (Prozor)	1951 Ljubinje)	364–658

There is a possibility that climate change can affect forests in B&H in a way that over time it could transform the entire forest systems by moving their spatial distribution and composition. This carries the burden of socio-economic and ecological consequences. Climate change that has occurred will not have the same impact on all forest ecosystems in Bosnia and Herzegovina. In support of this claim, the fact is that the survival of forest communities is not related only (or exclusively) to the average annual temperature in the area where the given community appears, which means that an increase in the average annual temperature will not be the only factor influencing the change.

In addition to average annual temperatures, other important elements include the distribution and the intensity of precipitation which should be analyzed together and in the interaction with the increase of average temperature, as well as with a number of other factors that appear in unpredictable periods and with intangible intensity. It can be noted that the changes in precipitation (+5 to −10%) would not have a drastic impact, as is the case with overlooked changes in average annual temperatures.

What has been a proven fact based on the research done in the world is that each region for which the changes are envisaged should be analyzed individually. This means that it could be expected that in the regions where climate changes are not anticipated there will be no change in the structure of forest ecosystems. Not all plant communities will react in the same way (some are at a higher altitude, a deeper pedological profile, with larger numbers of species and individuals, some are less sensitive, i.e. formed of more tolerant species, etc.), which means that it is necessary to analyze the reaction of each forest plant community separately.

The species that are in the center of their natural distribution will be more tolerant to climate change, while those near the edges (marginal populations) will be very vulnerable. In addition, the succession of species (their evolution) and the change in the structure of communities are related to the natural restoration of forests and are determined by the age of the trees. In some species (such as oaks), it is more than 100 years of age, and in some it is unrealistic to expect the changes in the existing vegetation over a period of less than one century (except in case of natural disasters). Finally, regarding all the changes and movements of forest communities, a number of other factors that influence the changes in forest ecosystems into account must be taken (changes in soil structure, changes in genetic resources and diversity, adaptability of species, etc.).

In addition to the above-mentioned threats, another significant threat to forest ecosystem is caused by increasing the number of forest fires. In some parts of B&H there is increased risk of forest fires caused by increased temperatures and changes in precipitation patterns, which requires the expansion of fire protection capacity. All these aspects (weather, pests, pathogens, fires) can, over a longer period of time, lead to a decrease in productivity and poorer forest conditions in B&H.

Therefore, it is indisputable that the productivity of forests in Bosnia and Herzegovina depends not only on the species and the location of the forest, but also on temperature and precipitation. Higher temperature limits adversely affect productivity, as radical temperatures limit growth. In addition, there are certain complex agents of stress in forests and forest systems: insects, diseases, droughts, floods, landslides, unplanned felling, fires, etc.

To put it briefly, the available data and research have been pointing out that climate change is a threat to all four macro-regions in B&H (ecological and vegetation areas). The Dinarides area, as a very important and rich center of endemic species in the Balkans, will be exposed to a particular threat. A wide range of different human activities impose threats for the rich plant and animal world. One of the significant consequences of global warming regarding ecosystems will be the shift of water supplies and the distribution of pests and diseases. The invasion of allochthonous species will increase, and those species which are more aggressive will push out the autochthonous species out of their habitats. It is currently not possible to accurately predict the success of the adaptation of many forest species to new habitats created by climate change (IPCC 2012).

Significant changes are expected in the genuses inhabiting the mountainous regions of Bosnia and Herzegovina, especially the migration of some woody species in the direction of mountain range Dinarides towards the northwest, with the possible local impoverishment of the flora. It can also be expected that the number of herbaceous species of the narrow ecological valence from the highest mountain areas which will not be able to adapt its area quickly enough will be reduced. It can also be considered that the highest mountainous areas in Bosnia and Herzegovina, from the altitudes of 1500 m and beyond, which is a subalpine zone, will be affected by climate changes.

For the future preservation of natural forests and their biological and genetic diversity it is of at most importance to prevent the occurrence of adverse consequences caused by climate change, primarily fire, insect and pathogenic attacks. At the same time, it is necessary to constantly investigate variability, test adaptability and give recommendations for the targeted use of reproductive material that will produce forests resistant to climate change.

31.4 Research Related to Adaptation to Climate Change

The use of appropriate reproductive material with the aim to make it adaptable to climate change is necessary. Through history, B&H has implemented ecological regionalization on the basis of which the transfer of forest reproductive material within the appropriate regions is recommended. The first steps were made by Đikić et al. (1965) and Fukarek (1970) who provided a detailed natural range of the most important forest species of trees: beech, fir and spruce. By developing the Ecological and vegetational reionization (Stefanović et al. 1983) the reionization of forest species was significantly improved. All studies were carried out only on the basis of bio-ecological characteristics. The variation determined by DNA markers was not done at that time which, in addition to the lack of the data from open field tests, should be the basis for the delineation of the regions (Ballian 2011). This is further reflected on nursery production (Ballian 2010).

Today, the research done on the basis of different types of biochemical and molecular markers is in use. In addition to genetic research, short-term tests are an important segment of research. Black pine, both from extreme habitats and from the most productive habitats, has been explored by setting short-term experimental plots in different environmental conditions (Mataruga et al. 2007, 2011, 2012). Studies have shown that there are significant differences between black pine populations, as well as individual mother trees (the half-sib lines), but there is no difference between the populations originating from extreme habitats and the most productive habitats. This research was a milestone for the melioration and reforestation operations in extreme habitats, bearing in mind that black pine is considered to be the most suitable species for the rehabilitation of the degraded habitats.

Studies on the effect of stress on plant material originating from B&H and the comparison with other European populations were carried out by Bolte et al. (2016). By investigating 1-year seedlings originating from Nevesinje (southern part of B&H) and comparing them with beech from other parts of Europe using the LP50 test, it was determined that the tested beech was in the group of medium-resistant populations.

The results of the study on the resistance of beech to dry conditions carried out by applying cavitation methods were shown at the final collection of the COST action of STREeSS and presented in the poster section under the title "Variation in xylem vulnerability to embolism and European beech from geographically marginal population". The study covered 15 populations from Europe. The tested populations from the northern part of B&H (Banja Luka), besides the population of Laveyron from France, are by far the most resistant beech populations in Europe based on cavitation research. This means that reproductive material from this population could be used for raising new forests in conditions of increased temperature according to the climate change scenarios.

Ballian (2011) provides the instructions for the delineation of seed zones in B&H, as well as the methodology of their separation through several steps that include the analysis of habitat conditions in certain areas, the analysis of genetic diversity by using biochemical and molecular markers and the analysis of the results obtained on experimental plots.

The investigations of the effects of climatic changes indirectly through the analysis of survival and phenologicalphases of domestic species in progeny tests were carried out by: Mataruga et al. (2010), Daničić et al. (2012), Ballian et al. (2015), Cvjetković et al. (2016; 2015a, b, c). At the moment, several domestic and introduced species significant for B&H forestry are being subjected to testing.

31.5 Activities Aimed at Mitigating the Effects of Climate Change

In order to mitigate and prevent climate change in B&H it is necessary to work on:

1. Stability of forest ecosystems

   Stable forest ecosystems are the guarantee for good adaptation to climate change and better adaptability. First of all, the stability of natural forests should be emphasized.

2. Legislation that deals with climate change in FB&H and RS, taking into account scientific achievements in this area.

   It is necessary to draft the specific legislation that will address the problem of climate change at the state and entity level, taking into account the specific situation in which B&H is at the moment.

3. Cross-sectoral dialogue in the fight against climate change

   Very often we have witnessed that there are no coordinated activities among sectors that are interested in combating with climate change.

4. Developing a network for monitoring climate

   It is necessary to start monitoring climate changes through the organization of permanent and temporary experimental plots and with systematic research at the laboratory level. All the obtained data would be unified into one database, and the data would be the basis for making timely and valid strategic decisions in the combat against climate change, as well as for the planned development of the FB&H and RS.

5. Creating expert teams for the fight against climate change

   Establishing expert teams that will manage databases and participate in drafting strategic documents in the fight against climate change. Scientists, politicians of different profiles, representatives of the non-governmental sector and representatives of local social-political communities should be included in the teams.

6. Creating a solid foundation for funding research at the level of all government ministries.

   Providing permanent funds for financing research projects related to climate change at the levels of state, entity and cantonal ministries, as well as the interested business entities.

7. Ensuring support through funds, fees, etc.

   Without the support of the local and wider social political community climate change combat cannot be approached, as certain administrative problems, as well as the problems of material support, might occur all the time. DPZ should ensure systematic financing, as well as specific intervention funds, and provide certain funding through certain types of taxes to combat climate change.

8. Providing cohesion funds at the level of the FB&H and RS.

To facilitate the fight against climate change consequences, in particular the fight of local populations, it is necessary to have an organized fund from which the damage resulting from climate change will be financed and to intervene in the activities carried out in the context of adaptation to climate change.

9. Ensuring the increase of forest surfaces

Increasing areas under forests through afforestation and the of conversion coppice forest into high valuable forests.

10. Encouraging the increase of areas under forests in the private sector

Re-establishing the Karst afforestation fund at B&H level, both the Federation of Bosnia and Herzegovina and the Republic of Srpska.

11. Constant media campaign

Actively conducting media campaigns ~~on~~ related to the importance of combating climate change in the coming period, with the use of all the media.

In addition, regarding the economic perspective, it is necessary to develop short-term and long-term operational plans for entities Federation of B&H and RS Forestry Program in which they need to:

- Develop a comprehensive and integrated methodology for assessing the impacts and sensitivity and provide cost estimates for effective adaptation.
- Improve the basic understanding and prediction of the impact of climate change throughout Bosnia and Herzegovina.
- Explain the expected impacts of climate change and the ozone layer on ecosystems and explore the ways to improve their effectiveness.
- Create an integral database of climate change in Bosnia and Herzegovina through a long-term and comprehensive activity by using wide data sets of high scientific value with the use of numerous models.
- Improve access to the existing data and integrate the data that are relevant to the adjustment, and inter alia through the inclusion of strong long-term monitoring of the quality and quantity of natural resources, biodiversity and other developments in ecosystems on the field.
- Using information systems should be promoted and its capabilities should be fully utilized, for example the link with the relevant Meteorological Data (Hydrometeorological Institute of B&H) or some of the European institutes (other links are also possible).
- It is necessary to provide an updated summary of climate change reports every 4–5 years.

- Together with the economy (forestry, agriculture, water management, energy, etc.), make a plan to support research on forest adaptation and business, services and industry.
- Improve knowledge related to the movement and availability of resources in Bosnia and Herzegovina, the region and Europe, including renewable energy sources.
- Promote cooperation, partnership and networking with the scientific community in the EU and the neighboring countries of the region, as well as with developing countries, and other key partners who are willing to help and share research results, work together on climate models and other methodological tools, in particular in the context of the development of the short-term (5-year) program of work on adaptation to climate change (UNFCCC).
- Provide support to forestry practice by providing guidance on the existing scientific knowledge and the adaptation measures and providing options and analyzing the costs and benefits of these options.

References

Ballian, D. (2010). Rasadnička proizvodnja. Savjetodavni sastanak oko ispunjavanja zakonskih obaveza vezanih za pregled rasadnika, sjemena i sjemenskih sastojina u cilju obezbjeđenja kvalitetnog materijala (sjemena i sadnica) za biološku reprodukciju šuma u FBiH, 16.12.2010. Sarajevo.

Ballian, D. (2011). Osnovni principi rajonizacije sjemenskih objekata za proizvodnju sjemenana genetičkim principima. *Radovi Hrvatskog društva za znanost i umjetnost, 12*(13), 18–41.

Ballian, D., Jukić, B., Kajba, D., & von Wüehlisch, G. (2015). Phenological variability of European beech (*Fagus sylvatica* L.) in the international provenance trial. *EUFORIA – European Forest Research and Innovation Area*. August 31st–4th September 2015, Rogla, Slovenia.

Bertović, S. (1975). Ekološko-vegetacijske značajke okoliša Zavižana u sjevernom Velebitu. *Glasnik za šumarske pokuse, 18*, 5–75.

Bolte, A., Czajkowski, T., Cocozza, C., Tognetti, R., De Miguel, M., Pšidová, E., Ditmarova, L., Dinka, L., Delzon, S., Cochard, H., Ræbild, A., De Luis, M., Cvjetkovic, B., Heiri, C., & Müller, J. (2016). Desiccation and mortality dynamics in seedlings of different European beech (*Fagus sylvatica* L.) populations under extreme drought conditions. *Frontiers in Plant Science*. https://doi.org/10.3389/fpls.2016.00751.

Cvjetković, B., Mataruga, M., Šijačić-Nikolić, M., Daničić, V., & Lučić, A. (2015a). Bud burst and height increment of Norway spruce (*Piceaabies* Karst.) in progeny tests in Bosnia and Herzegovina, International conference "Reforestation Challenges", Proceedings (pp. 251–259).

Cvjetković, B., Mataruga, M., Šijačić-Nikolić, M., Daničić, V., & Stojnić, S. (2015b). Survival of Norway spruce seedlings (*Piceaabies* (L.)Karst.) in progeny test "Srebrenica" in purpose of transfer of forest reproductive material of this species. The Third Republic of Srpska Biologist and Ecologist Symposium. *Book of abstracts* (pp. 136–137).

Cvjetković, B., Mataruga, M., Šijačić-Nikolić, M., Ivetić, V., Daničić, V., Stojnić, S., & Stojanović, M. (2015c). Norway spruce (*Picea abies* (L.) Karst.) seedlings survival in progeny test "Drinić". *Bulletin of Faculty of Forestry Banja Luka, 22*, 5–14.

Cvjetković, B., Mataruga, M., Šijačić-Nikolić, D.,. V., & Popović, V. (2016). Variability of Norway spruce morphometric characteristics in progeny tests in Bosnia and Herzegovina. *Builletin of Faculty of Forestry Belgrade, 113*, 11–35.

Daničić, V., Isajev, V., Mataruga, M., & Cvjetković, B. (2012). Variability of flowering and seed crops in a clone seed orchard of scots pine (*Pinuss ylvestris* L.). In *International scientific conference "Forests in the future – Sustainable use, risks and challenges"*, Belgrade, 4th–5th, October, Book of abstracts, p:99. Proceedings (pp. 441–450).

Đikić, S., Jovančević, M., & Panov, A. (1965). Principi i perspektive unapređenja proizvodnje šumskog sjemena u Bosni i Hercegovini, Šumarski fakultet i Institut za šumarstvo Sarajevo *Posebna izdanja* br. 3, 78.

Falk, W., & Hempelmann, N. (2013). Species favorability shift in Europe due to climate change, a case study for *Fagus sylvatica* L. and *Picea abies* (L.) Karst.based on an ensemble of climate models. Journal of Climatology Article ID 787250, 1–18.

Fukarek, P. (1970). Rasprostranjenje i raprostranjenosti bukve, jele i smrče na području Bosne i Hercegovine, Akad. nauka i umjetnosti BiH, radovi XXXIX, knjiga. *11*, 231–256.

IPCC. (2012). In C. B. Field, V. Barros, T. F. Stocker, D. Quin, D. J. Dokken, K. L. Ebi, M. Mastrandrea, K. Mach, G.-K. Plattner, S. Allen, M. Tignor, & P. Midgley (Eds.), *Summary for policymakers. Managing the risks of extreme events and disasters to advance climate change adaptation* (pp. 1–19). Cambridge: Cambridge University Press.

Kirschbaum, M. U. F. (2000). Forest growth and species distribution in a changing climate. *Tree Physiology, 20*, 309–322.

Mataruga, M., Isajev, V., Konstantinov, K., Mladenović-Drinić, S., & Daničić, V. (2007). Protein in seed and seedlings of selected austrian pine (*Pinus nigra* Arnold) trees as genetic markers tolerant to drought. *Genetika, 39*(2), 259–272.

Mataruga, M., Isajev, V., Balotić, P., Burlica, Č., & Cvjetković, B. (2010). Progeny tests of Norway spruce (*Picea abies Karst.*) in Bosnia and Herzegovina – Contribution to the European ex situ conservation. In *First Serbian congress future with forests, proceedings* (pp. 378–389).

Mataruga, M., Isajev, V., Daničić, V., & Cvjetković, B. (2011). The dynamics of germination and morphometrics properties of Austrian pine (*Pinusnigra* Arnold) saplings in terms of early indicators of tolerance toward the drought. *Genetika, 43*(1), 75–90.

Mataruga, M., Haase, D., Isajev, V., & Orlović, S. (2012). Growth, survival, and genetic variability of Austrian pine (Pinusnigra Arnold) seedlings in response to water deficit. *New Forests, 43*, 791–804.

Mataruga, M., Isajev, V., Orlović, S., Đurić, G., Daničić, V., Cvjetković, B., Balotić, P., & Ćopić, M. (2013). Program očuvanja šumskih genetičkih resursa Republike Srpske 2013–2025 godine. Ministarstvo poljoprivrede, šumarstva i vodoprivrede Republike Srpske (pp. 36–145).

Matić, S. (2011). Utjecaj stanišnih promjena i načina gospodarenja na sušenje obične smreke (*Picea abies* Karst.) u Hrvatskoj. *Croatian Journal of Forest Engineering, 31*(1), 7–17.

Nagel, T. A., Firm, D., Rozenbergar, D., & Kobal, M. (2016). Patterns and drivers of ice storm damage in temperate forests of Central Europe. *European Journal of Forest Research, 135*, 519–530.

Schlyter, P., Stjernquist, I., Bärring, L., Jönsson, A. M., & Nilsson, C. (2006). Asessment of the impacts of climate change and weather extremes on boreal forests in northern Europe, focusing on Norway spruce. *Climate Research, 31*, 75–84.

Second National Report to the UNFCCC. (2013).

Stefanović, V., Beus, V., Burlica, Č., Dizdarević, H., & Vukorep, I. (1983). *Ekološko-vegetacijska rejonizacija Bosne i Herzegovine* (pp. 1–49). *Posebna izdanja* br 18. Šumarski fakultet u Sarajevu.

Chapter 32
Conservation and Management of Romanian Forest Genetic Resources in the Context of Climate Change

Dragoş Postolache, Alexandru Lucian Curtu, Neculae Şofletea, and Flaviu Popescu

Abstract Forest genetic resources (FGR) of Romania harbour high genetic diversity although human impact, during the last two centuries, has strongly affected forest ecosystem composition and structure. Dynamic *in situ* conservation of FGR was the first choice approach to conserve the adaptation potential in natural populations of target tree species by selecting most representative and valuable populations. However, only 17 FGR growing in extreme environmental conditions have been selected and incorporated into the National Catalogue of Forest Genetic Resources. Establishment of new FGR in marginal tree populations that are close to the xeric limit of species' natural distribution should be extended. Almost half of tree populations selected for adaptation and included in the National Catalogue of Basic Materials for Production of Forest Reproductive Materials are represented by oak species, which is explained by lack of natural regeneration in the last decades. The shortage of native forest reproductive material of oak species (e.g. *Quercus*

D. Postolache (✉)
National Institute for Research and Development in Forestry "Marin Drăcea", Cluj-Napoca, Romania

University of Agricultural Sciences and Veterinary Medicine, Cluj-Napoca, Romania
e-mail: dragospostolache@yahoo.com

A. L. Curtu
Department of Forest Sciences, Transilvania University of Braşov, Braşov, Romania
e-mail: lucian.curtu@unitbv.ro

N. Şofletea
Faculty of Silviculture and Forest Engineering, Transilvania University of Braşov, Braşov, Romania
e-mail: nic.sofletea@unitbv.ro

F. Popescu
National Institute for Research and Development in Forestry "Marin Drăcea", Simeria, Romania
e-mail: flaviu.popescu@icashd.ro

M. Šijačić-Nikolić et al. (eds.), *Forests of Southeast Europe Under a Changing Climate*, Advances in Global Change Research 65,
https://doi.org/10.1007/978-3-319-95267-3_32

robur, Q. pedunculiflora, Q. pubescens) remains one of the biggest challenges in the conservation and management of rear edge oak populations in Romania. Molecular fingerprinting of FGR should constitute an important tool for genetic monitoring of evolutionary changes and to assist ecological reconstruction of fragmented tree populations. Collaboration with neighboring countries is an important aspect for dynamic conservation of FGR and their sustainable management.

Keywords Forest genetic resources · Climate change · Adaptation · Genetic monitoring · Romania

32.1 General Aspects

Climate is a major selective driver acting on forest tree species that have to cope with climate change by physiological adjustments in the short term and by optimizing their fitness in following generations.

In addition, there is general agreement that climate change is a reality, which has a growing impact on the European forests.

The Fifth Assessment Report (AR5) of the Intergovernmental Panel on Climate Change (IPCC 2013; Christensen et al. 2013) demonstrated consistent agreement with previous report (IPCC 2007; Parry 2007) and pointed out that the annual mean temperatures over all Europe will have a rising trend throughout the twenty-first century.

The last report of European Environmental Agency (EEA) indicate that in the last decade 2006–2015 the European land temperatures were approximately 1.5 °C warmer compared to the pre-industrial level (Marx et al. 2017).

Increasing frequency and severity of extreme drought events have been already observed in many forests from southern and eastern Europe, including Romania, (Bertini et al. 2011; Mátyás et al. 2010; Pravalie et al. 2014) that may cause negative effects on forest health, stability and productivity.

It is expected that forest tree species distribution will be modified by climate-induced biome shifts northward at the leading edge and retraction at the rear edge of species distribution with possible severe economic and social consequences (Hanewinkel et al. 2013).

While the frequency of high temperature and drought events increased, in the past two decades, at the rear range of species distribution, the frequency of events favorable for seedling recruitments have declined (Jump et al. 2009).

The rate at which temperature isotherms will shift across different areas of Earth's terrestrial landscapes vary significantly by local topography, being higher in plains such as temperate grassland (590 m year^{-1}), lower in temperate broadleaf and mixed forests (350 m year^{-1}) and negligible in mountainous biomes such as temperate coniferous forests (100 m year^{-1}) (Loarie et al. 2009).

Consequently, species will have to move at short distances on mountain slopes in order to track the rate of temperature changes, compared to flatter areas where species will have to migrate at much longer distances. In this context, mountainous areas will strongly mediate the consequences of temperature increase by facilitating species to move upwards (Loarie et al. 2009).

However, habitat fragmentation and land cover changes, resulted from the human activities, represent considerable obstacles to range shifts of tree populations and an impediment to gene flow among populations that may reduce the rate of adaptation (Davis and Shaw 2001).

Studies of range shifts of European tree species during the Quaternary, revealed by fossil and phylogeographic surveys (Hewitt 2004; Magri et al. 2006; Petit et al. 2002; Piotti et al. 2017; Postolache et al. 2017), provide details about species responses to past climate changes that shaped current population genetic diversity patterns.

The process of range shifts produced in the past by major climate changes were not simply the result of contraction and expansion at the range limits but have also involved adaptive evolutionary changes (Davis and Shaw 2001).

Understanding the postglacial history of European tree species may help to comprehend better the evolution of local climatic adaptation and also to identify favorable microhabitats (e.g. refugia) where species have persisted during past climate changes.

Ongoing global climate change points to the importance of studing adaptive mechanisms involved in local adaptation of natural tree populations in order to understand their capacity to adapt to rapid climate changes.

To get on with ongoing rates of climate warming, over the coming centuries, forest tree species will have to rely on the interplay of three mechanisms through *migration*, *phenotypic plasticity* and *adaptation through natural selection*; these options are not mutually exclusive (Davis and Shaw 2001; Jump and Penuelas 2005; Nicotra et al. 2010).

32.2 Challenges in Conservation of Standing Genetic Variation in Forest Genetic Resources From Romania

Historical forest management in Romania has strongly affected the forest ecosystem composition and structure with direct consequences on ecosystem functioning and stability.

Although the forest cover has increased in Romania since 1924 by 5% (by roughly 308,000 ha), the intensive management in the past disturbed forest composition with substantial decrease of deciduous oak forests, especially *Quercus pubescens*, *Q. robur* and *Q. pedunculiflora* and with an increase in proportion of Norway spruce forests (Munteanu et al. 2016).

Current forest disturbances, such as deforestation and forest degradation, were particularly high in the mountain regions of Eastern Carpathians (counties Suceava, Maramureş, Harghita, Covasna, Vrancea) and Southern Carpathians (counties Argeş, Vâlcea, Hunedoara), likely due to increased accessibility (Andronache et al. 2017).

The application of fractal analysis using the time-series of satellite remote sensing data revealed increased fragmentation of forest areas especially in North-Eastern and in South-West Oltenia region, mostly because of intensified rate of deforestations (Andronache et al. 2017).

Human impact such as deforestation and forest fragmentation, toghether with ongoing climate changes may amplify the rate of mortality and physiological stress of forest trees, especially at the rear edge of species distribution.

Standing genetic variation and population sizes will be playing a major role in maintaining vitality of forests at the rear-edge of the species distribution as being deprived of gene flow from better adapted populations and could consequently face risk of extinction (Kremer et al. 2014).

One of the best strategy to preserve the standing genetic variation of trees over long term is by maintaining evolutionary processes through dynamic conservation (*in situ* and dynamic ex situ) of forest genetic resources (FGR) (Lefevre et al. 2013).

The ultimate goal of FGR dynamic conservation is to preserve the adaptation potential of trees that will facilitate forests adaptation to climate change (Kelleher et al. 2015).

In situ conservation of FGR in Romania was the first choice approach to conserve adaptation potential in natural populations of target tree species by selecting most representative and valuable populations (Popescu et al. 2015).

Moreover, to mitigate the effect of climate change, 106 populations (1166.94 ha) of 21 tree species (19 are autochthonous species) have been selected for adaptation and included in the National Catalogue of Basic Materials for Production of Forest Reproductive Materials (Pârnuţă et al. 2012) (Table 32.1).

The number of populations and target species that have been selected for adaptation was based on the analysis of population performance in specific environmental conditions and also taking into consideration the economic importance of target species.

Almost half of selected populations for adaptation are represented by oak species, (more precisely, 58 populations of *Quercus* spp.) with a total area of 510 ha (Table 32.1). The large number of selected oak populations for adaptation is explained by lack of regeneration in the last decades.

Furthermore, large number of populations for adaptation have been selected for species of economic significance, such as Norway spruce (*Picea abies*) with 12 populations with total area of 267.7 ha, and beach (*Fagus sylvatica*) with 6 populations with total area of 117.9 ha (Table 32.1).

A number of 17 forest genetic resources have been identified growing under extreme environmental conditions and have been included in the National Catalogue of Forest Genetic Resources (Pârnuţă et al. 2011) (Table 32.2).

The most numerous populations have been selected for *Q. pedunculiflora* (5 populations with total area of 55.7 ha) and for *Robinia pseudoacacia* (5 populations with total area of 63.5 ha).

The predominant areas with forest tree species growing in extreme environmental conditions are located in wood steppe region and only few in high-alpine zones (e.g. *Picea abies*).

Pubescent oak (*Q. pubescens*) and grayish oak (*Q. pedunculiflora*) are main tree species of forest ecosystems in wood steppe region, particularly in southern part of Romania, where grayish oak occupies more drought sites.

Table 32.1 Number of selected populations for adaptation in The National Catalogue of Basic Materials for Production of Forest Reproductive Materials

No.	Species	No. Pop.	Area (ha)
Conifers			
1	*Abies alba* Mill.	1	20.5
2	*Picea abies* (L.) Karst.	12	276.7
3	*Pinus cembra* L.	3	7.4
4	*Pinus mugo* Turra	1	33.9
5	*Pinus nigra ssp. nigra* Arn.	1	26.5
6	*Pinus sylvestris* L.	1	57.5
7	*Taxodium distichum* (L.) Rich. (allochthonous species)	2	2.8
Broadleaves			
8	*Fagus sylvatica* L.	6	117.9
9	*Fraxinus excelsior* L.	3	9
10	*Fraxinus angustifolia* Vahl.	1	3.9
11	*Populus alba* L.	4	21.4
12	*Populus nigra* L.	2	4
13	*Prunus avium* L.	1	2
14	*Quercus cerris* L.	7	30.46
15	*Quercus frainetto* Ten.	4	41.14
16	*Quercus pedunculiflora* K.Koch	7	44.9
17	*Quercus petraea* (Matt.) Liebl.	8	63.74
18	*Quercus pubescens* Willd.	10	36.6
19	*Quercus robur* L.	22	293
20	*Robinia pseudoacacia* L. (allochthonous species)	5	31.1
21	*Tilia tomentosa* Moench.	5	42.5
	Total	**106**	**1166.94**

Table 32.2 Number of selected populations in extreme environmental conditions included in the National Catalogue of Forest Genetic Resources

No.	Forest tree species	No. of pop.	Area (ha)
1	*Fraxinus excelsior* L.	1	2.9
2	*Juglans nigra* L.	1	0.9
3	*Picea abies* (L.) Karst.	1	32.4
4	*Pinus sylvestris* L.	1	32.4
5	*Quercus pedunculiflora* K.Koch	5	55.7
6	*Quercus robur* L.	2	32.6
7	*Robinia pseudoacacia* L.	5	63.5
8	*Ulmus pumila* L.	1	2.1
	Total	**17**	**222.5**

Pedunculate oak (*Q. robur*) extends its distribution range far at the external limits of wood region but only rarely with grayish oak, as for example in Letea forest (Şofletea and Curtu 2007).

Additionally, comparative genetic analysis with molecular markers of pedunculate oak and graysh oak populations din not revealed a distinct genetic cluster for *Q. pedunculiflora* and authors suggested to consider it as a subspecies of *Q. robur*, but grayish oak tolerance to more drought climate conditions should be studied further and taken into consideration (Curtu et al. 2009).

Both graysh oak and pubescent oak populations that are growing in southern part of Romania are considered very important drought tolerant tree species that could mitigate the effect of climate changes (Apostol et al. 2016).

However, the natural regeneration in rear edge populations of oak species has decreased significantly in the last decades. Monitoring regeneration in the rear edge populations of oak species should be considered and implemented, as natural regeneration play an important role in demographic and genetic processes.

Rear edge populations are usually small and isolated populations that may harbour exceptional populations with very high levels of species genetic diversity (Hampe and Petit 2005).

Adaptation in populations located at the rear edge depends mostly on standing genetic variation within the local populations, because they are deprived of gene flow from better adapted populations (Davis and Shaw 2001).

On the other hand, higher gene-flow trhough human-aided migration between rear-edge and core populations might be very useful for adaptive evolution of central and leading edge populations in the context of clime change but only if natural regeneration in native population is absent or low (Fady et al. 2016).

The option that rear edge populations will migrate to track suitable conditions may be limited for species with long generation times and inhabiting highly fragmented landscape, as the case of pedunculate oak populations located at the southern edge of their distribution in Romania. In order to avoid extinction, these populations will have to cope with rapid temperature increase mainly through phenotypic plasticity and/or adaptive evolution.

At the national level there is a growing interest to identify, describe and include new forest genetic resources of pubescent oak and graysh oak in the National Catalogue of Forest Genetic Resources (Apostol et al. 2016; Budeanu et al. 2016).

Allochthonous species as black locust (*Robinia pseudoacacia*) has been successfully used in afforestation projects in steppe and wood steppe regions, especially for sand dune stabilization in south-western part of Romania.

Norway spruce (*Picea abies*) is the most dominant tree species in the alpine zones and fulfils important functions as protection against erosion.

32.3 Adaptive Management of Forest Genetic Resources to Climate Change

Southern Romania is already confronting with the problem of desertification and in many oak forests (e.g. *Q. robur*) the frequency of mast year events decreased significantly. Simulation results suggests that under extreme and moderate climate change scenarios, a considerable change in species composition occurred at large scales in the forests of Carpathian Mountains where the share of broadleaved species is expected to increase along with the growing abundance of more thermophile and/or drought-resistant species (e.g. *Q. pubescens*) (Bouriaud et al. 2015).

Additionally, a recent study based on various climatic scenarios show that one of possible effects would be changes in species distribution patterns (Barbu et al. 2016). In this context, it is necessary to re-evaluate the structure and the current distribution of FGR.

On the other hand, FGR located at the rear edge are expected to be more affected by global climate changes due to their small effective population size. However these FGR are of great scientific and practical interest, especially as they may contain specific adaptation to extreme environments.

One of the most difficult problem to solve in the management of rear edge populations of oak species (e.g. *Q. robur, Q. pedunculiflora, Q. pubescens*) remains the shortage of native forest reproductive material, but also the low number of FGR and forest stands designated for seed production (Apostol et al. 2016).

Establishment of new FGR and of new stands for forest reproductive material production in marginal populations should be accelerated to increase the resiliency of forests to forthcoming extreme droughts, especially in drought-threatened flatlands in southern Romania.

If the effective population size of rear edge populations is too small, the human-aided migration or establishment of ex situ conservation units in suitable areas should be taken into consideration for dynamic conservation of these threatened populations that may contain specific adaptations (Kelleher et al. 2015).

An adaptive management strategy to respond to climate change should rely on keeping natural regeneration or on using local seed source for artificial regeneration.

An additional adaptive strategy relies on selecting large FGR with multi-species mixtures with significant altitudinal or ecological variation to maintain a high adaptation potential.

Assessment and genetic monitoring of tree species genetic diversity with molecular markers can contribute with outstanding knowledge on species genetic structure that should be incorporated in new guidelines for conservation and management of FGR.

Previous national genetic studies performed with molecular markers are not yet integrated into the sustainable management of FGR, although various phylogeographic studies revealed genetic diversity hotspots in key economic species, such as oaks (Moldovan et al. 2010; Popescu and Postolache 2009) and silver fir (Postolache et al. 2016).

Effective long-term research and genetic monitoring should be instantly applied in small FGR located at the rear edge of the species' distribution range, in order to observe beforehand lack of adaptive capacity that might be vital for adaptive resource management (Filippos 2016; Mátyás 2010).

Several new European initiatives anchored on genetic monitoring of forest species include research activities on conservation and assessment of adaptive potential. The two European projects GenTree and LIFEGENMON are addressing the challenging task to develop guidelines for forest genetic monitoring, which will support better management and conservation of FGR.

Regional cooperation with neighboring countries in sustainable forest management and use of FGR is crucial in adaptation of forest species to climate change.

The European Information System on Forest Genetic Resources (EUFGIS) provides support to increase the level of cooperation and linkages between national FGR conservation programmes and research activities (Kelleher et al. 2015).

Last but not least the advanced cost-effective molecular approaches in forestry (e.g. genotyping-by sequencing, exome sequencing, amplicon sequencing, targeted gene panels) represent a powerful tool for obtaining answers on the role of natural selection in shaping patterns of genetic variation and detect environmental drivers of selection (Plomion et al. 2016). Genomic resources may increase the quality and amount of relevant data for identifying adaptive genetic variation in forest tree species.

32.4 Conclusions

Comprehensive genetic surveys of key economic tree species with molecular markers suggest that FGR in Romania harbour high genetic diversity that most probably was preserved by natural regeneration.

Maintaining high levels of genetic diversity through dynamic *in situ* and *ex situ* conservation of FGR is the first priority of national conservation strategy to support resilience of forests and mitigate the effects of climate change.

The adaptation of forest ecosystems to climate change can be facilitated by appropriate use of FGR and by increasing awareness of the importance of using forest reproductive material (FRM) of high-quality.

The National Catalogue of Basic Materials for Production of Forest Reproductive Materials includes a number of 106 populations of 21 tree species that have been selected for their adaptative value.

More attention should be devoted to forest tree populations growing under extreme environmental conditions and especially in semi-arid areas.

Permanent monitoring plots well distributed, especially in drought-threatened areas, should be established to detect early signs of lack of adaptive capacity and to elaborate guidelines for adaptive forest management.

Acknowledgements The research was supported by national program (DESFOR) of the Ministry of National Education and Scientific Research of Romania (PN16330201).

References

Andronache, I., Fensholt, R., Ahammer, H., Ciobotaru, A.-M., Pintilii, R.-D., Peptenatu, D., et al. (2017). Assessment of textural differentiations in forest resources in Romania using fractal analysis. *Forests, 8*, 54.

Apostol, E. N., Dinu, C. G., Apostol, B., Ciuvăţ, A. L., Lorenţ, A., Pleşca, I., et al. (2016). Importance of pubescent oak (Quercus pubescens Willd.) for Romanian forests in the context of climate change. *Revista de Silvicultură şi Cinegetică, 21*, 29–33.

Barbu, I., Barbu, C., Curcă, M., & Ichim, V. (2016). Adaptarea pădurilor României la schimbările climatice (pp. 378). Voluntari, Ilfov: Editura Silvica.

Bertini, G., Amoriello, T., Fabbio, G., & Piovosi, M. (2011). Forest growth and climate change: Evidences from the ICP-Forests intensive monitoring in Italy. *iForest-Biogeosciences and Forestry, 4*, 262.

Bouriaud, L., Bouriaud, O., Elkin, C., Temperli, C., Reyer, C., Duduman, G., et al. (2015). Age-class disequilibrium as an opportunity for adaptive forest management in the Carpathian Mountains, Romania. *Regional Environmental Change, 15*, 1557–1568.

Budeanu, M., Stuparu, E., & Tănăsie, Ş. (2016). Identificarea de noi resurse genetice forestiere de cvercinee cu adaptabilitate ridicată. *Revista de Silvicultură şi Cinegetică, 21*, 6.

Christensen, J. H., Kanikicharla, K. K., Marshall, G., & Turner, J. (2013). *Climate phenomena and their relevance for future regional climate change*. Cambridge: Cambridge University Press.

Curtu, A. L., Şofletea, N., Toader, A.-V., Enescu, M., Moldovan, C., & Chesnoiu, E.-N. (2009). Stejarul brumăriu: Specie sau unitate intraspecifică a stejarului pedunculat. *Revista pădurilor, 124*, 7.

Davis, M. B., & Shaw, R. G. (2001). Range shifts and adaptive responses to Quaternary climate change. *Science, 292*, 673–679.

Fady, B., Cottrell, J., Ackzell, L., Alía, R., Muys, B., Prada, A., et al. (2016). Forests and global change: What can genetics contribute to the major forest management and policy challenges of the twenty-first century? *Regional Environmental Change, 16*, 927–939.

Filippos, A. (2016). Conservation and monitoring of tree genetic resources in temperate forests. *Current Forestry Reports, 2*, 119–129.

Hampe, A., & Petit, R. J. (2005). Conserving biodiversity under climate change: The rear edge matters. *Ecology Letters, 8*, 461–467.

Hanewinkel, M., Cullmann, D. A., Schelhaas, M.-J., Nabuurs, G.-J., & Zimmermann, N. E. (2013). Climate change may cause severe loss in the economic value of European forest land. *Nature Climate Change, 3*, 203–207.

Hewitt, G. (2004). Genetic consequences of climatic oscillations in the quaternary. Philosophical transactions of the Royal Society of London. *Series B: Biological Sciences, 359*, 183–195.

Jump, A. S., & Penuelas, J. (2005). Running to stand still: Adaptation and the response of plants to rapid climate change. *Ecology Letters, 8*, 1010–1020.

Jump, A. S., Mátyás, C., & Peñuelas, J. (2009). The altitude-for-latitude disparity in the range retractions of woody species. *Trends in Ecology & Evolution, 24*, 694–701.

Kelleher, C., de Vries, S., Baliuckas, V., Bozzano, M., Frýdl, J., Gonzalez Goicoechea, P., et al. (2015). *Approaches to the conservation of Forest genetic resources in Europe in the context of climate change. European Forest Genetic Resources Programme (EUFORGEN)* (Vol. xiv, p. 3). Rome: Bioversity International.

Kremer, A., Potts, B. M., & Delzon, S. (2014). Genetic divergence in forest trees: Understanding the consequences of climate change. *Functional Ecology, 28*, 22–36.

Lefevre, F., Koskela, J., Hubert, J., Kraigher, H., Longauer, R., Olrik, D. C., et al. (2013). Dynamic conservation of forest genetic resources in 33 European countries. *Conservation Biology: The Journal of the Society for Conservation Biology, 27*, 373–384.

Loarie, S. R., Duffy, P. B., Hamilton, H., Asner, G. P., Field, C. B., & Ackerly, D. D. (2009). The velocity of climate change. *Nature, 462*, 1052–1055.

Magri, D., Vendramin, G. G., Comps, B., Dupanloup, I., Geburek, T., Gömöry, D., et al. (2006). A new scenario for the quaternary history of European beech populations: Palaeobotanical evidence and genetic consequences. *New Phytologist, 171*, 199–221.

Marx, A., Bastrup-Birk, A., Louwagie, G., Wugt-Larsen, F., Biala, K., Füssel, H.-M., et al. (2017). Terrestrial ecosystems, soil and forests. *European Environment Agency (EEA) Report, 1*, 30.

Mátyás, C. (2010). Forecasts needed for retreating forests. *Nature, 464*, 1271.

Mátyás, C., Berki, I., Czúcz, B., Gálos, B., Móricz, N., & Rasztovits, E. (2010). Future of beech in Southeast Europe from the perspective of evolutionary ecology. *Acta Silv Lign Hung, 6*, 91–110.

Moldovan, I. C., Sofletea, N., Curtu, A. L., Abrudan, I. V., Postolache, D., & Popescu, F. (2010). Chloroplast DNA diversity of oak species in Eastern Romania. *Notulae Botanicae, Horti Agrobotanici, Cluj-Napoca, 38*, 302–307.

Munteanu, C., Nita, M. D., Abrudan, I. V., & Radeloff, V. C. (2016). Historical forest management in Romania is imposing strong legacies on contemporary forests and their management. *Forest Ecology and Management, 361*, 179–193.

Nicotra, A. B., Atkin, O. K., Bonser, S. P., Davidson, A. M., Finnegan, E., Mathesius, U., et al. (2010). Plant phenotypic plasticity in a changing climate. *Trends in Plant Science, 15*, 684–692.

Pârnuţă, G., Stuparu, E., Budeanu, M., Scărlătescu, V., Marica, F. M., Lalu, I., et al. (2011). *Catalogul naţional al resurselor genetice forestiere* (pp. 5–522). Bucureşti: Editura Silvică.

Pârnuţă, G., Budeanu, M., Stuparu, E., Scărlătescu, V., Chesnoiu, E., Tudoroiu, M., et al. (2012). *Catalogul naţional al materialelor de bază pentru producerea materialelor forestiere de reproducere*. Bucureşti: Editura Silvică.

Parry, M. L. (2007). *Climate Change 2007: Impacts, adaptation and vulnerability: Contribution of Working Group II to the fourth assessment report of the Intergovernmental Panel on Climate Change* (Vol. 4). Cambridge: Cambridge University Press.

Petit, R. J., Brewer, S., Bordács, S., Burg, K., Cheddadi, R., Coart, E., et al. (2002). Identification of refugia and post-glacial colonisation routes of European white oaks based on chloroplast DNA and fossil pollen evidence. *Forest Ecology and Management, 156*, 49–74.

Piotti, A., Leonarduzzi, C., Postolache, D., Bagnoli, F., Spanu, I., Brousseau, L., et al. (2017). Unexpected scenarios from Mediterranean refugial areas: Disentangling complex demographic dynamics along the Apennine distribution of silver fir. *Journal of Biogeography, 44*, 1547–1558.

Plomion, C., Bastien, C., Bogeat-Triboulot, M.-B., Bouffier, L., Déjardin, A., Duplessis, S., et al. (2016). Forest tree genomics: 10 achievements from the past 10 years and future prospects. *Annals of Forest Science, 73*, 77–103.

Popescu, F., & Postolache, D. (2009). Variabilitatea genetică a populaţiilor de cvercinee din România, rezultat al interacţiunii dintre evoluţia postglaciară a vegetaţiei şi influenţele antropice. *Revista pădurilor, 124*, 49–54.

Popescu, F., Postolache, D., & Pitar, D. (2015). Aspecte privind conservarea şi managementul resurselor genetice forestiere din România. *Revista de Silvicultură şi Cinegetică, 21*, 13–17.

Postolache, D., Popescu, F., Pitar, D., Apostol, E., Iordan, A., Avram, A., et al. (2016). Origin, evolution and genetic structure of Silver fir stands of Romania evaluated through molecular markers. *Revista de Silvicultură si Cinegetică, 21*, 8–14.

Postolache, D., Popescu, F., Paule, L., Ballian, D., Zhelev, P., Farcas, S., et al. (2017). Unique postglacial evolution of the hornbeam (*Carpinus betulus* L.) in the Carpathians and the Balkan Peninsula revealed by chloroplast DNA. *Science of the Total Environment, 599*, 1493–1502.

Pravalie, R., Sîrodoev, I., & Peptenatu, D. (2014). Detecting climate change effects on forest ecosystems in Southwestern Romania using Landsat TM NDVI data. *Journal of Geographical Sciences, 24*, 815–832.

Şofletea, N., & Curtu, L. (2007). *Dendrologie*. Braşov: Editura Universităţii "Transilvania".

[illegible]

Chapter 33
Overview on Forest Genetic Monitoring (FGM) Including Case Studies on FGM for Two Species from Greece

Filippos A. Aravanopoulos, Paraskevi G. Alizoti, Nikolaos Tourvas, Ermioni Malliarou, Evangelia V. Avramidou, Ino-Vasileia Korompoki, Vasiliki-Maria Kotina, Evangelos Barbas, and Anna-Maria Farsakoglou

Abstract Forest tree genetic monitoring and conservation are of the utmost significance for the persistence of natural forests. Genetic conservation aspires to guard and maintain genetic diversity which is crucial for safeguarding species adaptive potential. Genetic monitoring is the quantification of time-based alterations in population genetic structure and diversity which expose processes that preserve genetic variation in nature, pioneering prognosis and assisting in defining instruments for forest genetic resource management. Genetic monitoring is founded on the geneecological approach and therefore the proposed indicators are linked to the evaluation of genetic variation, genetic drift, gene flow, mating systems and natural selection. The development of genetic baseline data for the application of genetic monitoring in forest conservation areas in Greece is presented as a case study.

Keywords Conservation genetics · Genetic monitoring · Forest trees · *Abies* · *Fagus* · Greece

---

F. A. Aravanopoulos (✉) · P. G. Alizoti ·
School of Forestry and Natural Environment, Aristotle University of Thessaloniki, Thessaloniki, Greece
e-mail: aravanop@for.auth.gr; alizotp@for.auth.gr

N. Tourvas · E. Malliarou · I.-V. Korompoki · V.-M. Kotina · E. Barbas · A.-M. Farsakoglou
Laboratory of Forest Genetics and Tree Breeding, Faculty of Agriculture, Forestry and Natural Environment, Aristotle University of Thessaloniki, Thessaloniki, Greece
e-mail: nikostourvas@gmail.com; ermionimalliarou@yahoo.gr; ikorompo@for.auth.gr; vmkotina@for.auth.gr; vbarbas@for.auth.gr; amfarsakoglou@gmail.com

E. V. Avramidou
Laboratory of Silviculture, Forest Genetics and Biotechnology, Institute of Mediterranean Forest Ecosystems and Forest Products Technology, Athens, Greece
e-mail: aevaggelia@yahoo.com

M. Šijačić-Nikolić et al. (eds.), *Forests of Southeast Europe Under a Changing Climate*, Advances in Global Change Research 65,
https://doi.org/10.1007/978-3-319-95267-3_33

33.1 Introduction

The protection of forest genetic resources is paramount today, but the need for this effort is the outcome of the notable forest devastation that has started during the Holocene and continues until present. In ancient Greece, it is known that recurrent incidents of deforestation, have taken place since the Neolithic era (van Andel et al. 1990). Currently, population extinction rate and species loss in the planet, could be 1000 times greater than background levels (UNEP 2015). Nevertheless, forests still present a sustainable basis of primary production and ecosystem services, while preserving biodiversity, especially the genetic legacy of natural ecosystems. Biological conservation, is employed to counteract species extinction and biodiversity loss. Forest genetic resource conservation, specifically signifies the science, management and policy actions that safeguard their sustained availability and survival in perpetuity (Aravanopoulos 2016; Fussi et al. 2016).

Until lately there has been a lack of use of genetic indicators in conservation (Laikre 2010). The discussion has progressed only during the current decade regarding the use of *genetic monitoring* for evaluating the status of forest genetic conservation. *Genetic monitoring* is the *quantification of time-based alterations in population genetic structure and diversity* (Aravanopoulos 2011). Discussion focuses on the use of an effective genetic monitoring system in real forest conditions (Aravanopoulos 2011) and on the advancement of genetic monitoring usages at different levels: local, regional and continental (Aravanopoulos et al. 2015).

In this chapter, the genetic monitoring concept is presented along with the first results of two case studies in Greece that provide genetic monitoring baseline data.

33.2 Genetic Monitoring of Forest Tree Populations

The objective of genetic monitoring is the assessment of the present status of genetic resources and the quantification of pertinent alterations at a chronological gradation in light of conserving long-standing adaptive evolutionary potential (Aravanopoulos 2011, 2016). Genetic monitoring goes beyond and above a mere analytical method, as by detecting time-based alterations in nature, underlying mechanisms may be deduced and their comparative magnitude can be appraised. It constitutes a timely discovery procedure that directs to management choices, intended to diminish probable damaging consequences before irreparable harm befalls. Therefore, genetic monitoring is a prognostic instrument and constitutes an approach to safeguard the protection of processes that preserve genetic variation in nature (Aravanopoulos 2011, 2016). The term has been forwarded more than 25 years ago (Allendorf and Ryman 1987), but has only lately warranted consideration (Schwartz et al. 2006; Kuparinen and Merila 2007). FGM was originally proposed by Namkoong et al. (1996), but at a national level only recently the necessity to conduct forest genetic monitoring programs was realized (e.g. Konnert et al. 2011; Geburek et al. 2010).

Assessment systems based on criteria, indicators and verifiers were suggested, but there is a lack of applied forest genetic monitoring studies, which is due to the intricacy of the purported systems.

Aravanopoulos (2011) attempted to ratify the genetic monitoring theory and methodology, highlighting that genetic monitoring is better investigated by a time-based approach than by approximating the response function of populations for specific traits across environments. Aravanopoulos (2011) employed one criterion for genetic monitoring: the conservation of genetic variation and adaptive evolutionary potential in nature. Genetic monitoring aims at biologically (or economically) keystone species, (the objective being avoidance of genetic variation loss), and endangered and/or rare species (the objective being restoration of genetic variation). Small populations were not considered as special cases besides rear edge populations due to anticipated stressful effects because of climate change (Aravanopoulos 2011, 2014; Fady et al. 2015; Hampe and Petit 2005). The scientific foundation for genetic monitoring was suggested to be the geneecological approach that considers as the chief evolutionary forces at microscale (and therefore as genetic indicators to be assessed), the restrictive (in the extent of genetic variation sense) forces of natural selection and genetic drift, as well as the expanding force of gene flow. The system is based on the assessment of a minimum set of very informative demographic and genetic parameters, minimum sample sizes, types and numbers of marker loci, frequency of monitoring, critical difference levels and thresholds between verifier measurements during temporal assessments (Aravanopoulos 2011).

Recently, a thematic study on indicators of genetic diversity assessing the evolutionary potential of trees was developed (FAO 2014; Graudal et al. 2014). "State" indicators (referring to the condition and status of aspects of biodiversity) could be employed as genetic monitoring indicators at the population level. The indicator "Trends in population condition" (corresponding to the "criterion" of the genetic monitoring approach presented above), is assessed by the diversity in adaptive traits and genes (Graudal et al. 2014). Indicator "demographic condition of selected populations" embraces the investigation of selection (i.e. indicator 1 of Aravanopoulos 2011), while indicator "genetic condition of selected populations" examines genetic drift and gene flow (i.e. indicators 2 and 3 of Aravanopoulos (2011).

In Europe, the development of a genetic monitoring system was advanced the Europena Forest Genetic Resources Network (EUFORGEN) since 2005 (Phase III), while in 2010 (Phase IV) the Genetic Monitoring Working Group was formed. The Working Group evaluated genetic monitoring approaches, techniques and methodologies, and recommended opportunities for creating a Pan-European genetic monitoring scheme (Aravanopoulos et al. 2015). The *geneecological approach* and the model of selected *indicators and verifiers* for population assessment were endorsed, based on selection, and genetic variation-mating system in a scheme concordant to that of Aravanopoulos (2011). The EUFORGEN system is also a "tight" system regarding the numbers of indicators and verifiers proposed. These methods (Aravanopoulos 2011; Aravanopoulos ct al. 2015; Graudal et al. 2014) include only 2–3 indicators and 7–11 verifiers; consequently, they are all schemes that can credibly be used in the field.

Lately, an EU funded project “LIFE for European Forest Genetic Monitoring System (LIFEGENMON)”, essentially provides a “proof of principle” application of forest genetic monitoring systems. The project applies genetic monitoring at a transect from South Germany to Central Greece. The project aims to produce strategies and procedures for genetic monitoring users (Fussi et al. 2016).

33.3 Develoment of Genetic Monitoring Baseline Data for The Protection of Hybrid Greek Fir (*Abies borisii-regis*) Genetic Resources in Greece

The adaptability of future forest tree generations relies on conserving biodiversity on every possible level, including genetic diversity. The Greek hybrid fir (*Abies borisii-regis*) is a keystone ecologically dominant species of biological and economical importance in Greece. *Abies borisii-regis* grows in pure or mixed stands in the presence of *Pinus nigra*, *Fagus sylvatica* and *Quercus* spp. in lower elevations. Because of its geographical distribution among the ecological niches of silver fir (*Abies alba*) and Greek fir (*Abies cephalonica*), *Abies borisii-regis* is a carrier of unique genetic adaptations and as such it is considered valuable material for conservation. Moreover, as a keystone species it serves a vital role in ensuring the stability of the ecosystems it forms.

The development of baseline genetic monitoring data in hybrid Greek fir has long been due. In an effort towards their development, 11 fluorescent microsatellite markers of nuclear DNA (f-nSSRs) were employed to estimate genetic variation parameters and composition of three age classes occurring simultaneously within a natural *Abies borisii-regis* stand in Kalipefki, Katerini (Region of Central Macedonia, Greece). The sampling included 250 adult individuals (70–100 years old), 200 juvenile plants from the regeneration (5–10 years old), and 400 seeds (20 open-pollinated families from 20 of the above 250 trees and 20 seeds per parental tree). DNA extraction was performed with the Macherey Nagel kit according to the manufacturer’s instructions. Genotyping was performed in an ABI3730xl sequencer using the GeneMapper software and results were analyzed using the GenAlEx software.

Results indicated that the percent of polymorphic loci was 100%. Other genetic diversity parameters such as mean observed heterozygosity ($H_O = 0.52$) and mean gene diversity ($H_E = 0.58$) were in agreement with the respective values found in similar studies conducted mainly for *Abies alba* (Krajmerová et al. 2015; Tourvas et al. 2017). Most of the variation (99%) was found within cohorts (Aravanopoulos et al. 2017; Tourvas et al. 2017). A comparative analysis of the results obtained for the cohorts and those obtained for natural regeneration, seeds and mature trees at a later time (genetic monitoring), will present valuable results regarding the maintenance of genetic diversity under changing climatic and environmental conditions. Awareness of the existing levels of genetic diversity and their temporal trends, being the main integral part of genetic monitoring, is expected to facilitate and promote the effort of genetic resources protection and management of fir forests.

33.4 Develoment of Genetic Monitoring Baseline Data for The Protecton of European Beech (*Fagus sylvatica*) Genetic Resources in Greece

Fagus sylvatica is an ecologically keystone species of biological and economical importance in Greece. The adaptive and evolutionary potential of future beech generations lies on the protection of its genetic diversity and the preservation of its adaptive potential. Currently, the genetic diversity present in the mature trees and the natural regeneration of a European beech stand located at the area of Arnaia (Chalkidiki, Region of Central Macedonia) has been investigated to derive baseline data for genetic monitoring purposes. The stand is a 90-years-old uniform beech forest, growing in an optimal environment at an altitude of 700 m a.s.l.

Two sub-populations (of 50 individuals each) consisting of the cohort of natural regeneration (two 10 m^2 sub-areas located at a distance larger than 100 m), and one population (50 individuals) representing the cohort of mature trees were studied using eight fluorescent microsatellite genetic markers (f-SSR). DNA extraction was performed with the Macherey Nagel kit. Multiple PCR reactions for SSR primers were performed in an Eppendorf Thermal Cycler PCR. Genotyping was performed in an ABI3730xl sequencer using the GeneMapper software and results were analyzed using the GenAlEx software for estimating genetic diversity parameters.

A total of eight SSR primers were used and all loci were polymorphic ($P = 100\%$). When the adult tree and regeneration cohorts were compared, generally similar results were found, but values were generally higher for the mature stand. The average number of effective alleles for mature trees was 5.94 and 5.89 for the regeneration. Expected heterozygosity for mature trees was 0.82 and observed heterozygosity 0.84, while for the regeneration it was 0.68 and 0.78 respectivelly. Most of the variation was found to reside within populations (99%). When the results of the two regeneration sub-populations were compared, the observed (H_O) and expected heterozygosity values (H_E) were 0.703 and 0.712 respectively for the first population, and 0.648 and 0.749 for the second one. Results indicated the presence of high genetic variability, comparable to that found in other studies concerning European beech populations (Aravanopoulos et al. 2017; Korompoki et al. 2017; Malliarou et al. 2016; Rajendra et al. 2014).

Comparison of the results obtained for the natural regeneration and mature trees at a later time (genetic monitoring), will yield valuable conclusions regarding the maintenance of genetic diversity under changing climatic and environmental conditions. The results will be also of value for the protection of beech genetic resources, while on a temporal basis they will yield important information on the time trends of genetic variability in beech ecosystems.

33.5 Conclusions

Forest genetic resource protection is the *sine qua non* for the maintenance of forest genetic diversity, considering the aggregate consequences of forest misuse and devastation that are aggravated by climate change. Genetic monitoring becomes imperative as the speed of environmental change increases and surpasses the timeframe of adaptation through evolution. New theoretical advances justify the use of the geneecological approach for genetic monitoring and the utilisation of a restricted number of very informative indicators (demographic parameters and genetic diversity estimates). Baseline levels of genetic diversity in two genetic monitoring studies currently underway in Greece that concern Greek hybrid fir (*Abies borisii-regis*) and European beech (*Fagus sylvatica*) indicate ample levels of contemporary genetic diversity that is an asset for the future adaptability of these populations under the ensuing climatic change scenarios in this region.

Acknowledgements Partial assistance from the European Union LIFE+ Project "LIFE for European Forest Genetic Monitoring System" is gratefully acknowledged.

References

Allendorf, F. W., & Ryman, N. (1987). Genetic management of hatchery stocks. In N. Ryman & F. M. Utter (Eds.), *Population genetics and fishery management* (pp. 141–159). Seattle: Univ. Washington Press.

Aravanopoulos, F. A. (2011). Genetic monitoring in natural perennial plant populations. *Botany, 89*, 75–81.

Aravanopoulos, F. A. (2014). Genomics of trees. In K. G. Ramawat, J. M. Merillon, & M. R. Ahuja (Eds.), *Tree biotechnology* (pp. 514–557). Boca Raton: CRC.

Aravanopoulos, F. A. (2016). Conservation and monitoring of tree genetic resources in temperate forests. *Current Forestry Reports, 2*, 119–129.

Aravanopoulos, F. A., Tollefsrud, M. M., Graudal, L., Koskela, J., Kätzel, R., Soto, A., Nagy, L., Pilipovic, A., Zhelev, P., Božič, G., & Bozzano, M. (2015). *Development of genetic monitoring methods for genetic conservation units of forest trees in Europe*. Rome: European Forest Genetic Resources Programme (EUFORGEN), Bioversity International 62 pp.

Aravanopoulos, F. A., Avramidou, E., Malliarou, E., Tourvas, N., Kotina, V., Korompoki, I., Ganopoulos, I., Alizoti, P., Barbas, E., Bekiaroglou, P., Hasilidis, P., Roussakis, G., Kiourtsis, F., & Fragiskakis, N. (2017). Forest genetic monitoring applied: First results from two FGM sites in Greece. In: *Proceeding of Forest Genetics: Health and Productivity under Changing Environments: A Joint Meeting of WFGA and CFGA*. University of Alberta, Edmonton, AB, pp. 40.

Fady, B., Cottrell, J., Ackzell, L., Alia, R., Muys, B., Prada, A., & Gonzalez-Martinez, S. C. (2015). Forests and global change: What can genetics contribute to the major forest management and policy challenges of the twenty-first century? *Regional Environmental Change*. https://doi.org/10.1007/s10113-015-0843-9.

FAO. (2014). *The state of the World's Forest genetic resources*. Rome: Commission on Genetic Resources for Food and Agriculture, Food and Agriculture Organization of the United Nations.

Fussi, B., Westergren, M., Aravanopoulos, F. A., Baier, R., Kavaliauskas, D., Finzgar, D., Alizoti, P., Bozic, G., Avramidou, E., Konnert, M., & Kraigher, H. (2016). Forest genetic monitoring genetic monitoring: An overview of concepts and definitions. *Environmental Monitoring and Assessment, 188*, 493. https://doi.org/10.1007/s10661-016-5489-7.

Geburek, T., Milasowsky, N., Frank, C., Konrad, H., & Schadauer, K. (2010). The Austrian forest biodiversity index: All in one. *Ecological Indicators, 10*, 753–761.

Graudal, L., Aravanopoulos, F. A., Bennadji, Z., Changtragoon, S., Fady, B., Kjaer, E. D., Loo, J., Ramamonjisoa, L., & Vendramin, G. G. (2014). Global to local genetic diversity indicators of evolutionary potential in tree species within and outside forests. *Forest Ecology and Management, 333*, 35–51.

Hampe, A., & Petit, R. J. (2005). Conserving biodiversity under climate change: The rear edge matters. *Molecular Ecology, 8*, 461–467.

Konnert, M., Maurer, W., Degen, B., & Katzel, R. (2011). Genetic monitoring in forests – Early warning and controlling system for ecosystemic changes. *iForest, 4*, 77–81.

Krajmerová, D., Paule, L., Zhelev, P., Voleková, M., Evtimov, I., Gagov, V., & Gömöry, D. (2015). Natural hybridization in eastern-Mediterranean firs: The case of *Abies borisii-regis*. *Plant Biosystems, 150*, 1–11.

Kuparinen, A., & Merila, J. (2007). Detecting and managing fisheries-induced evolution. *Trends in Ecology & Evolution, 22*, 652–659.

Laikre, L. (2010). Genetic diversity is overlooked in international conservation policy implementation. *Conservation Genetics, 11*, 349–354.

Namkoong, G., Boyle, T., Gregorious, H. R., Joly, H., Savolainen, O., Ratman, W., & Young, A. (1996). Testing criteria and indicators for assessing the sustainability of forest management: genetic criteria and indicators. *Centre for International Forestry Research*, Working Paper No. 10, Bogor; 1996.

Rajendra, K. C., et al. (2014). Subtle human impacts on neutral genetic diversity and spatial patterns of genetic variation in European beech (*Fagus sylvatica*). *Forest Ecology and Management, 319*, 138–149.

Schwartz, M. K., Luikart, G., & Waples, R. S. (2006). Genetic monitoring as a promising tool for conservation and management. *Trends in Ecology & Evolution, 22*, 25–33.

Tourvas, N., Malliarou, E., Avramidou, E., & Aravanopoulos, F. A. (2017). A comparative analysis of molecular genetic diversity of fir (*Abies* sp.) mature tree and regeneration cohorts in the frame of a genetic monitoring program. *In: Proceeding of 18th Pan-Hellenic Forest Science Conference*, Edessa, pp. 962–969.

UNEP. (2015). Biodiversity. http://www.unep.org/dgef/Biodiversity. Accessed 15 Dec 2016.

van Andel, T., Zangger, E., & Demitrack, A. (1990). Land use and soil erosion in prehistoric and historical Greece. *Journal of Field Archaeology, 17*, 379–396.

Korompoki, I. V., Kotina, V. M., Malliarou, E., Avramidou, E., & Aravanopoulos, F. A. (2017). Genetic diversity in natural regeneration populations of European beech using fluorescent microsattelite markers (f-SSRs). In: *Proceeding of 18th Pan-Hellenic Forest Science Conference*, Edessa, pp. 951–955.

Malliarou, E., Avramidou, E., Ganopoulos, I., & Aravanopoulos, F. A. (2016). Preliminary results of SSR genetic diversity in *Fagus sylvatica* in the frame of a genetic monitoring program. In*: Proceeding of 16th Pan-Hellenic Conference. Hellenic Scientific Society for Plant Genetics and Breeding*, Florina, p. 53.

Chapter 34
Climate Change, Biodiversity and Forest Genetic Resources: A Bulgarian Perspective

Petar Zhelev and Ina Y. Aneva

Abstract Climate change is predicted to have substantial effect on all components of biodiversity. The chapter presents a brief survey on the biodiversity in Bulgaria and its conservation. The impact of climate change on the biological diversity and on forests will provoke different and individual reactions of the species and ecosystems. These reactions will vary depending on the intrinsic biological characteristics of the species and ecosystems and on the particular environmental conditions. Biological diversity in Bulgaria is reviewed at different hierarchical levels, with particular reference to forest biodiversity and genetic resources. The high genetic diversity in the forest tree populations is a prerequisite for their successful adaptation to future environmental changes.

Keywords Forest ecosystems · Adaptation · Genetic diversity · Bulgaria

34.1 Introduction

Biodiversity conservation is one of the most important contemporary activities as a part of environmental protection in general. Its importance is outlined by the numerous legislative documents and decisions at international and national level. Biological diversity is defined and is studied and conserved at different hierarchical level – species, ecosystems, landscapes etc. As pointed by Noss (1990), different people have different understanding of biodiversity and it is difficult to find a definition, which will be at the same time simple, fully operational and clear. However, at operational level people are dealing with genetic, species and habitat diversity.

P. Zhelev (✉)
University of Forestry, Sofia, Bulgaria
e-mail: zhelev@ltu.bg

I. Y. Aneva
Institute of Biodiversity and Ecosystem Research, Bulgarian Academy of Sciences, Sofia, Bulgaria
e-mail: ina.aneva@abv.bg

M. Šijačić-Nikolić et al. (eds.), *Forests of Southeast Europe Under a Changing Climate*, Advances in Global Change Research 65,
https://doi.org/10.1007/978-3-319-95267-3_34

Forest ecosystems harbor a great deal of biological diversity, especially in countries and regions characterized by high percentage of forest cover, like Bulgaria. Climate change could have substantial effect on the biodiversity, and this effect could be very strong in regions with transitional climate, like Balkans. Different scenarios predict drier climate in the future, which could lead to substantial change in the biodiversity, and could affect forest genetic resources, too. Since the forests conserve mostly natural or semi-natural biodiversity, they should be a subject of special attention.

The reports of the Inter-Governmental Panel for Climate Change (IPCC) present evaluations about the potential increase of the global temperatures, which varies, according to different models and scenarios, from 1.8 to 4 °C until the end of twenty-first century (IPPC 1990, 1995, 2001, 2007, 2014).

How biodiversity will be affected by climate change could be understood if: (1) the existing biodiversity (and its functioning, desirably) is known, and (2) some scenario for climate change is available. Fortunately, there is a considerable amount of information concerning biodiversity of Bulgaria, at least at the species level, and there are different scenarios developed – pessimistic, neutral and optimistic ones – concerning the effect of climate change on biodiversity and on forests until the end of twenty-first century (Raev et al. 2011, 2015).

34.2 Brief Information About the Biodiversity in Bulgaria

34.2.1 General Information

Bulgaria is one of the European countries with high biodiversity. It results from the combination of several factors: southern latitude, diverse orography with altitude from 0 to almost 3000 m.a.s.l., various landscape, and influence of other biogeographic regions: Mediterranean region, Western Asia, the steppes, Alpine region and some others. Besides, the land remained practically unaffected by the last glaciations (Turill 1928; Polunin 1980). According to Kutzbach and Guetter (1986) the average temperatures in July during the last glaciation were slightly lower than today and the mountains were in most cases ice-free (Thiede 1974; Denton and Hughes 1981; Dawson 1992; Abrantes et al. 2012). Thus, during the last glaciation many species and communities persisted in Balkan Peninsula and in Bulgaria. Balkan Peninsula is considered as one of the main glacial refugia in Southern Europe (Birks and Willis 2008; Keppel et al. 2012), but new palaeobotanical evidence showed that the picture was more complicated and there were many more local micro-refugia in central and eastern Europe, which are called "cryptic refugia". All these circumstances contributed to the present-day biodiversity and to existence of many relic and endemic species in Balkans and in Bulgaria, in particular.

The biological diversity in Bulgaria has been summarized on several occasions. First, a National Strategy for Biodiversity Conservation was developed (Sakalian

1993). There were also reviews on particular groups, areas, or particular topics related to biodiversity, for example, Bondev (1991), Apostolova and Slavova (1997), Ganeva and Natcheva (2004), Sakalian (1999), but the last general review was the book Contemporary Status of Biodiversity in Bulgaria (Petrova et al. 2005).

There is information about the level of diversity at different hierarchical levels:

- Within species diversity (taxonomic and genetic diversity)
- Species diversity
- Habitat diversity
- Landscape diversity

34.2.2 Within-Species Diversity

The information about within-species diversity is modest, as compared to that concerning species diversity. The reason for that can be traced first in the fact that genetics is much younger than the classical biological sciences, like botany and zoology, and second, that evaluation of genetic diversity is usually not possible by field observations only, but rather requires laboratory analyses, which are sometimes expensive. Still, for many within-species diversity is limited to taxonomic diversity, i.e., subspecies, varieties and forms, but even this kind of diversity is relatively well known only in the most studied groups – higher plants and vertebrate animals.

There is a substantial information about the distribution of genetic variation in the tree species. The studies performed included application of phenotypic, biochemical and molecular markers. Traditionally, conifers received most attention (genera *Pinus*, *Picea*, *Abies*, *Taxus* and *Juniperus*) but broadleaved species attracted also the interest of experts, especially during the last two to three decades (see Alexandrov and Dobrev 2012, for review).

The information accumulated is valuable not only as a knowledge, but also as a background of strategies for close-to-nature use and conservation of forest tree genetic resources.

34.2.3 Species Diversity

The species diversity is relatively well-known in plants, fungi and vertebrates, but there are several groups of invertebrates that are still under study and the exact number of species is unknown. The last reviews (Petrova et al. 2005) are summarized in Table 34.1.

The last column of the table represents only an expert evaluation of the species occurring in the forest territories. It should be pointed out that the information about species diversity is not a final one, because in numerous complex and polymorphic

Table 34.1 Compiled approximate information about the biodiversity in Bulgaria in the more important groups of living organisms

Groups of organisms	Approximate number of species	Occurring in forests and forest territories (expert evaluation)
Fungi	4900	3000
Macromycetes	2200	2000
Lichenized fungu (Lichens)	710	600
Plants		
Algae (incl. Phytoplankton)	3350	500
Mosses (after Ganeva and Natcheva 2004, 2005)	754	500
Ferns and seed plants	4100	3000
Animals		
Lower invertebrates (after Golemanski et al. 2005)	4850	2000
Arachnids	1100	500
Acarids	1600	1000
Insects	20,900	15,000
Moluscs	445	–
Fishes	218 (107 freshwater)	–
Amphibians and reptiles	53	30
Birds	400	250 (about half of them nesting)
Mammals	100	80

From Petrova et al. (2005)

systematic groups the problematic issues are not completely solved. New species are being described, some of the existing ones are given lower taxonomic rank, or become synonyms. These processes are normal results of the taxonomic revisions, of the change in taxonomic and phylogenetic concepts and especially of the introduction of modern molecular genetic methods in systematics.

No separate information is presented about the arboreal species (trees, shrubs and semi-shrubs). They are about 400 species (Yurukov and Zhelev 2001; Tashev and Tsavkov 2008). Some of them deserve spècial attention because they are large trees, cover significant territories and play an important environment-formation role. Also, the forest communities formed by the tree species provide habitat for many other living organisms.

34.2.4 *Habitat Diversity*

Even though the habitats of a given species are always subject of interest from the point of view of its ecological requirements, the term gained additional meaning due to Council Directive 92/43/EEC of 21 May 1992 on the conservation of natural habitats and of wild fauna and flora (called briefly "The Habitat Directive"). Despite the numerous discussions and adjustments of the definitions of a habitat, the meaning used in the Habitat Directive (Davies and Moss 1997; Davies et al. 2004; Devilliers and Devilliers-Terchuren 1996) implies that plant terrestrial habitats practically coincide with the plant communities and their distinguishing is based on phytosciological criteria. Therefore, the terrestrial habitats and particularly, the forest habitats, often overlap with a syntaxon of certain rank, described by using the floristic approach (Braun-Blanquet 1928, 1964).

Large-scale studies on vegetation diversity were performed in 1960–1980s and considerable amount of information was gathered, however, mostly using the dominant approach of the Soviet school (Ramenski 1937; Sukachev 1972). Most of these results were summarized in the book of Bondev (1991). The floristic approach of Braun-Blanquet gained more popularity starting from 1980s onward. There is a comparatively detailed information about the vegetation of Balkan Peninsula (Horvat et al. 1974), and recently a detailed review on the classification of vegetation in Bulgaria was published (Tzonev et al. 2009). According to this review, until 2009 the syntaxa described in Bulgaria belong to 37 classes, 67 orders, 94 alliances, 218 associations, 48 sub-associations and 36 plant communities of lower rank. Since the studies using the floristic approach have increased substantially during the last 30 years, the number of plant associations is expected to increase with the studies of new, not studied territories. There is also a well-organized database about the vegetation in Balkans (Vassilev et al. 2016) that can be used in the studies of habitat diversity, too.

According to Meshinev and Apostolova (2005) all habitat units at different hierarchical levels are 977, and at habitat level they are 542, and 192 of them being of European importance. A special Handbook for identification of the habitats of European importance was issued (Kavrakova et al. 2009) containing 90 natural habitats included in NATURA 2000 network. The habitats that are typical for Bulgaria and were missing in the previous editions of the Interpretation Manual of European Union habitats were included in the last edition of the Manual (Anonymous 2013).

A special volume of the Red Data Book of Bulgaria (Bisserkov 2015) deals with rare habitats. It contains 165 habitat types, classified according to EUNIS Habitat classification, Palearctic Classification and Habitat directive. About one-third (56 habitats) of them are dominated by trees or shrubs and are typically forest territory.

The forest habitats are very important for conservation of the existing biodiversity. Most of them are also protected according to the Bird Directive (Council Directive 2009/147/EC on the conservation of wild birds).

There are 37 forest habitats included in the network NATURA 2000. Some of them are not typical forests, being dominated by shrub species and classified as "bush", or "heath", but they are situated on forest territories and gene conservation is related to these habitats, too. There are som habitats with high priority for conservation, most interesting of them being these with the presence of endemic species, like "95A0 High oro-Mediterranean pine forests", dominated by *Pinus peuce* or *P. heldreichii*.

The brief survey on the information available allows the conclusion that Bulgaria, like the other countries on Balkan Peninsula, is one of the countries with the highest biological diversity per unit area. The comparison with other European countries indicates that biodiversity in Bulgaria is conserved to a relatively good extent, which is caused to several factors, like: mountainous regions with less developed industries that can damage the environment, difficulties in application of mechanization in agricultural and forestry activities, relatively late industrialization, depopulation of some border areas as a result of migration to cities, state property on the land since the end of 1940s to the beginning of 1990s, which allowed rapid and relatively easy announcement of protected territories, and the activities taken by the state institutions for protection of the environment. The share of natural and virgin forests in Bulgaria is still relatively large (Raev 2006; Veen et al. 2010).

34.2.5 Landscape Diversity

Due to the variable orography and to some other supporting factors the landscape diversity in Bulgaria can be evaluated as high, too. It is a specific domain of research combining biological, agricultural and geographic sciences (and probably some more) and the studies are not so extensive and numerous like the studies of biological aspects of biodiversity. Landscapes of some protected territories were studied (Malinova 2007), and the existing landscape classifications and maps are being updated by the experts (Velchev et al. 1992, 2000; Todorov and Velchev 2014; Todorov et al. 2017). Koulov and Zhelezov (2016) considered the sustainable development of mountain regions in a landscape perspective.

34.3 Conservation of Biodiversity

Biodiversity conservation is regulated by Biological Diversity Act of Bulgaria (2002), which has undergone numerous amendments until 2013. Even though at the time of Biodiversity Act adoption Bulgaria was not still a member of EU, the European regulations were considered and incorporated in the Act and in its appendicaes. Figure 34.1 represents the protected plant species included in Appendices 2, 2a, and 3 of the Biodiversity Act. Only 1 fern species, 17 Angiosperms and six mosses are included in Appendix 2. Total 75 Angiosperms, 2 moss species and 10 fungi are included in Appendix 2a, and 4 Lycophytes, 13 Ferns, 6 Gymnosperms and 550 Angiosperms. The number of protected species is impressive and it indicates both that there are species needing special attention and measures for conservation, on the one hand, and that the state authorities take care for the nature conservation. Some beautiful examples of protected species and their habitats are presented in Fig. 34.2.

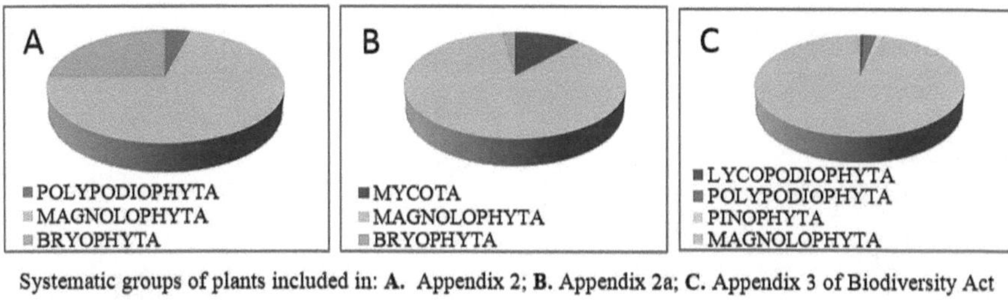

Fig. 34.1 Groups of plant and fungi protected by Biodiversity Act. Appendix 2 includes species protected according to the Directive 92/43/EC; Appendix 2a includes species that are protected in protected territories, and Appendix 3 includes species protected on the whole territory of the country

Table 34.2 presents the protected animals according to the appendices of Biodiversity Act. The fact that much more vertebrates than invertebrates are protected by the Biodiversity Act is due not to the fact that vertebrates are more endangered, but also that much more is known about them, in comparison with invertebrates. Even though not having legal power, the Red Data Book of the Republic of Bulgaria contains valuable information about the rare and endangered species and habitats. It is organized in three volumes. Volume 1 – Plants and Fungi (Peev 2015) includes 808 species, (6 Algae, 102 Bryophytes, 8 Ferns, 4 Gymnosperms, 539 Angiosperms and 149 Fungi). Volume 2 (Golemanski 2015) includes 51 invertebrates and 236 Vertebrates, the most numerous being the birds – 140 species, followed by fishes – 50 species. Volume 3 includes 165 natural habitats (see above). All volumes are available online (http://e-ecodb.bas.bg/rdb/en/).

A Red List of Vascular Plants (Petrova and Vladimirov 2009) preceded Red Data Book. It is coauthored by 28 researchers and contains 898 vascular plant species, classified according to their conservation status. The list contains 74 arboreal species – 8 Gymnosperms and 66 Angiosperms – many of them important forest trees (Table 34.3).

34.4 Climate Change and Biodiversity

It is difficult to make even an initial survey on the issues relating climate change and biodiversity. One search in Google using the word combination "climate change biodiversity" yields more than 60 million results. Even if the search is narrowed by using the combination "climate change forest resources" the results are still more than five million.

However, there are some trends that could be discussed in this respect, and particularly when the problem concerns smaller territory like Balkan Peninsula or Bulgaria.

Fig. 34.2 Examples of plant species of conservation importance, together with their habitats. (**a**) *Fritillaria gussichiae* (Deg. et Dörfl.) Rix in dry forests of subMediterranean type; (**b**). *Viola delphinantha* Boiss. on rocky limestone slopes with *Acer hyrcanum* Fisch. et C.A. Mey.; (**c**). *Lilium rhodopaeum* Delip. at the edge of a beech forest; (**d**). *Arctostapylos uva-ursi* (L.) Spreng.; (**e**). *Gentiana lutea* L. near *Pinus heldreichii* Christ forest and (**f**). *Ophrys cornuta* Steven near *Quercus pubescens* Willd. forest

There are numerous climate scenarios until the end of twenty-first century, which differ substantially in their end output, mostly concerning the increase of global temperature. However, the global averages cannot be very useful at local scale because significant differences among the different regions on the Earth are expected. In most scenarios, European countries are expected to experience more significant changes compared with other regions in the world, and one of the most affected countries will be these in the Mediterranean region, including the southern peninsulas. Generally, increase of temperatures and decrease of precipitations and humidity are predicted for South-eastern Europe, including in Bulgaria.

Table 34.2 Groups of animals protected by biodiversity act

Animal groups	Appendix 2 (2a)	Appendix 3
Vertebrates	182 (2)	443
Mammals	26	54
Birds	120 (2)	338
Reptiles	8	33
Amphibians	5	15
Fishes	23	3
Invertebrates	35	37
Crustaceans	1	
Insects	29	34
Gastropods	4	2
Bivalves	1	1

For the legal meaning of appendices of Biodiversity Act, see Fig. 34.1

34.4.1 Reaction of the Biodiversity to Climate Change

One of the serious problems arising in the prediction of reactions of different species to climate change is that in each community there are coexisting species that differ in their requirements to the environmental conditions and in their capacity to adapt to the changes in the environment. Therefore, co-occurring species in the moment could react differently to climate change and will eventually have different feats (Huntley 2007). A key inference concerning the past reactions of the species to climate change is that the reactions were individualistic for different species inhabiting common territory before the change (Graham and Grimm 1990; Huntley and Birks 1983; Huntley 1991).

IUCN (2003) determined seven categories of species according to their sensitivity to climate change and its consequences:

- Species with a distribution limited by physical barriers (high mountains, water basins, as well as high-mountain species with no space to migrate higher);
- Species with limited distribution (local endemics);
- Species with limited dispersal ability;
- Species that are particularly sensitive to extreme climate events;
- Species adapted to highly specific environmental conditions;
- Species, which have evolved in a close connection with other species;
- Species without abilities for flexible physiological reaction to climate variables.

The interglacial and postglacial migrations in Europe and in Balkans are relatively well documented (Willis 1994; Bennett et al. 1991; Huntley and Birks 1983; Mackintosh et al. 2011; Ohlemüller et al. 2012). For many tree species the postglacial migrations can be explain by a simple model. According to that model most species have colonized Northern Europe from several large glacial refugia, situated in the three southern peninsulas – Iberian, Italian and Balkan. However, this model cannot explain the mode of post-glacial migration of many species. There is an increasing evidence that many species persisted in "intermediate populations" in Middle Europe (Magri et al. 2006) and recolonization started from there. This scenario can explain the "Reid's paradox" (Clark et al. 1998) of the rapid plant migrations.

This brief mentioning of past migration underlines the complexity of prediction of ecosystems' reactions to future climate change, because the species constituting an ecosystem will probably react differently. The reaction of a species will depend, besides its biology and ecological requirements, on the place of its distribution where the impact of climatic factors will take place. The species does not react as an entity (with the exception of narrow endemics with single populations) but rather different populations will react, most probably these situated at the border areas of impact of climate factors. Huntley (2007) classified the species responses to climate change into several types: behavioral reactions, population dynamic reactions, adaptive genetic reactions, spatial reactions and macro-evolutionary reactions. Each one of these reactions could be more or less important depending on the species and the particular conditions and climate change impact.

34.4.2 Climate Change and Forest Genetic Resources

Genetic diversity, as a hierarchical component of the biological diversity has been studied to a lesser extent (see above). It is curious to cite Usher (2005) who pointed out that in the book about the global biodiversity of Groombridge (1992) there are about 240 pages dedicated to the species diversity, 80 pages – to the habitat diversity and only 6 pages – to the genetic diversity. However, the situation changed substantially since 1992 and today focusing on genetic diversity and its conservation is an integral part of each biodiversity conservation programme. In many plant and animal species and particularly in forest trees gene diversity is crucial for the adaptation potential of populations to the changing environment. Balkans are characterized by high level of genetic diversity (Griffiths et al. 2004; Weiss and Ferrand 2007). The phenotypic and ecotype diversity in Bulgarian tree species is relatively well studied (Dobrinov et al. 1982).

As pointed out above, the reactions to climate change of biological species will be individual. The most significant impact is expected on the tree species and forest in the lower altitudinal belt, and especially in the southern regions of the country. This concerns mostly oaks; however, there are also numerous other important tree species that do not form extensive stands, but are important from economic and environmental point of view. Here we could mention the genera *Ulmus, Fraxinus, Acer, Tilia, Platanus, Populus, Salix, Alnus, Sorbus* and many other important, albeit not so large, trees.

A very significant problem related to the lower altitudinal belt is related to coniferous plantations. Due to some historical and social reasons, a substantial part of the forests in the warmer parts of Bulgaria were destroyed, which had caused changes in the hydrological regime, erosion processes and other land degradation processes. Therefore, large-scale afforestations were performed in the 1960–1980s, with particularly large areas afforested in Rhodopes, on the southern slopes of Stara Planina and in the southernmost areas. Mostly *Pinus sylvestris* and *P. nigra* had been used in afforestations and currently significant area in the country is occupied by coniferous plantations. These plantations, often consisting of a single species (monocultures) are in many areas over-aged (i.e. more than 50-year old), lacking timely performed thinnings in the past and starting to suffer a decline. Another disadvantage of coniferous plantations in the lower altitudina belt is their succeptibility to forest fires. In many stands the indigenous broadleaved species are re-colonizing the area, after decline of the plantations (Gogushev 2010).

In most cases no records about the seed provenance exist and the origin of plantations is unknown. This is a serious problem, because without proper information about the provenance, the genetic factor cannot be precisely evaluated.

Alberto et al. (2013) in their extensive and thorough review underlined the importance of field experiments and predicted that adaptive response to changing environment will be most effective in the case of large populations with high genetic diversity, under strong selection pressure and with ecological opportunity for establishment of better adapted genotypes. These conclusions underline the importance of the classical forest genetic experiments – provenance trials (Matyas 1997).

There are numerous studies on the adaptation of forest trees and forest ecosystems to changing environment and particularly – to climate change (Muller-Starck and Schubert 2001; Savolainen et al. 2007; Koskela et al. 2007). Aitken and Bemmels (2015) recommended assisted gene flow, achieved by using multiple seed sources, in order to increase genetic diversity and adaptability. Kremer et al. (2012) postulated that gene flow in its classical sense is considered to have a positive effect on adaptation in response to changing environment. However, the authors underline that this positive effect depends on many other circumstances related mostly with the species distribution. Most authors underline the importance of maintaining high genetic diversity as a potential for future adaptation to the changing environmental conditions (Eriksson et al. 1993). The most recent comprehensive work on the adaptation of forest trees and their populations to climate change was the one that resulted from joint efforts of many European experts within the framework of EUFORGEN Programme (Koskela et al. 2007; Kelleher et al. 2015). It could be used as a guide for designing the strategies and measures in the particular countries, considering the specific regional issues.

34.4.3 Specific Issues for Bulgaria

Populations of most forest trees in Bulgaria possess relatively high genetic diversity (see Alexandrov and Dobrev 2012, for review), which is a good prerequisite for adaptation to future environmental changes. At the same time many tree species are here at the periphery of their natural distribution and this fact should be considered in the strategies and measures recommended (Ganopoulos et al. 2011; Fady et al. 2016). In this chapter it is difficult to formulate completely new and original ideas and recommendations, after the work of the mighty international group (Kelleher et al. 2015), but some specific Bulgarian issues could be underlined.

The main ideas and recommendations related to gene resources conservation can be summarized in the following main points:

- Using the existing gene conservation units to study the responses to climate change, and establishment of new ones, if necessary. Performing of regular monitoring on the responses;
- Maintaining of high biodiversity levels in the forest ecosystems and high genetic diversity in tree populations, including through promoting the natural regeneration and close to nature forest management;
- Broadening the species composition of the experiments and conservation units;
- Studying of assisted migration;
- Strengthening the international cooperation, including within Balkans;

As pointed above, most of these moments are present in the scientific papers and international documents, but they deserve short comments here.

There are enough in situ conservation units for most forest tree species in Bulgaria (Alexandrov and Dobrev, this issue), but practically all of them were selected exclusively for breeding purposes. Doubtless, they could be analyzed in view of climate change, but if necessary, some new conservation units could be added in order to conform with a preliminary developed experimental design to study the reactions to climate change. The high diversity of tree species in Bulgaria requires including more species in the conservation units and in the studies. As pointed by Lefévre (2007), the information about the "model" tree species has increased significantly, while less is known about non-model species.

The importance of maintaining high genetic diversity was outlined above. According to Eriksson et al. (1993) this leads to high additive variance in the populations and opportunity to flexible reaction and adaptation to the rapidly changing environment. Introducing the concept of Multiple Population Breeding Systems (Namkoong et al. 1988) could be an option in some species, for which accumulated breeding information is available. In natural populations the objective of maintaining high genetic diversity can be achieved at least partly by promoting natural regeneration through close-to-nature management. On the one hand, this approach will contribute to the conservation of biodiversity, and on the other hand, it will bring light about the response of the whole ecosystem to environmental change.

Due to high tree species diversity in Balkans and in Bulgaria, in particular, much more experiments will be necessary to eluciate the responses of different species to changing environment. This is not an easy task, because it is related to the availability of financial and expert human resources. Here we could find a link to the recommendation for strengthening the international cooperation in the field of forest genetic resources conservation. It is absolutely necessary to resume the participation of Bulgaria in the European Forest Genetic Resources Programme (EUFORGEN). This will promote exchange of valuable information and input new data to EUFGIS database (Kelleher et al. 2015). Studying the assisted migration will be much more effective at international scale, having in mind the relatively small geographic area of Bulgaria and the higher effectiveness of the international initiatives and efforts.

Acknowledgements The work on the paper was partially supported by the project INFORMED (INtegrated research on FOrest Resilience and Management in the mEDiterranean).

Appendix

Table 34.3 Arboreal species included in the red list of bulgarian vascular plants (Petrova and Vladimirov 2009)

Taxon	Family	Biodiversity act	IUCN category
Abies borisii-regis Mattf.	Pinaceae		LC
Acer heldreichii ssp. visianii Orph.	Aceraceae	Appendix 3	VU
Aesculus hippocastanum L.	Hippocastanaceae	Appendix 3	EN
Amygdalus webbii Spach	Rosaceae	Appendix 3	
Arbutus andrachne L.	Ericaceae	Appendix 3	CR
Arbutus unedo L.	Ericaceae	Appendix 3	CR
Arctostaphylos uva-ursi (L.) Spreng.	Ericaceae		VU
Astracantha aitosensis (Ivan.) Podl.	Fabaceae	Appendix 3	EN
Astracantha thracica (Griseb.) Podl.	Fabaceae	Appendix 3	VU
Calluna vulgaris L.	Ericaceae	Appendix 3	VU
Caragana frutex (L.) C. Koch	Fabaceae	Appendix 3	CR
Castanea sativa mill.	Fagaceae		EN
Celtis plachoniana K.I. Chr.	Ulmaceae		EN
Cercis siliquastrum L.	Fabaceae		NT
Chamaecytisus frivaldszkyanus (Degen) Kuzm.	Fabaceae		EN
Chamaecytisus danubialis (Velen.) Rothm.	Fabaceae		DD
Chamaecytisus kovacevii (Velen.) Rothm.	Fabaceae	Appendix 3	EN
Chamaecytisus neicheffii (Urum.) Rothm.	Fabaceae	Appendix 3	
Chamaecytisus ratisbonensis (Schaeff.) Rothm.	Fabaceae	Appendix 3	
Cistus salvifolius L.	Cistaceae	Appendix 3	EN
Clematis alpina (L.) mill.	Ranunculaceae	Appendix 3	VU
Crataegus microphylla C. Koch	Rosaceae	Appendix 3	CR
Crataegus orientalis pall. Ex M. Bieb.	Rosaceae		VU
Crataegus stevenii Pojark.	Rosaceae	Appendix 3	VU
Daphne blagayana Freyer	Tymeleaceae	Appendix 3	EN
Daphne cneorum L.	Tymeleaceae		EN
Daphne laureola L.	Tymeleaceae	Appendix 3	EN

(continued)

Table 34.3 (continued)

Taxon	Family	Biodiversity act	IUCN category
Daphne oleoides Schreb.	Tymeleaceae		LC
Daphne pontica L.	Tymeleaceae	Appendix 3	EN
Empetrum nigrum L.	Empetraceae	Appendix 3	EN
Ephedra campylopoda C.A.Mey.	Ephedraceae	Appendix 3	
Ephedra distachya L.	Ephedraceae	Appendix 3	VU
Erica arborea L.	Ericaceae	Appendix 3	VU
Eriolobus trilobata M. Roem.	Rosaceae	Appendix 3	CR
Ficus carica L.	Moraceae		LC
Frangula rupestris (Scop.) Schur	Rhamnaceae		VU
Fraxinus pallisiae Wilmott	Oleaceae		VU
Genista germanica L.	Fabaceae	Appendix 3	CR
Genista pilosa L.	Fabaceae	Appendix 3	EN
Hippophae rhamnoides L.	Elaeagnaceae	Appendix 3	CR
Hypericum androsaemum L.	Hypericaceae	Appendix 3	EN
Hypericum calycinum L.	Hypericaceae	Appendix 3	EN
Ilex aquifolium L.	Aquifoliaceae	Appendix 3	EN
Ilex colchica Poj.	Aquifoliaceae	Appendix 3	EN
Juniperus excelsa M. Bieb.	Cupressaceae	Appendix 3	VU
Juniperus sabina L.	Cupressaceae	Appendix 3	EN
Mespilus germanica L.	Rosaceae		LC
Myricaria germanica (L.) Desv.	Tamaricaceae	Appendix 3	CR
Ononis repens L.	Fabaceae	Appendix 3	CR
Pinus brutia ten.	Pinaceae	Appendix 3	CR
Pinus peuce Griseb.	Pinaceae		NT
Platanus orientalis L.	Platanaceae		NT
Pyrus bulgarica Khutath. & Sachok.	Rosaceae	Appendix 3	NT
Quercus coccifera L.	Fagaceae	Appendix 3	EN
Quercus hartwissiana Steven	Fagaceae	Appendix 2A	VU
Quercus mestensis Bondev & Gančev	Fagaceae	Appendix 3	CR
Quercus thracica Stef. & Nedjalkov	Fagaceae	Appendix 3	CR
Quercus trojana Webb	Fagaceae	Appendix 3	
Rhamnus alpinus L.	Rhamnaceae	Appendix 3	EN
Rhododendron myrtifolium Schott & Kotschy	Ericaceae	Appendix 3	EN
Rhododendron ponticum L.	Ericaceae	Appendix 3	VU
Ribes nigrum L.	Saxifragaceae	Appendix 3	CR
Salix caprea L.	Salicaceae	Appendix 4	
Salix pentandra L.	Salicaceae	Appendix 3	CR
Salix retusa L.	Salicaceae	Appendix 3	EN
Salix rosmarinifolia L.	Salicaceae	Appendix 3	CR
Salix xanticola Christensen	Salicaceae	Appendix 3	VU

(continued)

Table 34.3 (continued)

Taxon	Family	Biodiversity act	IUCN category
Satureja pilosa Velen.	Lamiaceae		LC
Spiraea crenata L.	Rosaceae	Appendix 3	CR
Spiraea hypericifolia L.	Rosaceae	Appendix 3	DD
Spiraea salicifolia L.	Rosaceae	Appendix 3	CR
Taxus baccata L.	Taxaceae	Appendix 3	EN
Tilia rubra DC.	Tiliaceae		LC
Vaccinium arctostaphyllos L.	Ericaceae	Appendix 3	EN

References

Abrantes, F., Voelker, A., Sierro, F. J., Naughton, F., Rodrigues, T., Cacho, I., Aritzegui, D., Brayshaw, D., Sicre, M.-A., & Batista, L. (2012). Paleoclimate variability in the Mediterranean region. In P. Lionello (Ed.), *The climate of mediterranean region: From past to future* (pp. 1–86). Amsterdam: Elsevier.

Aitken, S. N., & Bemmels, J. B. (2015). Time to get moving: Assisted gene flow of forest trees. *Evolutionary Applications, 9*, 271–290.

Alberto, F. J., Aitken, S. N., Alía, R., González-Martínez, S. C., Hänninen, H., Kremer, A., Lefévre, F., Lenormand, T., Yeaman, S., Whetten, R., & Savolainen, O. (2013). Potential for evolutionary responses to climate change – Evidence from tree populations. *Global Change Biology, 19*, 1645–1661.

Alexandrov, A., & Dobrev, R. (2012). State of forest genetic resources in Bulgaria. *Nauka za Gorata, 1-2*, 1–22.

Anonymous. (2000). National action plan for biodiversity conservation, S., MoEW, 50 pp. (in Bulgarian).

Anonymous. (2013). Interpretation manual of the European union habitats – EUR28. European Commission, 144 pp.

Apostolova I., & Slavova L.. (1997). Conspectus of plant communities in Bulgaria published in the period 1891–1995. S., 340 pp. (in Bulgarian).

Bennett, K. D., Tzedakis, P. C., & Willis, K. J. (1991). Quaternary refugia of north European tree species. *Journal of Biogeography, 18*, 103–115.

Biodiversity Act of Bulgaria. State Gazette, No 77/09.08.2002, Last amendment 2013.

Birks, H. J. B., & Willis, K. J. (2008). Alpines, trees, and refugia in Europe. *Plant Ecology and Diversity, 1*(2), 147–160.

Bisserkov V. (2015). Red data book of Bulgaria, vol. 3. Natural Habitats. S., IBER-BAS and MoEW.

Bondev I. (1991). The vegetation of Bulgaria. Map in scale 1:600000 with explanatory text. S., Sofia University "St. Kliment Ohridsky" Publishing House, 183 pp. (in Bulgarian).

Braun-Blanquet, J. (1928). *Pflanzensoziologie* (1st ed.). Berlin: Springer 330 pp.

Braun-Blanquet, J. (1964). *Pflanzensoziologie. Grundzüge der Vegetationskunde*. Wien: Springer 865 pp.

Clark, J. S., Fastie, C., Hurtt, G., Jackson, S. T., Johnson, C., King, G. A., Lewis, M., Lynch, J., Pacala, S., Prentice, C., Schupp, E. W., Webb, T., & Wyckoff, P. (1998). Reid's paradox of rapid plant migration. Dispersal theory and interpretation of paleoecological records. *Bioscience, 48*(1), 13–24.

Davies C. E., Moss D. 1997. EUNIS habitat classification. Final Report, European Topic Centre on Nature Conservation, Task 7.5.1., 307 pp.

Davies C.E., Moss D., Hill M. O. (2004). EUNIS habitat classification revised 2004. Raport to: European Environment Agency, European Topic Centre on Natyre Protection and Biodiversity, 307 pp.

Dawson, A. G. (1992). *Ice age earth: Late Quaternary geology and climate*. London: Routledge XX+293 pp.

Denton, G. H., & Hughes, T. J. (1981). *The last great ice sheets*. New York: Willey 484 pp.

Devilliers, P., J. Devilliers-Terchuren (1996) A classification of palearctic habitats. Nature and Environment 78, Council of Europe, Strasbourg, 194 pp.

Dobrinov I., Doykov G., Gagov V., 1982. Forest genepool in Bulgaira S., Zemizdat, 258 pp. (in Bulgarian).

Eriksson, G., Namkoong, G., & Roberds, J. H. (1993). Dynamic gene conservation for uncertain futures. *Forest Ecology and Management, 62*, 15–37.

Fady, B., Aravanopoulos, F., Alizoti, P., Mátyás, C., von Wühlisch, G., Westergren, M., Belletti, P., Cvjetkovic, B., Ducci, F., Huber, G., Kelleher, C. T., Khaldi, A., Kharrat, M. B. D., Kraigher, H., Kramer, K., Mühlethaler, U., Peric, S., Perry, A., Rousi, M., Sbay, H., Stojnic, S., Tijardovic, M., Tsvetkov, I., Varela, M. C., Vendramin, G. G., & Zlatanov, T. (2016). Evolution-based approach needed for the conservation and silviculture of peripheral forest tree populations. *Forest Ecology nad Management, 375*, 66–75.

Ganeva, A., & Natcheva, R. (2004). An overview of the Bulgarian bryophyte flora – Past, present and future. *Phytologia Balcanica, 10*, 17–21.

Ganeva A., Natcheva R., (2005). The moss flora of Bulgaria: Contemporary status of studies, conservation and future challenges. In: Petrova A., Vadimirov V., & Dimitrova D. (Eds.). Contemporary status of the biodiversity in Bulgaria – problems and prospects, S., BAS and Bulgarian Bio-platform: 69–74. (in Bulgarian).

Gogushev G. (2010). Syntaxonomic analysis of the oak forests in the Western Frontier Mountains. Ph.D. thesis, University of Forestry, Sofia, 185 pp.

Golemanski, V., Todorov, M., Pandurski, I., Georgiev, B., Uzunov, Y., Peneva, V., Konsulova, T., Kozuharov, D., Andreev, S., & Stoev, P. (2005). Biodiversity of lower invertebrates in Bulgaria: contemporary status, problems, prospects. In A. Petrova, V. Vladimirov, & D. Dimitrova (Eds.), *Contemporary status of the biodiversity in Bulgaria – problems and prospects. S., BAS* (pp. 105–128).

Golemanski V. (2015). Red data book of Bulgaria, vol. 2. Animals. S., IBER-BAS and MoEW.

Graham, R. W., & Grimm, E. C. (1990). Effects of global climate change on the patterns of terrestrial biological communities. *Trends in Ecology & Evolution, 5*, 289–292.

Griffiths, H. I., Krystufek, B., & Reed, J. M. (Eds.). (2004). *Balkan biodiversity: Pattern and process in the European hotspot*. Dordrecht: Springer 358 pp.

Groombridge, B. (Ed.). (1992). *Global biodiversity: Status of the Earth's living resources*. London: Chapman & Hall 614 pp.

Horvat, I., Glavač, V., & Ehlenberg, H. (1974). *Vegetation Südosteuropas*. Stuttgart: Gustav Fisher Verlag 768 pp.

Huntley, B. (1991). How plants respond to climate change: Migration rates, individualism and the consequences for plant communities. *Annals of Botany, 67*, 15–22.

Huntley B., 2007. Climatic change and the conservation of European biodiversity: Towards the development of adaptation strategies. Discussion paper presented at the First Meeting of the Group of Experts: "Climate change and biodiversity". Strasbourg, 58 pp.

Huntley, B., & Birks, H. J. B. (1983). *An atlas of past and present pollen maps for Europe: 0–13000 B.P.* Cambridge: Cambridge University Press 667 pp.

IPCC, 1990. Scientific assessment of climate change – Report of Working Group 1. Summary for Policymakers. IPCC/WMO/UNEP, UK Meteorological Office, Bracknell.

IPCC. (1995). *Climate change 1995: The science of climate change. Summary for policymakers*. Bracknell: IPCC/WMO/UNEP, UK Meteorological Office.

IPCC. (2001). *Climate change 2001: Synthesis report. Contribution of working groups I, II, and III to the third assessment report of the intergovernmental panel on climate change*. Cambridge: Cambridge University Press.

IPCC. (2007). In Core Writing Team, R. K. Pachauri, & A. Reisinger (Eds.), *Climate change 2007: Contribution of working groups I, II and III to the fourth assessment report of the intergovernmental panel on climate change*. Geneva: IPCC 104 pp.

IPCC. (2014). In Core Writing Team, R. K. Pachauri, & L. A. Meyer (Eds.), *Climate change 2014: Synthesis report. Contribution of working groups I, II and III to the fifth assessment report of the intergovernmental panel on climate change*. Geneva: IPCC 151 pp.

IUCN. (2003). *Guidelines for application of IUCN red list criteria at regional levels: Version 3.0. IUCN species survival commission*. Gland: IUCN.

Kavrakova V., Dimova D., Dimitrov M., Tzonev R., Belev T., Rakovska K., 2009. Handbook for identification of habitats with European importance in Bulgaria. 2nd updated edition, Sofia, WWF, Danube-Carpathian Program and "Green Balkans: federation, 131 pp.

Kelleher, C. T., de Vries, S. M. G., Baliuckas, V., Bozzano, M., Frýdl, J., Goicoechea, P. G., Ivankovic, M., Kandemir, G., Koskela, J., Kozioł, C., Liesebach, M., Rudow, A., Vietto, L., & Zhelev, P. (2015). *Approaches to the conservation of Forest genetic resources in Europe in the context of climate change. European Forest Genetic Resources Programme (EUFORGEN)*. Rome: Bioversity International xiv + 46 pp.

Keppel, G., Van Niel, K. P., Wardell-Johnson, G. W., Yates, C. J., Byrne, M., Mucina, L., Schut, A. G. T., Hopper, S. D., & Franklin, S. E. (2012). Refugia: Identifying and understanding safe havens for biodiversity under climate change. *Global Ecology and Biogeography, 21*(4), 393–404.

Koskela, J., Buck, A., & Teissier du Cros, E. (Eds.). (2007). *Climate change and forest genetic diversity: Implications for sustainable forest management in Europe*. Rome: Bioversity International 111 pp.

Koulov, B., & Zhelezov, G. (Eds.). (2016). *Sustainable Mountain regions: Challenges and perspectives in Southeastern Europe*. Basel: Springer.

Kremer, A., Ronce, O., Robledo-Arnuncio, J. J., Guillaume, F., Bohrer, G., Nathan, R., Bridle, J. R., Gomulkiewicz, R., Klein, E. K., Ritland, K., Kuparinen, A., Gerber, S., & Schueler, S. (2012). Long-distance gene flow and adaptation of forest trees to rapid climate change. *Ecology Letters, 15*(4), 378–392.

Kutzbach, J. E., & Guetter, P. J. (1986). The influence of changing orbital parameters and surface boundary conditions on climate simulations for the past 18,000 years. *Journal of the Atmospheric Sciences, 43*, 1726–1759.

Lefévre, F. (2007). Conservation of forest genetic resources under climate change: The case of France. In J. Koskela, A. Buck, & E. Teissier du Cros (Eds.), *Climate change and forest genetic diversity: Implications for sustainable forest management in Europe* (pp. 95–101). Rome: Bioversity International.

Mackintosh, A., Golledge, N., Domack, E., Dunbar, R., Leventer, A., White, D., Pollard, D., DeConto, R., Fink, D., Zwartz, D., Gore, D., & Lavoie, C. (2011). Retreat of the East Antarctic ice sheet during the last glacial termination. *Nature Geoscience, 4*, 195–202.

Magri, D., Vendramin, G. G., Comps, B., Dupanloup, I., Geburek, T., Gömöry, D., Latalowa, M., Litt, T., Paule, L., Roure, J. M., Tantau, I., van der Knaap, W. O., Petit, R. J., & de Beaulieu, J.-L. (2006). A new scenario for the quaternary history of European beech populations: Palaeobotanical evidence and genetic consequences. *New Phytologist, 171*, 199–221.

Malinova D. (2007). Typologic landscape structure of Bugarka Nature Park. Ph.D. thesis. S., University of Forestry, 171 pp.

Matyas, C. (1997). Climatic adaptation of trees: Rediscovering provenance tests. In P. M. A. Tigerstedt (Ed.), *Adaptation in plant breeding* (pp. 49–58). Dordrecht: Kluwer Academic Publishers.

Meshinev, T., & Apostolova, I. (2005). The habitats in Bulgaria. In: Petrova A., Vladimirov V., & Dimitrova D. (eds.) *Contemporary status of the biodiversity in Bulgaria – Problems and prospects, S., BAS and Bulgarian Bio-platform* (pp. 351–373). Sofia: Bulgarian Academy of Sciences.

Muller-Starck, G., & Schubert, R. (2001). *Genetic response of forest systems to changing environmental conditions*. Cham: Springer XX+366 pp.

Namkoong, G., Kang, H., & Brouard, J. S. (1988). *Tree breeding: Principles and strategies*. New York: Springer 175 pp.

Noss, R. (1990). Indicators for monitoring biodiversity: A hierarchial approach. *Conservation Biology, 4*(4), 355–364.

Ohlemüller, R., Huntley, B., Normand, S., & Svenning, J. C. (2012). Potential source and sink locations for climate-driven species range shifts in Europe since the last glacial maximum. *Global Ecology and Biogeography, 21*, 152–163.

Peev, D. (Ed.). (2015). *Red data book of the republic of Bulgaria. Vol. 1. plants and fungi*. Sofia: BAS and MoEW.

Petrova, A., & Vladimirov, V. (Eds.). (2009). Red list of Bulgarian vascular plants. *Phytologia Balcanica, 15*(1), 63–94.

Petrova, A., Vladimirov, V., & Dimitrova, D., 2005. Contemporary status of the biodiversity in Bulgaria – Problems and prospects, S., BAS and Bulgarian Bio-platform, 565 pp. (in Bulgarian).

Polunin, O. (1980). *Flowers of Greece and the Balkans – A field guide*. Oxford: Oxford University Press 605 pp.

Raev, I. (Ed.). (2006). *Environment and structure of the Norway spruce forests in the Rila Mountains*. Sofia-Moscow: Forest Research Institute, Bulgarian Academy of Sciences, Pensoft publ 194 pp.

Raev, I., Zhelev, P., Grozeva, M., Markoff, I., Velichkov, I., Zhiyanski, M., Georgiev, G., Miteva, S., & Alexandrov, V. (2011). *Program of measures for adapting the forests in the Republic of Bulgaria and mitigation of negative effect of climate change on forests*. Sofia: Executive Forest Agency, Ministry of Agriculture and Food 212 p. (in Bulgarian).

Raev, I., Alexandrov, V., & Tinchev, G. (2015). Assessment of drought related climate change impacts on forests in Bulgaria. *Silva Balcanica, 16*(1), 5–24.

Ramenski, L. G. (1937). Counts and description of vegetation (on the basis of the projective method). Moscow, Selkhozgiz, 100 pp. (In Russian).

Sakalian, M. (1993). National strategy for biodiversity conservation, vol. 1. S., MoE, 661 pp. (in Bulgarian).

Sakalian, M. (1999). Biological diversity in Central Balkan National Park. S., MoEW, 635 pp. (in Bulgarian.

Savolainen, O., Bokma, F., Knürr, T., Kärkkäinen, K., Pyhäjärvi, T., & Wachowiak, W. (2007). Adaptation of forest trees to climate change. In J. Koskela, A. Buck, & E. Teissier du Cros (Eds.), *Climate change and forest genetic diversity: Implications for sustainable forest management in Europe* (pp. 19–30). Rome: Bioversity International.

Sukachev, V. N. (1972). *The foundations of Forest typology and Biogeocoenology*. Leningrad: Nauka 420 pp. (in Russian).

Tashev, A. N., & Tsavkov, E. I. (2008). Medicinal plants of the Bulgarian dendroflora. *Phytologia Balcanica, 14*(2), 269–278.

Thiede, J. (1974). Glacial Mediterranean. *Nature, 276*, 680–683.

Todorov, N., & Velchev, A. (2014). *Landscapes of Bulgaria – Spatial structure*. Veliko Tarnovo: IVIS Publication House.

Todorov, N., Velchev, A., & Konteva, M. (2017). An update of the landndscape classification system in Bulgaria (based on the landscape map from 1992, scale 1:500 000). *Problems of Geography, 1–2*, 43–53.

Turill, W. B. (1928). *The plant life of the Balkan Peninsula*. Oxford: Clarendon Press 490 pp.

Tzonev, R. T., Dimitrov, M. A., & Roussakova, V. H. (2009). Syntaxa according to the Braun-Blanquet approach in Bulgaria. *Phytologia Balcanica, 15*(2), 209–233.

Usher, M. B. (2005). *Conserving European biodiversity in the context of climate change.* Strasbourg: Committee for the activities of the Council of Europe in the field of biological and landscape diversity 32 pp.

Vassilev, K., Pedashenko, H., Alexandrova, A., Tashev, A., Ganeva, A., Gavrilova, A., Gradevska, A., Assenov, A., Vitkova, A., Grigorov, B., Gussev, C., Filipova, E., Aneva, I., Knollová, I., Nikolov, I., Georgiev, G., Gogushev, G., Tinchev, G., Pachedjieva, K., Koev, K., Lyubenova, M., Dimitrov, M., Apostolova-Stoyanova, N., Velev, N., Zhelev, P., Glogov, P., Natcheva, R., Tzonev, R., Boch, S., Hennekens, S., Georgiev, S., Stoyanov, S., Karakiev, T., Kalníková, V., Shivarov, V., Russakova, V., & Vulchev, V. (2016). Balkan vegetation database: Historical background, current status and future perspectives. *Phytocoenologia, 46*(1), 89–95.

Veen, P., Fanta, J., Raev, I., Biriş, I.-A., de Smidt, J., & Maes, B. (2010). Virgin forests in Romania and Bulgaria: Results of two national inventory projects and their implications for protection. *Biodiversity and Conservation, 19*, 1805–1819.

Velchev, A., Todorov, N., Assenov, A., & Berouchashvili N. (1992). *A Landscape map of Bulgaria of the scale 1:500 000.* Annuaire de l'Université de Sofia «St. Kliment Ohridski», Fac. de Géologie et géographie, Liv. 2-Géographie, (*84*: pp. 85–106).

Velchev, A., Penin, R., Todorov, N., & Konteva, M. (2000). Landscape geography of Bugaria. S., Bulvest, 236 pp.

Weiss, S., & Ferrand, N. (Eds.). (2007). *Phylogeography of Southern European Refugia.* Dordrecht: Springer 377 pp.

Willis, K. J. (1994). The vegetation history of the Balkans. *Quaternary Science Reviews, 13*, 769–788.

Yurukov, S., & Zhelev, P. (2001). The woody flora of Bulgaria: A review. *Schweizerische Zeitschrift für Forstwesen, 152*(2), 52–60.

Chapter 35
Effect of Climate Change on the High-Mountain Tree Species and Their Genetic Resources in Bulgaria

Momchil Panayotov, Nickolay Tsvetanov, Evgeni Tsavkov, Georgi Gogushev, Peter Bebi, Petar Zhelev, and Stefan Yurukov

Abstract Climate change have the potential to strongly affect mountain coniferous forests in Bulgaria in several ways: (1) By directly affecting tree growth and the potential of trees to successfully cope with climate extremes; (2) By affecting disturbance events and regimes, which on their turn may trigger further habitat changes; (3) By facilitating migration of better adapted for the new climate condition species, which can outcompete other less adapted species and replace them and (4) By facilitating invasive species. In this chapter we provide data on the distribution of coniferous forests in Bulgarian mountains and short reviews of recent tree ring studies and studies on disturbance regimes. The tree ring data show the high importance of drought and other extreme climate events on high-mountain conifer species. This outlines that expected summer warming and temperature increase have the potential to strongly affect tree growth. The disturbance data shows the high importance of fires and windthrows, but also high number of snow damages and avalanches, which are not to be neglected. Insect outbreaks may be further facilitated if summer temperatures increase and allow species, which are currently limited by colder temperatures in higher altitudes, to affect also forests higher up. All these factors can act together and modify habitat quality and conditions and in this way put in risk species and genotypes with limited distribution and narrow growth niches. It is therefore necessary to take measures for ex-situ conservation of genotypes besides the well developed in-situ conservation in the network of protected areas in Bulgaria.

M. Panayotov (✉) · N. Tsvetanov · E. Tsavkov · P. Zhelev · S. Yurukov
University of Forestry, Sofia, Bulgaria
e-mail: panayotov.m@ltu.bg; nicktsvetanov@ltu.bg; etsavkov@ltu.bg; zhelev@ltu.bg; syurukov@ltu.bg

G. Gogushev
Regional Forestry Directorate – Blagoevgrad, Blagoevgrad, Bulgaria
e-mail: gogushev_g@abv.bg

P. Bebi
WSL Institute for Snow and Avalanche Research SLF, Davos, Switzerland
e-mail: bebi@slf.ch

M. Šijačić-Nikolić et al. (eds.), *Forests of Southeast Europe Under a Changing Climate*, Advances in Global Change Research 65,
https://doi.org/10.1007/978-3-319-95267-3_35

Keywords Mountain coniferous forests · Disturbance · Tree –ring studies · Climate change

35.1 Introduction

Mountain landscapes in Bulgaria cover about half of the country's territory. According to the criteria of the European Environmental Agency about 55,000 km^2 are mountains. Except the highest mountain areas above 2200 m a.s.l. in the Rila and Pirin Mountains and the highest parts of the ridges of Stara Planina, Vitosha, Belasitsa, Slavyanka and Osogovo, the rest of these areas are covered by forests, which are mostly coniferous or beech-dominated.

The mountain coniferous forests in Bulgaria occupy 1.05 million ha and dominate the landscape in the highest mountains in southwestern and southern Bulgaria and partly in Stara Planina mountains. Although most of them have long management history, there are also patches of old-growth forests of specific interest in terms of natural conservation and protection of the genotype richness. Special attention deserve the forests of Bosnian pine (*Pinus heldreichii* Christ) and Macedonian pine (*Pinus peuce* Griseb.) in the Pirin Mountains, which are among the best preserved ecosystems of these high-mountain Balkan tree species. Within the mountain coniferous forests, a limited number of deciduous tree species also occur. The typical ones are of the genera *Sorbus (S. aucuparia* L., *S. aria* (L.) Crantz), *Acer (A. pseudoplatanus* L., *A. heldreichii* Orph. ex Boiss., *A. platanoides* L., *A. hyrcanum* Fisch. et Mey.), *Betula (B. pendula* Roth), *Populus (P. tremula* L.), *Fraxinus (F. excelsior* L.), *Salix* (mostly *S. caprea,* but also a number of *Salix* species with limited distribution, like *S. lapponum* L., *S. waldsteiniana* Willd. and *S. pentandra* L.).

Beech forests (*Fagus sylvatica* L.) dominate the mountain landscapes of Stara Planina Mountain Range and have high participation in mixed beech-coniferous forests at 1200–1500 m a.s.l. in the other high mountain ranges.

Besides the typical coniferous species, beech and the accompanying above-mentioned deciduous species, there are also a few locations of additional interest such as the recently validated *Quercus protoroburoides* found at only two isolated locations on slopes in the the Rila mountains at about 1600 m a.s.l. (Tashev and Tsavkov 2017), populations of the above-mentioned *Acer heldreichii* (Alexandrov and Pandeva 2014), *Alnus viridis* (Chaix.) DC. (Roussakova 2015a) and even spots of some rare conifers like *Taxus baccata* L. and *Abies borisii-regis* Mattf. (Roussakova 2015b). There are also a number of shrub species, some of which are with limited local or global distribution (Assyov et al. 2012).

The Bulgarian mountain forests have high economic and ecological role. They are crucially important as the main source of high-quality wood, but also serve various ecosystem services, which are valued by society. Most of the drinking water catchments are in mountain forests or in the *Pinus mugo* shrub communities above the treeline. These forest also provide forest fruits such as blueberries (*Vaccinium* sp.) and raspberries (*Rubus* sp.) and commercially valued mushrooms mostly from *Boletus* sp. and *Cantharellus cibarius.* In addition, mountain resorts are located

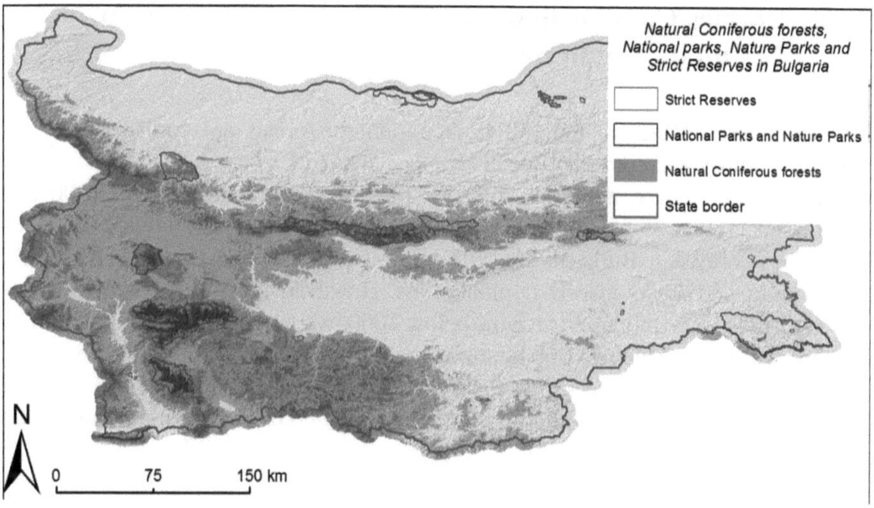

Fig. 35.1 Natural coniferous forests and extent of National Parks, Nature Parks and Strict forest reserves in Bulgaria. (Data source: Executive Forest Agency)

mostly within mountain coniferous forests which make them particularly important for the tourism economic sector. Most of the natural coniferous forests and big proportion of the beech forests in the Rila, Pirin and Stara planina Mountains are protected in National parks and reserves, while in the Rhodope Mountains this percentage is quite low. In Bulgaria there are 3 national parks, 11 natural parks and 55 strict reserves and the majority of them are located in mountains (Fig. 35.1).

In addition, much of the mountain forest territories are in NATURA 2000 areas. This network of protected areas provides solid foundation for the in-situ conservation of mountain forests, the species and genotype richness. However, there are numerous threats to mountain forests, mostly related to climate change. The species-specific and ecosystem-level responses to expected future climate remain highly unknown. There are high concerns that especially species with limited distribution may experience further shrinking of their habitats. This is particularly true for species at the subalpine zone, which may face pressure from more competitive species advancing from lower altitudes, but at the same time may be restricted to migrate higher up due to terrain (e.g. rocks) or climate constrains (e.g. severe winds in high mountains, snow drift and snow accumulation, avalanches, frosts) above the present treeline. Isolated old-growth *Pinus nigra* forests growing on steep sun-exposed rocky slopes in the Trigrad region in the Rhodope Mountains may also further loose territory due to high competition from advancing deciduous species, limited regeneration, and lack of possibilities to migrate because of lack of other suitable habitats in vicinity. In addition, invasive species and disturbances may dramatically change the present environmental conditions thus leading to different ecosystems from the presently found, which has the potential to negatively affect habitats with limited distribution.

35.2 Species Composition and Distribution of Bulgarian Mountain Coniferous Forests

The main tree species in the Bulgarian mountain forests are conifers from the *Pinaceae* family and common Beech (*Fagus sylvatica* L.).

Most (92%) of the natural coniferous forests are located between 1000 and 2200 m a.s.l. (Fig. 35.2) while only 1% are below 700 m a.s.l. About 39% of the coniferous forests in Bulgaria are with natural origin, while the rest are mostly *Pinus nigra* and *Pinus sylvestris* plantations. However, most of these plantations were created at low-mountain terrains below 1000 m a.s.l. Besides the pure coniferous forests there are also 64,000 ha mixed beech-conifer forests, of which 82% are natural. Above the treeline in the Pirin and Rila Mountains and isolated in Stara Planina Mountains there are also 19,300 ha *Pinus mugo* bush-type communities. Of

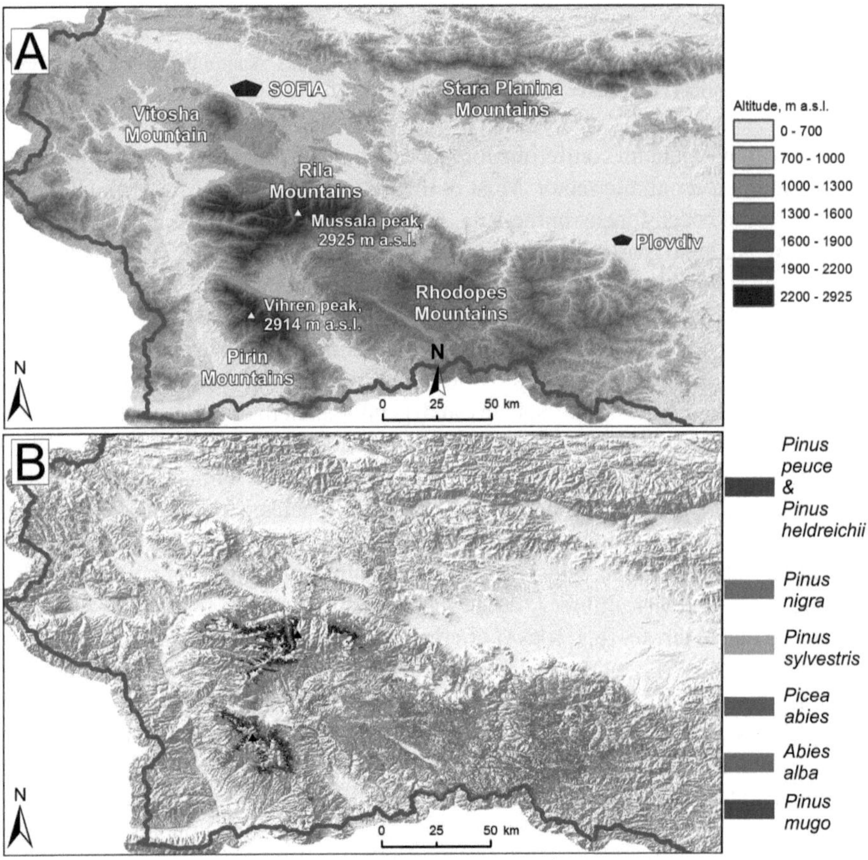

Fig. 35.2 Relief map of Central and Southwestern Bulgaria (**a**) and coverage of natural coniferous forests by dominant tree species (**b**). (Data source: Executive Forest Agency, processing following Panayotov et al. 2016b)

the natural coniferous forests, the largest territories are occupied by *Pinus sylvestris* (47%) followed by *Picea abies* (22%), beech-conifer (12%), *Pinus nigra* (8%), *Abies alba* (5%), *Pinus peuce* (2%), mixed coniferous without a clear dominant species (2%) and *Pinus heldreichii* (Panayotov et al. 2016a). *Taxus baccata* is found as single trees or small groups. The mountain bushes consisting of *Juniperus communis* and *Juniperus sabina* cannot be classified as forests but have important ecological role on steep and eroded slopes mostly above treeline.

Scots pine *(Pinus sylvestris* L.) forests occupy large territories mostly in the Rhodope, Pirin, Rila, Plana and Osogovo Mountains. Most of them are situated between 1000 and 1600 m a.s.l. and are below 120 years old. It is considered that this species increased its territory after the middle of the nineteenth century as a result of abandonment of pastures in the mountains and regeneration after fires in formerly spruce forests (Panayotov et al. 2016b). The spruce (*Picea abies* Karst) forests are situated mostly between 1300 and 1900 m a.s.l. (86%) in the Rhodope Mountains and Rila Mountains. In the other high mountains there are smaller patches or the species is mixed with *Pinus peuce* and *Pinus heldreichii* (the Pirin Mountains). About half of the spruce forests are between 80 and 120 years old, while 30% are above 120 years old.

The oldest and best preserved spruce forests are found in the high parts of the Rila and Rhodope Mountains. The *Pinus nigra* forests are situated mostly between 700 and 1300 m a.s.l. (75%). The lowest locations are about 500 m a.s.l., while the highest is at 1850 m a.s.l. (Tashev et al. 2013). Most of the Austrian pine forests are below 120 years old, but there are isolated patches with age of up to 400 years (Shishkova and Panayotov 2013). The Silver fir-dominated forests occupy 23,000 ha and are situated mostly between 1000 and 1600 m a.s.l. on shady locations (88%). The lowest location is at 400 m a.s.l. close to Kotel town in the Stara planina Mountain, while the highest are at about 1900 m a.s.l. These forests together with the mixed beech-fir (15,000 ha) and beech-fir-spruce (15,000 ha) are some of the most productive in Bulgaria with measured stock of more than 1400 $m^3.ha^{-1}$. These forests are situated mostly in the Rhodope, Rila, Stara Planina and Pirin Mountains. About 42% of the silver fir forests are with age between 120 and 200 years. The Macedonian pine (*Pinus peuce* Griseb.) forests occupy 18,700 ha but are of high importance for the subalpine zone in the Rila and Pirin Mountains. In the Stara planina Mountians there is only one natural location (the "Tsaritchina" reserve) of this species. Half of these forests are above 120 years old but there are well preserved patches with age of more than 400–500 years mostly in the Pirin Mountains (Panayotov and Yurukov 2007; Panayotov et al. 2010). The Bosnian pine (*Pinus heldreichii* Christ) forests occupy 1200 ha in the Pirin and Slavyanka Mountains where they form the treeline alone or together with *Pinus peuce*. They are of high conservational value and there are old-growth patches with age of more than 500–600 years (Panayotov et al. 2010; Rangelova and Panayotov 2013).

35.3 Climate Change Uncertainties and Potential Effects

Climate change has the potential to affect seriously mountain forests in Bulgaria. Amongst the conclusions of the report from the Intergovernmental panel on Climate Change (IPCC 2014) is the expectation that the effects of climate change in Europe will be highest in the southern regions and particularly the Mediterranean area, where there will be a decrease in available water resources and summer droughts. This can strongly affect the forests especially in the mountain ranges in southern Bulgaria.

Climate patterns in the Bulgarian mountains are determined by the combined influence of Mediterranean, Atlantic and Polar air masses (Brown and Petkova 2007). Annual precipitation varies from about 900–1400 mm. The Pirin, Belasista and Slavyanka Mountains, the southern parts of the Rhodopes Mountains and the southwestern parts of the Rila Mountains are under stronger influence of Mediterranean air masses and have a precipitation maximum in autumn and winter. Stara Planina, Osogovo and Vitosha Mountains and the central and northern parts of the Rila Mountains have a precipitation maximum at the end of spring and the beginning of summer. July to September is a generally dry period. Long summer droughts have occurred in some years (Raev et al. 2003) and strongly affected tree growth (Raev et al. 2003; Panayotov et al. 2013). Since the 1980s, there has been a trend of increasing temperatures (Grunewald et al. 2009) and decreasing snow cover (Brown and Petkova 2007; Nojarov 2012) in the Bulgarian mountain areas. The long-term records from Musala peak (2925 m a.s.l., 1933 to present days) show that between 2000 and 2015 were recorded seven of the ten warmest summer periods (June–August) and four of the ten warmest winters (December–March). These higher temperatures may have numerous effects on mountain ecosystems by changing precipitation patterns (i.e. snow vs. rain) and the duration of snow-cover in winter, causing higher drought stress in summer, providing growth advantage to some species while at the same time limiting other, and strongly affecting the potential for invasion of species from other ecosystems, biotic and abiotic disturbances (Panayotov et al. 2017).

Although there are efforts to partially fill the knowledge gaps by experimental work and modeling exercises, the wide range of potential effects of climate change causes high uncertainty in the understanding of potential effects on forest ecosystems. For example an applied project lead by the Balkani NGO (project "How to adapt forests to changes in climate – innovative contribution of NGO") used GIS modeling of future climate change scenarios for 2050 and 2100 and the Rivas-Martinez concept for bioclimatic classification to model the expected future climate and associated growth conditions in southwestern Bulgaria, including the Pirin mountains. The modeling showed potential dramatic changes in the climatically-suitable areas for moisture-demanding species such Norway spruce and Siver fir. They are expected to experience very high pressure from beech and pines (mostly *Pinus sylvestris*) which will be better adapted for the increased temperatures in these areas. The climatically optimal conditions for the Balkan endemic species

Pinus peuce will shift above present treeline (2200 m a.s.l.) up to about 2500 m a.s.l. However, at these altitudes the species will face problems with soil availability and other effects of mountain climate such as strong winds, frosts, snow accumulation and snow drift as well as disturbances such as avalanches. The combined effect could be that *Pinus peuce* will have limited opportunities for migration and facing drier and warmer climate and increased competition from better adapted species may lose much of its present-day territories. Similar situation could be expected for *Pinus heldreichii*, which has even narrower growth niche being well adapted to the very limited in distribution limestone and marble rocky terrains in the high parts of Pirin and Slavyanka Mountains. Higher temperatures may mean increased competition from advancing from below *Pinus nigra* and *Fagus sylvatica*. This modeling example shows that subalpine species with limited distribution could be threatened by further shrinking of the suitable habitats, which could limit their overall distribution. Such situation calls for emergent research actions to get deeper understanding and knowledge on the potential of rare species to handle future expected climate conditions and potentially for need of management decisions and actions to support the survival of species.

Another potential high effect of climate change is on the overall productivity and growth of tree species which may have dramatic effects on ecosystem services including production of biomass and carbon sequestration. Zlatanov et al. (2017) carried out modeling study for the mountain landscape in the valley of Mt. Perelik in the Western Rhodopes at altitudes between 1000 and 2100 m a.s.l., which is highly representative of the mountain areas dominated by the most important coniferous species (Norway spruce, Scots pine, Austrian pine, Fir) and beech in the Rhodopes Mountains, Rila Mountains and parts of the Pirin, Vitosha and Stara Planina Mountains. They used the stand-scale model PICUS, the landscape scale model LandClim to model forest growth, productivity and the capabilities to provide ecosystem services under selected regional climate simulations. The modeling showed that high impact of climate change might be expected. The most affected forests are expected to be Norway spruce-dominated forests where the participation of the main species will diminish and the general growing stock will significantly decrease (with up to 300 m^3/ha under the pessimistic scenarios). This is of high importance given the fact that large areas in the Rhodopes and Rila Mountains in Bulgaria are covered by this forest type and numerous other ecosystems services with high importance such as water-provisioning and tourism might be seriously affected. Beech forests at lower altitude are also expected to suffer loss of biomass (up to 200 m^3/ha below 1000 m a.s.l. under pessimistic scenarios) and abundance. However, at more mesic sites beech are expected to increase their share. Silver fir might be expected to increase its share, which is currently relatively low and also increase its growing stock in some of the scenarios. Most robust to climate change is predicted to be Austrian pine and Scots pine on sites with favorable conditions (i.e. sun-exposed mountain terrain above 1000 m a.s.l.). The general biomass quantity is expected to decrease at lower altitude sites. At higher-altitude sites it might initially increase due to better temperature conditions (in the case increased temperatures above 1800 m a.s.l.), but after 2070 decrease due to drought conditions and their negative impact of growth mostly on Norway spruce. Harsher impacts are expected on other ecosystem services at lower altitudes – there is expected loss in species richness, higher mortality and associated decrease of the ability of forests to provide soil stabilization. This is of great importance given the fact that most of the water catchments for provision of clean drinking water in Southern Bulgaria are in the mountain regions covered with the type of forests studied.

One of the key questions for the future of forests in high mountain areas is regeneration and survival of saplings. This could be highly dependent on numerous factors and while there is data on widely distributed in Europe species, the knowledge about locally important species on the Balkan Peninsula is limited. An example of a study to partially fill in such gaps is the carried out artificial drought experiment with saplings from *Picea abies*, *Pinus mugo*, *Pinus peuce* and *Pinus heldreichii* (Anev 2016; Anev and Tsvetkova 2018). Within it three artificial precipitation regimes were simulated*:* (1) Normal summer precipitation based on historical data from mountain stations for the period 1961–1990; (2) June-August drought and (3) Whole-summer drought with severe drought in August. In addition, the saplings were transported at approximately 400 m lower altitude thus simulating increased temperatures. The experiment showed that in drought situations *P. abies* saplings had maintained the level of their physiological processes unchanged, but this strategy lead to high mortality. *Pinus mugo* was strongly influenced especially by the long summer drought regime, but its survivorship remained high and stable. Both photosynthesis and transpiration of *P. peuce* and *P. heldreichii* were more sensitive to the long summer drought, in which they had lower survivorship. Based on their data the authors concluded that the limited potential for adaptation and competition of *P. abies* probably will decrease its range of distribution. The expected trends in climate change will probably lead to a further narrowing of the ecological and physiological comfortable zone for *P. peuce* and *P. heldreichii* in the mountainous regions of the Balkan Peninsula. *Pinus mugo* has the potential to cope with higher temperature and drought in the high mountain zones which means that in the lack of other limiting factors it has the potential to maintain its wide distribution at suitable habitats above 2200 m a.s.l. in Rila and Pirin Mountains and expand in Western Stara Planina Mountains.

35.4 A Look into the Past Growth of Trees as a Source of Information on the Potential of Tree Species to Cope with Extreme Climate

Tree rings are one of the most reliable proxies for past climate conditions and events and this is among the main reason for their wide use in the research of past environmental conditions. This also provides a good reference on the climate extremes that trees lived through and therefore on their capabilities of coping with such extremes.

In the last decades there were considerable efforts to use tree ring analysis to study past climate in Bulgaria.

The longest developed and published tree ring chronologies for Bulgaria (up to almost 800 years) were from *Pinus heldreichii* Christ from the Pirin Mountains (Panayotov et al. 2010; 2016b). They show high dependence of tree growth on summer temperature and moisture conditions, but also on winter climate. The developed tree-ring width (TRW) chronologies (e.g. Panayotov et al. 2010) show negative effect on growth from early summer droughts and positive effects of higher winter temperatures. This "mixed climate signal" partly hinders the use of TRW for climate reconstructions, but the longest TRW chronology was used to support the construction of an Atlas of mega droughts and pluvials for the European region for the last 2000 years (Cook et al. 2015). Maximum latewood density (MXD) chronology from *Pinus heldreichii* was also constructed (Trouet et al. 2012). It had strong summer temperature signal, demonstrating the importance of summer temperatures (July–August) on the latewood formation. The MXD chronology analysis showed common temperature variability in the Balkans, Italy, and the southern Carpathians and opposing temperature variability patterns over the British Isles and southern Scandinavia suggesting that the North Atlantic Oscillation (sNAO) is a main driver of the teleconnection between summer temperatures in southeastern versus northwestern Europe (Trouet et al. 2012). The tree ring studies of another species with high longevity in the subalpine forests in Bulgaria (*Pinus peuce* Griseb.) showed weaker climate signal, but again dependence mostly on summer temperature and moisture conditions (Panayotov et al. 2010). However, *Pinus peuce* tree rings with anomalous anatomy such as frost rings, light rings, unusually narrow tree rings have the potential to be used as markers of years with extreme climate (Panayotov and Yurukov 2007). Such tree ring features were also found in *Picea abies* and *Pinus sylvestris* trees from subalpine locations. An analysis of *Pinus sylvestris* tree ring cores from different altitudes showed that light rings, frost rings, tree rings with intra-annual density fluctuations (IADF) have high potential to be used to reconstruct extreme climate conditions which were not well presented in standard climate records (Panayotov et al. 2013). This study also demonstrated that the same feature types (e.g. light rings) may be formed under completely contrasting climate extremes at different altitudes – unusually cold summers in high altitude and unusually dry late summers in low altitude. The above-mentioned studies demonstrated that although summer drought was causing growth decreases mostly in low and mid-mountain locations, certain strong droughts had negative effects also in subalpine and treeline locations. A study of mortality in unmanaged Spruce forests carried out by Panayotov et al. (2016a) demonstrated that occasional summer drought could be among the most important factors initiating mortality at the period of high competition even in locations, which are normally not affected strongly by dry conditions. *Pinus nigra* tree ring width was also strongly affected by summer drought (Grozev and Nedelchev 1996; Shishkova and Panayotov 2013). A novel TRW chronology from *Pinus nigra* trees from the Trigrad region of the southern Rhodopes with length of almost 400 years was used for summer drought reconstruction. It showed several dry periods (1680–1687, 1715–1725, 1725–1730, 1765–1775, 1792–1797,

1830–1835) which add up to the famous droughts from the instrumental measurements (e.g. 1807–1908, 1928, 1945–1947, 1967, 1984, 1987–1988, 2000, 2007).

Unusually cold winters (e.g. 1929) or summers (e.g. 1949) also caused the production of narrow tree rings in subalpine species such as Norway spruce and Macedonian pine. In addition, frost rings clearly demonstrated that early summer frosts are occurring periodically and leave a permanent sign in tree rings due to anatomical changes caused by cell damages (Panayotov and Yurukov 2007; Panayotov et al. 2010). However, such events were not found to be associated with mass mortality.

The available up-to-date tree ring data from Bulgaria points out that summer drought has the potentially strongest negative effects on tree growth. This means that if climate change will be associated with more frequent droughts, then the growth of trees in many locations, particularly rocky and sun-exposed slopes will be limited. However further studies can contribute to better understand if certain genotypes are more resilient and can have better capabilities to cope with such conditions. This is particularly important for young trees and saplings, which could be more exposed to drought stress due to their root systems, which are still not so well developed.

35.5 Natural Dynamics and Disturbances in Bulgarian Mountain Coniferous Forests

Biotic and abiotic disturbances have the potential to strongly affect mountain forests and be among the main drivers of changes in them. They strongly affect species composition, structural properties, nutrient cycling, carbon storage, biodiversity and other ecosystem services (Oliver and Larson 1996; Foster et al. 1998; Bebi et al. 2009; Bradford et al. 2013; Thom and Seidl 2015; Vacchiano et al. 2015). The higher frequency of abiotic and biotic disturbances in coniferous forests in Europe during the last several decades (Schelhaas et al. 2003; Seidl et al. 2014) and the increasing desire to manage forests in a “close-to-nature” manner (Kraus and Krumm 2013) means that it is a necessity to improve the understanding of the Historic Range of Variability (HRV) (Landres et al. 1999; Veblen 2003; Kulakowski and Bebi 2004) and the probable Future Range of Variability (FRV) of these forests (Kulakowski et al. 2017). During the last decade the topic of natural disturbances in Bulgarian mountain coniferous forests was the focus of several studies, which contributed to the overall understanding of their role (Panayotov et al. 2011, 2015, 2017; Panayotov and Georgiev 2012).

The collected data show that natural disturbances are very important part of the dynamic of coniferous forests in Bulgaria. Historically most frequent were fires (39% of all reported abiotic disturbances) followed windthrows (31%) (Panayotov et al. 2017). Less data were related to damages caused by snow and ice load (6%) and avalanches (20%). From the biotic disturbances the most important are related

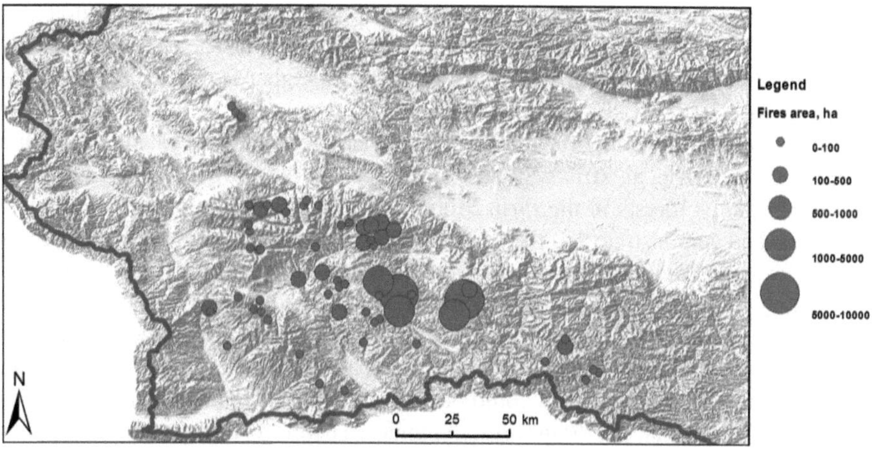

Fig. 35.3 Area of historically recorded forest fires in coniferous forests in mountain territories in Southwestern Bulgaria following Panayotov et al. (2017)

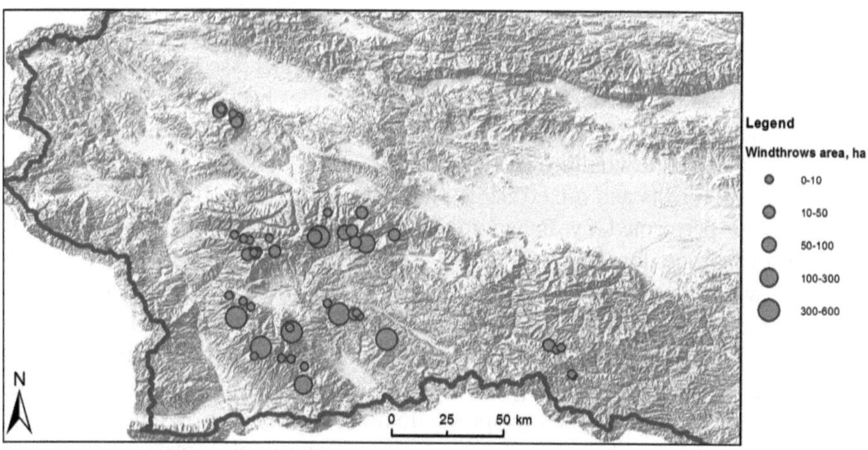

Fig. 35.4 Area of historically recorded windthrows in coniferous forests in mountain territories in Southwestern Bulgaria following Panayotov et al. (2017)

to insect outbreaks caused mainly by bark beetles (*Ips typographus, Ips acuminatus, Tomicus piniperda*), pine processionary moth (*Thaumetopoea pityocampa*), the pine sawflies *Neodpirion sertifer* and *Diprion pini*. Stressed trees are also affected by pathogenic fungi of which most important are *Heterobasidium annosum* in spruce and pine forests and recently *Sphaeropsis sapinea* in *Pinus nigra* plantations at lower elevation (Zafirov 2016).

The largest registered disturbances in mountain coniferous forests were fires (up to 9000 ha, Fig. 35.3) followed by windthrows (up to 3000 ha, Fig. 35.4). Fires affected all kinds of coniferous forests including Norway spruce-dominated, but were most frequent in Pine-dominated forests (Panayotov et al. 2017). There was no specific zone within the area of natural distribution of coniferous forests that was more affected, although there was slightly higher number of fires at altitudes between 1600 and 1900 m a.s.l. Several periods had higher frequencies: the 1880–1910, 1940–1950, and after 1990. These periods match the available statistics for the whole territory of Bulgaria and point to climatic reasons besides anthropogenic factors (Panayotov et al. 2017). Studies of fire scars in old-growth *Pinus heldreichii* and *Pinus peuce* forests in the Pirin Mountains show a number of fires in the last 500 years and demonstrate that these forests are characterized by periodic fire events (Rangelova and Panayotov 2013). This is also true for *Pinus nigra* forests, but there are still no studies on their fire history in Bulgaria. These data on fire occurrence in mountain coniferous forests in Bulgaria are still not full, but they clearly demonstrate the importance of fires. Having in mind the expected future frequent and stronger summer droughts, fires could become severe problem for many coniferous forests and be the main factor in forest dynamics. Therefore, there is need on further additional studies to better understand the role of fires in the forest dynamics of rare forest types (e.g. *Pinus nigra, Pinus heldreichii and Pinus peuce* forest) and plan potential activities and measures to insure the long-term stability of these ecosystems.

For the last 150 years there were registered at least 59 windthrows with area above 1 ha (Panayotov et al. 2017). There was a sharp increase in the number after 1960 (Fig. 35.5). The most affected zone was at altitudes between 1600 and 1900 m a.s.l. and the most affected type of forests was Norway spruce-dominated. However, there were also large windthrows (i.e. >300 ha) also in Macedonian pine *(Pinus peuce* Griseb.*)* forests and mixed Scots pine-Norway spruce forests. Available data and analyses performed by Panayotov et al.(2015) showed that uniform 120–200 year-old spruce-dominated forests have highest chances to be affected by stand-replacing windthrows while older and heterogeneous forests were affected mostly by gap-sized disturbances. This is very important having in mind that large proportion of the Bulgarian spruce forests are at that age group or will enter it in the next decades. In the unmanaged old-growth forest "Parangalitsa", which is one of the few examples of natural Norway-spruce dominated forests in mountain areas in Southern Europe, windthrows shaped the landscape-level forest structure and affected more than 20% of the forests within the last 150 years (Panayotov et al. 2011, 2015).

There was one large-scale bark-beetle outbreak caused by *Ips typographus* in the Bistrishko branishte Forest Reserve and Ofeliite region both in Vitosha Nature park (Dountchev et al. 2014; Panayotov and Georgiev 2012; Dountchev and Zhelev 2015). It started after a 60 windthrow (2001) and affected more than 200 ha of old (120–160 years old) spruce forest. These events were at least partly related to former human activities which were the reason for almost even-aged and simple-structured forests in the reserve (Panayotov and Georgiev 2012). However, climate changes coupled with aging of forests and potentially higher number of windthrows have the potential to considerably increase the damage from *Ips typographus* in future (Mirchev et al. 2011; Seidl et al. 2011; Temperli et al. 2013; Thom et al. 2013). Other insects known for causing considerable tree damages in coniferous

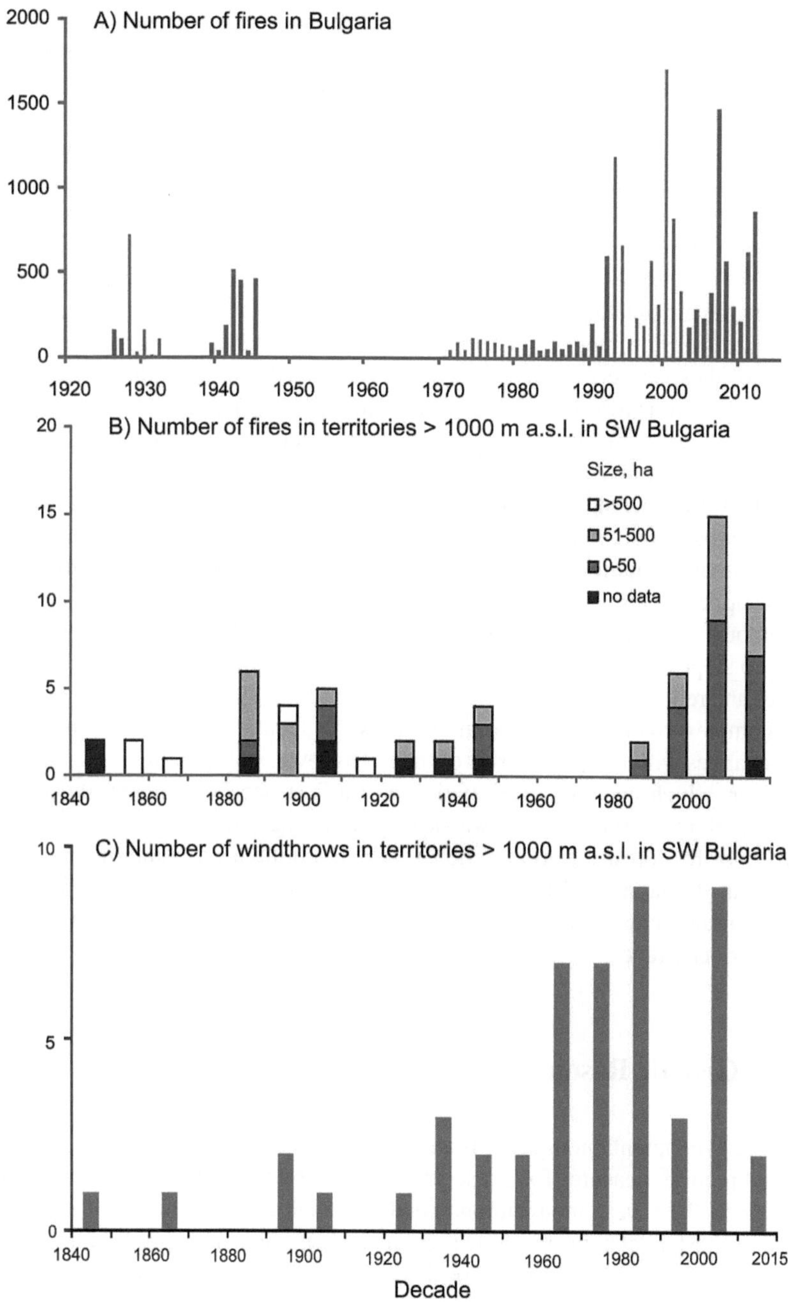

Fig. 35.5 Number of recorded forest fires in whole Bulgaria (Source: Executive Forest Agency reports) (**a**), historically documented fires in the mountains of Southwestern Bulgaria (Source: Panayotov et al. 2017) aggregated per decade (**b**) and historically documented windthrows in the mountains of Southwestern Bulgaria aggregated per decade. (Source: Panayotov et al. 2017) (**c**)

mountains are the pine processionary moth (*Thaumetopoea pityocampa*), which is currently attacking increasingly larger areas (up to 30,000 ha annually) in *Pinus nigra* and *Pinus sylvestris* forests at elevations up to 1400 m a.s.l. (Mirchev et al. 2011); the pine sawflies *Neodpirion sertifer* and *Diprion pini*; and bark beetles such as *Tomicus piniperda, Tomicus minor* and *Ips sexdentatus*, which also mostly attack pine forests and plantations at lower elevations (Mirchev et al. 2004). Several outbreaks of *Ips acuminatus* were reported in younger *Pinus sylvestris* forests in the Rhodopes Mountains after the extensive droughts in 1928 (Russkoff 1928) and 1945–1947 (Zashev 1950). In the last decade, this insect has been increasingly affecting pine plantations at lower altitudes. Increases in summer temperatures and in the frequency or intensity of droughts could result in conditions that support outbreaks of these insects in higher elevations and thus have unpredictable effects on natural coniferous forests and their composition (Raev et al. 2011).

Snow-related damages were mostly related to stem-breakage in young *Pinus sylvestris* and *Pinus nigra* forests. There were several reports for damages with more than 30,000 m^3 damaged wood stock the biggest of which was for the winter of 2015 (almost 1,000,000 m^3) (Panayotov et al. 2017). Is is questionable whether future climate conditions will increase the probability of wet snow occurrence, but if this happens, more such events could be expected.

Although avalanches are quite important disturbance factor for subalpine forests on steep slopes in the Pirin, Rila and Stara planina Mountains, there were only few studies and reports (Panayotov et al. 2017). Because avalanche events are dependent on a complex combination of weather and snow factors it is hard to predict if climate changes will increase the risk of them. However, sudden temperature increases in winter, which become rather typical in the last decades, almost always mean increase of the probability of wet snow avalanches, which can be particularly destructive for forests. Slab avalanches, which are the most common type of avalanches in Bulgaria are formed by strong winds and snow transport, which means that potential increase of the frequency of strong winds has the potential to modify also their occurrence.

35.6 Genetic Resources Conservation

The forests in mountainous regions are facing many challenges that treat their integrity and require measures for conservation of their genetic diversity. Besides being exposed to the severe mountainous conditions and climate change, they are subjected to substantial anthropogenic pressure because in many cases they are the only opportunity for livelihood for the local people, and are traditionally used as a source of fuel wood.

These circumstances underline the necessity for conservation efforts concerning the forest ecosystems and particularly, the natural ones. The studies have shown that natural populations in the mountainous regions possess high level of genetic diversity and in most cases are worthy of including in conservation programs. Information is available about the population diversity of *Fagus sylvatica* (Gömöry et al. 1999), *Pinus sylvestris* (Longauer et al. 1992; Zhelev et al. 1994; Doncheva et al. 2003), *Abies alba* Mill. (Bergmann and Gagov 2000), *Pinus peuce* (Zhelev and Tzarska 2009), *Pinus mugo* (Zhelev and Slavov 2002; Slavov and Zhelev 2004), *Pinus heldreichii* (Naydenov et al. 2005), *Pinus nigra* (Scaltsoyiannes et al. 2009).

The high genotype diversity of these species is conserved in their natural populations in the protected territories in the mountains of Bulgaria. These populations are protected in three national parks, several nature parks and many nature reserves (Alexandrov and Dobrev 2011). Also, forest seed production stands are managed in a way aimed toward conserving the existing genetic diversity. According to the report of Alexandrov and Dobrev (2011), for the main species represented in the mountain forests, the number of units for conservation (stands dominated by or with participation of the respective species) varies from 2 (*Taxus baccata*) to more than 1600 (*Pinus sylvestris*) and 2100 (*Picea abies*), with the mean area ranging from 4 to 26 ha. There are much more conservation units for broadleaved tree species – for example, only for *Fagus sylvatica* they are more than 3000, with an area more than 28,600 ha and mean area per unit 8.9 ha. However, few broadleaved species play important role in the formation of mountain forests. With the exception of beech, the others are not stand-forming broadleaves but rather have a scattered distribution.

Ex situ conservation efforts are related mostly with tree breeding, since the objects include many provenance and progeny tests, clone collections and seed orchards, especially concerning the conifers. Even though these activities resulted from gene conservation plans designed for a particular species and/or regions, they are being successfully incorporated into the ex situ gene conservation activities targeted to the high-mountain species. Number of ex situ conservation objects is increasing as is the number of accessions included. For *Abies alba* see Gagov et al. (*this volume*), and for the other relevant species, the information was summarized by Alexandrov and Dobrev (2011).

Genetic resources of some species, like *Pinus sylvestris*, are managed based on ecotype subdivision. Seed supply for afforestation is differentiated according to the ecotypes and results of provenance research, and there are seed orchards established especially for each particular ecotype: low-, middle- and high-mountain one (Gagov et al. 2003). Particular attention is given to the high-mountain ecotype of Scots Pine and to the zone of introgressive hybridization with *Pinus mugo*. Besides a special seed orchard for high mountain ecotype, there is also a seed orchard in Yundola designed for producing hybrid seeds for afforestations in the zone of alpine treeline (Dobrinov and Yurukov 1986). The conservation of *Pinus mugo* genetic diversity was reviewed by Alexandrov et al. (2011) but the authors put their study to a broader context and overlooked the genetic information available for Bulgarian populations.

Gene conservation activities in the mountain forest ecosystems are becoming an important part of the more complex and sophisticated, ecosystem-oriented nature conservation programmes. The conservation activities and measures should find place also in the regional plans for sustainable development in the mountainous regions, which will require coordination of the forestry and nature-conservation authorities with the local communities.

Acknowledgements Most of the studies reported in this chapter were performed within the framework of a Swiss-Bulgarian Project "Subalpine forest development in Bulgarian mountain forests under climate change" supported by Swiss National Science Fund (project IZEBZO143109). Part of the study was supported also by the project "Integrated Research on Forest Resilience and Management in the Mediterranean (INFORMED)".

References

Alexandrov, A., & Dobrev R. (2011). *The state of the world's forest genetic resources*. (Country Report Bulgaria, 60 pp). Rome: FAO.

Alexandrov, A., & Pandeva, D. (2014). Acer heldreichii. In *Enzyklopädie der Holzgewächse: Handbuch und Atlas der Dendrologie* (pp. 1–6).

Alexandrov, A., von Wühlisch, G., & Vendramin, G. G. (2011). Conserving the genetic diversity of *Pinus mugo* Turra. *Silva Balcanica, 12*(1), 5–11.

Anev, S. (2016). Physiological basis of growth, adaptation and survival. In M. Panayotov, P. Bebi, & S. Yurukov (Eds.), *Mountain coniferous forests in Bulgaria – structure and natural dynamics* (pp. 79–96). Sofia: University of Forestry ISBN: 978-954-332-146-9.

Anev, S., & Tsvetkova, N. (2018). Drought stress in four subalpine species: Gas exchange response and survivorship. *Russian Journal of Ecology, 49*(5), 422–427.

Assyov, B., Petrova, A., Dimitrov, D., & Vassilev, R., 2012. *Conspectus of the Bulgarian vascular flora. Distribution maps and floristic elements* (4th Revised and enlarged edition, 489 pp). Sofia: Bulgarian Biodiversity Foundation.

Bebi, P., Kulakowski, D., & Rixen, C. (2009). Snow avalanche disturbances in forest ecosystems—state of research and implications for management. *Forest Ecology and Management, 257*, 1883–1892.

Bergman, F., & Gagov, V. (2000). Genetische Diversität und Differenzierung von Tannenpopulationen der Balkanhalbinesel. In *Proceedings of the 9th International European Silver Fir Symposium in Skopje, Macedonia.*

Bradford, J. B., Jensen, N. R., Domke, G. M., & D'Amato, A. W. (2013). Potential increases in natural disturbance rates could offset forest management impacts on ecosystem carbon stocks. *Forest Ecology and Management, 308*, 178–187.

Brown, R. D., & Petkova, N. (2007). Snow cover variability in Bulgarian mountainous regions, 1931–2000. *International Journal of Climatology, 27*(9), 1215–1229.

Cook, E. R., Seager, R., Kushnir, Y., Briffa, K. R., Büntgen, U., Frank, D., Krusic, P. J., Tegel, W., van der Schrier, G., Andreu-Hayles, L., Baillie, M., Baittinger, C., Bleicher, N., Bonde, N., Brown, D., Carrer, M., Cooper, R., Čufar, K., Dittmar, C., Esper, J., Griggs, C., Gunnarson, B., Günther, B., Gutierrez, E., Haneca, K., Helama, S., Herzig, F., Heussner, K.-U., Hofmann, J., Janda, P., Kontic, R., Köse, N., Kyncl, T., Levanič, T., Linderholm, H., Manning, S., Melvin, T. M., Miles, D., Neuwirth, B., Nicolussi, K., Nola, P., Panayotov, M., Popa, I., Rothe, A., Seftigen, K., Seim, A., Svarva, H., Svoboda, M., Thun, T., Timonen, M., Touchan, R., Trotsiuk, V., Trouet, V., Walder, F., Ważny, T., Wilson, R., & Zang, C. (2015). Old world megadroughts and pluvials during the Common Era. *Science Advances, 1*(10), e1500561. https://doi.org/10.1126/sciadv.1500561.

Dobrinov, I., & Yurkov, S. (1986). Studies on the progeny of natural hybrids between *Pinus sylvestris* and *P. mugo*. *Scientific Works, Higher Institute of Forestry, Series Forestry, 30*, 123–138 (in Bulgarian).

Doncheva, N., Gagov, V., & Zhelev, P. (2003). Individual heterozygosity distribution in natural Scots pine (*Pinus sylvestris* L.) populations. *Genetics and Breeding, 32*(1–2), 61–67.

Dountchev, A., & Zhelev, P. (2015). Natural regeneration on the Norway spruce forests after large-scale natural disturbances in Bistrishko branishte reserve, Vitosha Mts. *Forestry Ideas, 21*(2), 293–305.

Dountchev, A., Tsvetanov, N., Zhelev, P., & Panayotov, M. (2014). Challenges for the conservation of the Norway spruce forests in Vitosha Nature Park after large-scale natural disturbances. *Ecologia Balkanica, 5*, 61–69.

Foster, D. R., Knight, D. H., & Franklin, J. F. (1998). Landscape patterns and legacies resulting from large, infrequent forest disturbances. *Ecosystems, 1*, 497–510.

Gagov, V., Zhelev, P., Evtimov, I., & Doncheva, N. (2003). Genetic structure of seed production stands and clonal seed orchards of Scots Pine. Project Report. University of Forestry, 145 pp (in Bulgarian).

Gömöry, D., Paule, L., Brus, R., Zhelev, P., Tomović, Z., & Gračan, J. (1999). Genetic differentiation and phylogeny of beech on the Balkan Peninsula. *Journal of Evolutionary Biology, 12*(4), 746–754.

Grozev, O., & Nedelchev, N., (1996). Comparative dendrochronological analysis of Austrian pine and Scots pine from the region of western Rhodopes. *Scientific works of the University of Forestry, XXXVII, Forestry*, 81–88 (in Bulgarian).

Grunewald, K., Scheithauer, J., Monget, J.-M., & Brown, D. (2009). Characterisation of contemporary local climate change in the mountains of southwest Bulgaria. *Climatic Change, 95*(3–4), 535–549.

IPCC. (2014). In R. K. Pachauri & L. A. Meyer (Eds.), *Climate change 2014: Synthesis report. Contribution of Working Groups I, II and III to the Fifth Assessment Report of the Intergovernmental Panel on Climate Change* (p. 151). Geneva: IPCC.

Kraus, D., & Krumm, F. (2013). *Integrative approaches as an opportunity for the conservation of forest biodiversity*. Freiburg: European Forest Institute.

Kulakowski, D., & Bebi, P. (2004). Range of variability of unmanaged subalpine forests. *Forum für Wissen, 2004*, 47–54.

Kulakowski, D., Seidl, R., Holeksa, J., Kuuluvainen, T., Nagel, T., Panayotov, M., Svoboda, M., Thorn, S., Vacchiano, G., Whitlock, C., Wohlgemuth, T., & Bebi, P. (2017). A walk on the wild side: Disturbance dynamics and the conservation and management of European mountain forest ecosystems. *Forest Ecology and Management, 388*, 120–131.

Landres, P. B., Morgan, P., & Swanson, F. J. (1999). Overview of the use of natural variability concepts in managing ecological systems. *Ecological Applications, 9*(4), 1179–1188.

Longauer, R., Zhelev, P., Paule, L., & Gömöry, D. (1992). The mating system, outcrossing rate and genetic differentiation of Scots pine (*Pinus sylvestris* L.) populations from Bulgaria. *Biologia (Bratislava), 47*(7), 539–547.

Mirchev, P., Georgiev, G. T., Tsankov, G. (2004). Economically important insect pests in the pine forests in Bulgaria. In *Proceedings of the Third Symposium on "Deliblato Sands"*, Pancevo, Serbia, pp. 223–228.

Mirchev, P., Georgiev, G. T., & Matova, M. (2011). Prerequisites for expansion of Pine processionary moth, *Thaumetopoea pityocampa* (Den. & Schiff.) in Bulgaria. *Journal of Balkan Ecology, 14*(2), 117–130.

Naydenov, K. D., Tremblay, F. M., Bergeron, Y., Alexandrov, A. H., & Fenton, N. (2005). Dissimilar patterns of *Pinus heldreichii* Christ. populations in Bulgaria revealed by chloroplast microsatellites and terpenes analysis. *Biochemical Systematics and Ecology, 33*(2), 133–148.

Nojarov, P. (2012). Variations in precipitation amounts, atmosphere circulation, and relative humidity in high mountainous parts of Bulgaria for the period 1947–2008. *Theoretical and Applied Climatology, 107*(1–2), 175–187.

Oliver, C. D., & Larson, B. C. (1996). *Forest stand dynamics*. New York: Wiley 544 pp.

Panayotov, M., & Georgiev, D. (2012). Dynamics in the *Ips typographus* outbreak following the 2001 windthrow in Bistrishko branishte reserve, Bulgaria. *Silva Balcanica, 13*(1), 38–48.

Panayotov, M., & Yurukov, S. (2007). Tree ring chronology from *Pinus peuce* in Pirin Mts and the possibilities to use it for climate analysis. *Phytologia Balcanica, 13*(3), 313–320.

Panayotov, M., Bebi, P., Trouet, V., & Yurukov, S. (2010). Climate signal in tree-ring chronologies of *Pinus peuce* and *Pinus heldreichii* from the Pirin Mountains in Bulgaria. *Trees – Structure and Function, 24*(3), 479–490.

Panayotov, M., Kulakowski, D., Laranjeiro Dos Santos, L., & Bebi, P. (2011). Wind disturbances shape old Norway spruce forests in Bulgaria. *Forest Ecology and Management, 262*(3), 470–481.

Panayotov, M., Zafirov, N., & Cherubini, P. (2013). Fingerprints of extreme climate events in *Pinus sylvestris* tree rings from Bulgaria. *Trees - Structure and Function, 27*(1), 211–227.

Panayotov, M., Bebi, P., Tsvetanov, N., Alexandrov, N., Laranjeiro, L., & Kulakowski, D. (2015). The disturbance regime of Norway spruce forests in Bulgaria. *Canadian Journal of Forest Research, 45*(9), 1143–1153.

Panayotov, M., Kulakowski, D., Tsvetanov, N., Krumm, F., Barbeito, I., & Bebi, P. (2016a). Climate extremes during high competition contribute to mortality in unmanaged self-thinning Norway spruce stands in Bulgaria. *Forest Ecology and Management, 369*, 74–88.

Panayotov, M., Tsvetanov, N., Gogushev, G., Tsavkov, E., Zlatanov, T., Anev, S., Ivanova, A., Nedelin, T., Zafirov, N., Aleksandrov, N., Dountchev, A., Vasileva, P., Shishkova, V., Stoyanov, B., Sotirova, N., Vatov, A., Bebi, P., & Yurukov, S., 2016b. *Mountain coniferous forests in Bulgaria – Structure and natural dynamics*. Sofia: University of Forestry, 332 pp. ISBN: 978-954-332-146-9.

Panayotov, M., Gogushev, G., Tsavkov, E., Vasileva, P., Tsvetanov, N., Kulakowski, D., & Bebi, P. (2017). Abiotic disturbances in Bulgarian mountain coniferous forests – An overview. *Forest Ecology and Management, 388*, 13–28.

Raev, I., Knight, C., & Staneva, M. (2003). *Drought in Bulgaria – A contemporary analogue for climate change – Natural, economical and social aspects of the dry period 1982–1994*. Sofia: BAS.

Raev, I., Zhelev, P., Grozeva, M., Markov, N., Velichkov, I., Zhiyanski, M., Georgiev, G., Miteva, M., Aleksander A., Trichkov, L., Bardarov, D., & Vasilev, N. (2011). *Programme of measures for adaptation of the forests in Republic of Bulgaria and mitigation the negative effects of the climate change on them*. Executive Forest Agency, Ministry of Agriculture, Food and Forests.

Rangelova, P., & Panayotov, M. (2013). Structure of old-growth *Pinus heldreichii* forests in Pirin Mountains. *Bulgarian Journal of Agricultural Science, 19*(2), 273–276.

Roussakova, V. (2015a). Green alder (*Alnus viridis*) mountain brush. In V. Bisserkov (Ed.), *Red data book of Bugaria* (Natural habitats, Vol. 3, pp. 233–235). Sofia: MoEW and IBER-BAS.

Roussakova, V. (2015b). King Boris's fir (*Abies alba* subsp. *borisii-regis*) forests. In V. Bisserkov (Ed.), *Red data book of Bugaria* (Natural habitats, Vol. 3, pp. 345–346). Sofia: BAS and MoEW.

Russkoff, M. (1928). Contribution to the study of damages by insects in our forests. *Gorski pregled (Forest review), 11–12*, 477–490 (in Bulgarian).

Scaltsoyiannes, A., Tsaktsira, M., Pasagiannis, G., Tsoulpha, P., Zhelev, P., Iliev, I., & Rohr, R. (2009). Allozyme variation of European Black (*Pinus nigra* Arnold) and Scots pine (*Pinus sylvestris* L.) populations and implications on their evolution: A comparative study. *Journal of Biological Research (Thessaloniki), 11*, 95–106.

Schelhaas, M.-J., Nabuurs, G.-J., & Schuck, A. (2003). Natural disturbances in the European forests in the19th and 20th centuries. *Global Change Biology, 9*(11), 1620–1633.

Seidl, R., Schelhaas, M.-J., & Lexer, M. J. (2011). Unraveling the drivers of intensifying forest disturbance regimes in Europe. *Global Change Biology, 17*(9), 2842–2852.

Seidl, R., Schelhaas, M.-J., Rammer, W., & Verkerk, P. J. (2014). Increasing forest disturbances in Europe and their impact on carbon storage. *Nature Climate Change, 4*, 806–810.

Shishkova, V., & Panayotov, M. (2013). Climate-growth relationship of *Pinus nigra* tree-ring width chronology from the Rhodope Mountains, Bulgaria. *Bulgarian Journal of Agricultural Science, 19*(2), 225–228.

Slavov, G. T., & Zhelev, P. (2004). Allozyme variation, differentiation, and inbreeding in populations of *Pinus mugo* in Bulgaria. *Canadian Journal of Forest Research, 34*(12), 2611–2617.
Tashev, A., & Tsavkov, E. (2017). Validation of the name *Quercus protoroburoides* (Fagaceae). *Phytotaxa, 308*(2), 232–238.
Tashev, A., Koev, K., Tashev, N., & Georgiev, S. (2013). New data on the vertical distribution of some arboreal species of the flora in Bulgaria. *Forestry Ideas, 19*(2), 201–207.
Temperli, C., Bugmann, H., & Elkin, C. (2013). Cross-scale interactions among bark beetles, climate change, and wind disturbances: A landscape modeling approach. *Ecological Monographs, 83*(3), 383–402.
Thom, D., & Seidl, R. (2015). Natural disturbance impacts on ecosystem services and biodiversity in temperate and boreal forests. *Biological Reviews, 91*(3), 760–781.
Thom, D., Seidl, R., Steyrer, G., Krehan, H., & Formayer, H. (2013). Slow and fast drivers of the natural disturbance regime in Central European forest ecosystems. *Forest Ecology and Management, 307*, 293–302.
Trouet, V., Panayotov, M., Ivanova, A., & Frank, D. (2012). A pan-European summer teleconnection mode recorded by a new temperature reconstruction from the northeastern Mediterranean (ad 1768–2008). *The Holocene, 22*(8), 887–898.
Vacchiano, G., Maggioni, M., Perseghin, G., & Motta, R. (2015). Effect of avalanche frequency on forest ecosystem services in a spruce–fir mountain forest. *Cold Regions Science and Technology, 115*, 9–21.
Veblen, T. T. (2003). Historic range of variability of mountain forest ecosystems: Concepts and applications. *Forestry Chronicle, 79*(2), 223–226.
Zafirov, N. (2016). Study of the health status of coniferous forests. In M. Panayotov, P. Bebi, & S. Yurukov (Eds.), *Mountain coniferous forests in Bulgaria – structure and natural dynamics* (pp. 137–150). Sofia: University of Forestry.
Zashev, B. (1950). On the biology and ecology of the engraver beetle *Ips acuminatus* Gyll in Bulgaria. *Annuaire of the Agricultural Academy, Faculty of Forestry, 3*, 356–361 (in Bulgarian).
Zhelev, P., & Slavov, G. T. (2002). Genetic differentiation of *Pinus mugo* in Bulgaria: Importance for gene conservation. In D. Temniskova (Ed.), *Proceedings of the Sixth National Conference of Botany*, pp. 467–472 (in Bulgarian).
Zhelev, P., & Tsarska, A. (2009). Genetic diversity in the Bulgarian populations of *Pinus peuce* Grsb. In: D. Noshad, E. W. Noh, J. King, & R.A. Sniezko (Eds.), *Breeding and genetic resources of five-needle pines. Proceedings of the Conference 2008* (pp. 10–16). Yangyang: Korea Forest Research Institute.
Zhelev, P., Longauer, R., Paule, L., & Gömöry, D. (1994). Genetic structure of indigenous Scots pine populations from Rhodopi Mountains. *Nauka za Gorata (Bulgarian Forest Science), 3*, 68–76.
Zlatanov, T., Elkin, C., Irauschek, F., & Lexer, M. (2017). Impact of climate change on vulnerability of forests and ecosystem service supply in Western Rhodopes Mountains. *Regional Environmental Change, 17*(1), 79–91.

Chapter 36
Climate Change Impacts on the Genetics of Post-Fire Regeneration and Reproductive Phenology

Filippos A. Aravanopoulos and Paraskevi G. Alizoti

Abstract Evidence regarding fast climatic change is constantly accumulating and has been linked to an ongoing increase of fire occurrence and reproductive phenology modifications, both being crucial factors in the evolution of forest tree populations. Studying the effects of forest fires in the gene pool, it was found that in basic genetic diversity parameters that reflect the magnitude of genetic variation in natural populations, no significant differences between post-fire and control populations were observed. Nevertheless, the genetic architecture of post-fire populations compared to control populations presented some differences in the frequencies of rare alleles, the occurrence of interspecific hybridization in the post-fire populations and the observation of genetic bottleneck effects, especially in isolated populations. Forest fires may not induce genetic erosion in forest tree populations per se however population isolation compounded by periods of high forest fire frequency and intensity, may lead to adverse consequences for the architecture of the genetic diversity of forest trees. Such consequences may be compounded by reduced natural regeneration due to the decreased percentage of sound seed, a result of the anticipated reduction of tree fertility and flowering synchronization, induced by climatic change.

Keywords Genetic diversity · Molecular markers · Forest fires · Seed set

36.1 Introduction

The evidence regarding the vast and fast climatic change is constantly accumulating. The year 2016 surpassed year 2015 as the year with the highest mean temperatures in 137 years of organized measurements (NOAA 2018). It was the third year in a row to set a new heat record and the fifth time the record has been broken within the twenty-first century. Global temperature is expected to rise about 1.8–4.0 °C during

F. A. Aravanopoulos (✉) · P. G. Alizoti
School of Forestry and Natural Environment, Aristotle University of Thessaloniki, Thessaloniki, Greece
e-mail: aravanop@for.auth.gr; alizotp@for.auth.gr

M. Šijačić-Nikolić et al. (eds.), *Forests of Southeast Europe Under a Changing Climate*, Advances in Global Change Research 65,
https://doi.org/10.1007/978-3-319-95267-3_36

the twenty-first century (IPCC 2013). Such phenomena have been linked to an ongoing increase of fire occurrence, intensity and severity, in the Mediterranean and in other areas with Mediterranean conditions (California, Chile, Australia etc.). Low elevation conifers have been especially affected in recent decades (Bariteau et al. 2003). Particularly important is the apparent change in the fire patterns in Mediterranean ecosystems. In recent years there has been a "divergence" of the fire's behavior from the established standards, that lies in: (a) increase of fire incidence and burned areas, (b) reduction of the interval between successive fires in affected areas, (c) increase of fire intensity, (d) occurrence of mega-fires, (e) increase of large fires in high forests mainly of black pine and beech, (f) severe reduction or even total destruction of the seed bank, and (g) depletion of regeneration capacity. Large fires affect the genetic resources due to the exhaustion, or destruction, of the seed bank (Aravanopoulos 2009). They degrade the soil, disrupt the biological cycles of the ecosystem and are able to lead to new pollution areas when pollutants bound to biomass are remobilized due to fire (Paliouris et al. 1995). Furthermore, post-fire restoration and ecosystem resilience may suffer from reduced natural regeneration due to the decreased percentage of sound seed, which in turn is the outcome of a reduction in tree fertility and flowering synchronization, also induced by climatic change.

36.2 Effects of Fires on the Genetic Diversity and Evolution of Forest Species: A Digression to Theory

Fires are one of the most important factors affecting the genetic evolution of species in fire-prone areas (Aravanopoulos et al. 2004), i.e. the process by which the genetic variation in populations is altered over time as a reaction to environmental change. A large part of forest species genetic variation today has evolved through their continuing interaction with the biotic and abiotic factors to which they have been exposed at least since the last glacial period and thereafter. The preservation of the genetic diversity and structure of forest species populations ensures their potential to adapt, by preserving unique clusters of genes that are "*sine qua non*" for the survival and reproduction of species, that is, ultimately for their presence in perpetuity.

However, there are indications today that the respective environmental stability that prevailed for more than 12,000 years has begun to be significantly disrupted. In fire-prone areas, the deterioration due to drought and the emerging climate change, as well as a series of specific anthropogenic factors, have exacerbated the phenomena of forest biological decline and require the direct protection of genetic resources. This is the case for instance, in the South and Eastern Mediterranean (Bariteau et al. 2003). Genes are the fundamental unit of biodiversity, the basic material of evolution and the primary source of planet's living organisms' diversity (Dobzhansky 1970). The genetic reserve of forest species, at least until the beginning of the twentieth century, was virtually intact. Evidently, if the genetic resources of natural populations are destroyed or significantly eroded, then the species adaptive potential and the capacity for genetic improvement are also jeopardized.

36.3 Regeneration After Fire: Effects of Fires on Genetic Diversity and Evolution of Forest Species

Despite an overwhelming understanding among most scientists that the genetic diversity is the foundation of sustainability and although it is generally reported that the effects of forest fires on the genetic stock of forest populations have negative results on genetic diversity (Ledig 1992), relevant case studies are not numerous and focus mostly on conifers. In a literature review for this chapter, about two-thirds of the relevant papers concerned the impact of forest fires on natural conifer populations (such as *Picea mariana, Picea glauca, Pinus halepensis, Pinus brutia, Sequoia semperiverns*). Fewer papers regarding woody perennial angiosperms (such as *Populus tremuloides, Betula maximowicziana*) have been identified. In these works, there is usually a comparison between the genetic parameters of post-fire populations and "control" populations, the latter being spatially proximal (within the fire area) mature populations that survived the fire (Aravanopoulos et al. 2004), population close but outside the fire area under normal forest management (e.g. Perry and Bousquet 2001; Rajora and Pluhar 2003), or populations with absence or low incidence of fires (e.g. Rogers 2000; Rajora et al. 2005; Baka and Aravanopoulos 2006; Uchiyama et al. 2006).

The study of the genetic constitution of post-fire populations and control populations was performed using biochemical or molecular markers (EST, isoenzyme, RAPD, SSR, STS). In cases where basic structural parameters reflecting the magnitude of genetic diversity (such as mean number of alleles per locus, polymorphic loci, observed heterozygosity, expected heterozygosity, genetic distance) were examined, there were generally no significant differences between post-fire populations and control populations (Rogers 2000; Perry and Bousquet 2001; Rajora and Rahman 2002; Rajora and Pluhar 2003; Rajora et al. 2005; Baka and Aravanopoulos 2006). However, when parameters reflecting the architecture of genetic diversity where examined (such as number of private alleles, effective population size, gene flow, linkage disequilibrium, Wahlund effect), a general reduction of such estimates was seen in the post-fire populations (Krauss 1997; England et al. 2002; Aravanopoulos et al. 2004; Uchiyama et al. 2006). Apart from the alterations in the genetic diversity architecture, significant changes were also observed in mating systems. For instance, the commonly sexually propagated *Populus tremuloides* appeared almost exclusively asexually propagated in post-fire populations during colonization of burnt areas (Romme et al. 1997). Also, interspecific hybridization may be favored in cases where hybrids are better adapted or present hybrid vigor to the post-pyric environment (Aravanopoulos and Panetsos 1998; Aravanopoulos et al. 2004). Subtle changes in the genetic composition of post-fire forest tree populations may also reflect true genic selection, besides the effects of genetic drift and/or incoming gene flow. There is evidence that oxidative stress and heat stress result in a re-organization of the natural regulation of the antioxidative defense system (Ray et al. 2012). Forest trees may adapt by accumulating heat shock proteins (HSP; Queitsch al. 2000) and generating reactive oxygen species (Ray et al. 2012). Differential plant response in the above processes, may lead to the manifestation of natural selection.

36.4 Regeneration After Fire: Two Case Studies with Opposing Results

Aravanopoulos et al. (2004) studied the genetics of the regeneration after fire in *Pinus brutia* using two areas (Kourteri, Lesvos island and Mytilenei, Samos island), with a fire event 25 years earlier. At the time of sampling there were: (a) a population of mature trees that survived the fire, (b) natural regeneration in cone maturity production. Genetic diversity was studied in two cohorts, mature "maternal" stand and post-fire regeneration in 11 loci using 25 individuals per population (Aravanopoulos et al. 2004).

In Lesvos, allele frequency did not differ significantly between the mature stand and the regeneration. However, in the regeneration one of the alleles that was present in the mature population was not found (*Pgm-1C*), while another allele presented a highly reduced frequency and it was characterized as "rare" in the regeneration (*Gdh-1A*). Moreover, new rare alleles (*Idh-1A, Lap-1B*) were found in the regeneration. The former in particular is species-specific for *Pinus halepensis*, hence it appears that natural hybridization between the naturally occurring *P. brutia* and the introduced *P. halepensis* is favored after fire. Genetic diversity in the two populations was significant and within the range observed in other populations (Gomez et al. 1999; Panetsos et al. 1998). There were no statistically significant differences between the genetic variation values of the young and the mature population. However, in the mature stand, an excess of heterozygotes was observed compared to the regeneration, indicative of a genetic bottleneck (Aravanopoulos et al. 2004).

In Samos, gene frequencies between the regeneration and the mature population did not show any significant difference. However, new rare alleles have also been observed in the young population (*Idh-1C, Gdh-1C*). It also appears in this case, that natural hybridization between the naturally occurring *P. brutia* and the introduced *P. halepensis* is favored after fire. Differences in the parameters of genetic diversity between the young and the mature cluster were not significant (Aravanopoulos et al. 2004). An excess of heterozygotes was observed in the mature stand compared to the regeneration, indicative of a genetic bottleneck (Aravanopoulos et al. 2004). These results agree with Rajora and Rahman (2002) and Rajora and Pluhar (2003) who carried out similar studies in *Picea mariana*. The allelic differences found in about 1/3 of the loci can be expected, especially when the seedbank is limited, or the population size of the young cluster after the fire is small, both increasing the chances of genetic drift (Aravanopoulos et al. 2004). Such results indicate earlier bottlenecks and a potential genetic drift may lead to changes in the architecture of genetic diversity and structure in post-fire populations especially in a restrictive island environment, in concordance to the results of Krauss (1997), England et al. (2002), and Uchiyama et al. (2006).

Baka and Aravanopoulos (2006) examined the effects of differential forest fire history in the genetic diversity of regeneration after fire in *Pinus halepensis* using a comparison between populations with high and low incidence of fires in terms of fire frequency and severity. They used a population series in a transect covering north and central Greece and spanning from a general NE (Vourvourou) to SW (Varibobi) direction, that presents significant differences in the frequency and severity of fires ranging from extremely high (Varibobi) to relatively low (Vourvourou) (Dimitrakopoulos 2004). The populations studied are large with abundant natural regeneration. Molecular population genetic diversity was investigated following Baka and Aravanopoulos (2000). A forest damage index (FDI) indicator was also defined, representing fire incidence and fire hazard based on the estimate of the average annual number of fires in each sampling area (Dimitrakopoulos 2004). *Pinus halepensis* populations presented high genetic variation, absence of inbreeding and Hardy-Weinberg equilibrium in areas with different fire incidence. Most of the molecular genetic variation (>90%) was found within the populations (Baka and Aravanopoulos 2000, 2002, 2006). Gene frequencies in areas with high and relatively low incidence of fires did not show statistically significant differences. High forest fire occurrence and damage (reflected FDI) was not correlated either positively or negatively with genetic diversity parameters (e.g. gene diversity–H_e, Fig. 36.1). Geographic and population genetic distances were not significantly correlated (Fig. 36.2); nevertheless, a genetic distance dendrogram (not shown) depicted congruence to general geographic loca-

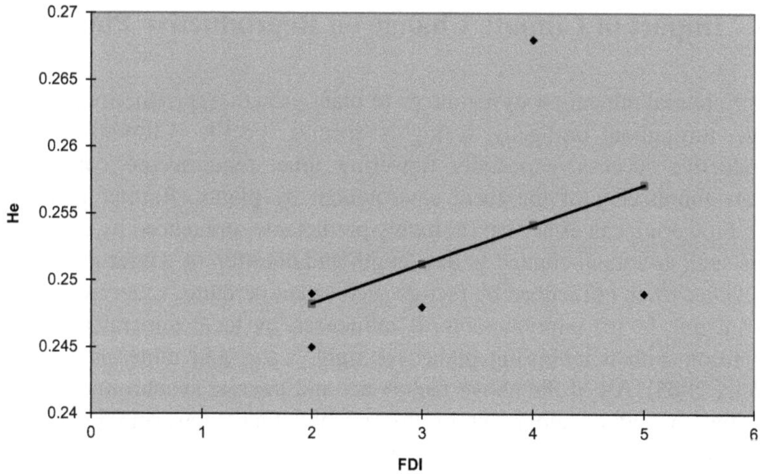

Fig. 36.1 Gene diversity

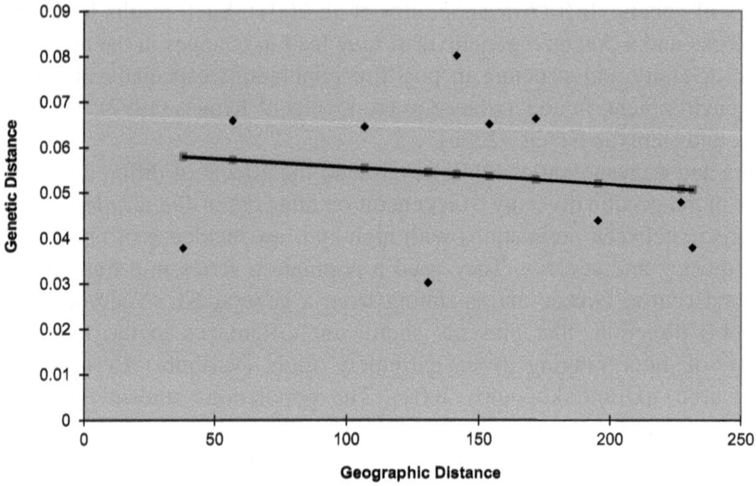

Fig. 36.2 Geographic and population genetic distances

tions. It is presumed that even if a population is repeatedly exposed to fires, it presents strong genetic mechanisms to cope with and to maintain high levels of genetic diversity, which do not differ significantly from a population that has not suffered a corresponding heavy burden. The results show that in areas where population sizes are large, natural distribution is continuous, populations are in a reproductive maturity and regeneration is abundant, the effects of frequent fires on the regeneration and the ensuing population do not lead to a reduction of genetic diversity and erosion of the species genetic resources.

36.5 Impact of Climate Change on Reproductive Phenology

The differential allocation of resources to plant growth, reproduction, defense and storage, throughout ontogeny, is highly species specific (Climent et al. 2008). Reproductive success, especially flowering time, requires the continuous and accurate monitoring of the local environment by plants (Bernier and Périlleux 2005). Flowering can either be: (a) highly predictable, influenced by environmental factors such as annual change in day length and duration of winter period, (b) less predictable, when influenced by factors such as temperature, water availability and light integral, or (c) unpredictable, if influenced by local mineral availability or competition with neighboring plants for light, water and minerals (Bernier and Périlleux 2005). All of the above factors act and interact synchronously, affecting the initiation and development of flowers and thus, flowering phenology and synchronization (Alizoti et al. 2010). Studies have demonstrated that reproduction is a costly process and that female flowering requires much more energy than the male one, due to the additional cost of seed and fruit production (Meagher 1988).

The costs, in terms of carbon, can be at least partially compensated by the increase of photosynthetic rates or by the photosynthesis of the reproductive structure itself. The level of costs required for reproduction would be most apparent in habitats with low resource availability or other stress conditions, due to reduced photosynthetic capacity (Reznick et al. 2000). In the Mediterranean, plant stress due to dehydration is closely related to high light and temperature conditions, factors that can enhance the negative effects of drought on the photosynthetic capacity of the plants due to stomata closure (Pereira et al. 2007). These conditions when interacting with the, typical for Mediterranean ecosystems, low nutrient availability, result in inhibition of uptake and transport of ions in the transpiratory stream (Gollan et al. 1992) and a decrease of nitrogen use efficiency (Pereira et al. 2007).

According to climate change projections, the climatic conditions are expected to worsen in southern Europe (high temperatures and drought), a region already vulnerable to climate change according to IPCC (2013)). Climate change is thus expected to become a critical factor for biodiversity loss in southern Europe, while according to EEA (2007) loss of almost the 25% of the plant species in Europe is projected by 2100. Moreover, climate change is expected to have a major impact on flowering of forest trees and its synchronization in the Mediterranean region. Alizoti et al. (2010) studied the flowering synchronization among 60 black pine clones, growing in a clonal seed orchard, during two consecutive years with contrasting inter-annual climatic conditions (1 year, with climatic conditions close to the long-term ones, and one xerothermic year, with conditions close to those expected due to climate change), and reported almost complete lack of male and female flowering synchronization among the seed orchard clones during the xerothermic year, as the male flowering and pollen shedding were completed for almost all clones prior to the receptive phase of the female conelets. Similar results have been reported by Sherry et al. (2011) who studied the reproductive phenology of six species under experimental climate warming and reported lagging of phenological responses. Thus, climate change may be proved detrimental for tree fertility and flowering synchronization in the Mediterranean region, with apparent consequences on the quantity and quality of the produced seed, as well as in the natural regeneration of forest trees in forest ecosystems due to the reduced percentage of sound seed.

36.6 Conclusions

Climate change may be linked to increased forest fire frequency, intensity and severity. Moreover, seed quantity and quality, as well as natural regeneration may also be adversely affected due to the reduced percentage of sound seed, a result of the anticipated reduction of tree fertility and flowering synchronization due to climate change.

However, the increase of forest fires does not cause genetic erosion phenomena in forest populations per se, as it has been shown by the genetic analyses of populations with different incidents of fires and populations where genetic diversity was investigated before and after the fire. Absence of inbreeding and absence of deviations from Hardy-Weinberg equilibrium were confirmed. In most cases, however, some qualitative differences especially in terms of rare alleles, interspecific hybridization and genetic bottlenecks were detected in post-fire populations, especially when populations were relatively isolated, particularly in island environments. Hence, while no immediate and major problems are expected in post-fire populations, their long-term future may be considered as still being under discussion.

References

Alizoti, P. G., Kilimis, K., & Gallios, P. (2010). Temporal and spatial variation of flowering among *Pinus nigra* Arn. clones under changing climatic conditions. *Forest Ecology and Management, 259*, 786–797.

Aravanopoulos, F. A. (2009). Wild fires as a factor for the erosion of the forest gene pool: Towards a genetic holocaust? In *Proceeding of 14th Pan-Hellenic forest science conference, Patras* (pp. 851–863).

Aravanopoulos, F. A., & Panetsos, K. P. (1998). Genetics and evolution of natural *Pinus brutia* Ten. populations in Lesvos island. *Geotechnical Science Issues, 9*, 10–19.

Aravanopoulos, F. A., Panetsos, K. P., & Scaltsoyiannes, A. (2004). Genetic structure of *Pinus brutia* stands exposed to wild fires. *Plant Ecology, 171*, 175–183.

Baka, M., & Aravanopoulos, F. A. (2000). Molecular genetics of *Pinus halepensis* populations. In *Proceeding of 8th Pan-Hellenic conference, Hellenic scientific society for plant genetics and breeding, Arta* (pp. 419–426).

Baka, M., & Aravanopoulos, F. A. (2002). Analysis of molecular genetic parameters in natural *Pinus halepensis* (Mill.) populations. In *Proceeding of 9th Pan-Hellenic conference, Hellenic scientific society for plant genetics and breeding, Thermi* (pp. 427–434).

Baka, M., & Aravanopoulos, F. A. (2006). Molecular genetic analysis of Aleppo pine populations originating from areas of differential forest fire history: Implications for reforestations and breeding. In *Proceeding of international conference of population genetics and genomics of forest trees, Madrid* (p. 112).

Bariteau, M., Alptekin, U., Aravanopoulos, F. A., et al. (2003). Les resources génétiques forestières dans le basin Méditerranéen. *Fôret Méditerranéenne, 24*, 148–158.

Bernier, G., & Périlleux, C. (2005). A physiological overview of the genetics of flowering time control. *Plant Biotechnology Journal, 3*, 3–16.

Climent, J., Prada, M. A., Calama, R., Chambel, M. R., Sánchez de Ron, D., & Alia, R. (2008). To grow or to seed: Ecotypic variation in reproductive allocation and cone production by young female Aleppo Pine (*Pinus halepensis*, Pinaceae). *American Journal of Botany, 95*(7), 833–842.

Dimitrakopoulos, A. (2004). Classification of Prefectures in Greece based on forest fire frequency and total burnt area during the period 1983–1997. *Science Annual Facility Forestry & Natural Environment*. Aristotle University of Thessaloniki, *40*, 215–226.

Dobzhansky, T. (1970). *Genetics of the evolutionary process*. New York: Columbia University Press.

EEA. (2007). *Europe's environment. 4th Assessment*. Copenhagen: EEA.

England, P. R., Usher, A. V., Whelan, R. J., & Ayre, D. J. (2002). Microsatellite diversity and genetic structure of fragmented populations of the rare, fire-dependent shrub *Grevillea macleayana*. *Molecular Ecology, 11*, 967–977.

Gollan, T., Schurr, U., & Schulze, E. D. (1992). Stomatal response to drying soil in relation to changes in the xylem sap composition of *Helianthus annuus*. I. The concentration of cations, anions, amino-acids, in and pH of, the xylem sap. *Plant, Cell & Environment, 15*, 551–559.

Gomez, A., Aravanopoulos, F. A., Alia, R., & Bueno, M. A. (1999). *Pinus halepensis* RAPD markers: Linkage and genetic diversity. In *Proceeding of international congress 'Applications of Biotechnology in Forest Trees', Vitoria, Spain, September 22–25, 1999*. 4 p.

IPCC – Intergovernmental Panel on Climate Change. (2013). In T. F. Stocker, D. Qin, G. K. Plattner, et al. (Eds.), *Climate change: The physical science basis, Fifth Assessment Report*. Geneva: IPCC.

Krauss, S. L. (1997). Low genetic diversity in *Persoonia mollis* (Proteaceae), a fire-sensitive shrub occurring in a fire-prone habitat. *Heredity, 78*, 41–49.

Ledig, F. T. (1992). Human impacts on genetic diversity in forest ecosystems. *Oikos, 63*, 87–108.

Meagher, T. R. (1988). Sex determination in plants. In D. J. Lovett & D. L. Lovett (Eds.), *Plant reproductive ecology: Pattern and strategies* (pp. 125–138). New York: Oxford University Press.

NOAA – National Oceanic and Atmospheric Administration. (2018). *NOAA National Centers for Environmental Information, State of the Climate: Global climate report for annual 2017*. http://www.noaa.gov/climate

Paliouris, G., Taylor, H. W., Wein, R. W., et al. (1995). Fire as an agent in redistributing fallout ^{137}Cs in the Canadian boreal forest. *Science Total Enviroment, 160/161*, 153–166.

Panetsos, K. P., Aravanopoulos, F. A., & Scaltsoyiannes, A. (1998). Genetic variation of *Pinus brutia* in islands of the north eastern Aegean Sea. *Silvae Genetica, 47*, 115–120.

Pereira, J. S., Chaves, M. M., Caldeira, M. C., & Correia, A. (2007). Water availability and productivity. In J. Morison & M. Morecroft (Eds.), *Plant growth and climate change* (pp. 119–145). London: Blackwell Publication.

Perry, D. J., & Bousquet, J. (2001). Genetic diversity and mating system of post-fire and post-harvest black spruce: An investigation using codominant sequence-tagged-site (STS) markers. *Canadian Journal of Forest Research, 31*, 32–40.

Queitsch, C., Hong, S. W., Vierling, E., & Lindquist, S. (2000). Heat shock protein 101 plays a crucial role in thermotolerance in *Arabidopsis*. *Plant Cell, 12*, 479–492.

Rajora, O. P., & Pluhar, S. A. (2003). Genetic diversity impacts of forest fires, forest harvesting, and alternative reforestation practices in black spruce (*Picea mariana*). *Theoretical and Applied Genetics, 106*, 1203–1212.

Rajora, O. P., & Rahman, M. H. (2002). EST and microsatellite DNA analysis of genetic effects of forest management practices and forest fires in black spruce (*Picea mariana*). In *X proc. plant and animal genome conference San Diego, CA, January 12–16* (p. 209).

Rajora, O. P., Mann, I. K., & Shi, Y. Z. (2005). Genetic diversity and population structure of boreal white spruce (*Picea glauca*) in pristine conifer-dominated and mixed wood forest stands. *Canadian Journal of Botany, 83*, 1096–1105.

Ray, P. D., Huang, B. W., & Tsuji, Y. (2012). Reactive oxygen species (ROS) homeostasis and redox regulation in cellular signalling. *Cellular Signaling, 24*, 981–990.

Reznick, D., Nunney, L., & Tessier, A. (2000). Big houses, big cars, superfleas and the cost of reproduction. *Trends in Ecology & Evolution, 15*, 421–425.

Rogers, D. L. (2000). Genotypic diversity and clone size in old-growth populations of coast redwood (*Sequoia sempervirens*). *Canadian Journal of Botany, 78*, 1408–1419.

Romme, W. H., Turner, M. G., Gardner, R. H., et al. (1997). A rare episode of sexual reproduction in aspen (*Populus tremuloides* Michx.) following the 1988 Yellowstone fires. *Natural Areas Journal, 17*, 17–25.

Sherry, R. A., Zhou, X., Gu, S., Arnone, J. A., III, Johnson, D. W., Schimel, D. S., Verburg, P. S. J., Wallace, L. L., & Luo, Y. (2011). Changes in duration of reproductive phases and lagged phenological response to experimental warming. *Plant Ecology and Diversity, 4*(1), 23–35.

Uchiyama, K., Goto, S., Tsuda, Y., et al. (2006). Genetic diversity and genetic structure of adult and buried seed populations of *Betula maximowicziana* in mixed and post-fire stands. *Forest Ecology and Management, 237*, 119–126.

Chapter 37
Watering Regime Influence on Sustainability of Floristic Structure of Special Nature Reserve "Obedska Bara

Vesna Nikolić and Dušan Jokanović

Abstract Obedska bara is the greatest flooded area of the river Sava in Serbia and one of the oldest protected natural ares in the world (protected since 1874). It is remain of the left riverbed of the river Sava. The main feature of this area is mixture of different types of wet sites (puddles, swamps, wet meadows, hygrophilous forests, etc.) that are occupied by different plant and animal species of a great national and international importance. Research in this nature reserve is related, before all, to influences between surface waters of the river Sava and Obedska bara on groundwater regime at wider area of this locality. Regime of fillfulling and emphasis of the water defines biocenosis development of this ecosystem. The research is related to period between 2002 and 2011 (10 years long) and obtained results define the state of this aquatic ecosystem.

Keywords Watering regime · Obedska bara · Aquatic ecosystem · Groundwater · Floristic structure

37.1 Introduction

The whole area of Ravni Srem is characterized, before all, with hygrophilous forest ecosystems presence that are under direct water influence. Alluvial or hygrophilous forests are highly dynamic ecosystems developed on flooded lowlands areas close to rivers (Hughes 2007). Some authors (Sperduto and Nichols 2004) define hygrophilous forests as variable plant associations present between aquatic and terrestrial environmental conditions in valleys of lowland rivers and river channels. These forest types can also be related to paraclimate associations whose floristic structure mainly depends on flooding duration and groundwater level (Franjić et al. 2006).

Knowing of watering regime in concrete site conditions, and relation between plant associations and extreme wet or dry biotope, is very important, not only for

V. Nikolić (✉) · D. Jokanović
Faculty of Forestry, University of Belgrade, Belgrade, Serbia
e-mail: vesna.nikolic@sfb.bg.ac.rs; dusan.jokanovic@sfb.bg.ac.rs

M. Šijačić-Nikolić et al. (eds.), *Forests of Southeast Europe Under a Changing Climate*, Advances in Global Change Research 65,
https://doi.org/10.1007/978-3-319-95267-3_37

normal growth and development of these associations, but also for suitable application of management measures. Among numerous ecological factors significant for forest ecosystems survival, one of the most important role plays water. Watering regime is evaluation of available water supply in the soil during longer period of time. It is expressed by the whole water quantity that enters the soil in many different ways, but it is also related to quantity of water loss during different processes (Burlica 1987).

Relation between water and forest is unconditional – forests namely influence to decreasing of precipitation quantity that turns into ground by interception and at the same time reduce soil humidity by transpiration. Forests have a great impact to water quality (Nikić et al. 2013) due to regulating pollutant content in the water, erosion and sedimentation processes, and finally they impact water temperature, as well (Vilhar and Fajon 2007).

Research of Pedunculate oak site characteristics and its watering regime in Ravni Srem introduces one of first steps in order to solve serious problem related to Pedunculate oak forests decline (Medarević et al. 2009; Letić et al. 2014; Nikolić 2017).

Previous ecological research found that water is one of the most significant ecological factors whose changes, such as surface and groundwater regime fluctuations, cause stability disorders and, as a result, decline of Pedunculate oak occurs. A big contribution related to above mentioned facts gave research of Mayer (1987, 1989, 1994, 1995, 1996) conducted in forest areas of north-western part of Croatia.

As for Ravni Srem, it is divided into Gornji Srem, that includes forest companies "Morovićand "Višnjićevo" and Donji Srem, consisted of "Klenak" and "Kupinovo". In the first half of twentieth century, defending embankment was built. In that way, influence of flooding to hygrophilous forests in Gornji Srem was disabled. Donji Srem is characterized with dominant influence of flooded water during high water level, and there is also established significant influence of groundwater due to capilar moving (Gajić and Karadžić 1991). Apart from Obedska bara, there is also another natural protected area in Serbia researched by some authors (Šijačić et al. 2011; Stanković et al. 2015b).

Bearing in mind that Pedunculate oak forests in Ravni Srem are of the greatest economic and ecological importance for Serbian forestry, it is necessary, before all, to prevent its decline, because this process affects serious loss of increment and causes many other damages. Numerous papers show that human activity is one of the main factors responsible for pedunculate oak decline.

In Russia, this phenomenon was mostly researched by Rahmanov (1984), who discovered existing of dead horizon dried by vegetation roots. As for Germany, papers related to Pedunculate oak decline originate from beginning of twentieth century in Westfalia and found gradual decline of oak trunks (Krahl-Urban et al. 1944). Different authors described process of oak decline in Poland (Siwecki 1989, 1993, 1995; Siwecki and Llese 1991; Bednarz 1993). Marcli (1966) connected intensive decline of Pedunculate oak forests in Romania to long duration of period without precipitation.

As for Pedunculate oak decline in Croatia and Serbia, it was firstly attributed to fungi and pests invasions, but later, it was concluded that interaction of many factors caused that (Vajda 1948; Harapin and Androić 1996; Prpić 1996; Donaubauer 1998; Steiner 1998; Fuhrer 1998; Siwecki and Ufnalski 1998; Thomas et al. 2002; Dubravac and Dekanić 2009; Dubravac et al. 2011).

Papers related to Pedunculate oak decline in alluvium of the Sava river treated this issue from many different angles: groundwater level and decline intensity (Dekanić 1974); oak pedunculate response to different watering conditions (Prpić 1976); trunk increment as indicator of habitat changes (Pranjić and Lukić 1989); ecological and biological factors (Prpić et al. 1994); problems related to silvicultural measures (Bobinac 2008; Bobinac et al. 1997).

It should be emphasized that site conditions and soil water content impact a lot to macroscopic (Jokanović et al. 2017, 2018) and microscopic (Jokanović and Nikolić 2017) wood structure of hygrophilous conifers. There are also other ecological factors related to structure of different plant parts (Jokanović et al. 2015; Veselinović et al. 2017; Mitrović et al. 2017).

37.2 General Information About Study Area

Obedska bara is situated on the long inundation area on the left bank of down flow of the river Sava. Area of Management Unit (MU) "Kupinske grede" is defined by form and position of special natural reserve "Obedska bara", Fig. 37.1. "Obedska bara" otherwise has existed more than 2.500 years and from morphological point of view, it is left riverbed of the river Sava. Because of its position within river Sava inundation, watering regime is directly related to the river whose high and low waters during centuries impacts to the form of this ecosystem, fulfilling it with water and caused also water emphasis. Complicated natural processes during thousands of years transformed water ecosystem of the lake into puddle-swamping ecosystem with tendency of its converting into wet meadows and into the land, finally. This

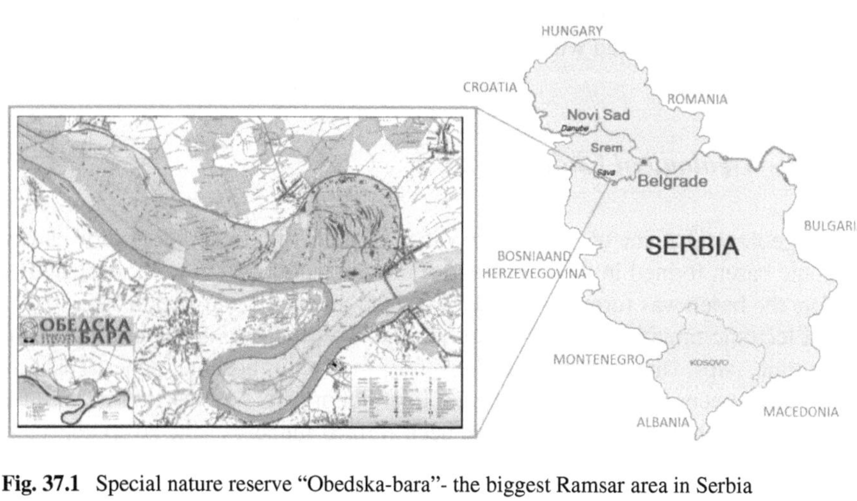

Fig. 37.1 Special nature reserve "Obedska-bara"- the biggest Ramsar area in Serbia

puddle-swamping ecosystem surrounds part of the river Sava inundation, named Kupinske grede, whose main features are an expressed micro relief with lots of small depressions and beams, then soils like fluvisoil, humofluvisoil and mold. These sites are occupied by hygrophilous vegetation *Fraxino angustifoliae-Quercetum roboris* (Gajić and Karadžić 1991). These sites are under permanent influence of middle and high waters of the river Sava, but they are also influenced by adding watering from Obedska bara whose waters, during low water levels of the river, drain through the alluvium and make high groundwater levels (Savić et al. 2013).

For each designated line within forestry complex are planned, depending on site and stand conditions, some goals and measures of management in the future that should provide converting of current forest condition into optimal related to participation and spatial distribution of forested and unforested areas, woody species and inside structure of the stand, duration of productive process, etc.

Management unit "Kupinske grede" is situated within special nature reserve "Obedska bara" in the second and third protecting zone. Based on these facts, forest management goals in this management unit are balanced with the Decision related to protection of the special nature reserve "Obedska bara".

Area of all fields that belong to this management unit is about 2.500 ha. Management unit is divided into 62 forest units whose areas range between 5.04 ha (forest unit 61) up to 596.18 ha (forest unit 62). An average area of forest unit is 39.97 ha. In this management unit, forests and forest soils include 2/3 from the whole area that is relatively high percent of forest cover of the management unit, bearing in mind its position, because a lot of swamps and puddles of "Obedska bara" belong to this management unit. Greater part of forests that belong to management unit are natural forests that occupy about 61.1% from the whole area. Lower part is occupied by forest cultures (4.2%). As for forest soil that potentially could be reforested includes relatively low percent in the management unit (30.71 ha or 1.2%). The other soil includes about 1/3 of the management unit area and it is consisted of ways, channels, different objects, etc. (Group of authors 2008).

As for other soil, it includes 17.93 ha or 0.7% from the whole area of this management unit. Other soil is not treated as forested area in spite of the fact that a great part of this soil is covered with the forest.

37.2.1 Geological Characteristics

From geological point of view, Vojvodina is southern and south-eastern part of Panonic basen formed in the beginning of miocen, about 25–30 million years ago. Before the basen was formed, this area was land, but at the end of paleogen, strong radial tectonic moving occured. Panonic mass was divided into blocks and at the same time, Alps, Dinaric Alps and Carpathian mountains became higher. This process caused occurence of Panonic basen. In northern parts of Dinaric Alps, close to border with Panonic mass, there have been formed tunnels of the rivers Sava and Drava (Josipović and Soro 2012).

During process of Panonic basen forming and a sea inside it, a long period of sedimentation started. Gradually, Panonic basen became completely separated and turns into a huge lake. Big Panonic lake gradually flowed and became divided into a few smaller lakes that leave in its deepest parts. These small lakes became much smaller and turn into big swamps. In that time, filling of the basen was the most intensive, because there was present a great quantity of sediments (Josipović and Soro 2012).

Next geological period, quartar, occured about two million years ago. In its first part, there were permanent changes of cold and warm periods. Climate changes caused vegetation changes and also participated in quartar remains forming (Ćurčić et al. 2002).

During geological history of Vojvodina, we can differ four phases: (1) phase related to lake; (2) phase related to puddle; (3) phase related to land; (4) phase related to ingression of water masses from northern parts and devastation of relief formed during the third phase (Bukurov 1952).

During later period of time, meandering of the rivers formed alluvial lowland areas (Bogdanović 1982), consisted of sand, gravel, clay and ash fractions.

37.2.2 Climate Characteristics

For climate characteristics of the forestry area Ravni Srem, data of meteorological measurements on climatology station Sremska Mitrovica, situated on 82 m height (φ 45°06N λ 19°33E) in period 2002–2011 has been used. During observed period, an average air temperature was 11.7 °C. Max annual temperature was 23,5 °C in 2003 in June and August, while min was recorded in February 2003 and it was −3,8 °C, (Diagram 37.1). Medium temperature during summer (June–August) was 21.5 °C, during spring (March–May) 12,2 °C, during autumn (September–November) 11,7 °C and during winter (December–February) 1,3 °C. An average air temperature during vegetation period (April–September) is 18,5 °C. July is the warmest month with an average temperature of 22.2 °C, while the coolest month is January with an average temperature of −0,03 °C. As for climate of this area, it is characterized with sharp cooling during winter that occurs under the influence of cold air mass from northern and eastern part of Europe. High air temperatures during January are a bit rare and it occurs under the influence of warm winds from west and south. The first cold days come in October and last to March, while the coolest are during January and February. Summer temperature is high and uniform.

Annual moving of relative air humidity has reverse relation with annual moving of air temperature. Relative air humidity is the greatest in winter period, especially in December and January (93% and 94%, respectively), and the lowest in April (56%). Annual fluctuations of relative air humidity range from 74% in 2008 and 2011 to 81% in 2004, 2005 and 2010. An average relative air humidity is 78%, while its medium value during vegetation period is 72%.

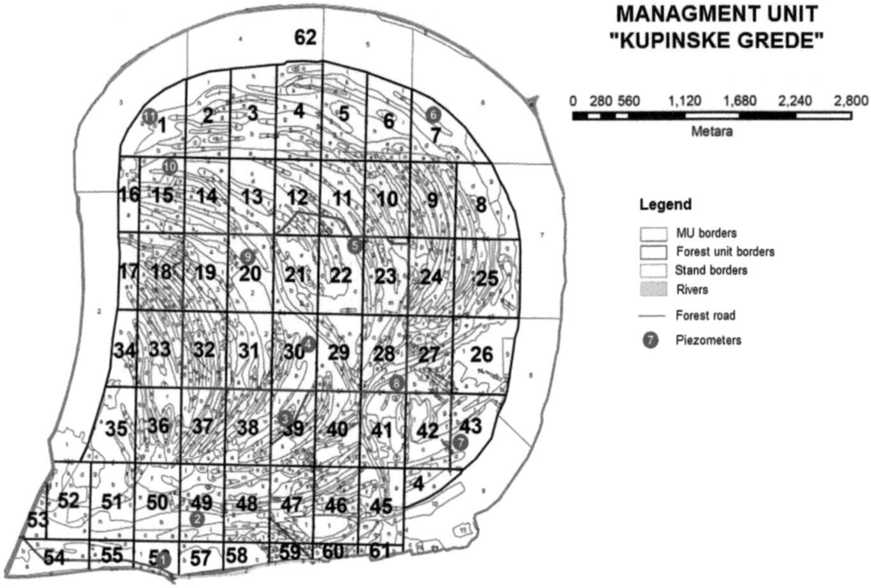

Diagram 37.1 Displaying of climate factors during 2002–2011

Based on observation during 10 years (2002–2011), there is about 627.4 mm of precipitation annualy. During the year, maximum occurs in June (78.7 mm), and minimum in January 38.2 mm (Diagram 37.1). As for precipitation distribution related to seasons, we have following situation: summer (32%), autumn (26%), spring (23%) and winter (19%). The average precipitation level during vegetation period was 351.7 mm.

37.2.3 Hydrographic and Hydrological Characteristics

Obedska bara is situated in south-eastern Srem (northern part of Serbia) on the left bank of down flow of the river Sava (Fig. 37.1). Its area is located between 44° 38′02″ and 44° 46′05″ of the northern latitude and between 19°47′16″ and 20° 03′30" of the eastern longitude. Obedska bara is arch-shaped like horseshoe between Obrež (on the west) and Kupinovo (on the east). Northern outside border includes soil less terrace 6–8 m high, while southern inside border is consisted of bank beams. Based on these beams, the whole inside part of the horseshoe was named "Kupinske grede".

The length of "Obedska bara" is about 13.5 km (arch length), an average width is between 500 and 700 m, while the whole area is about 7 km^2. The riverbed of "Obedska bara" is not fulfilled with water by the whole its width and length, and during dry periods, water is present just in deeper parts of "Obedska bara". The unique flow occurs just in periods of high water level. Kupinski part of the horseshoe is the closest to the 55th km of the river Sava, while Obreški part of "Obedska bara" is situated at 78th km of the river Sava. Horseshoe perforation is about 4 km long, while the whole horseshoe is about 5 km long. Kupinski part of the horseshoe is connected with the river Sava due to channel Vok, and Obreški due to channel Revenica. As for "Obedska bara"height above sea level, it is between 71 and 73 m, and alluvial lowland areas where it lays are between 71 and 76 m high.

Watering regime of "Obedska bara" is very complicated and directly connected with watering regime of the river Sava. That relation is less or more intensive depending on water level height. "Obedska bara" functions like accumulation – water comes there, changes the flow and finally emphasis. All three elements of balance mentioned above (flow, accumulation and water emphasis) are unconditionally related to numerous factors:

(a) "Obedska bara"is provided by water from the river Sava (through the channel Vok, channel Revenica, and alluvium), then from the back side, from ground waters from southern slopes of Fruška Gora, and finally from precipitation (about 650 mm);
(b) Water emphasis from "Obedska bara"is caused by decreasing of water level of the river Sava; through channel Revenica (a bit less); through channel Vok (the most of the water); underground through the porous alluvium; by transpiration from water surface (during summer up to 10 mm per day); by evapotranspiration (that is a significant loss, up to 20.000 m^3; (Milutinović and Vajda 2000).

On the upper ground, inside the horseshoe, there is situated Management Unit "Kupinske grede" (Fig. 37.2) managed by Public Enterprise (PE) "Vojvodinašume"

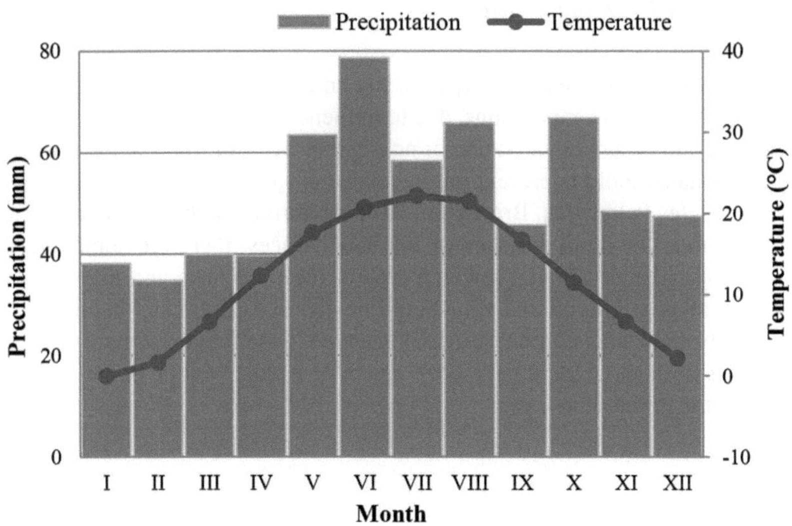

Fig. 37.2 Position of piezometers within MU "Kupinske grede"

that is surrounded from three sides by "Obedska bara" horseshoe, and from the forth side, it is surrounded by the river Sava.

This area is characterized with an expressed micro relief with a lot of small depressions (puddles) and higher grounds (beams). As for soil type, there are fluvisoils, humofluvisoils and mold. This area is site of vegetation type *Fraxino angustifoliae-Quercetum roboris* (Gajić and Karadžić 1991). The area of this management unit is about 2.478 ha, and about 61.1% belongs to lowland hard angiosperms of Ravni Srem forests. Great part of swamping and puddling areas are a basic ingredient of special natural area "Obedska bara". In northern part of forest complex in MU "Kupinske grede", there has been situated protected natural area "Debela gora", consisted of ash and oak forest old more than 200 years.

Implemented methodology supposes hydrological research, monitoring of watering regime of the river Sava, or better to say, watering regime of "Obedska bara", and also monitoring of groundwater level at the area of MU "Kupinske grede". In addition to this, there has been monitored the influence of watering regime on edaphic conditions and vegetation. Due to this, 11 piezometers have been installed on two bio indication lines – the first line was placed perpendicularly to the flow of Sava (includes piezometers 1–6) and another one was defined perpendicularly to the first (includes piezometers 7–11), Fig. 37.2. Apart from that, on the bridge over the channel Vok in Kupinovo hydrometric profile has been installed for fulfilling of "Obedska bara"monitoring.

Data used in the paper originates from scientific project "The influence of watering regime on characteristics of Pedunculate oak habitat in the area of Ravni Srem", that has lasted since 2000 (Nikić et al. 2010).

37.2.4 Soil Characteristics

There are a lot of different soil types in the small area: mold, existing on higher positions, while on lower positions, due to influence of ground waters, this soil type is degraded and occurs as carbonate or non-carbonate with some signs of waterlogging. Carbonated mold is present on the lowest position and it is rich with humus, and there is gley below that. Brown soil has been formed in the ground consisted of river alluvium. On small surfaces of the less terraces, there are some salt soils. Alluvial soils include alluvial lowland area of the river Sava and they have been formed on alluvial ground that still should be deposed. There are mainly two subtypes (sandy and loamy alluvial soil) and one type (waterlogged alluvial soil). In the zone of the horseshoe, there have been layers of puddle peat (2 m and more) that belongs to the group of low peat.

37.2.5 Horological and Phytogeographic Characteristics

There are three most important forest associations present in Donji Srem area – *Fraxinus angustifolia*; *Fraxinus angustifolia* and *Quercus robur*; *Fraxinus angustifolia, Quercus robur* and *Carpinus betulus*. Based on some papers (Jokanović et al. 2017), there has been determined species relations to a wide range of ecological factors, and its belonging to some areal types, as well. As for living plant forms present in these associations, the most dominant are phanerophytes with the greatest participation in *Fraxinus angustifolia*, *Quercus robur* and *Carpinus betulus* forest (about 42%). There are also a lot of hemicriptophytes. Based on plants relation to watering, mesophytes are dominant in all associations. There are also a lot of hygrophilous species, that are the most present in *Fraxinus angustifolia* association. As for soil acidity, all analyzed associations show neutrofil character, that is the most expressed in *Fraxinus angustifolia* association, while a bit lower neutrofiles content was recorded in another two associations. Based on plants relation to light regime, species present in these associations are between helyophytes and scyophytes. Relation to the temperature shows that the most frequent are mesotermic species. As for areal types in all associations, there is dominant participation of Middle-European and Euro-Asian areal type.

37.3 Implications of Hydrological Factors to Floristic Structure

In the area of "Obedska bara", that belongs to MU "Kupinske grede", relation between surface and groundwater level from one side and vegetation characteristics from the other side unconditionally exists. Watering regime affects distribution of the species present in all three height orders in forest associations. Waterlooging and adding flooded of some parts of microrelief cause occurrence of one plants type, while falling of the water level comes to plants alterations. Three edificators among woody species (*Fraxinus angustifolia, Quercus robur* and *Carpinus betulus*) are the most present, while its dominance directly depends on hydrological site conditions.

37.3.1 Monitoring of Surface and Groundwater Levels

Collected and analyzed data from experimental plots of researched area Obedska bara (MU "Kupinske grede") for period 2002–2011 show occurrence of high levels both of surface and groundwater that are in interaction with low water levels in some research years. These changes are related to water level of the river Sava fluctuations, that sometimes exceed bank height (76.51 m above sea level) and cause flooding of the whole area. High water levels were recorded during 2004, 2005, 2006, and 2010, while during other years, occurred just flooding of some low areas (74.47 m above sea level), which means that Sava did not leave its riverbed (Table 37.1, Diagram 37.2). During the researched period (2002–2011), we had to separate 2010 as representative of wet and 2011 as representative of dry period, respectively (Diagrams 37.3 and 37.4).

In addition to multiyear water level of the river Sava fluctuations, seasonal levels should be also mentioned. This implies occurrence of high water level in the beginning of the year in period January–June, and also at the end of the year, in November and December (Diagram 37.2). Absolute maximum of water level of the river Sava (77.05 m above sea level) was recorded in April 2005, while minimum (70.36 m above sea level) was found in July 2003. Number of the days with flooded water occurrence on the highest positions – beams (76.51 m above sea level) ranges from 9 days during 2005 to 19 days during 2010, while the overall number of the days with high water level during researched period is 60 days (Diagrams 37.2 and 37.3). Values on the lowest positions (74.47 m above sea level) on the researched area range from 7 days in 2003 to 150 days in 2010, while the overall duration of flooding on these localities is 565 days during 10 years long period. During 2002, 2003, 2007–2009 and 2011, there has not been recorded occurrence of flooding waters of the river Sava.

Water level in "Obedska bara"is followed by water level of the river Sava fluctuations, so fluctuations of high water levels are almost identical in the river and in "Obedska bara" (Diagram 37.2). From Diagram 37.2, it is obvious that high water levels in "Obedska bara" occured in 2004, 2005, 2006 and 2010. Decreasing of water level in "Obedska bara"was followed by water level of the river Sava, but just to overflow threshold on the channel Vok (71.43 m above sea level) where comes to its drainage. Below that level, "Obedska bara" loses water because of porous ground and evaporation.

Period of low water level in "Obedska bara" oscillates during period 2002–2011 going from 55 days in 2010 to 222 days in 2008 (Diagrams 37.2 and 37.3). Month with the lowest frequency of low water levels is March with 1.74% (24 days), while

Table 37.1 Extreme values of water level in period 2002–2011 at the area of "Obedska bara"

			2002	2003	2004	2005	2006	2007	2008	2009	2010	2011
Surface water	River Sava	*Min*	70.77	70.36	70.50	71.05	70.70	70.42	70.55	70.43	71.09	70.36
		Max	75.28	75.07	76.75	77.05	77.02	74.85	75.28	75.86	77.01	75.81
	Vok	*Min*	–	–	71.43	71.43	71.43	71.43	71.43	71.43	71.43	71.43
		Max	–	–	73.93	76.73	76.83	74.31	74.35	74.52	76.63	75.29
Ground water	P-1	*Min*	73.53	73.25	73.25	74.66	74.58	74.50	73.99	72.95	75.02	73.81
		Max	74.33	73.73	76.51	76.51	76.51	75.18	74.64	74.37	76.51	76.51
	P-3	*Min*	72.50	71.83	71.82	73.63	74.03	73.46	71.61	71.82	74.23	73.67
		Max	74.08	73.47	74.47	74.47	74.47	74.22	73.62	74.47	74.47	74.47

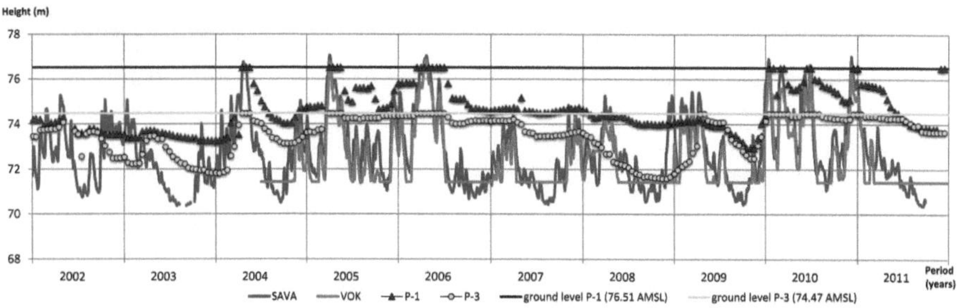

Diagram 37.2 Fluctuations of surface and groundwater levels in period 2002–2011 at the area of "Obedska bara"

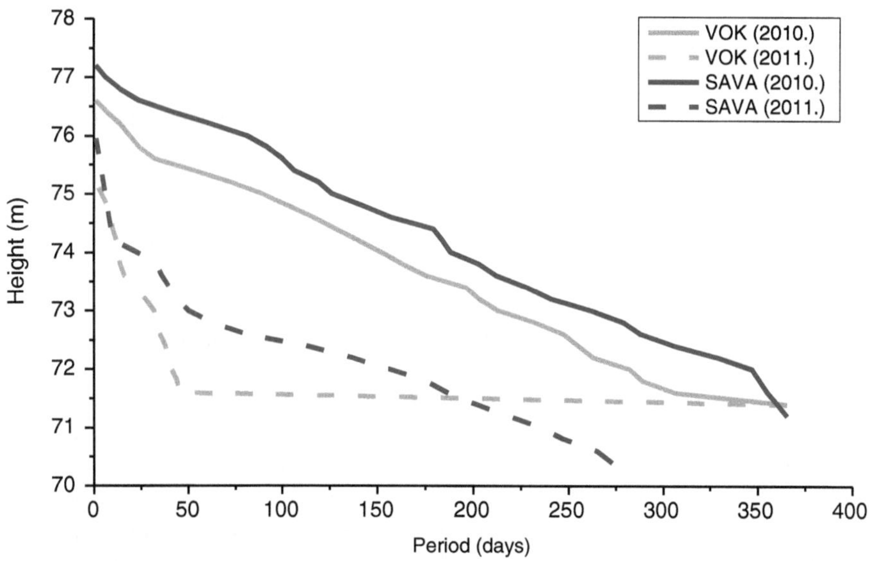

Diagram 37.3 Line of duration of surface waters level in 2010 and 2011 at the area of "Obedska bara"

August is a month with the most days of low water levels – 16.38% (226 days). As for groundwater levels, they mainly follow water level fluctuations in the river Sava and in "Obedska bara"going between 72.95 m above water level in 2009 and 76.51 m above sea level in 2004, 2005, 2006, 2010 and 2011 for higher localities situated on the beams (Table 37.1, Diagram 37.2). On lower positions of the study area in depressions (P-3, 74.47 m above sea level), groundwater levels oscillate between 71.61 m above sea level in 2008 and 74.47 m above sea level recorded during period 2004–2006 and 2009–2011. Bearing in mind above mentioned, wet years (2004, 2005, 2006 and 2010) with high water level of the river Sava provide high

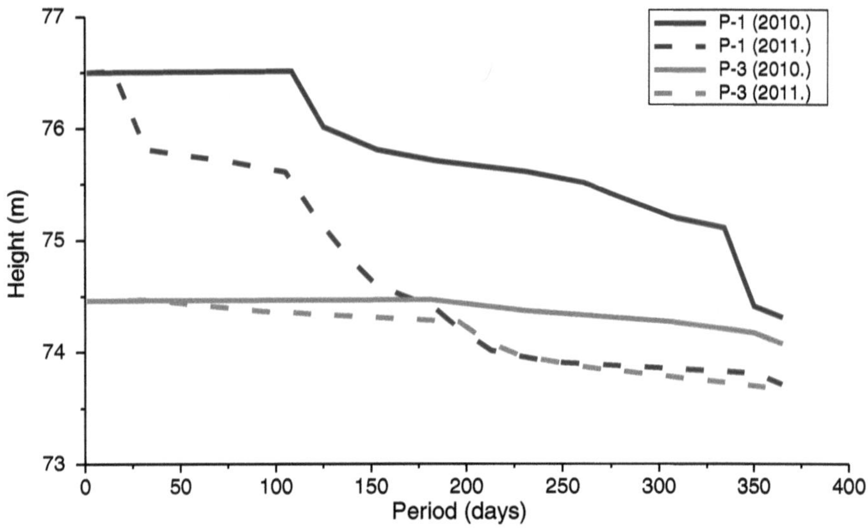

Diagram 37.4 Line of duration of groundwater level in 2010 and 2011 at the area of "Obedska bara"

groundwater levels that are over reference level (up to 2.0 m in depth). These conditions enable growth and development of hygrophilous forest species (Letić et al. 2006; Nikolić et al. 2012).

During dry years, such as 2002, 2003, 2007, 2008, 2009 and 2011, with low water level of the river Sava, groundwater levels form different conditions depending on micro relief.

In depressions (P-3), groundwater level oscillates in the range 0–2.0 m, apart from two periods (2003–2004 and 2008–2009), that are shown on Diagram 37.2. Groundwater levels lower than 2.0 m (for this location, it is 74.47 m above sea level) were recorded from half of January till the beginning of March with duration of 61 days and in period August–December in 2003 with duration of 153 days. This trend with low water levels continues in 2004 during January till the half of February with duration of 46 days.

In the second period (2008–2009), water levels under reference depth occur from May in 2008 till the beginning of March in 2009 with duration of 245, and 59 days, respectively (Diagrams 37.2 and 37.4). On higher positions, on beams (P-1), groundwater levels oscillate much more compared to reference level and form longer time periods that are characterized with not enough water content for growth and survival of present plant associations in the river Sava inundation.

During research period, there have been observed two more expressed and two not so significant periods with water level under reference level (for this locality, it is 74.51 m above sea level). In 2002 and 2003, till the half of March in 2004, groundwater level oscillated below 2.0 m (Diagram 37.2). Duration period of low groundwater level was 804 days. In another half of 2004, from the beginning of August till the half of November, during 106 days, occurred low groundwater levels. Another significant period related to low groundwater level occurred from the beginning of February 2008 till the end of December in 2009 with duration of 701 days. During 2011, there were recorded values of low groundwater levels from the beginning of July to the end of December with duration of 184 days (Diagrams 37.2 and 37.4).

Analysis of the whole observed period shows that water level of the river Sava defines its relation to watering regime in "Obedska bara". Bottom height of the channel Vok (71.43 m above sea level) and bank height of the river Sava (76.51 m above sea level) are the most important benchmark levels. As for area of "Kupinske grede", by high water levels of the river Sava, that area becomes the bottom, and mirror of "Obedska bara" integrates into water mirror of the river Sava. Conditions like these influence biocenosis forming whose features are directly influenced by environmental conditions of this puddle-swamp ecosystem. Detailed research related to above mentioned, as initial base, is shown in the edition "Ecology of Obedska bara" with much information about ceno-ecological relations in this area. Based on this, interaction between biotic and abiotic factors is of the greatest importance for defining water ecosystem present in "Obedska bara" area.

The main primary production holder in this ecosystem is phytoplankton. There are over 360 taxons of phytoplankton in "Obedska bara", but on the other side, there are no many individuals within the population. As a result, phytoplankton biomass is pretty small (0.007–8.015 mg/dm^3) because of influence of different factors.

Trophic degree of "Obedska bara"is related to production degree in time and many other processes are involved in this such as mineralization, transpiration, etc. Bearing in mind above mentioned, aquatic system "Obedska bara" is not able to degrade produced organic matter. As a result, it comes to creating plenty of semi-degraded mater which causes forming of peats. It is well-known that trophic degree directly depends on a wide range of different factors, so it means that quality of the water from the river Sava plays a significant role in that process as autochthonous factor. Waters of the lower flow of the river Sava are pretty burdened with a lot of toxic substances (Veljković and Jovičić 2010). It limits occurrence of microphytes (alochthonous factor), while developed biomass of macrophytes, especially located in marshes during process of bio-recovery, creates a plenty of these substances due to its capacity. As for using of nutrients from the water, macrophytes in these marshes ecosystem are competitive to phytoplankton (Stojanović et al. 2008). In addition to that, macrophytes occupy water area of "Obedska bara"making the area a bit narrower for microphytes development. Generally, due to numerous papers, "Obedska bara" position is defined as oligo-mezotrophic in the zone of free water area, while litoral zone, at the edge of "Obedska bara", is eutrofic-distrofic system with dominant semiaquatic plant associations belonging to *Scipreto-Phragmitetum and Phragmiteto- Salicetum.*

37.3.2 Monitoring of Floristic Structure

In the area of management unit "Klenak", that is also situated in Donji Srem, Jurišić (2015) established that, in cenoecological group consisted of Pedunculate oak and hornbeam, among species present in the first height order, the most dominant is Pedunculate oak, but within this cenotype, in some of researched stands, oak pedunculate was repressed by Turkey oak and hornbeam. Although these sites are very suitable for oak pedunculate development, in the most drained soils belonging to automorphic, far away from flooding zone, oak pedunculate gradually ceased to be dominant in the first height order (Jurišić 2015). After oak pedunculate, as the most dominant species occurs hornbeam. As for undergrowth order, within the same cenoecological group, plants from this height order are outside influence of flooding. Specific features of this area – there is no waterlogging under influence of raising and stagnation of groundwater on the surface, there is also no atmospheric precipitation and direct inundation from the riverbeds – cause a significant differentation between species, especially in undergrowth.

Within ceno-ecological group of hygrophilous forests of black alder in management unit "Klenak", the same author found dominant presence of *Fraxinus angustifolia, Salix cinerea* and *Amorpha fruticosa* in undergrowth. Forest association of ash and oak pedunculate in the area of MU "Kupinske grede", characterized with hygrophilous site conditions similar to these present in hygrophilous association of black alder, has a similar floristic structure in undergrowth with dominant presence of *Fraxinus angustifolia* and *Amorpha fruticosa*.

Within forest association of black willow and poplar in MU "Klenak", Jurišić (2015) in undergrowth order separated *Amorpha fruticosa* by its presence. The same author concluded that key factors responsible for variation of species occurrence within researched flooded forest sites area: alternation of flooding duration; distance from the riverbed; height above sea level. He also separated, based on ecological model related to adaptibility of researched dendroflora to inundation duration, *Acer campestre, Carpinus betulus* and *Fraxinus angustifolia*. These species are consisted of very similar ecological optimum, but it was also shown that *Fraxinus angustifolia* has significantly wider ecological valence and greater tolerance on anaerobic conditions compared to *Carpinus betulus* and especially to *Acer campestre*.

As for phytocenological research in researched area ("Obedska bara", MU "Kupinske grede"), there have been collected 28 phytocenological footages within forest units number 13, 14, 18, 19 and 20, and they also have been collected on 5 experimental plots in associations of ash, ash and oak pedunculate, and ash, oak pedunculate and hornbeam, respectively (Jokanović et al. 2018). Species determination has been performed by using Flora of Serbia (Stevanović 2012).

In association of *Fraxinus angustifolia* with *Carex sp,* there has been found presence of monodominant *Fraxinus angustifolia* in a wide alluvium of the river Sava, inside forest complex of oak pedunculate, where these associations are located in depressions where surface water stagnates during a longer period of time. Mentioned association is drier type of monodominant associations of *Fraxinus angustifolia* and it is, from floristical point of view, the richest association that belongs to *Alnion-*

glutinosae forests. It is mainly present on alpha and beta-gley with reduced horizon depth between 60 and 100 cm (Tomić 2004; Tomić and Rakonjac 2013).

As for floristic structure of *Carici remotae-Fraxinetum angustifoliae* association, the most dominant is *Fraxinus angustifolia* as edificator in the wood order with grade of presence 5. Apart from it, there are also *Ulmus minor*, that is a bit less present, while in the order of bushes, the most dominant is *Amorpha fruticosa*. In the order of undergrowth, there are numerous species: *Rubus caesius, Carex remota, Amorpha fruticosa, Aristolochia clematitis, Fraxinus angustifolia,* etc.

Pedunculate oak and ash forests are the wettest part of oak pedunculate associations and they are closely related to monodominant forests of *Fraxinus angustifolia* from *Alnion glutinosae* forest. This forest includes partly flooded part of alluvium, but also partly its unflooded part that is under the influence of groundwater. As for floristic structure, the most dominant are hygrophilous species (Tomić 2004). In the wood order, the most dominant are *Fraxinus angustifolia* and *Quercus robur*, then follows *Ulmus minor* and *Acer tataricum*. In this association, *Quercus robur* is completely competetive to *Fraxinus angustifolia,* which means that development in this forest characterizes a bit drier site conditions than in previous one. In the order of bushes, the most dominant is *Amorpha fruticosa*, while *Ulmus minor* and *Cornus sanguinea* follow. In the order of undergrowth, the most dominant is *Rubus caesius,* and apart from it, there are also: *Quercus robur, Fraxinus angustifolia, Ulmus minor, Amorpha fruticosa, Acer tataricum, Glechoma hederacea, Clematis integrifolia,* etc.

Three-dominant association consisted of Pedunculate oak, *Fraxinus angustifolia* and *Carpinus betulus* is present in flooding area of the rivers Danube, Sava and Velika Morava. It is transitional community between wet (*Fraxinus angustifolia* and *Quercus robur*) and drier (*Quercus robur* and *Carpinus betulus*) associations. The first time, this association was described nearby Obedska bara (Tomić 2004). In the wood order, apart from *Carpinus betulus, Fraxinus angustifolia* and *Quercus robur*, there is significant presence of *Ulmus minor* and *Acer tataricum*. As for the order of bushes, the most dominant are *Cornus sanguinea* and *Acer tataricum*. In the lowest order consisted of undergrowth, there are following species: *Convallaria maialis, Glechoma hederacea, Acer tataricum, Crataegus monogyna, Amorpha fruticosa, Galium palustre, Cynanchum vincetoxicum, Quercus robur, Lysimachia nummularia, Cardamine pratensis, Rubus caesius, Fraxinus angustifoila, Cornus sanguinea, Ulmus minor*, etc.

37.4 Conclusions

Research of watering regime of special nature reserve "Obedska bara" opened numerous questions about complicated relations that are present in this puddle-swamping ecosystem. During research period 2002–2011, a lot of data has been collected and it was attributed and analyzed. Based on the data, we can conclude the following:

Water level of the river Sava is characterized by significant fluctuations. In the zone of "Obedska bara", these fluctuations are between 70.36 m above sea level

(2005) and 77.05 m above sea level (2003), when the river Sava leaves its riverbed and floods wide area of "Obedska bara". As a result, "Kupinske grede" becomes the bottom of the river Sava, while wider area becomes an integral part of the flow of the river Sava.

Seasonal fluctuation of the Sava water level, defined with occurrence of high water levels during winter and spring, occurs mainly in period January–July, but also in November and December. Duration of flooded water levels on higher ground – beams goes from 0 to 19 days, while on lower positions, such as depressions, it lasts between 7 and 150 days.

Water level in "Obedska bara" is under direct influence of water level of the river Sava. Connection between "Obedska bara" and the river Sava is defined by height on the channel Vok (71.43 m above sea level), that serves for charging of the riverbed by the water, while through the height of the river bank (76.61 m above sea level) comes to converting of "Obedska bara" mirror to the mirror of the river Sava. Draining of "Obedska bara" by emphasis of the water through the channel Vok defines conditions related to isolation degree of the area compared to the river and it is between 55 days (2010) and 222 days (2008). Seasonal fluctuations of the "Obedska bara" isolation from the river goes between 1.74% (24 days) in March and 16.38% (226 days) in August for research period (2002–2011).

Watering regime of the river Sava and of "Obedska bara", as well, has the greatest influence on trophic degree of this aquatic ecosystem, but it also impacts to trophic relations of other ecosystems.

Bearing in mind this, the most significant factor of surface and groundwaters dynamics is the river Sava that defines contact periods and periods of isolation between the river and "Obedska bara", so this changes a lot the influence of autochthonous and alochthonous factors of this water ecosystem, and it affects the ecological capacity of "Obedska bara".

As for floristic structure of researched area, the most dominant forest associations are these consisted of ash; ash and oak pedunculate; ash, oak pedunculate and hornbeam. Association consisted of ash and oak pedunculate is the wettest, while this consisted of oak pedunculate and hornbeam is the driest. Association consisted of ash, oak pedunculate and hornbeam is transitional between above two mentioned. Floristic structure of researched associations in all height orders clearly determinates local environmental conditions.

References

Bednarz, Z. (1993, October 18–20). Water deficit limit tree-ring widths of the oak (Quereus robur L.) in the Niepolomice forest, Southern Poland. In Proceedings of the Vth Symposium In A. Szujecki, & P. Paschais (Eds.), *The protection of forest ecosystems. Forest ecosystems versus climate change* (Bialowieza, Vol. 82, pp. 57–68).

Bobinac, M. (2008). Treatment of hardwoods forests in the area of Ravni Srem. *Monographs 250 years of Ravni Srem forestry*. Public company Vojvodinašume – management unit Sremska Mitrovica, Petrovaradin (pp. 137–146).

Bobinac, M., Grbić, P., Janjatović, G. & Abjanović, Z. (1997). Thinning measures in young stands of *Quercus robur* and *Fraxinus excelsior* at the area of management unit "Sremska Mitrovica". *Šumarstvo* 4–5. Belgrade, pp. 33–43.

Bogdanović, Ž. (1982). Hydrological problematic of Srem, *Doctoral dissertation*. Faculty of nature and mathematics, University of Novi Sad.

Bukurov, B. (1952). About stratigraphy of quaternary deposits of Vojvodina from dr V. D. Laskarev, *Geološki Anali Balkanskog Poluostrva*, knjiga XIX, pp. 1–16.

Burlica, Č. (1987). Physical soil properties. *Forestry encyclopedia*, Sv. III, Zagreb, pp. 479–481.

Ćurčić, S., Đuričić, J., & Marjanović, V. (2002). *Municipality of Sremska Mitrovica, geography monographs*. Novi Sad: University of Novi Sad, Faculty of Nature and Mathematics.

Dekanić, I. (1974). Influence of level and fluctuations of groundwater on pedunculate oak decline. *(Quercus robur L.). Šum. list, 99*(7–10), 267–280.

Donaubauer, E. (1998). Die Bedeutung von Krankheitserregern beim gegenwärtigen Eichensterben in Europa – eine Literaturübersicht. *European Journal of Forest Pathology, 28*, 91–98.

Dubravac, T., & Dekanić, S. (2009). Structure and dynamics of the harvest of dead and declining trees of pedunculate oak in the stands of Spačva forest from 1996 to 2006. *Šumarski list, 133*(7–8), 391–405.

Dubravac, T., Dekanić, S., & Roth, V. (2011). Damage dynamics and structure of the oak tree crowns in stands in microelevations and microdepressions – Results from the monitoring on permanent experimental plots. *Šumarski list*, Zagreb, 125. *Posebni broj, 28*, 74–89.

Franjić, J., Škvorc, Ž., & Čarni, A. (2006). Distribution of *Crataegus nigra* Waldst. et Kit.in Croatia and its importance for forest edge vegetation formation. *Šumarski list, 1–2*, 3–8.

Führer, E. (1998). Oak decline in Central Europe: Asynopsis of hypotheses. *USDA Forest Service General Technical Report, NE-247*, 7–24.

Gajić M. & Karadžić D. (1991). Floristic structure of Ravni Srem with a special insight on Obedska bara. Faculty of Forestry in Belgrade and MU "Sremska Mitrovica". Belgrade – Sremska Mitrovica.

Group of authors. (2008). *General management plan for Srem forestry area*. Sremska Mitrovica: MU "Sremska Mitrovica".

Harapin, M., & Androić, M. (1996). Decline and protection of pedunculate oak forest. In D. Klepac (Ed.), *Oak pedunculate in Croatia* (pp. 227–256). Zagreb: Center for scientific work Vinkovci and forests of Croatia, HAZU.

Hughes, F. M. R. (2007). *The importance of flooding regimes in the conservation of floodplain forests*. Madrid: International Symposium on River Restoration.

Jokanović, D., & Nikolić, V. (2017). The width of tracheids of the taxodium trees on alluvial sites in Serbia, *Šumarstvo* 3–4. *Belgrade*, 189–198.

Jokanović, D., Vilotić, D., Mitrović, S., Miljković, S., Rebić, M., Stanković, D., & Nikolić, V. (2015). Correlations between the anatomical traits of Gymnocladus canadensis Lam. in heartwood and sapwood of early- and latewood zones of growth rings. *Archives of Biological Sciences, 67*(4), 1399–1404.

Jokanović, D., Vilotić, D., Nikolić, V., Nonić, M., Devetaković, J., & Stanković, D. (2017). Latewood proportion inside growth rings by bald cypress stems in Serbia. *Fresenius Environmental Bulletin, 26*(12A), 7925–7930.

Jokanović, D., Vilotić, D., Nikolić, V., Šijačić-Nikolić, M., Lakušić, B., & Jović, Đ. (2018). Growth rings width of bald cypress stems from two alluvial sites in Serbia. *Fresenius Environmental Bulletin, 27*(1), 306–312.

Josipović, J. & Soro, A. (2012). Groundwaters of Vojvodina, *Monographs*, Institute for water-power engineering "Jaroslav Černi" Belgrade.

Jurišić, B. (2015). Diversity of vascular flora of lowland forests of Ravni Srem. *Doctoral dissertation,* Biological Faculty of Belgrade University, pp. 1–490.

Krahl-Urban, J., Llesh, J., & Schwf.rtfeger, F. (1944). Das Eichensterben im Forstamt Flellefeld. *Zeitsehr. f.d. gesamte Forstwesen, 76/70*, 70–86.

Letić, L. J., Ivanišević, P., & Rončević, S. (2006). Watering regime influence on poplar growth. *Bulletin of Faculty of Forestry, University of Belgrade – Faculty of Forestry, Belgrade, 93*, 105–119.

Letic, L. J., Savić, R., Nikolić, V., & Lozjanin, R. (2014). Effects of reclamation works on the state of pedunculate oak forests in Ravni Srem. *Šumarstvo, 3–4*, 117–130.

MarclI, G. (1966). *Studial Cauzelor si al Metodclor de Prevenlre si Combatere a Uscarii Stejarului.* Bucharest: Centrul de Doeumentare Tehniea pentru Economia Forestiera 582.

Mayer, B. (1987). Regime of groundwater and surface water in the soils of lowland forests of basen Pokupski in period 1981–1986. *Final study case,* Forestry Institute Jastrebarsko – Zagreb, pp. 1–139.

Mayer, B. (1989). Ecological importance of underground and surface water regime for lowland forests of basen Pokupski. *Doctoral dissertation*, Faculty of Forestry, Zagreb.

Mayer, B. (1994). Effects of soil watering dynamics, groundwater, precipitation and defoliation on seasonal dynamics of radial increment and decline of pedunculate oak (*Quercus robur* L.) in Varoški lug. *Forestry Institute, 29*(1), 83–102.

Mayer, B. (1995). Underground and surface water in the lowland forest Turopoljski lug in period 1989–1993. *Forestry Institute, 30*(1), 47–73.

Mayer, B. (1996). Hydropedological relations in the area of lowland forests of basen Pokupski. *Forestry Institute, 31*(1/2), 37–89.

Medarević, M., Banković, S., Cvetković, Đ., & Abjanović, Z. (2009). Problem of forest dying in Gornji Srem. *Šumarstvo, 61*, br. 3–4–str. 61–73.

Milutinović R. & Vajda LJ. (2000). Obedska bara and growing of hydroenergetic system Đerdap – I, *Vodoprivreda* 183–185, god. 32, JDP, Beograd, pp. 229–236.

Mitrović, S., Jokanović, D., Vilotić, D., Miljković, D., Veselinović, M., & Stanković, D. (2017). Stomata characteristics of two Paulownia species under different conditions of light. *Fresenius Environmental Bulletin, 26*(3), 1876–1882.

Nikić, Z., Letić, L., Nikolić, V., & Filipović, V. (2010). Procedure for underground water calculation regime of pedunculate oak habitat in plain Srem. *Bulletin of Faculty of Forestry, University of Belgrade – Faculty of Forestry, Belgrade, 101*, 125–138.

Nikić, Z., Srećković Batoćanin, D., Burazer, M., Ristić, R., Papić, P., & Nikolić, V. (2013). Conceptual model of mildly alkaline water discharging from the Zlatibor ultramafic massif, western Serbia. *Hydrogeology Journal Official Journal of the International Association of Hydrogeologists, 21*(5), 1147–1161.

Nikolić, V., Letić, L. J., & Savić, R. (2012). Effects of watering regime of the river Sava on aquatic ecosystem of Obedska bara, Proceedings *Melioracije 12*, University of Novi Sad, Faculty of Agriculture, Department for water management, pp. 23–29.

Nikolić, V. (2017). Watering regime influence on site characteristics of pedunculate oak (*Quercus robur* L.) in Ravni Srem. Doctoral dissertation, Faculty of Forestry, Belgrade University.

Pranjić, A., & Lukić, N. (1989). Increment of pedunculate oak stems as indicator of site changes. *Glas.šum. pokuse, 25*, 79–94.

Prpić, B. (1976). Response of pedunculate oak (*Quercus robur* L.) from two different sites on different wetting conditions. *Šum. List., 100*(3), 117–123.

Prpić, B. (1996). Devastation of pedunculate oak forests. In D. Klepac (Ed.), *Quercus robur in Croatia* (pp. 273–298). Zagreb: HAZU Center for scientific work Vinkovci and Croatia forests.

Prpić, B., Seletković, Z., & Žnidarić, G. (1994). Ecological and biological causes of pedunculate oak (*Quercus robur* L.) stems dying in the lowland forest Turopoljski lug. *Glas. šum. pokuse, 30*, 193–222.

Rahmanov, V. V. (1984). *Hydroclimatic rule of forests (in Russian), Leshnaya promyshlennost.* Moscow: Forest Industry Publ 204.

Savic, R., Ondrasek, G., Bezdan, A., Lj, L., & Nikolic, V. (2013). Fluvial deposition in groyne fields of the middle course of Danube river. *Journal Tehnički vjesnik/Technical Gazette, 20*(6), 979–983.

Šijačić-Nikolić, M., Vilotić, D., Veselinović, M., Mitrović, S., & Jokanović, D. (2011). Bald cypress (*Taxodium distichum* (L.) Rich.) in the area of protected nature reserve Veliko ratno ostrvo. *Bulletin of Faculty of Forestry, Belgrade, 103*, 173–184.

Siwecki, R. (1989). A decline of oak forests caused by abiotic factors and attempts at biological research on this syndrome. *Arboretum Korniekie, 34*, 161–169.

Siwecki, R. (1993). Global climatic change and oak decline. In A. Szuji'cki & P. Pascilalis (Eds.), *Proe. of 5th Symp. the Proteetion of Forest Ecosystems – Forest Ecosystems Versus Climate Change, Bialowieza, 18–20 Oct., 1993* (pp. 69–83). Warszawa: IBL.

Siwecki, R. (1995). Syndrom zamierama dcjbow na Plycie Krotoszyriskiej. Quereus sp. OKL Goluchc)w. pp. 10–14.

Siwecki, R. & Llese, W. (1991). Oak deeline in Europe. In *Proceedings of the International IUFRO Symposium*, Kornik, Poland, 15–18 May 1990. Poznan: Panstwowe Wydawnietwo Rolnieze i Lesne, p. 360.

Siwecki, R., & Ufnalski, K. (1998). Review of oak stand decline with special reference to the role of drought in Poland. *European Journal of Forest Pathology, 28*, 99–112.

Sperduto, D. D., & Nichols, W. F. (2004). *Natural communities of New Hampshire*. New Hampshire: The NH Natural Heritage Bureau and The Nature Conservancy.

Stanković, D., Jokanović, D., Veselinović, M., Letić, Lj., Jović, Đ., &Karić, D. (2015). Zinc concentration in woody and herbaceous plants at Kosmaj area, Fresenius Environmental Bulletin, Freising-Germany *Fresenius Environmental Bulletin, 24*(11), 3672–3675.

Stanković, D., Ivetić, V., Ocokoljić, M., Jokanović, D., Oljača, R., & Mitrović, S. (2015b). Manganese concentration in plants of the protected natural resource, Kosmaj, in Serbia. *Archives of Biological Sciences, Belgrade, 67*(1), 251–255.

Steiner, K. C. (1998). A decline-model interpretation of genetic and habitat structure in oak populations and its implications for silviculture. *European Journal of Forest Pathology, 28*, 113–120.

Stevanović, V. (2012). *Classification of living forms in Serbian flora, "Flora SR Srbije I"* (pp. 39–42). Belgrade: Serbian Academy of Science and Art.

Stojanović, V., Pavić, D., & Mesaroš, M. (2008). The use of natural assets of reeds marshland in Vojvodina in view of sustainable development. *Zbornik Matice srpske za prirodne nauke Novi Sad, 115*, 109–116.

Thomas, F. M., Blank, R., & Hartman, G. (2002). Abiotic and biotic factors and their interactions as causes of oak decline in Central Europe. *Forest Pathology, 32*, 277–307.

Tomić, Z. (2004). Forest phytocenology. *Faculty of Forestry, Belgrade University*.

Tomić, Z., & Rakonjac, L. J. (2013). *Forest plant associations of Serbia*. Belgrade: University "Singidunum", Faculty for applied ecology "Futura" and Institute of Forestry.

Vajda, Z. (1948). *Effects of climate fluctuations on oak pedunculated decline from lowland forests in Posavina and Donja Podravina area*. Zagreb: Institute for forestry research of Forestry Ministry of Republic of Croatia 154.

Veljković, N., & Jovičić, M.. (2010). Long-term trend of water quality of aquatic area of Sava, 39, Conference about using and protection of water "Water 2010", Serbian society for water protection and Institute for waterpower engineering "Jaroslav Černi", Divčibare, Srbija, pp. 69–72.

Veselinović, M., Vilotić, D., Mitrović, S., Čule, N., Stanković, D., Jokanović, D., & Madzgalj, J. (2017). Air pollutant effects on chlorenchyma cell and chloroplasts of Douglas-fir (Pseudotsuga menziesii (Mirb.) Franco) needles. *Fresenius Environmental Bulletin, 26*(3), 1974–1979.

Vilhar, U. Š., & Fajon. (2007). Vpliv gozda in gozdnogojitvenih ukrepov na hidrološki režim vodozbirnega območja. In M. Kovač (Ed.), *Gozd in voda* (Gozdarski inštitut Slovenije) (pp. 16–21). Ljubljana: Zavod za gozdove Slovenije.

Concluding Remarks and Recommendations

According to the FAO (Global Forest Resources Assessment 2015), percentage of land area under the forests in the selected countries of Southeast Europe is high: Slovenia 62.0%, Montenegro 61.5%, Bosnia and Herzegovina 42.8%, Bulgaria 35.2%, Croatia 34.3%, Greece 31.5%, Serbia 31.1%, Romania 29.8%, indicating significant forest genetic resources within the overall biodiversity of this area.

Climate models predict that an increase in air temperature and precipitation regime changes will adversely affect the trees, with the majority of species that will exceed their ecological optimality. Adaptation to new environmental conditions will require a long period of time during which the forest genetic resources would be exposed to high pressure.

If forest genetic resources are viewed as inherited by the previous generations in order to preserve them for future generations, then it is clear that the activities on their conservation and sustainable utilization must be developed by science and profession, legally defined and coordinated globally, and implemented regionally and locally.

The first step in conservation of forest genetic resources presents the protection of biodiversity. According to the FAO (Global Forest Resources Assessment 2015), it can be noted that forest area designed for conservation of biodiversity in countries of Southeast Europe is relatively small, but it has increased in the period of 25 years (1990–2015).

Therefore, on the basis of presented state, legal frameworks and concrete activities on conservation of forest genetic resources (FGR) in the selected countries of Southeast Europe, we decided to define overall objectives for FGR conservation, which are common to all countries:

1. Research and education on FGR conservation in Southeast Europe;
2. Strategic, legal and program framework for FGR conservation;
3. International and cross-sectoral cooperation for FGR conservation;
4. Capacity building and information dissemination in FGR conservation;

M. Šijačić-Nikolić et al. (eds.), *Forests of Southeast Europe Under a Changing Climate*, Advances in Global Change Research 65,
https://doi.org/10.1007/978-3-319-95267-3

5. Implementation and monitoring of regional priorities in FGR conservation.

Within the **first overall objective** (*Research and education on FGR conservation in Southeast Europe*), the specific objectives and priorities were defined and concrete measures were proposed.

The **first specific objective** is systematization of available information on FGR, in order to provide a basis for state assessment, monitoring, conservation and sustainable use. The priority is systematization of available information on FGR within *in situ and ex situ* conservation programs. Within this priority, the following measures have been proposed:

- Develop a synthesis report on the state of FGR involved in programs of *in situ* conservation;
- Conduct an assessment of the vulnerability of rare and endangered species of forest trees involved in programs of *in situ* conservation to endangering factors;
- Conduct an impact assessment of different management systems on FGR conservation in accordance with climate change;
- Assess the existing system of forest species provenance regions in relation to needs for climate change adaptation and mitigation;
- Determine the status of forests in private ownership from the aspect of genetic resources conservation;
- Develop a synthesis report on the state of FGR within the existing network of seed orchards, provenance trials, progeny tests, clone archives and genetic collections, as well as report on previous results in the area of genetic preservation.

The **second specific objective** is determination of research priorities within *in situ* and *ex situ* conservation of FGR. The following measures have been proposed:

- Develop a long term program of research on: endangering factors influence on FGR; the adaptability of forest tree species to environmental changes; the influence of different management systems on FGR conservation;
- Develop guidelines for priority species identification and involvement in the programs of intensive research on climate change impacts on FGR;
- Develop guidelines for exploring the possibility of applying climate scenarios and models for the needs of assessing the vulnerability of FGR and the potential of tree species and forest types for carbon sequestration;
- Develop guidelines for establishing the priority list of tree species and their traits for inclusion in intensive research programs of *ex situ* conservation and breeding;
- Develop a long term program of research within: the seed orchards network, provenance trials network, progeny tests network; the clone archives, genetic collections network, research on genetic preservation and according to the synthesis report on genetic resources state and previous results.

The **third specific objective** is improvement of formal and informal education in conservation of FGR. The first priority is improving formal education in conservation of forest genetic resources and introducing specific themes into curricula at all levels of education. Within this priority, the following measures have been proposed:

- Develop guidelines for introducing specific themes on FGR conservation into curricula of primary and secondary education;
- Develop a synthesis report on the representation of FGR conservation topics in curricula of higher education with defined measures for improvement;
- Develop national and regional databases of experts' capacities and potentials in the area of FGR conservation.

The second priority is improving informal education in conservation of FGR and cooperation with civil sector. Within this priority, the following measures have been proposed:

- Develop national and regional databases of civil sector capacities for educational and dissemination activities in the area of forest genetic resources conservation;
- Capacity assessment of protected natural areas managers and forest owners and users for implementation of FGR conservation programs and definition of improvement measures;
- Develop a long term program of integration of research, experts and civil society capacities with the aim of improvement in the area of FGR conservation education.

Within the **second overall objective** (*Strategic, legal and program framework for FGR conservation*), the specific objectives and priorities were defined and concrete measures were proposed.

The **first specific objective** is strategic framework improvement for FGR conservation. The first priority is improvement of the strategic goals of forestry development in Southeast Europe with emphasis on the holistic approach to forest management. Within this priority, the following measures have been proposed:

- Review existing national forestry development strategies with the aim of aligning with the new EU Forestry Strategy, emphasizing the holistic approach to forest management;
- Develop national strategies of FGR conservation in all countries of Southeast Europe and incorporate objectives and measures into the regional strategy.

The second priority is improvement of the strategic frameworks in the environmental protection of Southeast Europe in order to improve the status of FGR conservation. Within the second priority, the following measures have been proposed:

- Assess the representation of FGR conservation topics in strategic frameworks for environmental protection and define guidelines for improvement;
- Establish an *ad hoc* regional expert group for monitoring of FGR conservation status in the strategic framework for environmental protection.

The **second specific objective** is legal framework improvement for FGR conservation. The first priority is improvement of legal framework in the area of *in situ* conservation of forest genetic resources, with following measures:

- To harmonize the legal frameworks in the management of forest reproductive material with the expected intensity of changes in environmental conditions and the needs of assisted migration of genetic material in accordance with the intensity of climate change;
- Assess the existing legal framework for regions of provenance of forest tree species in relation to the needs of adjusting and mitigating the consequences of changes in the environmental conditions;
- To improve existing by-law framework with the aim of integration of forest genetic resources conservation into the system of regular forest management.

The second priority is improving the status of FGR conservation in the legal framework for the protection of nature and the environment. The following measures have been proposed:

- Assess the representation of FGR conservation themes in the legal framework for environmental protection and nature protection and define the guidelines for improvement;
- Establish an *ad hoc* regional expert group for monitoring of FGR conservation status in the strategic framework for environmental and nature protection.

The **third specific objective** is program framework improvement for FGR conservation, within which three priorities were defined. The first priority is improvement of program framework for *in situ* conservation of FGR, with following proposed measures:

- Develop a long term regional program of *in situ* conservation of FGR;
- Develop separate programs of genetic conservation for forest tree species in natural protected areas and at specific stands;
- Develop programs and guidelines for production of high quality forest reproductive material.

The second priority is improvement of program framework for *ex situ* conservation of FGR. The following measures have been proposed:

- Develop a long term regional program of *ex situ* conservation of FGR;
- Develop guidelines for the adequate sampling of genetic material in order to establish experimental plantations and gene banks and to preserve the adaptive variability of forest tree species;
- Develop a breeding program for forest tree species of economic importance.

The third priority is improvement of program framework for integral approach to conservation of FGR. The proposed measures within this priority are:

- Develop a long term program for integral approach to conservation of FGR, based on the combination of *in situ* and *ex situ* methods and inter-sectoral cooperation;
- Develop guidelines for introduction of Multiple Population Breeding System (MPBS) conservation method as an efficient approach to climate change mitigation;

- Develop guidelines for establishment and management of gen-ecological zones within forest ecosystems with the aim of adaptability improvement of forest species.

Within the **third overall objective** (*International and cross-sectoral cooperation for FGR conservation*), two specific objectives were defined with priorities and concrete measures.

The **first specific objective** is international cooperation improvement for FGR conservation. The first priority is continuation of the active participation of national institutions and organizations in international processes and initiatives related to the conservation of FGR. Within this priority, the following measures have been proposed:

- Adopt and actively implement all recommendations of the MCPFE process in all countries and at the regional level;
- Continue active participation in the EUFORGEN Program and adopt all recommendations for improving national strategic and legal frameworks;
- Develop a long-term action plan for the participation of national institutions in international processes and fulfill international recommendations and guidelines.

The second priority is monitoring and implementation of international initiatives and standards in the field of mapping and evaluation of ecosystem services of FGR. The following measures have been proposed:

- Establish an *ad hoc* regional expert group for monitoring international movements in the field of mapping and evaluation of ecosystem services of FGR;
- Encourage institutional and expert involvement in international networks for evaluating ecosystem services;
- Develop guidelines for the introduction of international recommendations for evaluating ecosystem services in national and regional strategic and legal frameworks.

The **second specific objective** is improvement of cross-sectoral cooperation for FGR conservation.

The first priority is improving cross-sectoral cooperation in order to strengthen the capacity and promote the concept of conservation of FGR, with following measures:

- Develop a long-term program for the integration of research, expert and civil society capacities in order to improve the promotion of the concept of conservation of forest genetic resources;
- Establish an *ad hoc* regional expert group from representatives of the forestry sector, nature protection, and environmental protection, legal and economic experts in order to strengthen the capacity for conservation of forest genetic resources.

The second priority is improvement of cross-sectoral cooperation in order to develop a system of ecological-economic evaluation of forest genetic resources. The following measures have been proposed:

- Develop guidelines for networking of experts of different profiles in order to develop a system of ecological and economic evaluation of forest genetic resources;
- Establish an *ad hoc* regional expert group from representatives of different sectors in order to develop and implement the system of ecological and economic evaluation of FGR.

Within the **fourth overall objective** (*Capacity building and information dissemination in FGR conservation*), two specific objectives were defined.

The **first specific objective** is capacity building for FGR conservation, with two priorities. The first priority is capacity building of state and public sector for FGR conservation and the second priority is strengthening the capacities of the private and civil sector in conservation of FGR. Within the first priority, three measures have been proposed:

- Develop a long-term training program for representatives of the state and public forestry sector in order to strengthen the capacity to conserve FGR;
- Prepare educational materials and improve the knowledge of employees in state administration and public enterprises about the necessity of harmonizing forest management with goals of FGR conservation;
- Provide equipment and improve the skills of employees in public enterprises for the application of modern software packages for modeling ecosystem processes in relation to changes in environmental conditions.

Within the second priority, following measures have been proposed:

- Develop a long-term training program for representatives of the private and civil sector in order to strengthen the capacity to conserve FGR;
- Prepare educational materials and improve the knowledge and skills of representatives of the private and civil sector on the importance of FGR conservation;
- Identify all stakeholders in the process of programming and planning the conservation of forest genetic resources and actively involve them.

The **second specific objective** is raising public awareness and disseminating information in the conservation of FGR. Within this specific objective, the priority is awareness raising and information dissemination in the *in situ* and *ex situ* conservation of FGR. The following measures have been proposed:

- Develop informative and educational material on regional networks of conservation areas and the importance of *in situ* conservation of FGR;
- Integrate forest conservation areas and conservation programs into educational and tourist activities of protected natural areas;

- Map *ex situ* conservation points and develop educational and information program for acquaintance with this form of preservation of genetic variability of forest species;
- To promote *ex situ* conservation of FGR through good practice from all countries of Southeast Europe;
- Develop an internet portal on the importance of FGR with the presentation of all *in situ* and *ex situ* conservation capacities.

Within the **fifth overall objective** (*Implementation and monitoring of regional priorities in FGR conservation*), two specific objectives were defined, with priorities and proposed measures.

The **first specific objective** is implementation and monitoring of *in situ* and *ex situ* FGR conservation regional priorities. The first priority is monitoring of the impact of factors that threaten FGR and the creation of programs with adequate measures for their suppression. The following measures have been proposed:

- Comparatively analyze the degree of: genetic variability and the vulnerability of forest tree populations to disease and damage; forest tree species populations genetic variability and vulnerability to pollution (water, air, soil);
- Develop bio-indicators to monitor the impact of pollution on the diversity of forest species with emphasis on sensitive species or species endangered by the pollution;
- Develop guidelines for the inclusion of vulnerability assessment and adaptability of forest tree species into the regular forest management planning process.

The second priority is improving the system of *in situ* and *ex situ* conservation and achieving forest management in accordance with conservation principles. Within this priority, following measures have been proposed:

- Establish networks of conservation areas of forest tree species within protected natural areas and include them in management systems;
- Develop a system for monitoring of the state of FGR within the network of conservation areas;
- Introduce modern software packages for modeling ecosystem processes in relation to environmental changes in the system of forest management planning;
- Develop guidelines for harmonization of the management system with the highest degree of utilization of natural processes and mechanisms of self-regulation of forest ecosystems in order to conserve genetic variability;
- Develop guidelines for including the approach of mapping and evaluating ecosystem services of forest genetic resources in regular management systems.
- Establishment of new seed orchards from the species that have the largest share in the total volume of forests at the regional level, the species that have the highest economic importance for regional forestry;
- Establish provenance trials of the species which have not been covered by this method of conservation so far and include them in international networks;
- Establish progeny tests in field and laboratory conditions, and also as a preselection for the establishment of seed orchards;

- Conservation of existing gene pool of endangered species in clone archives, as well as applying genetic preservation methods.

The third priority is application of biotechnology in the conservation of FGR. The proposed measures within this priority are:

- Include methods of molecular genetics for identification of genome of endangered species in the process of development of conservation program for FGR;
- Develop efficient marker systems depending on forest tree species;
- Include MAS selection in the programs of "molecular breeding" for relations between phenotypic traits and gene expression;
- Use of biomarkers and genetic markers for determining resistance to diseases;
- Introducing *in vitro* techniques and techniques of a new generation of plant biotechnology.

The **second specific objective** is implementation and monitoring of the regional priorities for forest genetic resources conservation in Southeast Europe. Within this specific objective, the first priority is monitoring system for implementation of the regional priorities for conservation of FGR. The following measures have been proposed:

- Establish a cross-sectoral expert group to monitor the implementation of regional priorities;
- Develop a system of measurable indicators for monitoring the implementation of regional priorities;
- Develop a long-term program for financing the conservation of forest genetic resources and identify possible sources of financing;
- Monitor development and apply indicators to implement the Sixth MCPFE goal.

The second priority is evaluation system for implementation of the regional priorities for conservation of FGR. Within this priority, following measures have been proposed:

- Develop a detailed system of evaluation criteria for short-term and long-term evaluation;
- Perform an annual evaluation in order to improve / revise the regional objectives;
- Develop a system for assessing the economic efficiency of individual priorities and measures.

The implementation of the proposed measures, as the basis for achieving the specified objectives of conservation of forest genetic resources, should be initiated at the level of each country individually, and after a certain period, it will lead to the improvement of the state of genetic resources in the forests of South-East Europe.

MIX
Papier aus verantwortungsvollen Quellen
Paper from responsible sources
FSC® C105338

If you have any concerns about our products,
you can contact us on
ProductSafety@springernature.com

In case Publisher is established outside the EU,
the EU authorized representative is:
Springer Nature Customer Service Center GmbH
Europaplatz 3, 69115 Heidelberg, Germany

Printed by Libri Plureos GmbH
in Hamburg, Germany